I0057369

Recent Progress in Atmospheric and Climate Science

Editor: Clara Farrow

R CALLISTO REFERENCE

www.callistoreference.com

Callisto Reference,
118-35 Queens Blvd., Suite 400,
Forest Hills, NY 11375, USA

Visit us on the World Wide Web at:
www.callistoreference.com

© Callisto Reference, 2019

This book contains information obtained from authentic and highly regarded sources. Copyright for all individual chapters remain with the respective authors as indicated. All chapters are published with permission under the Creative Commons Attribution License or equivalent. A wide variety of references are listed. Permission and sources are indicated; for detailed attributions, please refer to the permissions page and list of contributors. Reasonable efforts have been made to publish reliable data and information, but the authors, editors and publisher cannot assume any responsibility for the validity of all materials or the consequences of their use.

ISBN: 978-1-64116-194-7 (Hardback)

Trademark Notice: Registered trademark of products or corporate names are used only for explanation and identification without intent to infringe.

Cataloging-in-Publication Data

Recent progress in atmospheric and climate science / edited by Clara Farrow.
    p. cm.
Includes bibliographical references and index.
ISBN 978-1-64116-194-7
1. Atmospheric physics. 2. Atmosphere. 3. Climatology. I. Farrow, Clara.
QC861.3 .R43 2019
551.5--dc23

Table of Contents

Preface

Atmospheric sciences refer to the study of the Earth's atmosphere and its various processes, as well as the factors affecting it. It integrates varied fields like aeronomy, climatology and meteorology. Climatology or climate science deals with the study of climate. Modern climatology encompasses the areas of oceanography and biogeochemistry. Climatic studies using analog techniques are fundamental to the understanding of short-term weather forecasting. Some of these techniques are the Pacific decadal oscillation (PDO), El Niño–Southern Oscillation (ENSO), the Interdecadal Pacific Oscillation (IPO), the Madden–Julian Oscillation (MJO), etc. Climate models are crucial for the study of weather and climate systems as well as for the projection of future climates. This book aims to shed light on some of the unexplored aspects of atmospheric and climate sciences and the recent researches in these fields. The various sub-fields along with technological progress that have future implications are glanced at. Researchers and students involved in these fields will be assisted by this book.

The information contained in this book is the result of intensive hard work done by researchers in this field. All due efforts have been made to make this book serve as a complete guiding source for students and researchers. The topics in this book have been comprehensively explained to help readers understand the growing trends in the field.

I would like to thank the entire group of writers who made sincere efforts in this book and my family who supported me in my efforts of working on this book. I take this opportunity to thank all those who have been a guiding force throughout my life.

Editor

Comparison of two successive versions 6 and 7 of TMPA satellite precipitation products with rain gauge data over Swat Watershed, Hindukush Mountains, Pakistan

Muhammad Naveed Anjum,[1,2,3]* Yongjian Ding,[1,2] Donghui Shangguan,[1] Adnan Ahmad Tahir,[4,5] Mudassar Iqbal[6] and Muhammad Adnan[1]

[1] State Key Laboratory of Cryospheric Science, Cold and Arid Regions Environmental and Engineering Research Institute, Chinese Academy of Sciences, Lanzhou, P.R. China
[2] Key Laboratory of Inland River Ecohydrology, Cold and Arid Regions Environmental and Engineering Research Institute, Chinese Academy of Sciences, Lanzhou, P.R. China
[3] University of Chinese Academy of Sciences, Beijing, P.R. China
[4] Institute of Earth Surface Dynamics, Faculty of Geosciences and Environment, University of Lausanne, Lausanne, Switzerland
[5] Department of Environmental Sciences, COMSATS Institute of Information Technology, Abbottabad, Pakistan
[6] Key Laboratory of Land Surface Process and Climate Change in Cold and Arid Regions, Cold and Arid Regions Environmental and Engineering Research Institute, Chinese Academy of Sciences, Lanzhou, P.R. China

*Correspondence to:
M. Naveed Anjum, State Key Laboratory of Cryospheric Science, Cold and Arid Regions Environmental and Engineering Research Institute, Chinese Academy of Sciences, 320 Donggang West Road, Lanzhou 730000, Gansu, China.
E-mail: naveedwre@lzb.ac.cn

Abstract

Swat watershed in Hindukush Mountains was selected for the evaluation of Tropical Rainfall Measuring Mission (TRMM) Multi-satellite Precipitation Analysis (TMPA) Versions 6 and 7 using rain-gauge data. Agreement between the satellite and gauge estimates was good at monthly scale but poor at daily and seasonal (monsoon and westerlies, particularly for monsoon) scales. Over and underestimations were observed at foothills and high-altitude areas, respectively. Although, bias was still present, but overall performance of TMPA-V7 was improved compared with TMPA-V6. Bias corrected-TMPA-V7 estimates were better than corrected-TMPA-V6. Basin average bias-corrected precipitation estimates of TMPA-V7 were estimated at daily, monthly, westerlies, monsoon, and annual scales [means (mm) 1.60, 48, 190, 173 and 582, respectively] for the period 1998–2014. Results suggest that regional bias correction of satellite-precipitation products is critical and can yield to the substantial improvement in capturing the precipitation.

Keywords: Swat Watershed; Hindukush Mountains; satellite precipitation; bias correction

1. Introduction

Precipitation is the key parameter for various applications and disciplines related to water resources. Getting accurate precipitation data is thus crucial for local, regional and global hydrologic predictions. However, acquisition of precipitation data is often limited to ground-based observations, but usually this traditionally available information suffers from low spatial and/or temporal coverage, particularly in the Hindukush-Himalayan (HKH) mountainous region.

At present, availability of remotely sensed data provides the information of global precipitation distribution with high spatiotemporal resolution (Hu *et al.*, 2014). Several satellite-based precipitation products have been operationally available to the researchers. Among these products, Tropical Rainfall Measuring Mission (TRMM) Multi-satellite Precipitation Analysis (TMPA) products are widely used for various hydrological studies. Owing to high orientation, the performance of these products is expected to vary from area to area (Prakash *et al.*, 2014). Thus, it is essential to evaluate the performance of these products using locally observed ground-data before their applications in a specific area.

The latest version 7 (V7) of TMPA products is available since 1998. In contrast to the previous version 6 (V6), algorithms of V7 have been improved and additional datasets are incorporated. For the broadest usage and applications of TMPA precipitation products, a number of studies have been made to evaluate these products at regional to global scales (e.g. Vernimmen *et al.*, 2012; Mashingia *et al.*, 2014; Liu, 2015). As for the HKH region, Nair *et al.* (2009) evaluated the TMPA-V6 product using gauge-based data in Western Ghats Mountains in India for the period of 1998–2004. They concluded that TMPA-V6 does not capture well the precipitation over the study area. Xue *et al.* (2013) evaluated TMPA-V6 and V7 products using rain-gauge data in the Wangchu Basin, Bhutan. They concluded that TMPA-V7 products have significant improvements compared with the TMPA-V6 in terms of accuracy. Some studies have enumerated the similarities and differences between TMPA-V6 and V7 (e.g. Chen *et al.*, 2013; Prakash *et al.*, 2014; Zulkafli *et al.*, 2014). These studies showed that V7 agreed well with ground-based precipitation data than V6.

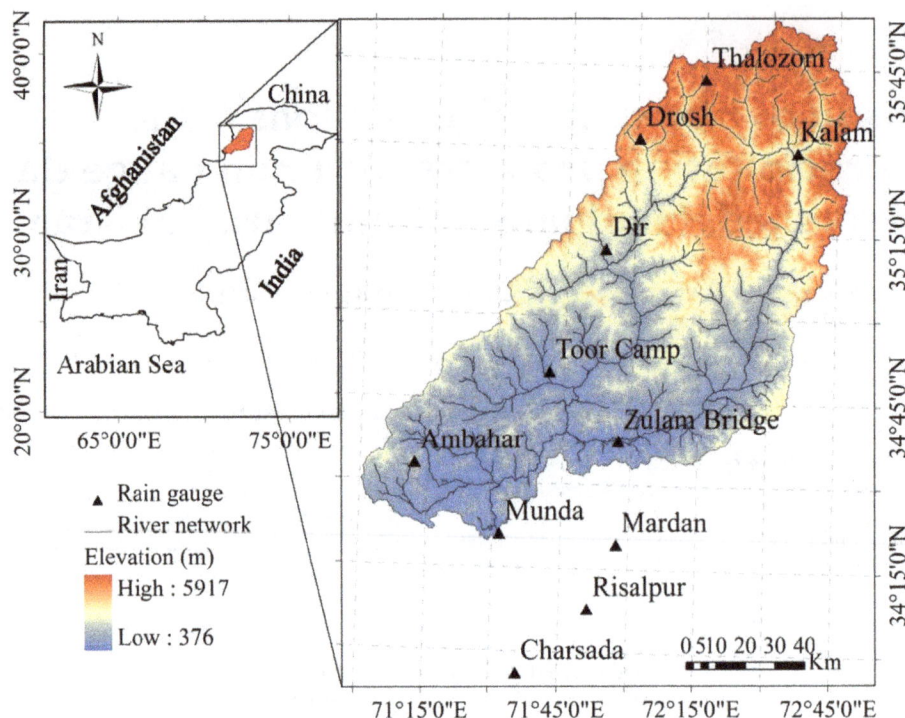

Figure 1. Location and the hydro-meteorological network of the Swat River Watershed in the northwest of Pakistan.

Nevertheless, no particular study has been conducted to evaluate the performance of TMPA products in the north-western mountainous region of Pakistan, which consists of diverse topography and precipitation distributions. Therefore, an attempt was made to evaluate the error characteristics of two widely used high-resolution TPMA products over Swat Watershed. There are two main rainy seasons in this region: summer precipitations due to monsoon currents during July–September, and winter precipitations due to disturbances in the mid-latitude westerlies during January–March (Wang et al., 2011). Accurate estimation of precipitation during these seasons is crucial for operational flood monitoring and prediction in northern areas, which are considered as most flood-prone areas of Pakistan. Thereby, the specific objectives of this study are to (i) evaluate the widely used and globally available TMPA products (V6-V7) and quantify the errors associated with these two successive versions in Hindukush Mountainous range with varying precipitation climatology; (ii) assess how much they differ during daily annual scales and also to assess the improvements in the upgraded version (V7) relative to its predecessor version (V6).

2. Case study specifications and datasets

This experiment is performed using 8 years of available rain-gauge data (1999–2006) over Swat River Watershed (drainage area = 14 039 km^2) located in the north-western Pakistan (Figure 1). Source of this river is in Hindukush Mountains, from where it flows through the Kalam Valley and Swat District. The elevation within the watershed varies from 376 m a.s.l. to 5,917 m a.s.l. (south-north). The average annual precipitation over the watershed varies from 300 to 980 mm. This area falls in the monsoon and westerlies belt. Heavy precipitations occurred merely under the interaction of westerly wave and intensified monsoon trough. Considerably high precipitation occurs only during westerlies as well as in the monsoon season. Heavy isolated precipitations in summer are often caused by the orographic lifting of the monsoon air mass arriving from south to south-east direction (Wang et al., 2011). Hourly datasets from total 15 automatic rain-gauge stations were collected and accumulated for daily precipitation considering UTC because TMPA-estimates were available at UTC. Eight-years (1999–2006) precipitation data of nine stations, used for evaluation of satellite estimates, were selected after sensitivity analysis of observed precipitation records. Similarly precipitation data from two stations, 'Munda' for the period of 2000–2005 and 'Dir' for the period of 2002–2009, were selected for the validation of bias correctors. All selected automatic-gauging stations were located within, and around the study area and their datasets were considered as ground truth for evaluation of satellite-precipitation products. Pakistan Meteorological Department (PMD), Water and Power Development Authority (WAPDA), and Irrigation Department (ID) of Khyber Pakhtunkhwa (KPK), Pakistan, provided the observed datasets.

Satellite-based precipitation estimates of TMPA-3B42 (V6-V7) products were obtained from the website of Goddard Earth Sciences Data and

Information Services Centre (http://mirador.gfsc.nasa.gov) which are freely available. Detailed information about these products can be found in Huffman *et al.* (2007) and Xue *et al.* (2013). Daily TMPA-3B42 satellite-precipitation products (V6-V7) at 0.25°×0.25° resolutions were used in this study.

3. Methodology

The performance of TMPA-V6 and V7 was evaluated on point (gauging-station) and basin levels at daily, monthly, seasonal (monsoon and westerlies) and annual timescales. Precipitation time-series of January–March were considered for westerlies while July–September for monsoon season. Gauge-based basin average precipitation was estimated using the Thiessen polygons approach [with ArealRain extension in ArcView (Petras, 2001)]. Satellite-based basin average precipitations were estimated by averaging values of all pixels that lie within the watershed. Pearson's correlation coefficient (CC), mean error (ME), mean absolute error (MAE), root mean square error (RMSE) and relative bias (BIAS) were used to evaluate the precision of satellite-based precipitation products. Detailed information about statistical indices can be found in Mashingia *et al.* (2014). The formulas of statistical indices are given below:

$$CC = \frac{\sum_{i=1}^{n}(PG_i - P\overline{G})\left(PS_i - P\overline{S}\right)}{\sqrt{\sum_{i=1}^{n}\left(PG_i - P\overline{G}\right)^2} \times \sqrt{\sum_{i=1}^{n}\left(PS_i - P\overline{S}\right)^2}} \quad (1)$$

$$ME = \frac{1}{n}\sum_{i=1}^{n}\left(PS_i - PG_i\right) \quad (2)$$

$$MAE = \frac{1}{n}\sum_{i=1}^{n}|PS_i - PG_i| \quad (3)$$

$$RMSE = \sqrt{\frac{1}{n}\sum_{i=1}^{n}\left(PS_i - PG_i\right)^2} \quad (4)$$

$$BIAS = \frac{\sum_{i=1}^{n}\left(PS_i - PG_i\right)}{\sum_{i=1}^{n}PG_i} \times 100 \quad (5)$$

where n represents the total amount of rain-gauge or satellite precipitation data, PG_i and PS_i represent the ith values of gauge and satellite precipitation, respectively; and $P\overline{G}$ and $P\overline{S}$ are the mean values of gauge and satellite precipitation estimates, respectively.

For more detailed evaluation and to estimate correspondence between the TMPA-based and gauge-based precipitation observations, four additional categorical statistical measures were adopted: false alarm ratio (FAR), probability of detection (POD), critical success index (CSI) and equitable threat score (ETS), details of categorical statistical measures are given in Mashingia *et al.* (2014). Thresholds of 0.5–25 mm were adopted to measure the ability of both TMPA products to capture precipitation occurrences at different intensities. Perfect values were considered as 1 for FAR, and 0 for each of POD, CSI and ETS. Formulas of categorical statistics are given by:

$$FAR = \frac{F}{H + F} \quad (6)$$

$$POD = \frac{H}{H + M} \quad (7)$$

$$CSI = \frac{H}{H + M + F} \quad (8)$$

$$ETS = \frac{H - Ar}{H + M + F - Ar} \quad (9)$$

Ar, represents the random hits that could occur by chance and is given by:

$$Ar = \frac{(H + M)(H + F)}{H + M + F + Z} \quad (10)$$

where H represents the hits (event forecast to occur, and did occur), F shows false alarms (event forecast to occur, but did not occur), M represents misses (event forecast not to occur, but did occur), and Z stands for correct negatives (event forecast not to occur, and did not occur).

4. Results and discussions

4.1. Evaluation results

Table 1 shows the statistical error characteristics of both TMPA products. At daily scale, both TMPA (V6 and V7) products did not show a good agreement with the rain-gauge data on point and basin level. The correlation coefficients (CC) were low (0.25–0.30 for V6–V7, respectively, on basin level and the best values of 0.18–0.26 on point level for V6–V7, respectively) and values of statistical errors were high [ME = 0.28–0.19; MAE = 2.31–2.23; RMSE = 5.19–5.29 for V6–V7, respectively, on basin level and the best values of ME= −0.21–(−0.03); MAE=1.90–1.73; RMSE=6.12–6.24 for V6–V7, respectively, on point level]. Both products showed a significant bias compared with the gauge data (18.5–12.9% for V6-V7, respectively, on basin level and the best values of −11.14% – (−1.66%) for V6-V7, respectively, on point level). At monthly timescale, both products showed a good correlation with the gauge data. Monthly data comparison exhibited the best values of CC as 0.66–0.71 for V6–V7, respectively, on point level and 0.71–0.73 for V6–V7, respectively, on basin level. But at seasonal (monsoon and westerlies) scale, both products (V6–V7) did not show a good agreement with the gauge data both on

Table I. Statistical error characteristics of TMPA (V6 and V7) precipitation estimates at different temporal and spatial scales.

Station name		Ambahar		Charsada		Drosh		Kalam		Mardan		Risalpur		Thalozom		Toor Camp		Zulam Bridge		Basin ave.	
TMPA	product	V7	V6	V7	V6	V7	V6	V7	V6	V7	V6	V7	V6	V7	V6	V7	V6	V7	V6	V7	V6
CC	Daily	0.06	0.10	0.14	0.10	0.16	0.15	0.26	0.18	0.18	0.13	0.17	0.12	0.14	0.10	0.13	0.09	0.21	0.15	0.30	0.25
	Monthly	0.52	0.64	0.58	0.66	0.50	0.55	0.72	0.72	0.68	0.59	0.54	0.51	0.66	0.60	0.61	0.50	0.71	0.64	0.73	0.71
	Monsoon	0.06	0.05	0.21	0.11	0.02	0.07	0.12	0.13	0.21	0.18	0.21	0.11	-0.01	-0.05	0.24	0.12	0.12	0.10	0.24	0.16
	Westerlies	0.06	0.12	0.12	0.14	0.17	0.13	0.23	0.23	0.24	0.12	0.17	0.19	0.18	0.15	0.12	0.12	0.25	0.24	0.34	0.31
ME	Daily	0.33	0.66	-0.03	0.38	0.61	0.84	-0.93	-1.13	0.06	-0.26	0.12	-0.21	-0.05	-0.21	0.46	0.60	0.69	0.77	0.19	0.28
	Monthly	9.99	19.7	-0.76	11.6	18.6	25.2	-28.2	-34.3	1.99	-8.58	3.64	-7.04	-1.56	-6.85	13.9	17.8	21.0	22.6	6.11	8.63
	Monsoon	0.96	1.18	-0.96	-0.13	1.63	1.79	0.59	0.72	-0.69	-1.18	0.02	-0.68	1.12	1.03	1.45	1.68	1.17	1.18	1.15	1.28
	Westerlies	-0.49	0.08	-0.10	0.02	0.05	0.34	-2.26	-2.78	-0.14	-0.92	-0.48	-1.18	-1.10	-1.50	-0.34	-0.54	0.07	0.18	-0.64	-0.66
MAE	Daily	1.73	1.90	2.45	2.89	2.42	2.59	3.41	3.39	2.83	2.75	2.61	2.50	3.07	3.08	2.09	2.28	2.51	2.66	2.23	2.31
	Monthly	25.5	26.1	22.8	23.7	34.5	36.4	42.8	53.8	31.6	36.4	35.6	34.1	30.8	33.9	27.2	32.6	31.5	31.6	22.9	24.6
	Monsoon	1.33	1.46	3.99	4.81	1.90	2.06	2.21	2.32	5.05	4.91	4.10	3.98	2.17	2.19	2.25	2.54	3.79	3.89	2.07	2.18
	Westerlies	2.98	3.22	3.05	3.12	3.45	3.69	5.47	5.22	3.14	2.88	3.52	3.06	4.84	4.85	3.26	3.14	3.29	3.41	3.34	3.36
RMSE	Daily	6.24	6.12	8.48	8.77	6.49	6.65	8.72	8.63	9.63	9.43	9.17	8.87	9.77	9.73	7.17	7.35	8.17	8.45	5.29	5.19
	Monthly	34.1	32.6	43.8	42.4	46.2	47.0	60.8	72.2	47.7	53.3	52.5	52.8	43.4	46.5	36.5	42.2	42.9	42.2	31.5	31.8
	Monsoon	3.92	4.10	10.85	11.5	3.27	3.78	4.21	4.58	12.95	12.9	11.15	11.1	4.06	4.24	6.35	6.95	10.38	10.4	3.81	3.94
	Westerlies	8.43	8.37	9.42	9.07	8.83	8.96	12.80	12.07	9.85	9.32	11.69	10.84	13.58	13.42	9.64	9.43	9.41	9.55	7.56	7.41
BIAS	Daily	40	80.64	-1.66	25.39	52.1	71.88	-34.6	-42.3	3.75	-14.6	7.6	-13.2	-2.74	-11.1	45.1	59.1	56.3	62.5	12.5	17.9
	Monthly	40	79.90	-1.66	25.45	52.1	71.69	-34.6	-42.1	3.75	-14.2	7.59	-13.7	-2.74	-11.0	45.1	59.2	56.3	60.5	13.1	18.7
	Monsoon	370	453	-31.3	-13.0	711	783	50.8	62.5	-19.4	-33.2	0.64	-25.4	161	144	226	262	71.1	71.4	132.6	147.3
	Westerlies	-24.7	4.06	-5.16	1.22	2.53	15.83	-48.3	-59.3	-6.30	-42.3	-19.3	-46.8	-30.9	-41.9	-15.9	-25.1	2.94	7.99	-22.4	-23.3

point and basin levels. As shown by CC, the precision of both products during monsoon season was reduced (0.16–0.24 for V6–V7, respectively, on basin level and the best values of 0.11–0.21 for V6–V7, respectively, on point level). During this season, both products overestimated the precipitation at most of the stations and basin level. However, the performance of both versions was slightly improved during westerlies compared with the monsoon season (shown by improved values of CC). Both TMPA products underestimated the precipitation compared with the gauge observations during westerlies precipitations. Results indicate that about 10% overestimations during monsoon and about 4% underestimations during westerlies precipitations were reduced in TMPA-V7.

Figure 2(a–i) shows the spatial distribution [adopting inverse distance weighting (IDW) interpolation] of average annual, westerlies and monsoon precipitation estimates of gauge and TMPA products. Gauge-based maps of average annual and westerlies precipitation showed as increasing pattern towards to north-east but monsoon map indicated an increasing pattern towards the south to south-east of the watershed (Figure 2(a–c)). From seasonal to annual scale, both TMPA products failed to capture the spatial patterns and magnitudes of precipitation over the watershed (Figure 2(d-i)). Compared with TMPA-V6, somewhat of these spatial precipitation patterns were captured by TMPA-V7. Figure 2(j–q) shows the results of evaluations conducted to estimate the occurrence of precipitation at different intensities. Results show that the precipitation detection skill of both satellite-precipitation products, on point and basin levels, decreases with the increase of precipitation intensity. Values of FAR were increased, while the scores of POD, ETS and CSI decreased with the increase of thresholds. This means that both TMPA products are less skilful to detect the occurrence and magnitude of intense precipitation events.

Figure 3(a) shows the comparison of average annual gauge and TMPA-based precipitation estimates at all stations. It is clear from this comparison that in lower altitude areas agreement between the TMPA-V7 and gauge observations are higher than that of the TMPA-V6. Nevertheless, both satellite products overestimated (69.6% by TMPA-V6 and 48.4% by TMPA-V7) the precipitation at stations on altitudes between 500 and 1500 m a.s.l. and underestimated (26.73% by TMPA-V6 and 19% by TMPA-V7) at stations above an altitude of 2000 m a.s.l. This error variability shows a significant geo-topographical dependent distribution of both satellite precipitation products. In low altitude areas, the overestimation by TMPA products was mainly ascribed to evaporation of precipitation, because of the warm atmosphere in lower areas. While the underestimation of precipitation over high-altitude areas (cooled atmosphere) was mainly ascribed to topology and orographic effects. Overall, the TMPA-V7 performed well as compared with TMPA-V6. Our results are consistent with the already published findings of Chen et al. (2013).

4.2. Correction of satellite-precipitation estimates

Results showed that the upgraded TMPA-V7 product performed slightly better than TMPA-V6. Nevertheless, significant bias was still present in TMPA-V7. Based on the findings, an effort was made to remove the bias of both TMPA products, to compare both versions after bias adjustment. In this regard, several time scales (daily, monthly, and seasonal) were considered for bias correction. Still high correlations between gauge data and TMPA products were found at the monthly scale, so bias correction factors at monthly timescale were developed within the study area. Previously, many researchers have adopted a monthly bias correction factor for the adjustment of satellite-based precipitation data (Vernimmen et al., 2012; Arias-Hidalgo et al., 2013). In this study, we adopted the methodology of Arias-Hidalgo et al. (2013) for the estimation of bias correctors. For that objective, the average monthly precipitation values of both TMPA products were compared with gauge-based observations. Equation (11) shows the developed relationship between rain gauge observations and their corresponding satellite-based precipitation estimates at the monthly timescale.

$$\mathrm{GP}_{i,m} = f_{i,m} \times \mathrm{TMPA}_{i,m} \qquad (11)$$

where $f_{i,m}$ is monthly bias factor at the ith rain-gauge. $\mathrm{TPMA}_{i,m}$ is original satellite-based monthly precipitation (mm/month) at the ith rain-gauge during the month m, $\mathrm{GP}_{i,m}$ is total precipitation at the ith rain-gauge and the month m.

In order to assess the validity of corrected satellite-precipitation estimates over Swat Watershed, RMSE and relative bias of corrected data were calculated. Table 2(a) shows the bias correction factors, estimated RMSE, and relative BIAS of adjusted data. Results showed that the bias-adjusted TMPA-V7 product was quite comparable with gauge-based data. Therefore, by adopting IDW approach, the bias correction factors for TMPA-V7 were spatially distributed across the whole watershed resulting in a distributed map of bias correctors. The corresponding bias correction factors for 'Munda' and 'Dir' gauging stations were estimated from that map. Before bias correction, statistical error characteristics of both validation stations were also calculated. The calculated values were CC = 0.17 – 0.17; ME = 0.74 – 0.93; MAE = 2.41 – 2.39; RMSE = 7.87 – 7.06, and bias = 66 – 94% for Munda and Dir stations, respectively. As expected, significant bias (90% at Munda and 105% at Dir station) was reduced after correction. Thus, the developed correction factors are considered valid for whole watershed and adjacent similar watersheds.

As various hydrological models use only daily precipitation data, thus monthly corrected precipitation estimates were disaggregated to daily scale, to make them available at daily time-scale for various hydrological modelling studies. For that, temporal disaggregation

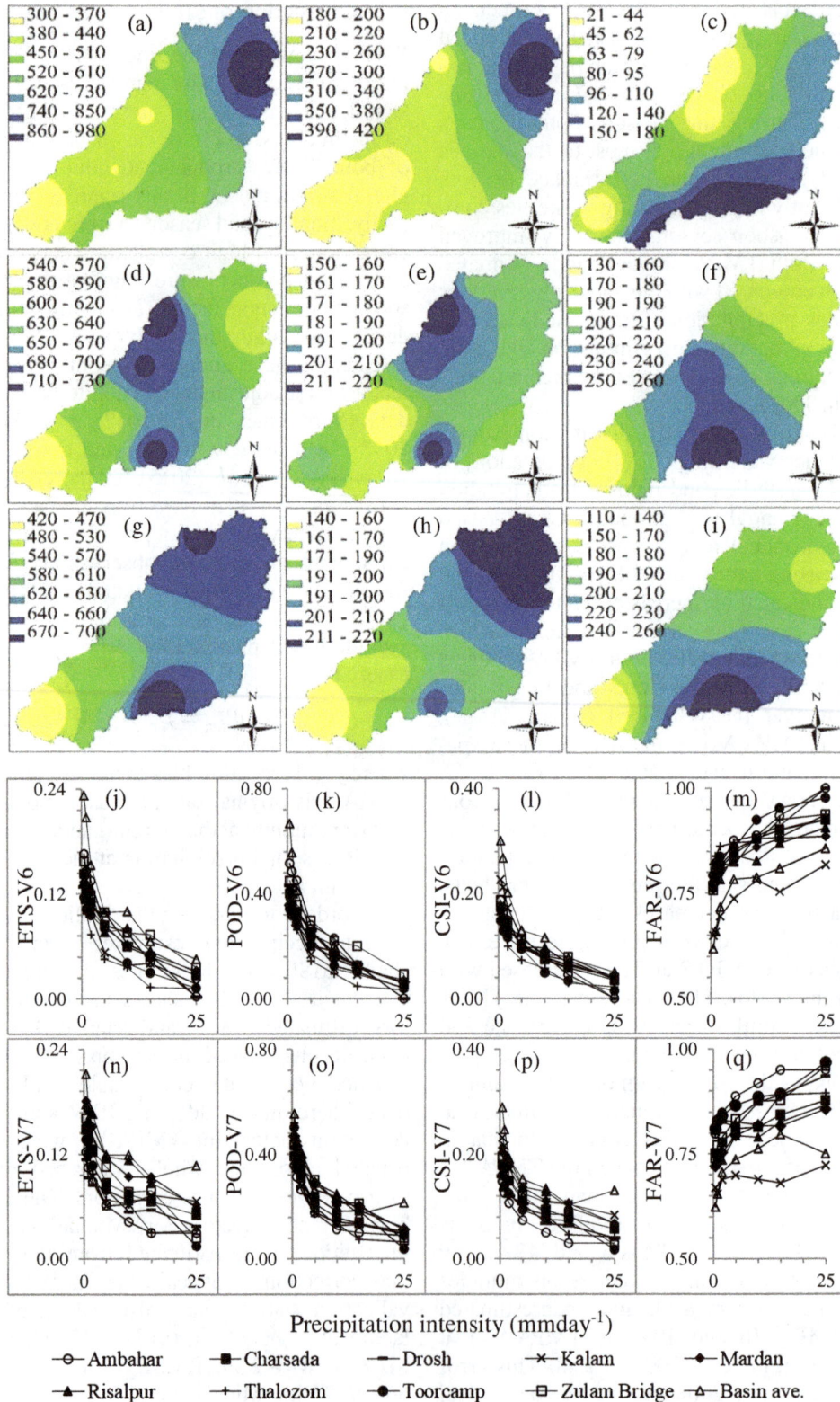

Figure 2. Spatial patterns of average annual, westerlies and monsoon precipitations of gauge based (a–c), TMPA-V6 (d–f), and TMPA-V7 (g–i), respectively. Binary analysis (j–q) of both products for precipitation occurrence using ETS, POD, FAR, and CSI for a range of thresholds.

coefficients (k_i) were derived from gauge-based daily precipitation time series as follows:

$$k_{i,d,m} = \frac{P_{i,d,m}}{TP_{i,m}} \tag{12}$$

where $k_{i,d,m}$ is temporal disaggregation coefficient at the ith rain-gauge, for the day d of month m. $P_{i,d,m}$ is cumulative precipitation at the ith rain-gauge on the day d of month m (mm day^{-1}), and $TP_{i,m}$ is cumulative precipitation at ith rain-gauge during the month m

Table 2. (a) Bias correction based on monthly correction, gauge measurements vs. TMPA (V6 and V7) precipitation estimates. (b) Basin wide mean daily-annual values of gauge and TMPA precipitation estimates.

(a)

Validation station name	Gauge data, annual rainfall (mm year⁻¹)	Original TMPA (V7) Annual rainfall (mm year⁻¹)	rBias (%)	RMSE (mm year⁻¹)	Original TMPA (V6) Annual rainfall (mm year⁻¹)	rBias (%)	RMSE (mm year⁻¹)	Monthly bias corrector (V7)	(V6)	Corrected TMPA (V7) Annual rainfall (mm year⁻¹)	rBias (%)	RMSE (mm year⁻¹)	Corrected TMPA (V6) Annual rainfall (mm year⁻¹)	rBias (%)	RMSE (mm year⁻¹)
Ambahar	300	420	40	22.3	536	78.9	26.7	0.74	0.59	309	3.3	20	318	6.1	17.9
Charsada	633	537	−15.3	28.2	685	8.07	32.8	1.22	0.93	653	3.2	26.1	634	0.1	32.4
Drosh	428	650	52.1	31	730	70.7	34.1	0.65	0.6	426	−0.5	23.7	439	2.6	22.7
Kalam	979	640	−34.6	48.7	567	−42.1	60.2	1.6	1.62	988	1	37	918	−6.3	51.2
Mardan	637	661	3.8	22.7	534	−16.2	30.5	1.17	1.3	692	8.6	22.5	695	9.1	26.9
Risalpur	575	619	7.6	21.4	491	−14.7	27.3	0.96	1.24	592	3	21.3	609	5.8	25.2
Thalozom	683	664	−2.7	30.5	601	−12	35.9	1.06	1.13	707	3.5	30.2	679	−0.6	35.2
Toor Camp	370	537	45.1	26.2	584	57.9	32.5	0.67	0.58	361	−2.4	20.7	341	−7.9	24
Zulam Bridge	448	700	56.3	26.4	719	60.5	26.9	0.68	0.67	476	6.3	16.7	479	7	15.7

(b)

		Means over the period of 1999–2006				Means over the period of 1998–2014	
		TMPA-V6 (mm)		TMPA-V7 (mm)		TMPA-V7 (mm)	
Resolution	Gauge (mm)	Original	Corrected	Original	Corrected	Original	Corrected
Daily	1.51	1.78	1.54	1.69	1.52	1.77	1.60
Monthly	46	55	47	52	46	54	48
Westerlies	262	201	170	204	182	213	190
Monsoon	80	198	168	186	162	194	173
Annual	552	656	564	625	556	650	582

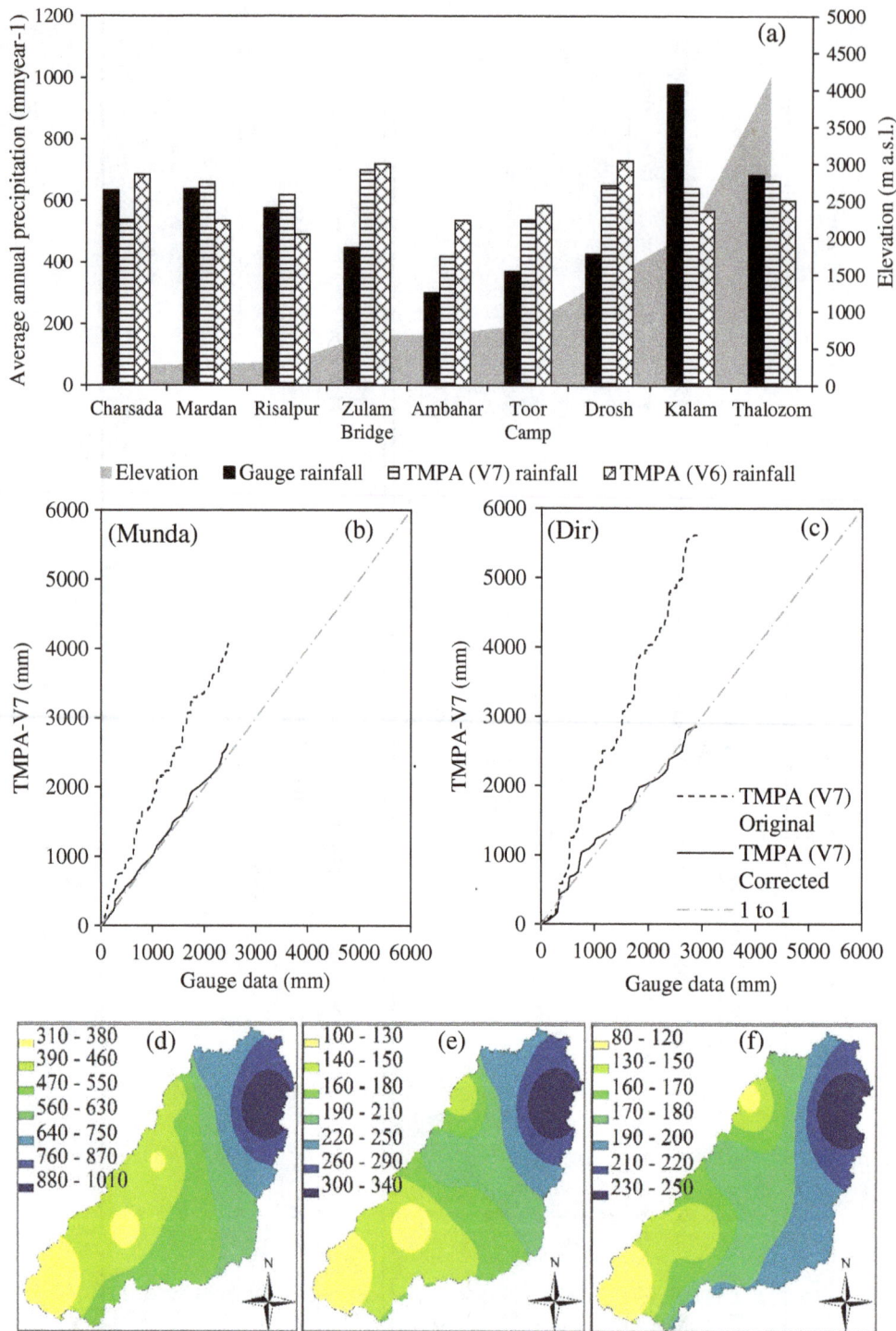

Figure 3. (a) Average annual precipitation of the rain–gauges and TMPA (V6 and V7) estimates. (b–c) Accumulative precipitation of gauge observations against the original and corrected daily TMPA (V7) precipitation estimates at both validation stations. (d–f) Spatial patterns of corrected TMPA–V7 average annual, westerlies and monsoon precipitations, respectively.

(mm month^{-1}). $k_{i,d,m}$ were then applied back to the corrected TMPA-V7 data to estimate the daily corrected precipitation estimates. Equation (13) shows the final expression.

$$\text{TMPA}_{\text{corr},i,d} = k_{i,d,m} \times \text{TMPA}_{\text{corr},i,m} \quad (13)$$

where $\text{TMPA}_{\text{corr},i,d}$ is the disaggregated, corrected daily TMPA-V7 precipitation data at location i (mm day^{-1}) for month m.

Daily synthetic and gauge-based precipitation datasets were quite comparable with each other. To check the effectiveness of corrected daily TMPA-V7 precipitation values, double mass curves at both validation stations were developed between the gauge-based and the original and corrected daily satellite-based precipitation estimates. Synthetic daily data agreed well with observed data, as shown in Figure 3(b–c). By considering the validity of correction factors, corresponding correction factors for each grid centre of TMPA-V7 was estimated to calculate the corrected basin-wide precipitation estimates (mean values are given in Table 2b) over the Swat Watershed, for the period of 1998–2014. Spatial patterns of corrected basin average annual and seasonal precipitation estimates were plotted as shown in Figure 3(d–f). The spatial patterns of corrected average annual and westerlies TMPA-V7 precipitation estimates (Figure 3(d–e)) were quite comparable with the gauge-based patterns. But the corrected monsoon estimates were still unable to capture spatial pattern of precipitation over Swat Watershed (Figure 3(f)).

5. Summary and conclusions

This study was conducted to evaluate the performance of TMPA-V6 and V7 precipitation products over Swat River Watershed in Hindukush Region. Both satellite-based products have been evaluated by using rain-gauge network on point and basin levels. Daily, monthly, seasonal (westerlies and monsoon) and annual precipitation time-series were analysed. Findings of this study are summarized and concluded as following:

1. Both TMPA-V6 and V7 failed to capture the spatial pattern of precipitation on annual and seasonal scale in the Swat Watershed.
2. Daily and seasonal time series of both TMPA products showed very low correlation (CC < 0.35 for daily and seasonal precipitations both on point and basin levels) with the observed precipitations. However, agreement between satellite and gauge precipitations was good at monthly scale with CC > 0.70 for basin and CC > 0.50 for point level.
3. Both TMPA products tend to overestimate precipitation events over 500–1500 m a.s.l. altitude areas, particularly overestimations were quite significant during monsoon season. On the other hand, underestimations were observed over high-altitude (above 2000 m a.s.l.) areas.
4. Skill of satellite-based products to detect the intense precipitation events decreases with the increase of precipitation intensity. Overall, the performance of TMPA-V7 was better than TMPA-V6 product. Basin average estimates show that biases were improved by 4% and 10% in TMPA-V7 during westerlies and monsoon precipitation, respectively.
5. The adjusted monthly TMPA-V7 precipitation estimates showed better agreement with gauge data compared with adjusted monthly TMPA-V6 estimates. Also, the synthetic daily precipitation estimates were quite comparable with rain gauge observations.

By using the developed areal bias-correction factor, the corrected daily, monthly, westerlies, monsoon and annual time-series (with mean (mm) values of 1.60, 48, 190, 173 and 582, respectively) of TMPA-V7 products were estimated for the period of 16-years (1998–2014).

The spatial pattern of corrected-TMPA-V7 product on annual scale was quite comparable with gauge observed spatial pattern. Although the spatial pattern of precipitation was captured well by the corrected-TMPA-V7 westerlies estimates but still the precipitation magnitudes were not captured well. Unfortunately, both the magnitudes and spatial patterns were not captured well by the corrected-TMPA-V7 monsoon estimates.

Findings of this study suggest that direct utilizations of both TMPA-V6 and V7 products are unreliable at daily-annual scales. However, bias corrected monthly, annual and synthetic daily TMPA-V7 estimates have a good potential for hydrological applications. Future research may focus on the evaluation of V7 for other seasons and also the application of adjusted data within a distributed hydrological modelling framework for precipitation-runoff simulations.

Acknowledgements

This work was supported by the Ministry of Science and Technology of China (Grant: 2013CBA01808); the National Natural Science Foundation of China (Grant: 41130638 and 41271082); and CAS-TWAS President's Fellowship programme. The authors are grateful to the WAPDA, KPK ID and PMD for providing the data. The authors declare no conflict of interests.

References

Arias-Hidalgo M, Bhattacharya B, Mynett AE, Griensven AV. 2013. Experiences in using the TMPA-3B42R satellite data to complement rain gauge measurement in the Ecuadorian coastal foothills. *Hydrology and Earth System Sciences* **17**: 2905–2915, doi: 10.5194/hess-17-2905-2013.

Chen S, Hong Y, Cao Q, Gourley JJ, Kirstetter PE, Yong B, Tian Y, Zhang Z, Shen Y, Hu J. 2013. Similarity and difference of the two successive V6 and V7 TRMM multisatellite precipitation analysis performance over China. *Journal of Geophysical Research [Atmospheres]* **118**: 13060–13074.

Hu Q, Yang D, Li Z, Mishra AK, Wang Y, Yang H. 2014. Multi-scale evaluation of six high-resolution satellite monthly rainfall estimates over a humid region in China with dense rain gauges. *International Journal of Remote Sensing* **35**: 1272–1294.

Huffman GJ, Adler RF, Bolvin DT, Gu GJ, Nelkin EJ, Bowman KP, Hong Y, Stocker EF, Wolff DB. 2007. The TRMM multisatellite precipitation analysis (TMPA): quasi-global, multiyear, combined-sensor precipitation estimates at fine scales. *Journal of Hydrometeorology* **8**: 38–55.

Liu Z. 2015. Comparison of precipitation estimates between Version 7 3-hourly TRMM Multi-Satellite Precipitation Analysis (TMPA) near-real-time and research products. *Atmospheric Research* **153**: 119–133.

Mashingia F, Mtalo F, Bruen M. 2014. Validation of remotely sensed rainfall over major climatic regions in Northeast Tanzania. *Physics and Chemistry of the Earth* **67-69**: 55–63.

Nair S, Srinivasan G, Nemani R. 2009. Evaluation of multisatellite TRMM derived rainfall estimates over a western state of India. *Journal of the Meteorological Society of Japan* **87**: 927–939, doi: 10.2151/jmsj.87.927.

Petras I. 2001. AreaRain.ave Areal Precipitation Calculation using Thiessen Polygons. AS11813.zip. ESRI, New York, Redlands, USA.

Prakash S, Gairola RM, Momin AM. 2014. Comparison of TMPA-3B42 versions 6 and 7 precipitation products with gauge-based data over

India for the Southwest Monsoon Period. *Journal of Hydrometeorology* **16**: 346–362.

Vernimmen RRE, Hooijer A, Mamenun Aldrian E, Van AIJM. 2012. Evaluation and bias correction of satellite rainfall data for drought monitoring in Indonesia. *Hydrology and Earth System Sciences* **16**: 133–146.

Wang SY, Davies RE, Huang WR, Gillies RR. 2011. Pakistan's two-stage monsoon and links with the recent climate change. *Journal of Geophysical Research, [Atmospheres]* **116**: D16114, doi: 10.1029/2011JD015760.

Xue X, Hong Y, Limaye AS, Gourley JJ, Huffman GJ, Khan SI, Dorji C, Chen S. 2013. Statistical and hydrological evaluation of TRMM-based Multi-satellite Precipitation Analysis over the Wangchu Basin of Bhutan: Are the latest satellite precipitation products 3B42V7 ready for use in ungauged basins? *Journal of Hydrology* **499**: 91–99.

Zulkafli Z, Buytaert W, Onof C, Manz B, Tarnavsky E, Lavado W, Guyot JL. 2014. A comparative performance analysis of TRMM 3B42 (TMPA) versions 6 and 7 for hydrological applications over Andean-Amazon river basins. *Journal of Hydrometeorology* **15**: 581–592.

Application of a developed distributed hydrological model based on the mixed runoff generation model and 2D kinematic wave flow routing model for better flood forecasting

Hongjun Bao,[1]* Lili Wang,[1]* Ke Zhang[2,3] and Zhijia Li[2,3]

[1] National Meteorological Centre, China Meteorological Administration, Beijing, China
[2] College of Hydrology and Water Resources, Hohai University, Nanjing, China
[3] State Key Laboratory of Hydrology-Water Resources and Hydraulic Engineering, Hohai University, Nanjing, China

*Correspondence to
H. Bao, National Meteorological
Centre, China Meteorological
Administration, Beijing, 100081,
P. R. China.
E-mail: baohongjun@cma.gov.cn
L. Wang, National
Meteorological Centre, China
Meteorological Administration,
Beijing 100081, P. R. China.
E-mail: wangll@cma.gov.cn

Abstract

Due to the specific characteristics of semi-arid regions, this paper aims to establish a Grid-and-Mixed-runoff-generation-and-two-dimensional-Kinematic-wave-based distributed hydrological model (GMKHM-2D model) coupling a mixed runoff generation model and 2D overland flow routing model based on kinematic wave theory for flood simulation and forecasting with digital drainage networks. Taking into consideration the complex runoff generation in semi-arid regions, a mixed runoff generation model combining saturation-excess mechanism and infiltration-excess mechanism is developed for grid-based runoff generation and 2D implicit finite difference kinematic wave flow model is introduced for routing to solve depressions water storing for grid-based overland flow routing in the GMKHM-2D model. The GMKHM-2D model, the GMKHM-1D model with coupling the mixed runoff generation model and one-dimension kinematic wave flow model, the Shanbei model and the Xin'anjiang model were employed to the upper Kongjiapo basin in Qin River, a tributary of the Yellow River, with an area of 1454 km^2 for flood simulation. Results show that two grid-based distributed hydrological models and the Shanbei model perform better in flood simulation and can be used for flood forecasting in semi-arid basins. Comparing with GMKHM-1D, the flood peak simulation accuracy of the GMKHM-2D model is higher.

Keywords: flood forecasting; GMKHM-2D model; mixed runoff generation model; 2D kinematic wave flow model; Shanbei model; Xin'anjiang model

1. Introduction

Hydrological events always vary temporal and spatial distribution for nonlinear and interaction of hydrological process. Distributed hydrological model can describes hydrological mechanism and rainfall-runoff response with spatial precipitation input and the underlying surface information with GIS and RS technology. Coupled to the opening up of the technological feasibility of making such models possible has been the recognition and development of a demand for more distributed predictions as outlined by Freeze and Harlan (1969), Beven and O'Connell (1982), Abbott et al. (1986), Bathurst and O'Connell (1992). There are now a number of distributed hydrological models that are being used regularly for practical application, such as System Hydrological European (SHE) (Beven and O'Connell, 1982; Bathurst and O'Connell, 1992; Refsgaard and Storm, 1995), and the Institute of Hydrology Distributed Model (Beven, 1987; Calver and Wood, 1995). However, these models are not applied for flood operational forecast (Abbott et al., 1986). The Distributed Model Intercomparison Project (DMIP) was conducted to provide insights into the flood simulation

and forecast capabilities of 12 distributed hydrological models and a widely used lumped hydrological model (Smith et al., 2004a). Result shows that calibrated distributed hydrological models can perform better than calibrated lumped hydrological models, and distributed hydrological models should be coupled with conceptual rainfall-runoff model and distributed flow routing model for flood forecast purpose (Smith et al., 2004b).

From the last part of the 20th century, a large number of distributed hydrological models or semi-distributed hydrologic models (Beven and Kirkby, 1979; Singh, 1995; Yu, 2000; Todini and Ciarapica, 2001; Yang et al., 2002; Xiong and Guo, 2004; Jia et al., 2005; Xu and Cheng, 2010; Ye et al., 2010; Xue et al., 2013) have been developed for flood simulation and flood forecast. Grid-based distributed hydrological models, which consider the spatial variability, have become one of the most important tools for the present hydrological investigations (Abbott and Refsgaard, 1996; Vieux, 2001). Large number of distributed hydrological models such as Distributed Hydrology-Soil-Vegetation Model (DHSVM) model (Wigmosta et al., 1994), TOPographical Kinematic Approximation and Integration (TOPKAPI) model (Ciarapica and Todini, 2002; Liu et al., 2005),

TOPgraphy based hydrological MODEL (TOP-MODEL, Beven and Kirkby, 1979) have been developed and applied in practical use. The runoff generation of the models is based on saturation-excess mechanism (Bao, 2006; Li *et al.*, 2006; Wang *et al.*, 2007; Yao *et al.*, 2009, 2012; Bao *et al.*, 2010, 2011). These models perform well in humid regions and semi-humid regions, but do poorly in semi-arid basins and arid basins as the runoff generation mechanism is different in such regions (Bao *et al.*, 2016, 2017). The hydrological regime in semi-arid basins is extreme and highly variable, where flash floods from a single large storm can exceed the total runoff from a sequence of years (Wheater *et al.*, 2008). In recent years, some distributed hydrological models based on infiltration-excess runoff mechanism such as the Grid and Holtan method based distributed hydrological (Grid–Holtan) model (Bao *et al.*, 2016) have been developed for flood forecasting in semi-arid basins. Actually, there are saturation-excess runoff and infiltration-excess runoff in the semi-arid basin. The process of runoff generation on a hillslope with a complex topography should be calculated such that the process of flow concentration is adequately represented (Liu *et al.*, 2004). Depressions storing water is usually ignored because of using D8 method in sinks filled module of distributed hydrological process modeling (Govindaraju *et al.*, 1992; Singh, 1996, 2001; Singh and Frevert, 2002; Mohammadian *et al.*, 2004; Tayfur and Singh, 2006, 2007; Wang, 2010; Wang *et al.*, 2010b; Ye *et al*, 2013).

In this paper, a conceptual mixed runoff generation model combining saturation-excess runoff and infiltration-excess runoff is introduced for runoff generation calculation, 2D implicit finite difference kinematic wave model is applied to solve catchment sinks storing water for grid-based concentration of overland flow routing and 1D implicit finite difference kinematic wave model is used for channel routing. A Grid-and-Mixed-runoff-generation-and-two-dimensional-Kinematic-wave-based distributed hydrological model (GMKHM-2D model) coupling the mixed runoff generation model and 2D overland flow routing model based on kinematic wave theory was developed for flood simulation and forecast with digital drainage networks. In the GMKHM-1D model, 1D kinematic wave model is applied for flow routing and the modules of runoff generation and channel routing is the same to the GMKHM-2D model'. The GMKHM-2D model, the GMKHM-1D model, the Shanbei model and the Xin'anjiang (Zhao, 1983; 1992; Zhao *et al.*, 1995) model were applied in the upper reaches of the Qin River above Kongjiapo station, a tributary of the Yellow River, to investigate the reliability and improvement of the developed model.

2. Study basin

In order to investigate the applicability of the GMKHM-2D model, it was applied to the upper reaches of the Qin River above Kongjiapo station, a tributary of the Yellow River. The GMKHM-1D model, the Shanbei model and the Xin'anjiang model were also applied for comparison. The control area of Kongjiapo station is $1454\,km^2$ and is located between latitudes $36.18°N$ and $37.00°N$ and longitudes $111.95°E$ and $112.56°E$ (Figure 1). The basin average annual precipitation is $544\,mm$, 67.8% of which is within the period of the flood season. Three rainfall stations and one hydrological station are available in the Kongjiapo basin.

3. The GMKHM model

The GMKHM-2D model and the GMKHM-1D model are based on raster or grid data structures and use identical square DEM cells as primary computational elements for rainfall-runoff modeling, and adopt grids for the DEM as computational elements for distributed rainfall-runoff modeling with each element consisting of a runoff generation component and a grid-based flow routing component.

3.1. The mixed runoff generation model

Single point runoff generation mechanism is classified into, saturation-excess in humid regions and infiltration-excess in arid regions (Zhao, 1983). The mechanism of runoff is the mixture of both in semi-humid and semi-arid regions. The Xin'anjiang model, based on saturation-excess runoff generation, has established flood forecasting efficiency by long-term use in humid regions. However, in semi-arid region, the efficiency of the model is not always ideal. Therefore, a mixed runoff generation model with coupling the Green–Ampt infiltration-excess model (Green and Ampt, 1911) to the Xin'anjiang model runoff generation formation was developed in this paper.

According to the relation of soil moisture (W), water field capacity (W_T) and precipitation intensity (i) with infiltration rate (f), we have the following governing equations.

If $i > f$, $W < W_T$,

$$R_S = i - f, R_G = 0 \qquad (1)$$

If $i \leq f$, $W > W_T$,

$$R_S = 0, R_G = i \qquad (2)$$

If $i > f$, $W > W_T$,

$$R_S = i - f, \ R_G = f \qquad (3)$$

If $i < f$, $W < W_T$,

$$R_S = 0, \ R_G = 0 \qquad (4)$$

In which R_S is a surface runoff, R_G is groundwater runoff. The four equations are the complete runoff formula at a point of soil layer. The surface and groundwater runoff in saturated areas is calculated through the

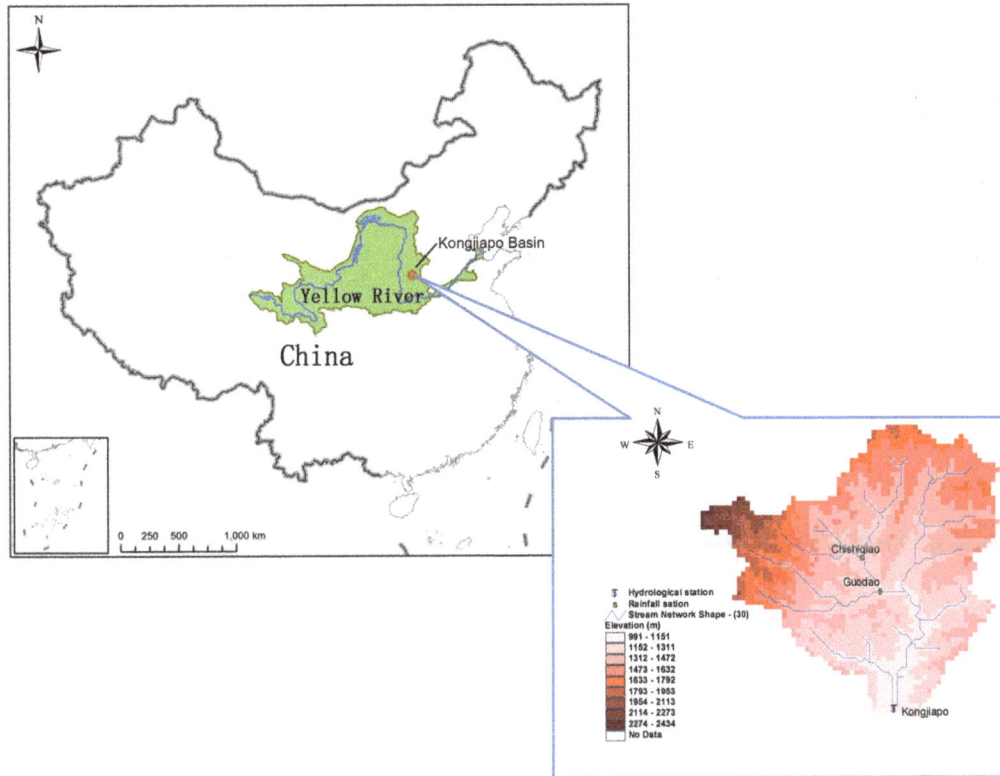

Figure 1. Sketch of the upper reaches of the Qin River above Kongjiapo station based on digital drainage networks.

Equations (2) and (3). The infiltration-excess runoff is calculated by Equation (1). The runoff formation on repletion of storage is opposite to the runoff formation in infiltration-excess, but the combination of both is the complete.

The Shanbei model (Zhao, 1983) is the simplest method simulating the infiltration-excess runoff. In the loess plateau, the runoff can be calculated by the infiltration curve:

If $i > f$,

$$R = R_S = i - f \qquad (5)$$

If $i < f$,

$$R = 0 \qquad (6)$$

During the process of a flooding event, the saturation area is varying. all precipitation is runoff ,which called the excess storage runoff in the saturation area. The runoff in unsaturated area is infiltration-excess overland flow. As flowing downhill through the saturation area, it can be as return flow, subsurface flow or groundwater flow. Therefore, the part of infiltration-excess overland runoff (the ratio is β) could be mixed with the runoff from saturation area. The other part (the ratio is $1-\beta$) as overland runoff directly flows to channel network. The area ratio generation infiltration-excess overland flow is considered as parameter (i_{mf}). Therefore, the mixed runoff generation model can be summarized as follows;

If $PE = P - E > 0$, then there is runoff, otherwise, not. Let R is the runoff amount calculated through the Xin'anjiang model. Then the runoff area is $FR = R/PE$ and unsaturated is $1 - FR$. A parabolic equation, which represents the area distribution of infiltration rate in the unsaturated area, defines as

$$\delta = 1 - \left(1 - \frac{f}{f_{mm}}\right)^{E_f} \qquad (7)$$

Where δ is the area in which infiltration rate is less than f and E_f is a parameter. $f_{mm} = (1 + E_f)f_m$. Here f_m is the point maximum of f and f_{mm} is the average maximum of f. The actual infiltration f can be calculated by the Green–Ampt infiltration model. i_{rs} is runoff amount generation from the unsaturated area:

If $PE \geq f_{mm}$ then

$$i_{rs} = \left(PE - f_{mm}\right)\left(1 - FR\right) i_{mf} \qquad (8)$$

If $PE < f_{mm}$ then

$$i_{rs} = \left\{PE - \frac{f_{mm}}{E_f + 1}\left[1 - \left(1 - \frac{PE}{f_{mm}}\right)^{E_f + 1}\right]\right\}$$
$$\times (1 - FR) i_{mf} \qquad (9)$$

The amount of runoff ($i_{rs}\beta$) flows through the saturated area and the amount $i_{rs}(1 - \beta)$ directly flow to channel networks.

If $PE + AU + \frac{i_{rs}}{FR}\beta < SMM$ then

$$R_S = \left\{PE + \frac{i_{rs}}{FR}\beta - SM + S + SM \right.$$
$$\left. \times \left[1 - \left(PE + \frac{i_{rs}}{FR}\beta + AU\right)/SMM\right]^{1+E_f}\right\} FR \qquad (10)$$

If $PE + AU + \frac{i_{rs}}{FR}\beta \geq SMM$ then

$$R_S = \left(PE + \frac{i_{rs}}{FR}\beta - SM + S\right)FR \qquad (11)$$

The water storage in the free water reservoir is

$$S = S + PE + \frac{i_{rs}}{FR}\beta - \frac{R_s}{FR} \qquad (12)$$

The other runoff generation and separation of runoff calculations are same with that of the Xin'anjiang model. In which, SM and SMM are respective the basin mean of free water storage capacity and free water storage capacity.

3.2. Routing model based on 2D kinematic wave

The influence of microtopography on the hydrological response of natural basins is significant (Singh, 2002). Some depressions in the overland storing water for some time during rainfall after ponding occurs. This water is retained on the surface and is ultimately evaporated or infiltrated (Cappelaere et al., 2003).

To solve this problem, this paper introduced 2D implicit finite difference kinematic wave model for overland flow routing. The errors contained in the DEM data are corrected with ANUDEM method (Hutchinson, 2004; Zhang et al., 2005). Once the retention storage in a grid cell is filled by excess rainfall, overland flow occurs.

Figure 2(a) is the partly original DEM data of the upper reaches of the Qin River above Kongjiapo station. From the figure, the elevation of the central grid is the lowest in that of all grids. The elevation of the central one ought to be added with D8 method in sinks filled module of most distributed hydrological modeling, which results in grid flow can be out of the central grid with single direction (shown in Figures 2(b) and (c)). However, for the contribution for the hydrograph of the basin outlet, depressions water storing capability is neglected in this hydrological modeling process. In this paper, two-directional model (Singh, 2002) was introduced for taking into consideration sinks water storing. The sinks in DEMs were calculated as a reservoir, so the flow can routing out of the central grid while the water stage being above that of the other grid (shown in Figure 2(d)).

The Saint–Venant equations with the continuity equation and momentum equation were used to describe the governing equation of overland flow. Overland flow model based on kinematic wave theory can perform enough precision in flood simulation and forecasting (Singh, 2002; Rui et al., 2008; Rui and Jiang, 2010). In the GMKHM-2D model, 2D implicit finite difference kinematic wave model was applied overland flow routing. Expressed as the two partial differential equations in the form of Equations (13)–(15) (Taky et al., 2009).

The governing equation for the overland flow is a 2D continuity equation:

$$\frac{\partial h}{\partial t} + \frac{\partial q_x}{\partial x} + \frac{\partial q_y}{\partial y} = r_e \qquad (13)$$

Where, h is the depth of overland flow, q_x is the unit discharge in the x-direction, q_y is the unit discharge in the y-direction, $r_e = (P_e - f)$, is excess rainfall rate, P_e is the net rainfall rate and f is the infiltration rate.

Momentum equation:

x-direction:

$$\frac{\partial u}{\partial t} + u\frac{\partial u}{\partial x} + v\frac{\partial u}{\partial y} = g\left(S_{0x} - S_{fx} - \frac{\partial h}{\partial x}\right) \qquad (14)$$

y-direction:

$$\frac{\partial u}{\partial t} + u\frac{\partial v}{\partial x} + v\frac{\partial v}{\partial y} = g\left(S_{0y} - S_{fy} - \frac{\partial h}{\partial y}\right) \qquad (15)$$

Where $S_{o(x,y)}$ is the land surface slopes in the x- and-y-directions, $S_{f(x,y)}$ is the friction slopes in the x- and y-directions, u, v are respectively the average velocity in the x- and- y-directions, in units m/s, g is the acceleration of gravity, in units m/s^2.

In accordance with the kinematic wave theory, ignoring the inertia term and the pressure term in the Saint–Venant equation, Equations (14) and (15) can be simplified to obtain kinematic wave equation, as following:

$$S_{fx} = S_{0x} \qquad (16)$$

$$S_{fy} = S_{0y} \qquad (17)$$

Where S_{ox} and S_{oy} are the land surface slopes in the x- and y-direction, respectively, and are calculated from DEM data. S_{fx} and S_{fy} are the friction slopes in the x- and y-direction, respectively.

The unit discharge in the x-direction q_x is defined using a general resistance form:

$$q_x = \alpha_x h^\beta \qquad (18)$$

Similarly, in the y-direction q_y:

$$q_y = \alpha_y h^\beta \qquad (19)$$

For the Manning equation β is a constant (Maidment, 1993), and the terms α_x, α_y are defined as:

$$\alpha_x = \frac{1}{n}|S_{0x}|^{1/2}\frac{S_{0x}}{|S_{0x}|} \qquad (20)$$

$$\alpha_y = \frac{1}{n}|S_{0y}|^{1/2}\frac{S_{0y}}{|S_{0y}|} \qquad (21)$$

$$\beta = 5/3 \qquad (22)$$

Where n is the Manning roughness coefficient. The terms $(S_{fy}/|S_{fy}|)$ and $(S_{fx}/|S_{fx}|)$ in Equations (20) and (21) are used to determine the direction of the flow.

(a)

1466	1502	1525	1469	1578
1536	1460	*1437*	1527	1584
1506	1535	1445	1490	1578

(b)

1466	1502	1525	1469	1578
1536	1460	*1445. 1*	1527	1584
1506	1535	1445	1490	1578

(c)

2	3	3	4	5
1	1	*3*	5	6
3	1	3	4	5

(d)

1466	1502	1525	1469	1578
1536	1460	*1437*	1527	1584
1506	1535	1445	1490	1578

(e)

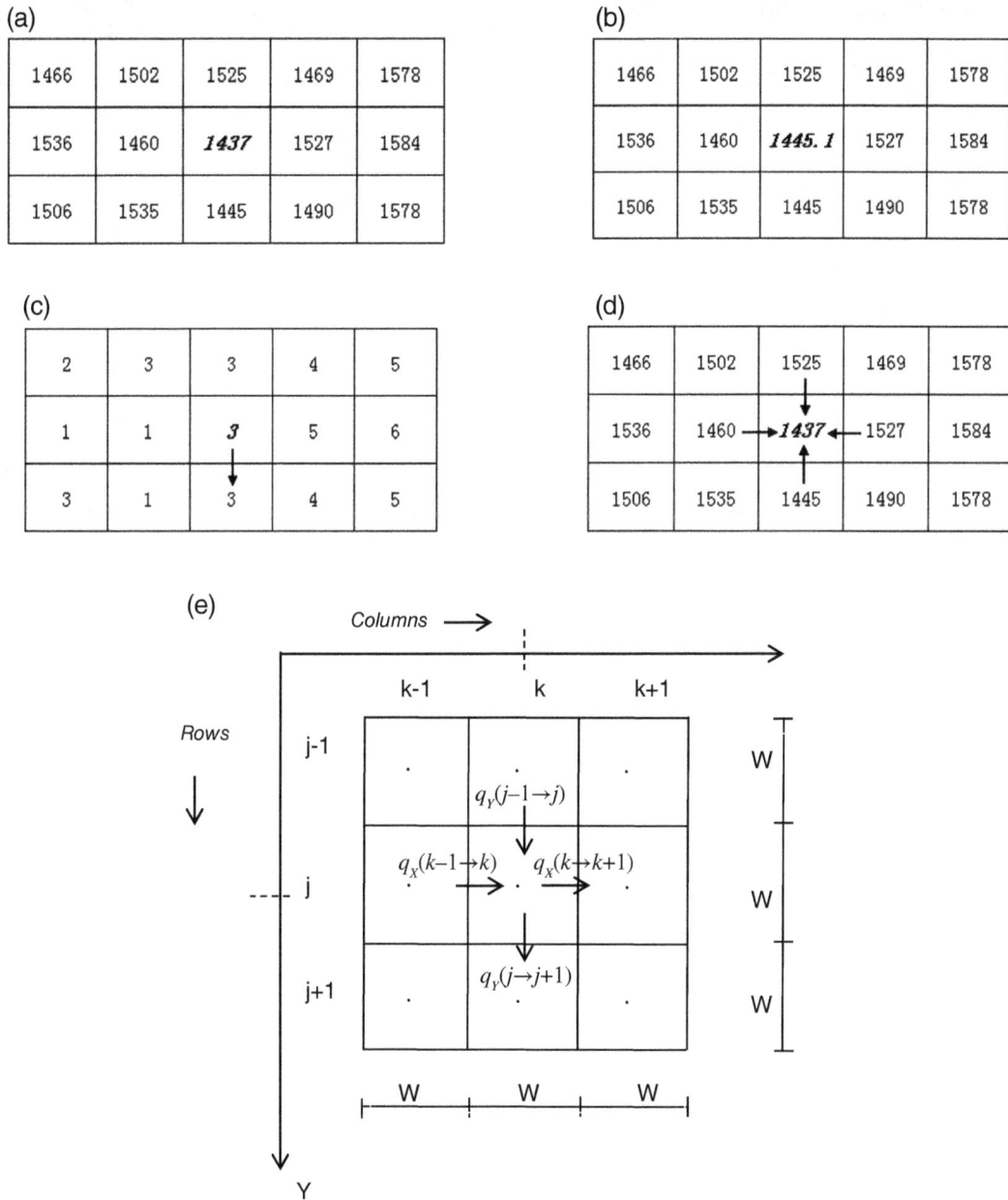

Figure 2. (a) The original DEM data. (b) Filled DEM data with D8 method. (c) Flow direction with D8 method. (d) Two-dimensional flow direction with the original DEM data. (e) Sketch of DEM grid in two-dimensional kinematic wave model.

The GMKHM-2D model used a grid where each grid cell is assumed a homogenous unit with one representative value for any hydraulic or hydrological parameter. Figure 2(e) is the sketch of DEM grid in two-dimension kinematic wave model.

For a grid unit, its principle of mass conservation can be described that the variation of the grid cell water volume is the difference between inflow and outflow in the dt time.

Calculation of h in Equations (18) and (19) is as follows.

For x or y-direction, the continuity equation can be simplified based on kinematic wave theory.

$$B\frac{\partial h}{\partial t} + \frac{B\sqrt{S_0}}{n}\frac{\partial}{\partial x}\left(h^{5/3}\right) = P_e(x,t) \quad (23)$$

where B is water width, L is channel length, $P_e(x,t)$ is the net rainfall rate.

Through to discretize Equation (12) with the four-point implicit finite difference scheme, the following equation can be shown.

$$\left(h^{5/3}\right)_{i+1}^{j+1} - \left(h^{5/3}\right)_{i}^{j+1} + D_1\left(h_{i+1}^{j+1} + h_i^{j+1}\right) = D_2 \quad (24)$$

where

$$\begin{cases} D_1 = \frac{n}{\sqrt{S_0}}\cdot\frac{\Delta x}{2\Delta t\theta} \\ D_2 = \frac{n\Delta x}{2\theta B\sqrt{S_0}}\left(\left(P_e\right)_{i+1}^j + \left(P_e\right)_i^j\right) \\ \quad -\frac{1-\theta}{\theta}\left(\left(h_{i+1}^j\right)^{5/3} + \left(h_i^j\right)^{5/3}\right) + \frac{n\Delta x}{2\theta\Delta t\sqrt{S_0}}\left(h_{i+1}^j + h_i^j\right) \end{cases} \quad (25)$$

For j = 0, the following equations can be shown.

$$\begin{cases} \left(h_1^1\right)^{5/3} - \left(h_0^1\right)^{5/3} + D_1\left(h_1^1 + h_0^1\right) = D_2\Big|_0^0 \\ \left(h_2^1\right)^{5/3} - \left(h_1^1\right)^{5/3} + D_1\left(h_2^1 + h_1^1\right) = D_2\Big|_1^0 \\ \cdots\cdots \\ \left(h_N^1\right)^{5/3} - \left(h_{N-1}^1\right)^{5/3} + D_1\left(h_N^1 + h_{N-1}^1\right) = D_2\Big|_{N-1}^0 \end{cases}$$

$$(26)$$

Where $N = L/\Delta x$, is iteration number. According to Equations (24) and (25), D_1 and D_2 can be calculated. Substituting the upstream boundary condition into Equation (26), we get the following equation.

$$\left(h_1^1\right)^{5/3} + D_1 h_1^1 = \beta_1 \qquad (27)$$

Where

$$\beta_1 = D_2\Big|_0^0 + \left(h_0^1\right)^{5/3} - D_1 h_0^1 \qquad (28)$$

Assuming the calculated runoff depth is z, so the following equation can be shown.

$$f(x) = D_1 x - \beta_1 + (z)^{5/3} \qquad (29)$$

h_1^1 can be calculated by bifurcated approach. In a similar way, according to Equation (26), runoff depth of every grid unit can be obtained and every grid discharge can be calculated by Manning formula in the studied basin.

Flow direction is based on friction slopes in the kinematic wave model. S_{0x} and S_{0y} can be calculated by DEM data.

The unit discharge in the x-direction between grid cell $(j, k-1)$ and (j, k).

If $S_{0x}^t(k-1 \rightarrow k) \geq 0$, then

$$q_x^t(k-1 \rightarrow k) = \frac{1}{n(j,k-1)}\left[h^t(j,k-1)\right]^{5/3}$$
$$\times \left[S_{0x}^t(k-1 \rightarrow k)\right]^{1/2} \qquad (30)$$

If $S_{fx}^t(k-1 \rightarrow k) < 0$, then

$$q_x^t(k-1 \rightarrow k) = \frac{-1}{n(j,k-1)}\left[h^t(j,k-1)\right]^{5/3}$$
$$\times \left[-S_{0x}^t(k-1 \rightarrow k)\right]^{1/2} \qquad (31)$$

Similarly, $q_x^t(k \rightarrow k+1)$, $q_y^t(j \rightarrow j+1)$, $q_y^t(j-1 \rightarrow j)$ can be calculated (Figure 2(e)).

3.3. Routing model based on 1-D kinematic wave

One-dimensional implicit finite difference kinematic wave model is used for channel routing in the GMKHM-2D model and the GMKHM-1D model. Compared with routing model based on 2D kinematic wave, 1-D kinematic wave routing model is simplified as the single direction. Therefore, the governing equation and momentum equation for 1-D overland flow can be written as

$$\frac{\partial h}{\partial t} - \frac{\partial q}{\partial x} = r_e \qquad (32)$$

Table 1. Parameter value of runoff generation in the GMKHM-2D model, the GMKHM-1D model and the Xin'anjiang model in Kongjiapo basin.

No.	Parameter	Physical description	Value
1	K_E	Ratio of potential evapotranspiration to pan evaporation	0.82
2	B	Distribution exponent of tension water capacity	0.30
3	C	Evapotranspiration coefficient of deeper layer	0.10
4	W_{UM}	Tension water storage capacity of upper layer	10.00
5	W_{LM}	Tension water storage capacity of lower layer	80.00
6	I_M	Ratio of impervious area to the total area of the basin	0.01
7	E_X	Distribution exponent of free water storage capacity	1.50
8	K_G	Outflow coefficient of free water storage to groundwater flow	0.30
9	K_I	Outflow coefficient of free water storage to interflow	0.40
10	C_G	Recession constant of groundwater storage	0.90
11	C_I	Recession constant of lower interflow storage	0.70
12	C_S	Recession constant of channel network storage	0.10
13	Lag	Lag time	0.00
14[a]	fc	Steady infiltration rate	8.60
15[a]	b_f	Distribution curve exponent of infiltration capacity	0.07
16[a]	fp	Ratio of infiltration-excess area to the total area of the basin	0.85

[a]Represents the parameter of infiltration-excess runoff generation in the GMKHM-2D model and the GMKHM-1D model.

Table 2. Parameter value of Green–Ampt model.

Soil type	Effective hydraulic conductivity	Soil suction in the wetting front	Saturated moisture content
Clay	0.03	31.63	0.475
Sand	11.78	4.95	0.436
Loam	0.34	8.89	0.436

$$S_0 - S_f = 0 \qquad (33)$$

Calculation steps of numerical solution for 1D kinematic wave overland flow routing model are similar to that of 2D kinematic wave (from Equation (23) to Equation (31)), and same to that of 1D kinematic wave channel routing model with replacing h, q (Equation (32)) for channel.

According to the global DEM data and the land use data resolution of $30'' \times 30''$ provided by USGS (2015), the Kongjiapo basin was extracted. The types of land use are reclassified into four main types: forest, grassland, cropland, and water, which respond to 0.1, 0.17, 0.035 and 0.0 of roughness coefficient value respectively (Yao *et al.*, 2012). The interflow and groundwater runoff concentration calculations are same with that of the Xin'anjiang model.

Table 3. Simulated results of flood events in Kongjiapo station.

	Flood No.	Precipitation (mm)	Peak flow (m³/s)	Relative error of flood volume (%)				Relative error of flood peak (%)				Peak time error (hour)				Nash–Sutcliffe efficiency value			
				G2	GI	S	X	G2	GI	S	X	G2	GI	S	X	G2	GI	S	X
Calibration	19580802	28.9	280	−10.7	−10.8	−9.2	26.1	6.5	17.1	−20.8	−25.3	0	1	−1	−2	0.82	0.78	0.82	0.23
	19640713	28.5	264	−12.9	−13.1	6.5	37.5	−5.7	10.1	−15.1	−34.9	1	2	−2	−3	0.81	0.77	0.81	0.21
	19750725	68.6	205	14.7	20.1	15.3	59.7	10.1	22.3	−20.1	−10.3	0	1	−1	−1	0.75	0.69	0.70	0.66
	19810815	82.7	380	−11.1	−11.2	11.2	81.3	−10.1	−10.4	−13.6	−23.1	−1	0	1	1	0.78	0.83	0.79	0.11
	19820801	72.1	214	17.5	22.1	21.1	60.8	7.2	13.2	−13.2	−12.7	0	1	−1	−1	0.72	0.70	0.70	0.71
	19850913	116.3	1026	−15.1	−13.8	−17.5	−37.2	7.1	10.1	−13.8	−73.2	0	0	−1	−4	0.94	0.91	0.85	0.06
	19880813	79.9	613	−10.1	−9.9	−13.6	23.3	−10.5	−9.7	16.3	−37.0	1	1	1	3	0.75	0.78	0.71	0.42
Validation	19930803	147.5	2070	−5.8	−6.5	−9.0	−38.7	0.47	5.1	−11.9	−48.2	0	0	1	2	0.95	0.99	0.88	0.17
	19960804	59.3	269	−3.9	6.5	9.6	9.5	−10.2	−9.8	11.0	−11.1	2	3	−2	−4	0.87	0.86	0.81	0.63
	20010726	50.1	390	−30.2	−28.3	33.4	53.1	4.8	18.8	13.2	−72.9	0	1	−1	−1	0.81	0.83	0.79	0.38
	The mean absolute value			13.2	14.2	14.6	42.7	7.3	12.7	14.9	34.9	0.5	1	1.2	2.2	0.82	0.81	0.79	0.36

GI, the GMKHM-1D model; G2, the GMKHM-2D model; S, the Shanbei model; X, the Xin'anjiang model.

4. Experiments and analysis application of models

4.1. Application and comparison with the GMKHM-1D model, the Shanbei model and the Xin'anjiang model

In order to verify the applicability of the GMKHM-2D model, it was applied to the upper reaches of the Qin River above Kongjiapo station for flood simulation. The GMKHM-1D model, the Shanbei model and the Xin'anjiang model were used for comparison. Ten representative flood events from 1958 to 2001 are selected. Flood events from 1958 to 1988 were applied for models calibration and the other three events were used to verify the models in the studied basin. In this paper, the time step is 1 h due to the data constraints in Kongjiapo basin (Cundy and Tento, 1985; Wang *et al.*, 2010a). Table 3 is simulated results of 10 representative flood events in Kongjiapo basin. The statistic values of models results consist of the relative error of flood volume, the relative error of flood peak discharge, the peak time error and the Nash–Sutcliffe efficiency value (Schaefli and Gupta, 2007). The Shanbei model and the Xin'anjiang model parameters are calibrated with artificial optimization method based parameters physical characteristics and historical flood data (Zhao, 1983). For better comparison, the parameters of saturation-excess runoff generation are unified in the GMKHM-2D model, the GMKHM-1D model and the Xin'anjiang model, and the parameters of infiltration-excess runoff generation in the GMKHM-2D model agree with that in the GMKHM-1D model and the Shanbei model. The routing parameter is Muskingum coefficient, which is valued 0.2 in the Xin'anjiang model and the Shanbei model. The runoff generation parameters of GMKHM-2D model, the GMKHM-1D model, the Shanbei model and the Xin'anjiang model can be seen in the Tables 1 and 2.

For the GMKHM-2D model and the GMKHM-1D model, most peak flows of simulated flood events were overestimated, probably because of the theory of runoff generation mechanism, since the two grid-based distributed hydrological models assumes the generation of overland flow as precipitation is over the infiltration rate. For the Xin'anjiang model, all of flood peak flows the simulated events were underestimated. The main reason is that the runoff generation calculation model is based on saturation-excess mechanism in the Xin'anjiang model, but infiltration-excess runoff generation is the main runoff in Kongjiapo basin. It caused that the Xin'anjiang model performed poor for peak occur time and flood volume simulation and Nash–Sutcliffe efficiency value. Therefore, the mean value of peak discharge simulated by the GMKHM-2D model and the GMKHM-1D model is better obviously than that of the Xin'anjiang model. From Table 3, for the relative error of flood volume, the peak time error and the Nash–Sutcliffe efficiency value, the two grid-based distributed hydrological models and the Shanbei model performed better than the Xin'anjiang model. This proves the advantage of mixed runoff generation mechanism in simulating flood hydrograph, especially peak flows simulation in semi-arid basin.

4.2. Comparison of the GMKHM-2D model and the GMKHM-1D model

For the advantages of GIS technology, DEM and excess infiltration mechanism, the GMKHM-2D model and the GMKHM-1D model performed well in simulating observed flood thin hydrograph in Kongjiapo basin. The two distributed hydrological models obtained the closed accuracy in the relative error of flood volume and the Nash–Sutcliffe efficiency value of all flood events, but the GMKHM-2D model performed significantly better in flood peak discharge and peak time simulation (Figure 3(a)). The reason is depressions storing water in the basin overland is neglected because of using single direction with D8 method in sinks filled module of hydrological process modeling, which results in flood peak discharge overestimating and peak time advances from zero to three hours. It is serious

(a)

(b)

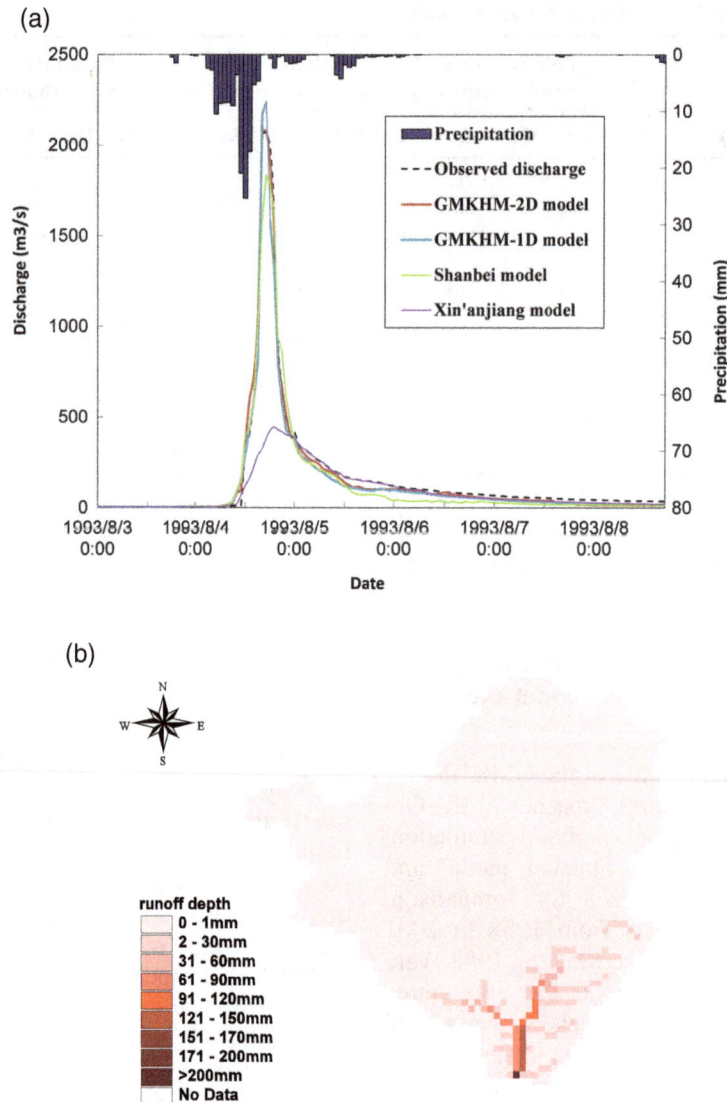

Figure 3. (a) Model results of FloodNo.19930803. (b) The peak (19th) time step simulated spatial map of the GMKHM-2D model runoff depth for FloodNo.19580802.

for flood simulation and forecasting in semi-arid basin. In the GMKHM-2D model, 2D implicit finite difference kinematic wave model was introduced to solve basin depressions storing water for grid-based concentration of overland flow routing, so the relative error of flood peak discharge and the peak time error is 7.3%, 0.5 h, respectively. Distributed hydrological models can represent the spatial map of the hydrological processes, which is one of the most representative feature in distributed hydrological modeling. From Figure 3(b), it can be clear that the peak (19th) time step simulated spatial map of the GMKHM-2D model runoff depth for FloodNo.19580802 is very reasonable and close to observation. Wherever, the relative error of flood peak discharge and the peak time error simulated by the GMKHM-1D model is 12.7%, 1 h, respectively. It proves the reliability and improvement of the developed GMKHM-2D model.

5. Conclusions

The GMKHM-2D model coupling a mixed conceptual runoff generation model and 2D overland flow routing model based on kinematic wave theory was developed for flood simulation and forecasting. The mixed runoff generation model with coupling the Green–Ampt infiltration-excess model to the Xin'anjiang model runoff generation formation on repletion of storage was applied for runoff generation calculation in semi-arid regions. Two-dimensional implicit finite difference kinematic wave model was introduced to solve depressions storing water of basins overland for grid-based concentration of overland flow routing. In order to investigate the reliability and improvement of the GMKHM-2D model, the GMKHM-1D model, the Shanbei model and the Xin'anjiang model were also employed for comparison in the Kongjiapo basin.

Depressions storing water of overland flow routing was taken into consider, so the GMKHM-2D model perform higher accuracy, especially in flood peak discharge and peak time simulation in semi-arid basin. Compared with the Xin'anjiang model, advantages of the two grid-based distributed hydrological models show obviously flood simulation and forecasting. However, compared with the Xin'anjiang model, five parameters are added to the GMKHM-2D model. The further research should focus on the parameter physical meanings and on how to estimate the parameters according to the physical aspect such as soil and land cover information.

Acknowledgements

This work was funded by the National Natural Science Foundation of China (Grants No. 51509043, 41105068, 91537211 and 51679061), the National Key Research and Development Program of China (Grants No. 2016YFC0402702, 2016YFC0402701 and 2016YFC0402705, the 1st Young Talents Program Project of China Meteorological Administration (2014–2017) and the 3rd Young Talents Program Project of China Meteorological Administration (2016–2019)).

Reference

Abbott MB, Refsgaard JC. 1996. *A Discussion of Distributed Hydrological Modelling. Distributed Hydrological Modelling.* Springer: Netherlands; 255–278.

Abbott MB, Bathurst JC, Cunge JA, O'Connell PE, Rasmussen J. 1986. An introduction to the European Hydrological System-Systeme Hydrologique Europeen, "SHE", 1: History and philosophy of a physically-based, distributed modelling system. *Journal of Hydrology* **87**(1–2): 45–59.

Bao, H. J. 2006 Research on the application of flood forecasting and scheduling model in the basin of Yishusi. ME dissertation,. Hohai University, Nanjing.

Bao HJ, Wang LL, Li ZJ, Zhao LN, Zhang GP. 2010. Hydrological daily rainfall-runoff simulation with BTOPMC model and comparison with Xin'anjiang model. *Water Science and Engineering* **3**(2): 121–131.

Bao HJ, Zhao LN, He Y, Li ZJ, Wetterhall F, Cloke HL, Pappenberger F, Manful D. 2011. Coupling ensemble weather predictions based on TIGGE database with Grid-Xinanjiang model for flood forecast. *Advances in Geosciences* **29**: 61–67.

Bao HJ, Wang L, Li ZJ, Yao C. 2016. A distributed hydrological model based on Holtan runoff generation theory. *Journal of Hohai University (Natural Sciences)* **44**(4): 340–346 (in Chinese).

Bao HJ, Li ZJ, Wang LL, Huang XX, Yao C, Qu CY, Zhao LQ, Zhang QH. 2017. Flash flood Forecasting Method Based On Distributed Hydrological Models In A Small Basin And Application. *Torrential Rain and Disasters* **36**(2): 92–99 (in Chinese).

Bathurst JC, O'Connell PE. 1992. Future of distributed modeling: the systeme hydrologique European. *Hydrological Processes* **6**: 265–277.

Beven K. 1987. Reply to "Comments on "On subsurface stormflow: prediction with simple kinematic theory for saturated and unsaturated flows" by Keith Beven". *Water Resources Research* **23**(4): 749.

Beven KJ, Kirkby MJ. 1979. A physically based, variable contributing area model of basin hydrology/un modèle à base physique de zone d'appel variable de l'hydrologie du bassin versant. *Hydrological Sciences Bulletin* **24**(1): 43–69.

Beven K J, O'Connell P E. 1982.On the role of physically-based distributed modelling in hydrology[R]. Institute of hydrology. NO.81.

Calver A, Wood WL. 1995. The Institute of Hydrology distributed model. In *Computer Models of Watershed Hydrology*. Water Resources Publication: Highlands Ranch, Colorado; 809–846.

Cappelaere B, Vieux BE, Peugeot C, Maia A, Séguis L. 2003. Hydrologic process simulation of a semiarid endoreic catchment in Sahelian West Niger. 2. Model calibration and uncertainty characterization. *Journal of Hydrology* **279**(1–4): 244–261.

Ciarapica L, Todini E. 2002. TOPKAPI: a model for the representation of the rainfall-runoff process at different scales. *Hydrological Processes* **16**(16): 207–229.

Cundy TW, Tento SW. 1985. Solution to the kinematic wave approach to overland flow routing with rainfall excess given by Philip's equation. *Water Resources Research* **21**(8): 1132–1140.

Freeze RA, Harlan RL. 1969. Blueprint for a physically-based, digitally-simulated hydrologic response model. *Journal of Hydrology* **9**: 237–258.

Govindaraju RS, Kavvas ML, Tayfur G. 1992. A simplified model for two-dimensional overland flows. *Advances in Water Resources* **15**: 133–141.

Green W, Ampt G. 1911. Studies on soil physics, part 1, the flow of air and water through soils. *The Journal of Agricultural Science* **4**(1): 1–24.

Hutchinson,M F. 2004. ANUDEM version 5.1 User Guide. Centre for Resource and Environmental Studies. The Australian National University, Canberra, pp. 3–26.

Jia YW, Wang H, Wang JH, Luo XY, Zhou ZH, Yan DH, Qin DY. 2005. Development of verification of a distributed hydrologic model for the Yellow river basin. *Journal of Natural Resources* **20**(2): 300–308.

Li, Z. J., Cheng, Y., and Xu, P. Z. 2006. Application of GIS-based hydrological models in humid watersheds. Water for Life: Surface and Ground Water Resources, In Proceedings of the 15th APD-IAHR & ISMH, 685–690. Madras.

Liu QQ, Chen JC, Li JC, Singh VP. 2004. Two-dimensional kinematic wave model of overland-flow. *Journal of Hydrology* **29**: 28–41.

Liu Z, Mlv M, Todini E. 2005. Flood forecasting using a fully distributed model: application of the TOPKAPI model to the Upper Xixian catchment. *Hydrology and Earth System Sciences* **9**(4): 347–364.

Maidment DR. 1993. *Hydrology Handbook.* McCraw-Hill: New York.

Mohammadian A, Tajrishi M, Azad FL. 2004. Two-dimensional numerical simulation of flow and geo-morphological processes near headlands by using unstructured grid. *International Journal of Sediment Research* **19**(4): 258–277.

Refsgaard JC, Storm B. 1995. MIKE SHE. In *Computer Models of Watershed Hydrology*. Water Resources Publications: Highlands Ranch, Colorado; 809–846.

Rui XF, Jiang CY. 2010. Review of research of hydro-goemorpjological processes interaction. *Advance in Water Science* **21**(4): 444–449 (in Chinese).

Rui XF, Yu M, Liu FG, Gong XL. 2008. Calculation of watershed flow concentration based on the grid drop concept. *Water Science and Engineering* **1**(1): 1–9.

Schaefli B, Gupta HV. 2007. Do Nash values have value? *Hydrological Processes* **21**(15): 2075–2080.

Singh VP. 1995. Watershed modeling. In *Computer Models of Watershed Hydrology*, Singh VP (ed). Water Resources Publications: Highlands Ranch, Colorado; 1–22.

Singh VP. 1996. *Kinematic Wave Modeling in Water Resources: Surface Water Hydrology.* Wiley: New York.

Singh VP. 2001. Kinematic wave modelling in water resources: a historical perspective. *Hydrological Processes* **15**(4): 671–706.

Singh VP. 2002. Is hydrology kinematic? *Hydrological Processes* **16**(3): 667–716.

Singh VP, Frevert D. 2002. *Mathematical model of small watershed hydrology and applications.* Water Resources Publications: Littleton, Colorado.

Smith MB, Georgakakos KP, Liang X. 2004a. The distributed model intercomparison project (dmip). *Journal of Hydrology* **298**(1): 1–3.

Smith MB, Seo DJ, Koren VI, Reed SM, Zhang Z, Duan Q, Moreda F, Cong S. 2004b. The distributed model intercomparison project (dmip): motivation and experiment design. *Journal of Hydrology* **298**(1): 4–26.

Taky A, Mailhol JC, Belaud G. 2009. Using a furrow system for surface drainage under unsteady rain. *Agricultural Water Management* **96**: 1128–1136.

Tayfur G, Singh VP. 2006. Kinematic wave model of bed profiles in alluvial channels. *Water Resources Research,* **42**: W06414. https://doi .org/10.1029/2005WR004089.

Tayfur G, Singh VP. 2007. Kinematic wave model for transient bed profiles in alluvial channels under nonequilibrium conditions. *Water Resources Research* **43**: W12412. https://doi.org/10.1029/ 2006WR005681.

Todini E, Ciarapica L. 2001. The TOPKAPI model. In *Mathematical Models of Large Watershed Hydrology*, Singh VP, Frevert DK (eds). Water Resources Publications: Littleton.

U.S. Geological Survey (USGS) GTOP30. 2005. http://edc.usgs.gov/ products/elevation/gtopo30/gtopo 30.html (accessed 20 May 2010).

Vieux, B. E. 2001. Distributed hydrological modeling using GIS. Dordrecht, the Netherlands. Kluwner Academic Publishers.

Wang, L. L. 2010. Study on Grid and Exceed-Infiltration Runoff Mechanism Based Hydrologic Models and Comparison Application. PhD Dissertation. Nanjing: Hohai University.

Wang LL, Li ZJ, Bao HJ. 2007. Application of hydrological models based on DEM in the Yihe basin. *Journal of Hydrologic Engineering* **37**(S1): 417–422.

Wang LL, Li ZJ, Bao HJ. 2010a. Application of developed grid-GA distributed hydrologic model in semi-humid and semi-arid basin. *Transactions of Tianjin University* **16**(3): 209–215.

Wang LL, Li ZJ, Bao HJ. 2010b. Development and comparison of grid-based distributed hydrological models for excess-infiltration runoffs. *Journal of Hohai University (Natural Sciences)* **38**(2): 123–128. https://doi.org/10.3876/j.issn.1000-1980.2010.02.001.

Wheater H, Sorooshian S, Sharma KD. 2008. *Hydrological Modeling in Arid and Semi-arid Areas*. Cambridge University Press: New York.

Wigmosta MS, Vail LW, Lettenmier DP. 1994. A distributed hydrology vegetation model for complex terrain. *Water Resources Research* **30**: 1165–1679.

Xiong LH, Guo SL. 2004. Effects of the catchment runoff coefficients on the performance of TOPMODEL in rainfall runoff modeling. *Hydrological Processes* **18**: 1823–1836.

Xu ZX, Cheng L. 2010. Progress on studies and applications of distributed hydrological models. *Journal of Hydrologic Engineering* **41**(9): 1009–1017 (in Chinese).

Xue X, Hong Y, Limaye AS, Gourley JJ, Huffman GJ, Khan SI, Dorji J, Chen S. 2013. Statistical and hydrological evaluation of TRMM-based Multi-satellite Precipitation Analysis over the Wangchu Basin of Bhutan: are the latest satellite precipitation products 3B42V7 ready for use in ungauged basins? *Journal of Hydrology* **499**(13–14): 91–99.

Yang DW, Herath S, Musiake K. 2002. A hillslope-based hydrological model using catchment area and width functions. *Hydrological Sciences Journal* **47**(1): 49–65.

Yao C, Li ZJ, Bao HJ, Yu ZB. 2009. Application of a developed Grid-Xin'anjiang model to Chinese watersheds for flood forecasting purpose. *Journal of Hydrologic Engineering* **14**(9): 923–934. https:// doi.org/10.1061/(ASCE)HE.1943-5584.0000067.

Yao C, Li Z, Yu Z, Zhang K. 2012. A priori parameter estimates for a distributed, grid-based xinanjiang model using geographically based information. *Journal of Hydrology* **468-469**(6): 47–62.

Ye A, Duan Q, Zeng H, Li L, Wang C. 2010. A distributed time-variant gain hydrological model based on remote sensing. *Journal of Resources and Ecology* **1**(3): 222–230.

Ye A, Duan Q, Zhan C, Liu Z, Mao Y. 2013. Improving kinematic wave routing scheme in Community Land Model. *Hydrology Research* **44**: 886–903.

Yu Z. 2000. Assessing the response of subgrid hydrologic processes to atmospheric forcing with a hydrologic model system. *Global and Planetary Change* **25**(1–2): 1–17.

Zhang CX, Yang QK, Duan JJ. 2005. A method to build high quality DEMs-ANUDEM method. *Chinese Agriculture Science Bulletin* **21**(12): 411–215.

Zhao RJ. 1983. *Watershed Hydrological Model: Xin'anjiang Model and Shanbei Model*. China WaterPower Press: Beijing (in Chinese).

Zhao RJ. 1992. The Xin'anjiang model applied in China. *Journal of Hydrology* **135**(1–4): 371–381. https://doi.org/10.1016/ 0022-1694(92)90096-E.

Zhao, R. J., Liu, X. R., & Singh, V.P.1995. The Xinanjiang model. In Proceedings of the Oxford Symposium on Hydrological Forecasting Iahs Publications, 135(1), pp. 371–381.

Intraseasonal rainfall variability in the Bay of Bengal during the Summer Monsoon: Coupling with the ocean and modulation by the Indian Ocean Dipole

Siraput Jongaramrungruang,[1,2,3] Hyodae Seo[2]* and Caroline C. Ummenhofer[2]

[1]*Trinity College, University of Cambridge, UK*
[2]*Physical Oceanography Department, Woods Hole Oceanographic Institution, MA, USA*
[3]*Now at Division of Geological and Planetary Sciences, California Institute of Technology, CA, USA*

*Correspondence to:
H. Seo, 266 Woods Hole Road,
MS#21, Woods Hole, MA
02543, USA.
E-mail: hseo@whoi.edu*

Abstract

The Indian Summer Monsoon rainfall exhibits pronounced intraseasonal variability in the Bay of Bengal (BoB). This study examines the intraseasonal rainfall variability with foci on the coupling with sea surface temperatures (SST) and its interannual modulation. The lagged composite analysis reveals that, in the northern BoB, SST warming leads the onset of intraseasonal rainfall by 5 days. Latent heat flux is reduced before the rain event but is greatly amplified during the rainfall maxima. Further analysis reveals that this intraseasonal rainfall-SST relationship through latent heating is strengthened in negative Indian Ocean Dipole (IOD) years when the bay-wide local SST is anomalously warm. Latent heat flux is further increased during the intraseasonal rainfall maxima leading to strengthened rainfall variability. The moisture budget analysis shows this is primarily due to stronger low-level moisture convergence in negative IOD years. The results provide important predictive information on the monsoon rainfall and its active/break cycles.

Keywords: Bay of Bengal; intraseasonal; monsoon; IOD

1. Introduction

During boreal summer (June to September, JJAS), strong southwesterly monsoon winds carry moisture from the tropical Indian Ocean onto the Indian subcontinent and Southeast Asia, producing heavy rainfall events in nearby countries (Figure 1(a)), the period known as the Indian Summer Monsoon (ISM) (Webster *et al.*, 1998; Wang, 2005). As the ISM precipitation accounts for almost 90% of the total annual rainfall, it exerts a significant impact on agriculture and livelihoods of over a billion people. One of the most important characteristics of the ISM is its intraseasonal variability of deep convection and precipitation (Lawrence and Webster, 2001; Goswami, 2005; Goswami and Ajayamohan, 2001, Figure 1(b)), for which the wet and dry conditions occur intermittently every 2–3 weeks (Vecchi and Harrison, 2002). The intraseasonal rainfall variability accounts for nearly 80% of the total variability in the Bay of Bengal (BoB) in JJAS (Figure 1(c)). Since the rainfall variability in the BoB is strongly coupled with the upper ocean processes, we expect that the sea surface temperature (SST) and surface heat fluxes significantly influence the characteristics of the intraseasonal rainfall variability (e.g. Lawrence and Webster, 2001). Indeed, a number of observational and modeling studies have examined the intraseasonal relationship between SST and rainfall, showing that the intraseasonal SST warming precedes the deep convection and heavy precipitation (Sengupta

et al., 2001; Vecchi and Harrison, 2002; Fu *et al.*, 2003; Seo *et al.*, 2014; Xi *et al.*, 2015).

On the interannual time scale, tropical Indian Ocean SST substantial variability associated with the Indian Ocean Dipole (IOD). The IOD is characterized by an anomalous warming or cooling in the southeastern Indian Ocean, accompanied by an SST anomaly of the opposite sign in the western Indian Ocean, during the negative or positive phase of the IOD, respectively (Saji *et al.*, 1999; Murtugudde and Busalacchi, 1999; Webster *et al.*, 1999, Figure 3). The impact of the IOD on the ISM rainfall has been well established (e.g. Ashok *et al.*, 2001, 2004; Gadgil *et al.*, 2004; Ihara *et al.*, 2007; Ummenhofer *et al.*, 2011). As the IOD switches its phase, the change in the background SST modulates the strength of the ocean-atmosphere coupling, intraseasonal convection, and thus the monsoon circulation over the BoB (Ajayamohan *et al.*, 2008). The IOD also modulates the well-known teleconnection between the El Niño-Southern Oscillation (ENSO) and the ISM (Ashok *et al.*, 2004; Ihara *et al.*, 2007; Ummenhofer *et al.*, 2011).

A key question of the study is how the intraseasonal rainfall variability and its coupling with the BoB ocean are modulated by the IOD. In this study, the impact of ENSO is not rigorously considered, as previous studies demonstrated that ENSO tends to be relatively uncorrelated with summertime intraseasonal convective variability in the BoB (Lawrence and Webster, 2001; Ajayamohan *et al.*, 2008). Nevertheless, additional

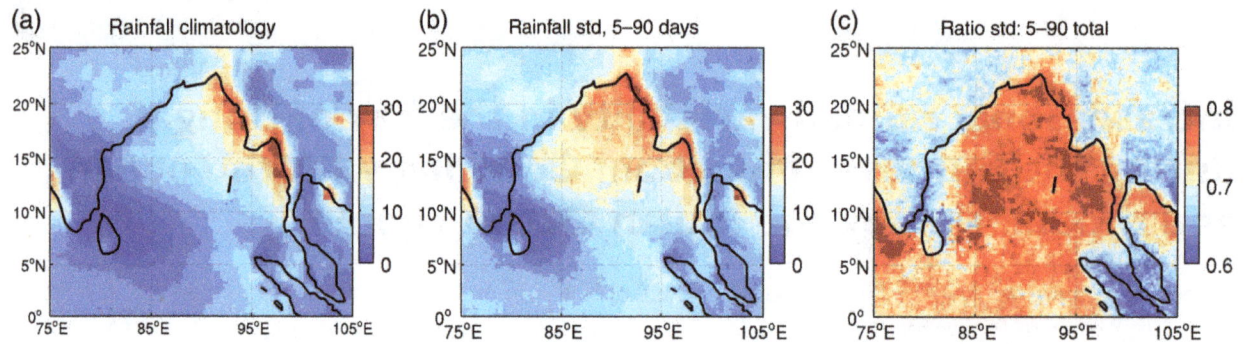

Figure 1. (a) JJAS rainfall climatology (mm day^{-1}) from TRMM 3B42 (1998–2015). (b) SD of the 5–90 day filtered JJAS rainfall, and (c) the ratio between the SD of 5–90 day filtered JJAS rainfall to SD of the unfiltered rainfall.

analysis has been carried out that isolates the IOD influences from that of ENSO, and the results are provided in the Supporting information.

The article is organized as follows: Section 2 describes the data and methodology. Section 3 presents the results investigating the intraseasonal air–sea interaction in Section 3.1 and its interannual variability in Section 3.2, followed by the atmospheric moisture budget analysis in Section 3.3. Section 4 is a summary and discussion of implications.

2. Data and methodology

We use the daily precipitation estimate from the Tropical Rainfall Measuring Mission (TRMM) version 3B42 on a 0.25° × 0.25° grid for 1998–2015 (Huffman *et al.*, 2007). We also use the Global Precipitation Climatology Project (GPCP) 1° daily rainfall product version 1.2 (Huffman *et al.*, 2001) for the period of 1997–2015. For the SST, the 0.25° daily NOAA Optimum Interpolation SST analysis (Reynolds *et al.*, 2007) is used. The latent heat (LH) flux for the same period is taken from the daily objectively-analyzed air-sea fluxes (OAFlux) version 3 (Yu and Weller, 2007). The column-integrated moisture budget analysis is based on the National Centers for Environmental Prediction-National Center for Atmospheric Research reanalysis dataset (Kalnay *et al.*, 1996). The time period of the analysis is restricted to 1998–2015 due to the availability of TRMM rainfall data. To investigate the intraseasonal variations of oceanic and atmospheric properties, the data are filtered using 5–90-day Butterworth band-pass filtering (e.g. Xi *et al.*, 2015).

3. Analysis and results

3.1. Intraseasonal SST-LH-rainfall relationship

To examine the evolution of the ocean-atmosphere fields in association with the intraseasonal rainfall variability, this study adopts the approach by Xi *et al.* (2015) and extends the analysis to 2015. Specifically, lagged composite anomalies for SST, wind speed (WS), and rainfall are constructed for ±25 days around

the onset of the heavy intraseasonal rainfall events. Heavy or extreme rainfall events are defined as when the intraseasonal precipitation anomaly averaged over the BoB (15°–23°N, 85°–95°E) exceeds one standard deviation (SD). If two intraseasonal rainfall events take place <10 days apart, they are considered as one event. This way, we obtained a total of 83 heavy intraseasonal rain events for the period of JJAS 1998–2015.

Figure 2(a) shows the zonally averaged (85°–95°E) lagged composite evolutions of the intraseasonal anomalies against the intraseasonal heavy rainfall peaks. The SST warming emerges north of 10°N about 10–15 days before the onset of the precipitation, with the maximum SST found at 15°N about approximately 5-days before the peak. This pre-convection period is quiescent, with anomalously negative WS and LH anomalies leading to the warming of the sea surface. The transition from the positive to negative SST anomaly occurs on Day 0 at 18°N and around Day 2 north of 18°–22°N. This period corresponds to the arrival of the intraseasonal convective anomaly, accompanied by the maximum positive LH and WS anomalies (Figure 2(b)). The enhanced moisture transfer to the atmosphere then facilitates the extreme precipitation events. During and after the peak rainfall, the heat loss from the sea surface through evaporation and upper ocean mixing results in SST cooling that lasts for >+10 days (Sengupta and Ravichandran, 2001; Sengupta *et al.*, 2001). These northward propagating intraseasonal anomalies and their phase relationships are driven by the monsoon intraseasonal variability inherent to the summer atmospheric circulation over the ISM region (e.g. Goswami, 1998; Jiang *et al.*, 2004). However, recent studies (e.g. Fu *et al.*, 2003, 2007) also suggest the leading role of the SST anomalies and air–sea interaction in the northward propagation of the intraseasonal disturbances in rainfall and clouds over the ISM region (Yasunari, 1979, 1980; Sikka and Gadgil, 1980). This is discussed in the following section.

3.2. Modulation of the intraseasonal ocean-atmosphere coupling by the IOD

How does this intraseasonal air-sea coupling vary interannually in association with the IOD? To answer this,

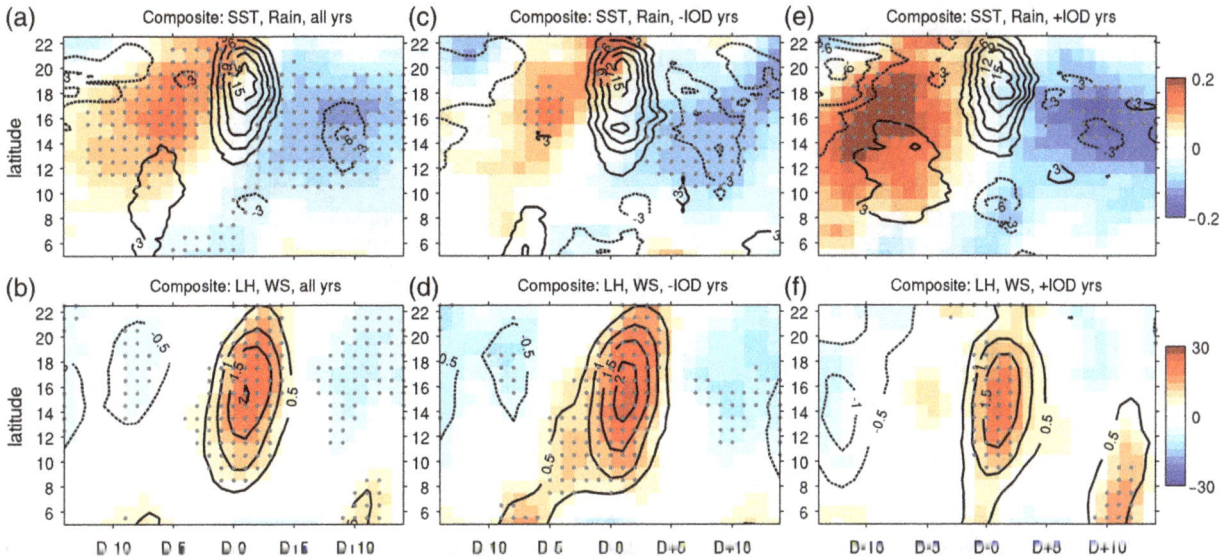

Figure 2. Lagged composite evolutions of the 5–90 day filtered (a) SST (color, °C) and rainfall (contours, mm day^{-1}) and (b) LH (shading, Wm^{-2}, positive out of the ocean) and WS (ms^{-1}, contours) associated with the extreme intraseasonal rainfall events. The extreme intraseasonal rainfall events are defined as when the normalized 5–90 day rainfall averaged over 15°–23°N, 85°–95°E exceeds +1 SD. From this definition, a total of **83** extreme intraseasonal rainfall events are identified for the whole period of 1998–2015. During the +IOD (−IOD) years, there are **22 (26)** heavy rain events. Gray dots denote 90% significance level based on a two-sided t-test.

conditional lagged composite anomalies are calculated following the phase of the IOD. The years in which negative or positive phases of IOD take place were identified using the JJAS Dipole Mode Index (DMI) following Saji *et al.* (1999). The JJAS DMI index is defined as the difference of the area-averaged JJAS SST anomalies between the western (10°S–10°N, 50°–70°E) and the southeastern (10°S–EQ, 90°–110°E) equatorial Indian Ocean (Figure 3). Negative and positive IOD years are defined as those when JJAS DMI > 0.75 SD and DMI < −0.75 SD, respectively (e.g. Cai *et al.*, 2013). With such criteria, a total of six negative IOD years (1998, 2001, 2005, 2010, 2013, and 2014) and five positive IOD years (2003, 2007, 2008, 2012, and 2015) are obtained (Table 1). Among these years, the year 2015 is classified as El Niño, while 1998 and 2010 are La Niña according to the Climate Prediction Center Oceanic Niño Index. Thus, the composite analysis is repeated for the 'pure' IOD years where IOD years that coincided with ENSO years are excluded in the analysis. The result for the latter is, in general, similar to the former (the supplementary figures), which is consistent with the finding from Ajayamohan *et al.* (2008, 2009). Therefore, to keep the sample size not too small for the significance testing, the analysis in this study is based on the IOD index. The statistical significance of the composite anomalies is assessed by a Student's t-test. Unless otherwise noted, the gray dots in the figures will denote the areas of 90% confidence level.

The spatial distributions of the composite SST anomalies during −IOD and +IOD years are shown in Figure 3. In addition to the reversal of the east-west SST contrast on the equator in the opposite phases of

the IOD, the composite SST fields show that the BoB experiences moderately warm (cold) SST anomaly in −IOD (+IOD) years. As the local SST condition is strongly coupled to the intraseasonal rainfall variability in the BoB through LH, one would expect a distinctive SST–rainfall relationship over BoB in different IOD years. This is hinted at in Figure 4, which compares the SD of the JJAS intraseasonal rainfall for −IOD and +IOD years. It shows that the intraseasonal rainfall variability is more pronounced during the −IOD years, especially over the northern BoB. An examination of the difference in rainfall variability for the pure IOD years yields very similar results (Figure S1). The enhancement of the intraseasonal rainfall variability motivates us to re-examine the composite evolution of the SST, LH, and WS in relation to intraseasonal rainfall separately for −IOD and +IOD years.

The results are shown in Figures 2(c)–(f). Across the BoB (north of 10°N), the pre-convection SST warming and the post-convection SST cooling tend to be noticeably weaker in −IOD years compared to +IOD years. Both the composites show the northward propagating SST anomalies, but the SST warming peaks at about 5–6 days before the peak in −IOD years, as opposed to 10–15 days prior in +IOD years. Ajayamohan *et al.* (2008) showed that, during the −IOD years, the pre-convection SST warming is weaker, and the intraseasonal SST variability has a shorter period north of 10°N (their Figure 13). The corresponding LH and WS anomalies are, however, much more pronounced prior to and during the peak rainfall events in −IOD years, with the clearer northward propagating characteristics from the equatorial southeastern Indian Ocean toward the BoB (Ajayamohan *et al.*, 2008).

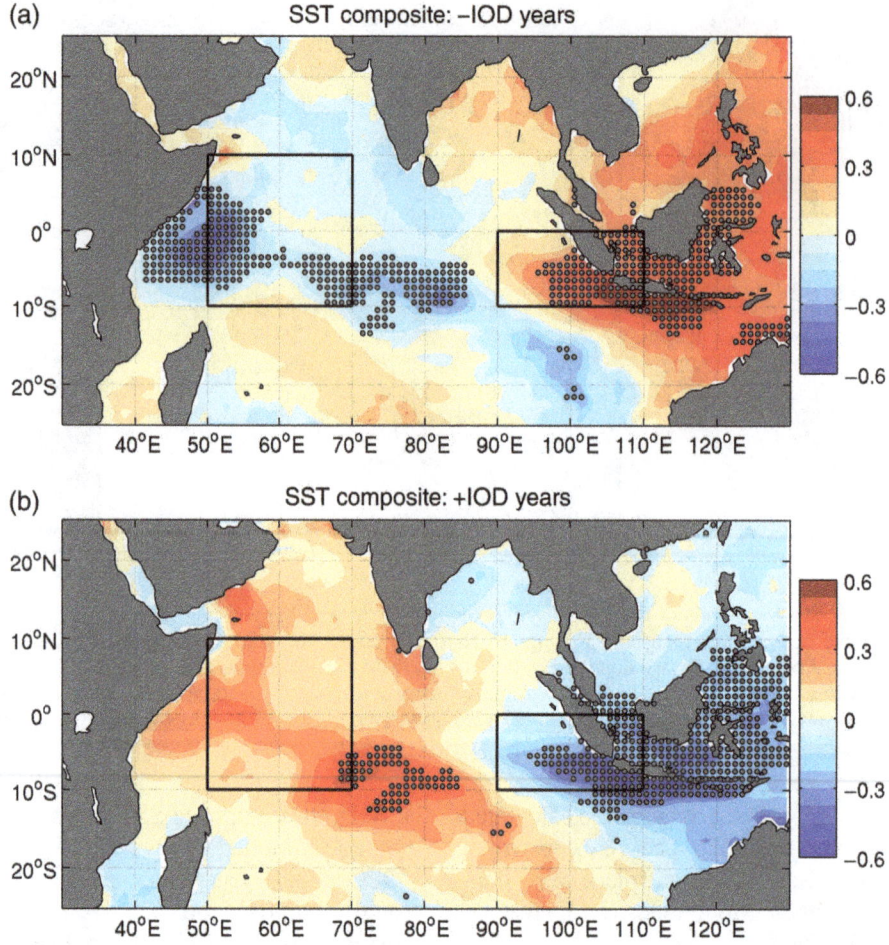

Figure 3. JJAS SST anomaly patterns associated with the (a) −IOD and (b) +IOD years. The boxes indicate the western (10°S−10°N, 50°−70°E) and southeastern (10°S−10°N, 50°−70°E) equatorial Indian Ocean used to construct the JJAS DMI (Saji et al., 1999). Gray dots denote 95% significance level in SST anomaly based on a two-sided t-test.

Recalling that the Bay-wide local SST anomaly during the −IOD years is weakly positive (Figure 3), the amplified LH and WS anomalies indicate that the reduced SST intraseasonal variability is a result of stronger heat flux damping, which is in turn conducive to stronger intraseasonal rainfall events. In contrast, in +IOD years, the LH and WS anomalies associated with the rainfall events are weaker, and the northward propagating characteristics are less coherent. This difference in the level of intraseasonal rainfall variability is consistent with Figure 3. The same analysis is repeated for the pure IOD years, showing that the results remain largely similar (Figure S2).

3.3. Moisture budget analysis

The difference of the characteristics of the intraseasonal rainfall events during the −IOD and +IOD years is further examined via column-integrated moisture budget analysis (Hsu and Li, 2012; Hsu et al., 2013; Xi et al., 2015),

$$\frac{\partial q}{\partial t} = -\left(\vec{V}_H \cdot \vec{\nabla}_H q\right) - \left(q \nabla \cdot \vec{V}_H\right) - \frac{\partial}{\partial p}\left(\omega q\right) - \frac{Q_2}{L}$$

(1)

Table 1. The chosen +IOD and −IOD years for the period of 1988–2015. +IOD (−IOD) years are defined in this study as the years when the JJAS DMI is greater than (less) than +0.75 SD (−0.75 SD). The JJAS DMI is defined as the difference of the JJAS SST anomaly between the western tropical Indian Ocean (50°−70°E, 10°S−10°N) and tropical southeastern Indian Ocean (90°−110°E, 10°S−0°N). The years with an asterisk denote the IOD years that coincide with ENSO based on JJAS ENSO index (http://www.cpc.ncep.noaa.gov/products/analysis_monitoring/ensostuff/ensoyears.shtml).

IOD+	IOD−
2003	1998*
2007	2001
2008	2005
2012	2010*
2015*	2013
	2014

where q is specific humidity, $\vec{V}_H$ the horizontal velocity vectors, $\vec{\nabla}_H$ the horizontal gradient operator, ω the pressure vertical velocity, Q_2 the atmospheric apparent moisture sink (Yanai et al., 1973; Johnson et al., 2015), and L the latent heat of condensation. The rate of change of the moisture on the left-hand side of Equation (1)

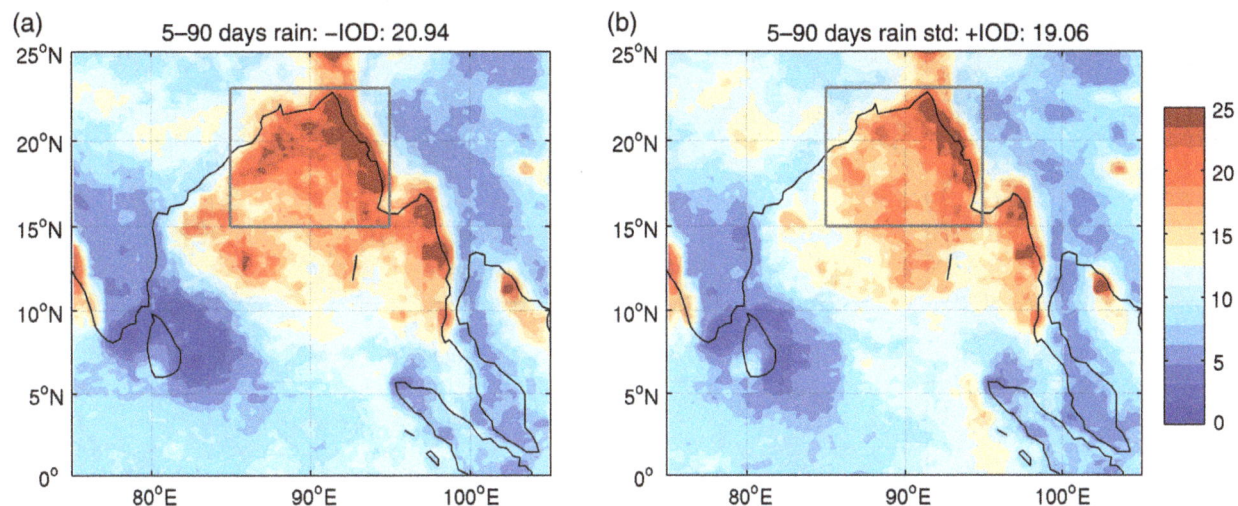

Figure 4. Map of mean SD of the 5–90 day filtered rainfall (mm day^{-1}) for the −IOD and +IOD years. Mean SD value averaged over the gray box is denoted in the title.

is balanced by the terms on the right. The first term is the horizontal moisture advection, and the second and the third are the horizontal moisture convergence and vertical moisture flux, respectively, which together represent the vertical moisture advection. The last term represents the loss of moisture out of the air column as condensation into the rain. In this analysis, this term is obtained as the residual, and thus it also includes other processes that are not fully resolved from the daily data as well as numerical errors due to finite differencing. Each term in Equation (1) is integrated from 1000 to 300 hPa.

The composite evolutions of each term averaged during the peak rainfall period are shown in the top panel of Figure 5, as well as those for the −IOD and +IOD years in the middle and lower panels. The total moisture rate is close to zero (not shown), indicating the quasi-equilibrium state during the peak of the heavy rainfall event. The greatest source of moisture in the atmospheric column over the BoB is the horizontal convergence of moisture, which is largest at the lower level near the sea surface (Figure S3). The vertical flux also adds additional moisture to the air column but is of secondary importance. In the northern BoB, the moisture increase through the horizontal convergence is mostly balanced by the moisture loss through condensation and rainfall. This moisture addition and removal processes are most pronounced in the northern BoB. The horizontal advection also accounts for the loss of moisture but occurs over the broad area over the BoB.

Evidently, the horizontal advection terms do not vary appreciably during the different phases of the IOD (Figures 5(e) and (i)), especially in the northern BoB, where the intraseasonal rainfall variability is most pronounced. In this region, the principal balance during the −IOD years is between the increased horizontal moisture convergence and the increased condensational loss (Figure 5). The increase in moisture convergence is most pronounced in the lower atmosphere

below 700 hPa with the peak at 925 hPa (Figure S3), suggesting the critical role of greater moistening of the lower troposphere by the evaporation during the −IOD years. In contrast, the moisture convergence in +IOD years is weakened mostly in the lower level. The enhanced vertical moisture flux does not exhibit large changes. Hence, the stronger low-level moisture convergence associated with the warmer SST and increased LH flux in −IOD years (Figure 2(d)) leads to the stronger intraseasonal precipitation events in the northern BoB (Kemball-Cook and Wang, 2001; Jiang *et al.*, 2004).

The role of SST rise in the intraseasonal moisture convergence is further illustrated in Figure 6 showing the lag correlation between the SST and moisture convergence over the northern BOB. For all years (black), the SST and moisture convergence are correlated most significantly when the former leads (lags) the latter by about 7 days (10 days). This quadrature phase relationship implies coherent intraseasonal variability with a period of approximately 40 days. Furthermore, the lag correlation is enhanced during the −IOD years (orange) compared to +IOD years (blue). This lends further support to the notion that the SST warming during the pre-convection period is important for the intensity of intraseasonal rainfall during the −IOD years.

4. Conclusions and discussions

This study examines the coupling of the intraseasonal rainfall events with the SST in the BoB and its interannual modulation of the intraseasonal rainfall variability. Based on the lagged composite analysis, it is shown that the ocean–atmosphere coupling on the intraseasonal time scale is significant, especially in the northern BoB, with LH playing a key role as the communicator. On average, the maximum warming in SST precedes the peak in precipitation by approximately

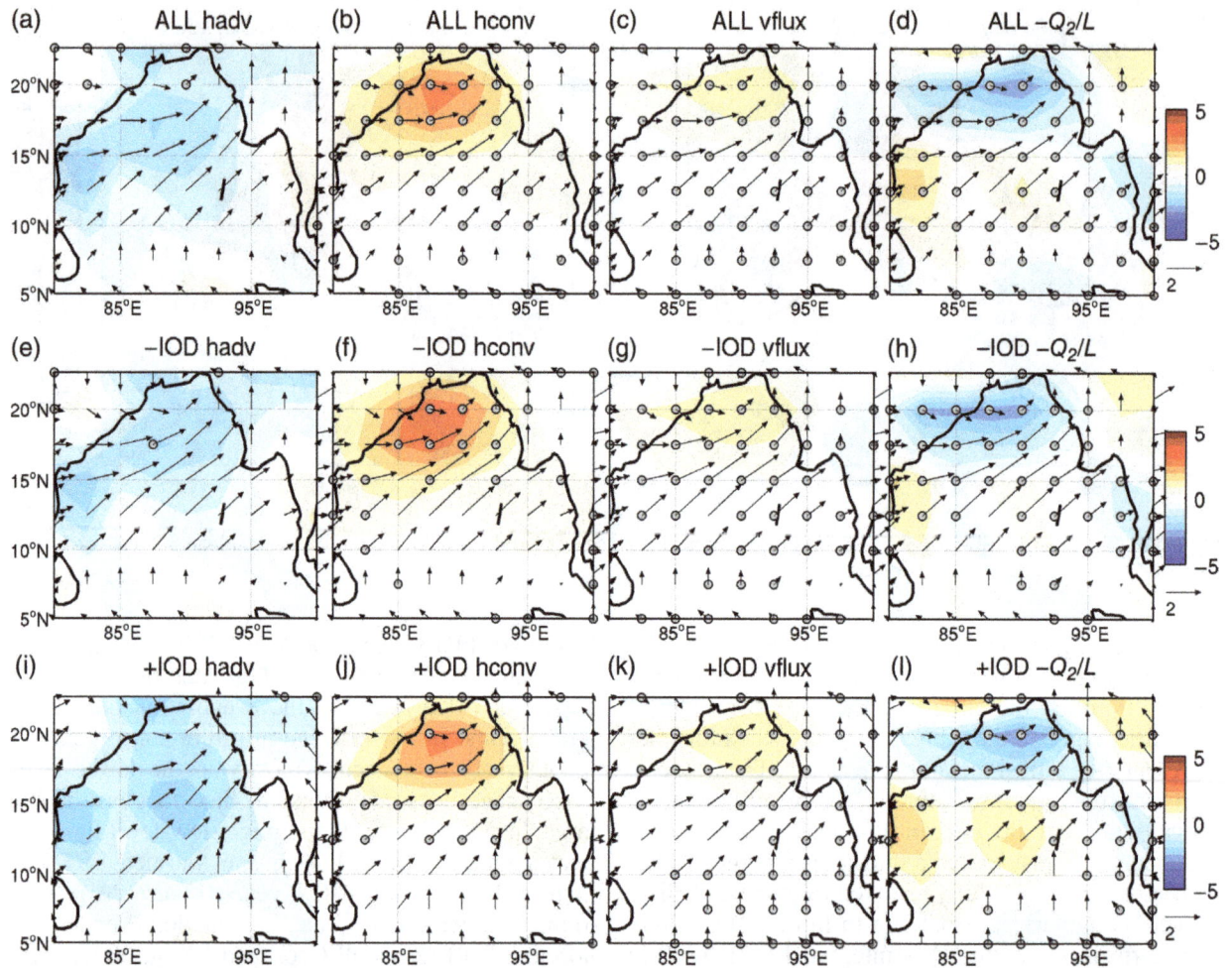

Figure 5. The column-integrated (1000–300 hPa) moisture budget terms (kg m^{-2} day^{-1}) averaged during Day 0–2, showing (from left to right) horizontal moisture advection, horizontal moisture convergence, vertical moisture flux, and the moisture loss due to condensation. Vectors are composite anomalies of the 5–90 day filtered 1000 hPa winds. Gray dots denote the 90% significance level of the composite anomalies.

5 days in the northern BoB. In addition, the coherent northward propagating anomalies of SST and LH are observed, suggesting that they are coupled to the northward propagating convection and clouds over the ISM region.

The intraseasonal ocean-atmosphere coupling is strongly modulated by the IOD. With the use of the JJAS DMI, a total of six years, out of the 1998–2015 TRMM data period, are identified as −IOD years, and five as +IOD years. Despite the relatively small number of the chosen IOD years due to the short satellite data record, the clear distinction in the ocean-atmosphere coupling patterns in different phases of the IOD can be observed. In −IOD years when the BoB SST is anomalously warm, the intraseasonal variability of LH and WS are enhanced, facilitating the stronger moisture transfer to the atmosphere, leading to strengthened intraseasonal rainfall variability. This is supported by the column-integrated moisture budget analysis showing that, indeed in −IOD years, the horizontal moisture convergence, the primary source of moisture during the intraseasonal rainfall peak, is significantly strengthened in the lower atmosphere (below 750 hPa)

over the northern BoB. The resultant stronger vertical moisture advection predominantly balances the increased rainfall variability in −IOD years in the northern BoB.

The same analysis is repeated using the GPCP 1° daily rainfall dataset to confirm the robustness of the finding based on the TRMM 3B42. Overall, the key conclusion of the study is not sensitive to the choice of the dataset. For example, the elevated level of intraseasonal rainfall variability in the BoB in the −IOD years compared to the +IOD years. Therefore, the intraseasonal SST-rainfall coupling in the northern BoB and its modulation by the IOD is significant and robust. The fact that SST leads rainfall during the summer monsoons provides important predictive information on the onset of the monsoon rainfall and its active/break cycles over the BoB and surrounding land areas. This relationship varies with the large-scale mode of variability in the IOD, with implications for predictions on the interannual time scale. In particular, −IOD years exhibit enhanced intraseasonal rainfall variability over Bangladesh and into Myanmar, while +IOD years show increased variability over northern India.

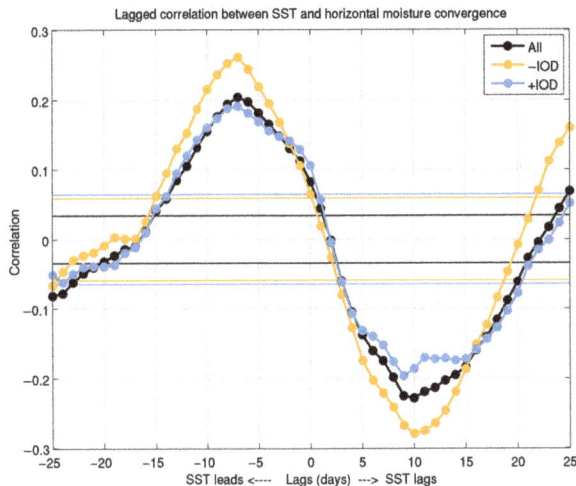

Figure 6. Lagged correlations between SST and horizontal moisture convergence averaged over the northern BoB (85°–95°E, 15°–25°N), color-coded to represent the correlations during all, –IOD and +IOD years. The horizontal lines represent the lower and upper 95% confidence bounds for each case.

Acknowledgements

This project is part of SJ's Summer Student Fellowship (SSF) program at Woods Hole Oceanographic Institution (WHOI) sponsored by the National Science Foundation Research Experience for Undergraduates Program (NSF-REU). HS acknowledges grants from the Office of Naval Research (N00014-15-1-2588) and National Oceanic and Atmospheric Administration (NA15OAR4310176). CCU acknowledges support from the National Science Foundation under AGS-1304245. The authors thank two anonymous reviewers for their constructive comments, which helped to substantially improve the manuscript.

Supporting information

The following supporting information is available:

Figure S1. As in Figure 3, but for pure IOD years. Mean SD value averaged over the gray box is denoted in the title.

Figure S2. As in Figure 2, but for pure IOD years. During the pure +IOD (pure –IOD) years, there are 18 (16) heavy rain events.

Figure S3. Vertical profiles of the horizontal moisture convergence (10^{-3} kg m^{-2} day^{-1}) averaged during the peak rainfall event for all (black), +IOD (blue) and –IOD (orange) years.

Figure S4. As in Figure 3 but using the GPCP 1° daily rainfall estimate from 1997 to 2015. Mean SD value averaged over the gray box is denoted in the title.

References

Ajayamohan RS, Rao SA, Yamagata T. 2008. Influence of Indian Ocean Dipole on poleward propagation of boreal summer intraseasonal oscillations. *Journal of Climate* **21**: 5437–5454.

Ajayamohan RS, Rao SA, Luo J-J, Yamagata T. 2009. Influence of Indian Ocean Dipole on boreal summer intraseasonal oscillations in a coupled general circulation model. *Journal of Geophysical Research* **114**: D06119.

Ashok K, Guan Z, Yamagata T. 2001. Impact of the Indian Ocean Dipole on the relationship between the Indian monsoon rainfall and ENSO. *Geophysical Research Letters* **28**: 4499–4502.

Ashok K, Guan Z, Saji NH, Yamagata T. 2004. Individual and combined influences of the ENSO and Indian Ocean Dipole on the Indian summer monsoon. *Journal of Climate* **17**: 3141–3155.

Cai W, Zheng X-T, Weller E, Collins M, Cowan T, Lengaigne M, Yu W, Yamagata T. 2013. Projected response of the Indian Ocean Dipole to greenhouse warming. *Nature Geoscience* **6**: 999–1007.

Fu X, Wang B, Li T, McCreary JP. 2003. Coupling between northward-propagating, intraseasonal oscillations and sea surface temperature in the Indian Ocean. *Journal of Atmospheric Science* **60**: 1733–1753.

Fu X, Wang B, Waliser DE, Tao L. 2007. Impact of atmosphere–ocean coupling on the predictability of monsoon intraseasonal oscillations. *Journal of Atmospheric Science* **64**: 157–174.

Gadgil S, Vinayachandran PN, Francis PA, Gadgil S. 2004. Extremes of the Indian summer monsoon rainfall, ENSO and equatorial Indian Ocean oscillation. *Geophysical Research Letters* **31**: L12213.

Goswami BN. 1998. Interannual variations of Indian summer monsoon in a GCM: external conditions versus internal feedbacks. *Journal of Climate* **11**: 501–522.

Goswami BN. 2005. South Asian monsoon. In *Intraseasonal Variability in the Atmosphere-Ocean Climate System*, Lau WKM, Waliser DE (eds). Praxis Springer: Chichester, UK; 19–61.

Goswami BN, Ajayamohan RS. 2001. Intraseasonal oscillations and interannual variability of the Indian summer monsoon. *Journal of Climate* **14**: 1180–1198.

Hsu P-C, Li T. 2012. Role of the boundary layer moisture asymmetry in causing the eastward propagation of the Madden-Julian Oscillation. *Journal of Climate* **25**: 4914–4931.

Hsu P-C, Li T, Murakami H, Kitoh A. 2013. Future change of the global monsoon revealed from 19 CMIP5 models. *Journal of Geophysical Research – Atmospheres* **118**: 1247–1260.

Huffman GJ, Adler RF, Morrissey MM, Curtis S, Joyce R, McGavock B, Susskind J. 2001. Global precipitation at one-degree daily resolution from multi-satellite observations. *Journal of Hydrometeorology* **2**: 36–50.

Huffman GJ, Adler RF, Bolvin DT, Gu G, Nelkin EJ, Bowman KP, Hong Y, Stocker EF, Wolff DB. 2007. The TRMM multi-satellite precipitation analysis: quasi-global, multi-year, combined-sensor precipitation estimates at fine scale. *Journal of Hydrometeorology* **8**: 38–55.

Ihara C, Kushnir Y, Cane MA, De la Pena VH. 2007. Indian summer monsoon rainfall and its link with ENSO and Indian Ocean climate indices. *International Journal of Climatology* **27**: 179–187.

Jiang X, Li T, Wang B. 2004. Structures and mechanisms of the north-ward propagating boreal summer intraseasonal oscillation. *Journal of Climate* **17**: 1022–1039.

Johnson RH, Ciesielski PE, Ruppert JH Jr. 2015. Sounding-based thermodynamic budgets for DYNAMO. *Journal of Atmospheric Science* **72**: 598–622.

Kalnay E, Kanamitsu M, Kistler R, Colling W, Deaven D, Gandin L, Iredell M, Saha S, White G, Wollen J, Zhu Y, Chelliah M, Ebisuzaki W, Higgins W, Janowiak J, Mo KC, Ropelewski C, Wang J, Leetmaa A, Reynolds R, Jenne R, Joseph D. 1996. The NCEP/NCAR 40-year reanalysis project. *Bulletin of the American Meteorological Society* **77**: 437–470.

Kemball-Cook SR, Wang B. 2001. Equatorial waves and air– sea interaction in the boreal summer intraseasonal oscillation. *Journal of Climate* **14**: 2923–2942.

Lawrence DM, Webster PJ. 2001. Interannual variations of the intraseasonal oscillation in the South Asian summer monsoon region. *Journal of Climate* **14**: 2910–2922.

Murtugudde R, Busalacchi AJ. 1999. Interannual variability of the dynamics and thermodynamics of the tropical Indian Ocean. *Journal of Climate* **12**: 2300–2326.

Reynolds RW, Smith TM, Liu C, Chelton DB, Casey KS, Schlax MG. 2007. Daily high-resolution-blended analyses for sea surface temperature. *Journal of Climate* **20**: 5473–5496.

Saji NH, Goswami BN, Vinayachandran PN, Yamagata T. 1999. A dipole mode in the tropical Indian Ocean. *Nature* **401**: 360–363.

Sengupta D, Ravichandran M. 2001. Oscillations of Bay of Bengal sea surface temperature during the 1998 summer monsoon. *Geophysical Research Letters* **28**: 2033–2036.

Sengupta D, Senan R, Goswami BN. 2001. Origin of intraseasonal variability of circulation in the tropical central Indian Ocean. *Geophysical Research Letters* **28**: 1267–1270.

Seo H, Subramanian AC, Miller AJ, Cavanaugh NR. 2014. Coupled impacts of the diurnal cycle of sea surface temperature on the Madden-Julian Oscillation. *Journal of Climate* **27**: 8422–8443.

Sikka DR, Gadgil S. 1980. On the maximum cloud zone and the ITCZ over Indian longitude during southwest monsoon. *Monthly Weather Review* **108**: 1840–1853.

Ummenhofer CC, Sen Gupta A, Li Y, Taschetto AS, England MH. 2011. Multi-decadal modulation of the El Niño-Indian monsoon relationship by Indian Ocean variability. *Environmental Research Letters* **6**: 034006.

Vecchi GA, Harrison DE. 2002. Monsoon breaks and subseasonal sea surface temperature variability in the Bay of Bengal. *Journal of Climate* **15**: 1485–1493.

Wang B. 2005. Theory. In *Intraseasonal Variability in the Atmosphere-Ocean Climate System*, Lau WKM, Waliser DE (eds). Praxis Springer: Chichester, UK; 307–360.

Webster PJ, Magana VO, Palmer TN, Shukla J, Romas RA, Yanai M, Yasuniari T. 1998. Monsoons: processes, predictability, and the prospects for prediction. *Journal of Geophysical Research* **13**: 14451–14510.

Webster PJ, Moore AM, Loschnigg JP, Leben RR. 1999. Coupled ocean-atmosphere dynamics in the Indian Ocean during 1997-98. *Nature* **401**: 356–360.

Xi J, Zhou L, Murtugudde R, Jiang L. 2015. Impacts of intraseasonal SST anomalies on precipitation during Indian summer monsoon. *Journal of Climate* **28**: 4561–4575.

Yanai M, Esbensen S, Chu J-H. 1973. Determination of bulk properties of tropical cloud clusters from large-scale heat and moisture budgets. *Journal of Atmospheric Science* **30**: 611–627.

Yasunari T. 1979. Cloudiness fluctuation associated with the Northern Hemisphere summer monsoon. *Journal of the Meteorological Society of Japan* **57**: 227–242.

Yasunari T. 1980. A quasi-stationary appearance of 30–40 day period in the cloudiness fluctuation during summer monsoon over India. *Journal of the Meteorological Society of Japan* **58**: 225–229.

Yu L, Weller RA. 2007. Objectively analyzed air-sea heat fluxes (OAFlux) for the global ice-free oceans. *Bulletin of the American Meteorological Society* **88**: 527–539.

4

Mesoscale modeling for the rapid movement of monsoonal isochrones

Vinay Kumar* and T. N. Krishnamurti

Department of Earth, Ocean and Atmospheric Science, Florida State University, Tallahassee, FL, USA

*Correspondence to:
V. Kumar, Department of Earth,
Ocean and Atmospheric Science,
Florida State University,
Tallahassee, FL 32306, USA.
E-mail: vkumar@fsu.edu*

Abstract

The progresses of fast moving 2013 monsoonal isochrones were simulated with WRF model at 25 km and compared with slower moving isochrones of 2014. A large number of sensitivity experiments were performed by enhancing of soil moisture, stratiform rain up to 25%, their combination and reducing soil temperature by 12%, over 5 grid points to the north of the isochrones over a relatively dry soil layer. The modified soil parameters and startiform rain ahead of isochrones enhances the population of the buoyant elements, divergent flows of the local Hadley circulation normal to the isochrones; resulting in a fast movement of the isochrones.

Keywords: monsoonal isochrones; dry soil layer; buoyant elements

1. Introduction

Northward extension of ITCZ reaches the southern tip of India (around May–June at 7°N) and augmentation of the first rainfall over Indian landmass is known as the arrival of Indian summer monsoon (Rao, 1976). Evolution of Indian summer monsoon includes onset, active-break phases, and thus a progress of rainfall from southernmost tip (Trivandrum, Kerala) to the northwestern end of India (Jaisalmer, Rajasthan). The movement of rain bands from southeast to northwest direction is locally termed as the onset of monsoonal isochrones or simply isochrones (Chang *et al.*, 2004). Once the monsoon onset is officially declared, Indian meteorologists watch its day to day northward progress over Indian subcontinent. Movement of monsoonal isochrones (northern limit of monsoon) over Indian region is governed by symmetric and antisymmetric heating/circulation (Gill, 1980), heat source and sink (Krishnamurti and Ramanathan, 1982), pressure gradient across landmasses (Ananthakrishnan *et al.*, 1968), reversal of temperature gradient over land and sea (Yanai *et al.*, 1992), meridional passage of the ISO wave (Wang, 2005), stratiform rain (Krishnan *et al.*, 2011; Samir and Sikka, 2013), and land surface processes (Yeh *et al.*, 1984). From all abovementioned parameters, we examined the importance of soil moisture in layer-1 (SM), stratiform rain (SR) from nonconvective clouds and soil temperature in layer-1 (ST) using WRF model sensitivity for the assessment of the fast movement of isochrones especially during June 2013 and for a comparison of slower moving isochrones of June 2014.

Sensitivity of SM, SR and ST were noted to be a significant factor for controlling hydrology and land–atmosphere interactions (Alfieri *et al.*, 2008; Krishnamurti *et al.*, 2012). Over the monsoonal region, more than 70% of the rain is SR (Krishnan *et al.*, 2011). In Krishnamurti *et al.* (2012), the following scenario was noted from observations and modeling: An isochrone, as it is advancing north is in fact a new family of clouds parallel to the parent isochrone. The nonconvective rains from cloud anvils of the parent isochrone enhance the soil moisture to assist the formation of a new line of clouds. These anvils, accompanied by nonconvective rains, of the parent isochrone often extend to the north of the parent isochrone. This region to the immediate north of the isochrone generally carries very warm surface and surface air temperatures in the planetary boundary layer (PBL), such conditionally unstable environment helps in the formation of new cumulus cloud lines. Keeping such hypothesis in background and predicting fast moving isochrones of June 2013 was an appealing experiment.

2. Model configurations, experiments and datasets

We used the following default physics options of WRF: long-wave radiation scheme – Rapid Radiative Transfer Model (RRTM); short-wave radiation scheme – Dudhia scheme; surface physics – Monin–Obukhov and Janjic scheme; land surface model – five-layer thermal diffusion; PBL scheme – Mellor–Yamada–Janjic (MYJ) turbulent kinetic energy (TKE) PBL; convection scheme the so-called Kain–Fritsch (new eta) scheme; and an explicit moisture scheme – WRF six-class graupel scheme (WSM6). We ran WRF model (with nonhydrostatic approach) with a single domain, horizontal resolution of 25 km, 27-vertical levels, 6-h initial and boundary conditions from FNL, and 16 days continuous run while model's lateral boundary conditions were updated at every 6 h. For 3-km resolution WRF with nonhydrostatic approach, Goddard microphysics option and cloud resolving model was opted.

From observations, it is noted that a couple of days after the passage of the onset isochrone the value of SM (volumetric cubic meters of water in 1 m³ of soil) jump up from 0.15 to 0.35 of saturation in the forecasts, which accounts to around 13.5% increment of SM. This awareness was used in the design of the proposed sensitivity experiments for the enhancements of SM and SR. However for ST, observations showed that the temperature generally decreased 297 K from 300 K (2 to 2.5 °C), which is 7.5% reduction of ST. Somehow for 2013, in WRF experiments we had to enhance SM (10 cm), SR by 25% and reduce ST (10 cm) by 12%, on the 5 grid points to the north of 1 June isochrone. For comparison, same experiments were performed for June 2014 also. In a series of experiments, we enhanced SM and SR by 15, 20 and 25%, and these are respectively labeled as SM15/SR15, SM20/SR20 and SM25/SR25 respectively; combination of enhanced SM and SR (labeled as SMSR25) and reduced the ST by 12% (ST12). A comprehensive sensitivity study was carried out using modified values of SM25, SR25, SMSR25 and ST12 for both June 2013 and June 2014.

The objective criteria to plot the lines of isochrones for WRF control simulation and experiment simulation are as follows: (1) after onset of monsoon over Kerala, we plotted the isochrones for 2, 5, 8, 11, 14 and 16 June connecting the lines which has rainfall 2.5 mm and more for two consecutive days; and (2) We face the situation of multiple isochrones lines during 14 and 16 June 2014, then the isochrones line is drawn closet to the previous day position.

For observational datasets, we used daily rainfall (mm) from TRMM3B42; monthly and daily datasets of volumetric soil water layer-1 (SWVL1, soil layer 0–7 cm, unit m³ m⁻³), low cloud cover (LCC, unit 0–1), soil temperature level-1 (STL1, soil layer 0–7 cm, unit deg-K), and total column water vapor (TCWV, unit kg m⁻²) from ERA Interim (Dee *et al.*, 2011).

3. Prediction of fast moving isochrones: a sensitivity of soil parameters and clouds

The simulation of the passage of isochrones during June 2013 was a challenge, because the monsoonal isochrones (solid green-line) covered India completely within first 16 days of June, after onset of southwest monsoon (Figure 1). Red dashed lines show the climatological isochrones, those cover India in 45 days (1 June to 15 July). Monsoonal isochrones of 2013 recorded the fastest movement of isochrones in almost last 100 years of Indian summer monsoonal rainfall. Monsoonal isochrones of 2013 were always ahead than the normal isochrones over western side of the India while monsoonal isochrones were always lagging behind till June

10 and strangely within 2 days (15–16 June) monsoonal isochrones covered all India.

First, we enhanced the SM and SR by 15%, while ST reduced by 7.5% and found that the northward movement of monsoonal isochrones in WRF experiments is almost similar to the control experiment (Figure S1). Furthermore, we gave a shot to enhance SM and SR by 20% (Figure S2); here the northward movements of isochrones were relatively faster than Figure S1, but could not simulate rainfall over northwestern India. In all these simulations, monsoonal rainfall did not reach to perennial desert region, Madhya Pradesh and western Utter Pradesh, which shows the land surface model in WRF has dry bias over this region. Another reason may be the prevailing soil is extremely dry over northwest India. Is there something more stored in soil moisture and SR for the fast monsoonal isochrones of 2013?

Finally, we enhanced the SM and SR by 25% and ST reduced by 12% in WRF experiments. Figure 2 shows accumulated rainfall from TRMM3B42, control experiment, SM25, SR25, SMSR25, ST12 experiments for 2–16 June. TRMM3B42 rainfall does not show northward movement of rainfall for 2 June 2013, but from 5–16 June, TRMM3B42 rainfall was always ahead of the IMD isochrones (black line). During 2013, isochrones (green lines) moved faster from 8 to 12 June as compared to the climatological isochrones (red dashed lines, Figure 1). In these experiments through 8 June, the predicted isochrones (red lines) and the IMD isochrones (black lines) were moving northward together over most of its length, except in the eastern flanks where the predicted isochrones moved faster (Figure 2). During 11 and 14 June, the northern flank of the isochrones did not progress, whereas eastern flank one moved ahead of the official IMD isochrones. Rainfall distribution for 14 June from TRMM3B42, control, SM25, SR25, SMSR25 and ST12 shows a band of maximum rainfall, to the immediate south of central India. This band of maximum rainfall between the Arabian Sea and the Bay of Bengal at 20°N is an interesting aspect related to the SM, SR and ST. The central region of India is a region of strong land–atmosphere coupling and thus a favorable source of increased precipitation (Koster *et al.*, 2004). This feature is attributed to the fast movement of monsoonal isochrones from 14 June onward (see Figures S3 and S4). Monsoonal isochrones in the SM25 experiment show (over several regions of India) a faster motion as compared to the control experiment from WRF model.

However, there was a large gap of 3–4 longitude between the observed and the predicted isochrone over Northwest India on 16 June. To look minutely the northern extent of the monsoonal isochrones on 16 June, a zoomed picture is shown at the bottom of Figure 2. We compared WRF control (CNTL) simulation with other WRF simulation experiments. Various noteworthy differences in rainfall found over Rajasthan region, western Utter Pradesh, Haryana and northern Madhya Pradesh. In terms of magnitude of rainfall all the designed WRF experiments do better than CNTL

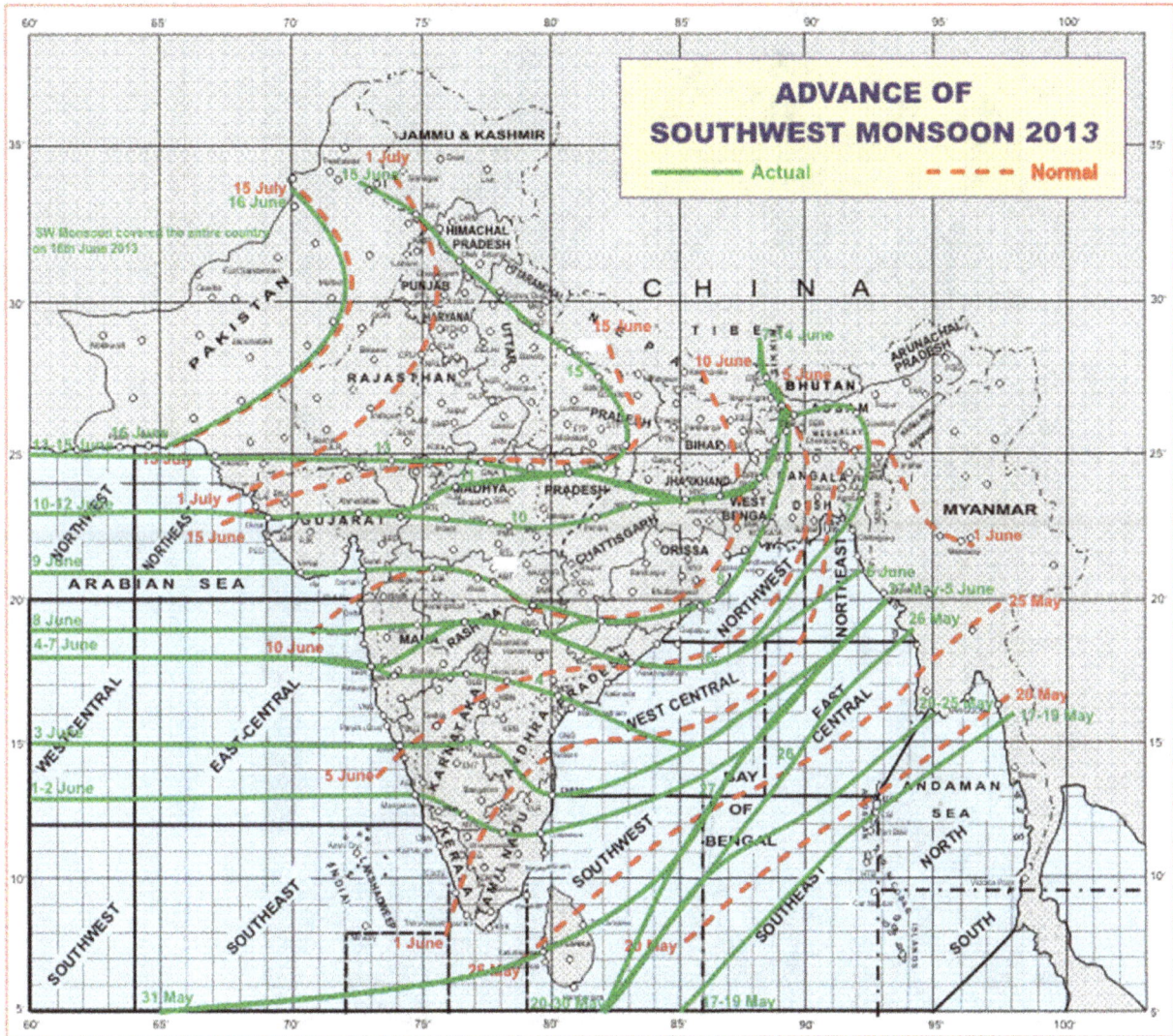

Figure 1. The IMD onset isochrones for year June 2013.

over Uttarakhand, western Utter Pradesh and Himachal Pradesh. SRSM25 experiment shows the best results among all. To compare slow versus fast moving monsoonal isochrones, we compare WRF simulation experiments between 2013 and 2014. In case of year 2014, IMD monsoon isochrones moved faster than CNTL simulation (Figure S5). For 2014 also, we enhanced SM and SR by 25% and ST reduced by 12% ahead (5 grids points) of the isochrones, model was able to capture rainfall variability with TRMM in all experiments (Figure 3). After 11 June, monsoonal isochrones in WRF experiments simulation were ahead of the CNTL simulation. But, 2014 being a drought year, so in soil there were not much soil moisture in June (Figure 4(a)). On applying same methodology (as in case of June 2013) simulated monsoonal isochrones of 2014 are in sacrosanct with observed isochrones (Figure S3). Amazingly a gap along eastern coast of Indian landmass in the Bay of Bengal is left void in all the WRF experiments of 2014, which were completely filled in all the WRF experiments of 2013. Presence of this gap from 8 June onward may be one of the indications of a

slower motion of monsoonal isochrones. These experiments show the dependency of monsoonal isochrones over soil characteristics (moisture and temperature) and stratiform rainfall. Largely, these experiments show the fast movement of June 2013 modeled isochrones compared favorably with the observed isochrones over the interior of India.

4. Spatial variability of soil parameters, clouds and circulation features

4.1. Observational features of soil parameters, clouds and water vapor

Figure 4 shows several observational fields (from ERA Interim), SWVL1 (upto 10 cm below the earth's surface) (Figure 4(a)–(c)); LCC (Figure 4(d)–(f)); STL1 (Figure 4(g)–(i)), and TCWV (Figure 4(j)–(l)) for June 2012, 2013 and 2014. SWVL1, LCC and TCWV being enhanced over central India around 20°N latitude, while STL1 showed lower values for June

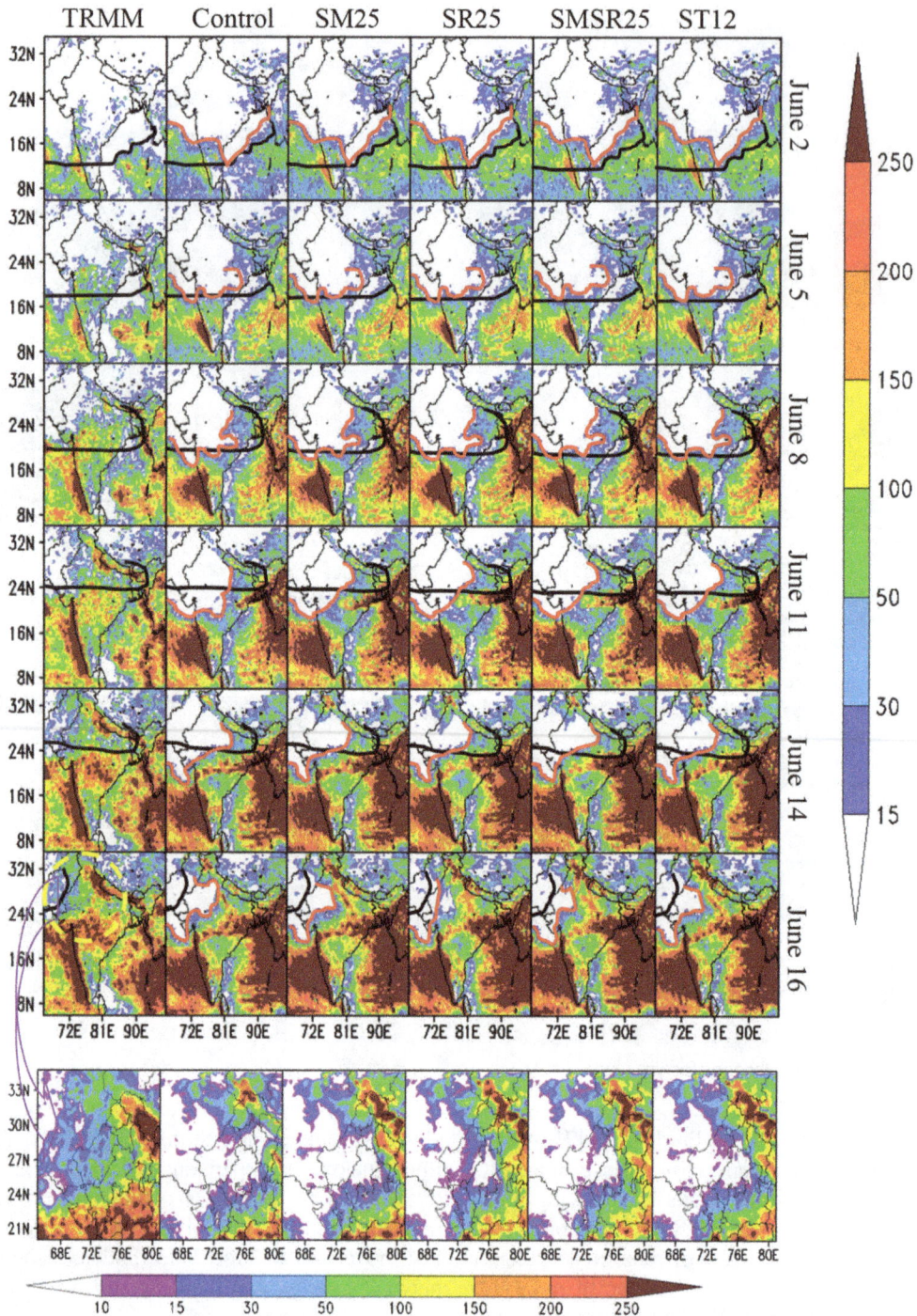

Figure 2. Daily accumulated rainfall (mm) from for 2, 5, 8, 11, 14 and 16 June 2013 from TRMM3B42 and WRF-ARW model sensitivity experiments (control, enhancing soil moisture by 25%, enhancing stratiform rain by 25%, enhancing soil moisture and stratiform rain by 25% and reducing soil temperature by 12% ahead of the 1 June 2013 onset isochrone). Black line shows the IMD isochrones while red line shows predicted isochrones. Lower panel shows the zoom out part of north-western India for 16 June.

2013. SWVL1 is increased from 0.15 to 0.3 $m^3\,m^{-3}$, LCC increased from 0.1 to 0.3 unit, STL1 dropped down from 307 to 304 K and TCWV increased from 40 to 60 $kg\,m^{-2}$. This was one of the reasons why we reduced soil temperature from 7.5 to 12% in case of soil temperature experiment for year 2013. The increased soil moisture and clouds were clearly related to increased rainfall over central India during 14 and 16 June. The line diagrams are plotted for the

rectangular box (76–78°E, 12–15°N) over southern India for daily values of SWVL1 (Figure 4(m)), LCC (Figure 4(n)) and STL1 (Figure 4(o)) from 15 May to 30 June 2013. These daily values show sudden increase of volumetric soil water, smooth growth of low cloud cover and drop of soil temperature from 2 June onward. The variability of above parameters confirms the role of the enhanced soil moisture, nonconvective rain and a reduction of soil temperature during the start of summer

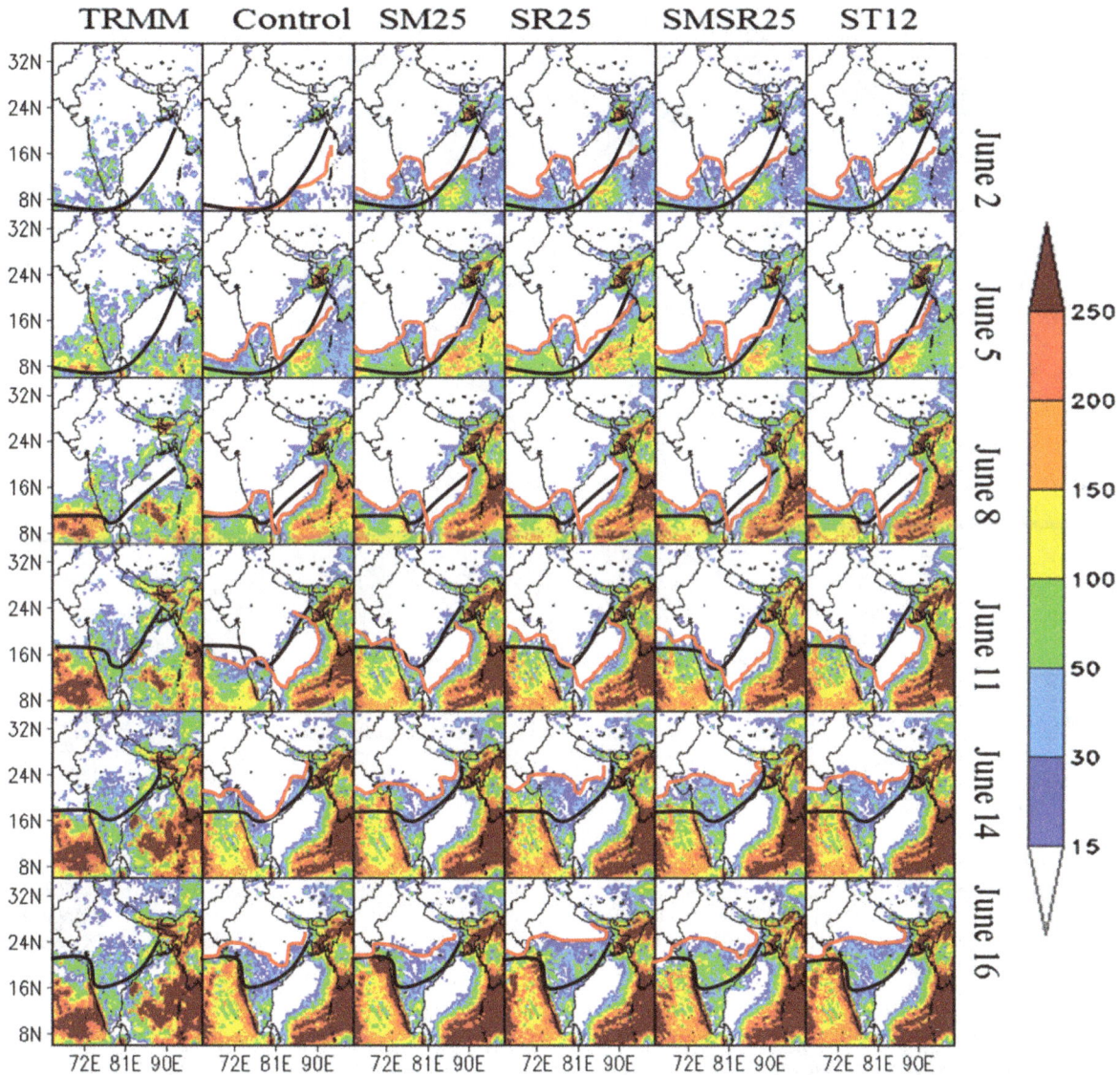

Figure 3. Daily accumulated rainfall (mm) from for 2, 5, 8, 11, 14 and 16 June 2014 from TRMM3B42 and WRF-ARW model sensitivity experiments (control, enhancing soil moisture by 25%, enhancing stratiform rain by 25%, enhancing soil moisture and stratiform rain by 25% and reducing soil temperature by 12% ahead of the 1 June 2014 onset isochrone). Black line shows the IMD isochrones while red line shows predicted isochrones.

monsoon season of 2013. Furthermore, Figure S3 and S4 show the gradual advancement of soil parameters and nonconvective cloud over central India and their crucial role in land surface modeling in WRF model.

4.2. Observational and model predicted local divergent circulation, Hadley circulation and Buoyancy field

Local divergent circulations and rotational fields at 200 mb gave an added insight for the fast movement of onset isochrones during June 2013. Figure 5 shows the velocity potential and divergent field at 200 mb from observations (Figure 5(a)) and model experiment SM25 (Figure 5(b)). The divergent circulation, this feature increased visibly from 2 to 16 nonconvective rain, which shows a divergent wind along the north and northwest (Figure 5(a)). We examined day by day,

observed, directions and amplitudes of the divergent circulations from 2 to 16 June and noted that their orientation and speeds were consistent with the steering of the isochrones toward the northwestward direction. On comparing these observed-based results with the model experiment SM25, the divergent flow was reasonably close to observations over the central longitudes of India confirming a northward steering. However, rotational parts of winds (streamline flows) were almost parallel to the onset isochrones (figure not shown), while rotational flow does not steer the isochrone. Prediction from model shows a stronger divergent circulations field June (Figure 5(b)). The model results did not carry a direction of divergent flow to the northwest direction as a result the model isochrone did not extend up to northwestern India. Figure 5(c) is based on observations, and shows the difference of the velocity potential between Kerala (southernmost end of India) and

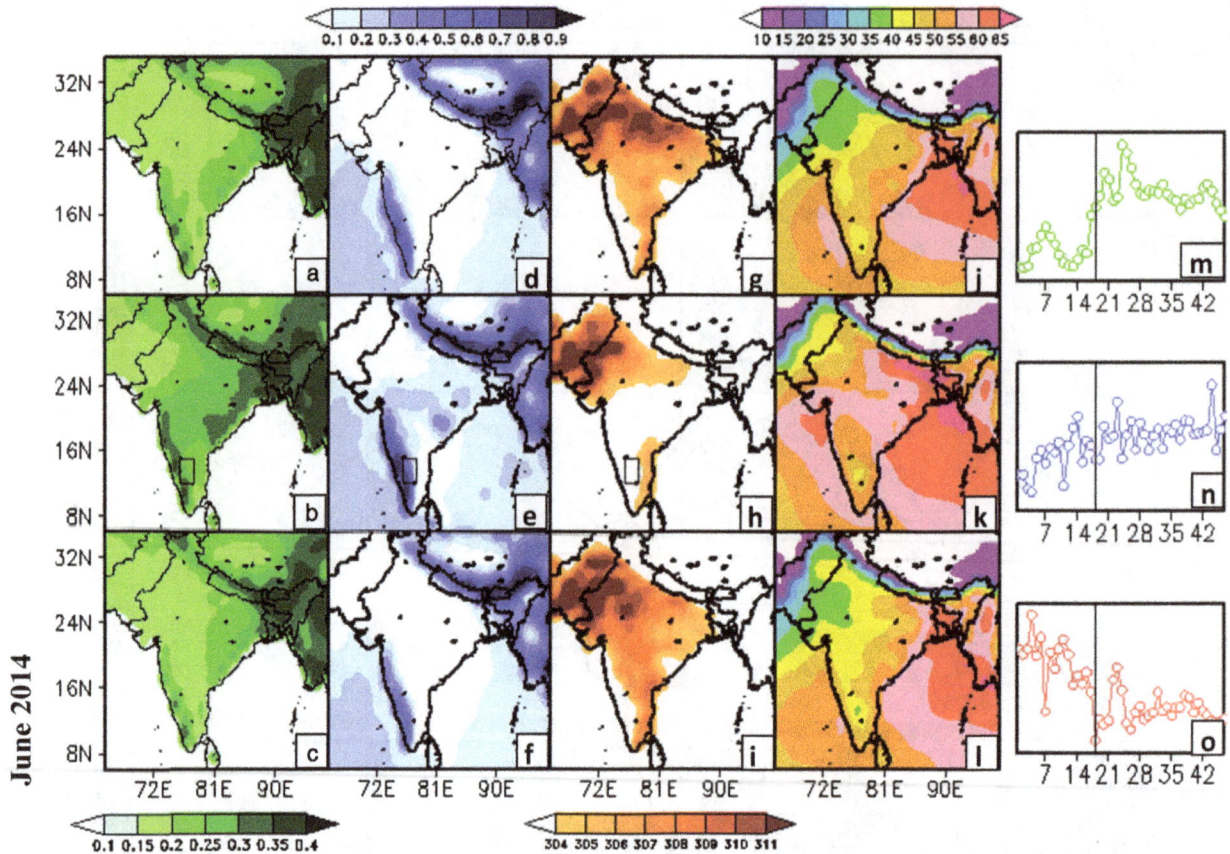

Figure 4. Volumetric soil water layer-1 (unit m^3 m^{-3}): (a) June 2014, (b) June 2013, (c) June 2012; low-level cloud cover (unit 0–1): (d) June 2014, (e) June 2013, (f) June 2012; soil temperature layer-1 (unit deg K): (g) June 2014, (h) June 2013, (i) June 2012; total water vapor column (unit kg m^{-2}): (j) June 2014, (k) June 2013, (l) June 2012. Line diagrams show averaged (over rectangular box in each figure of June 2013) daily variation for 15 May to 30 June 2013, (m) daily volumetric soil water layer-1, (n) low-level cloud cover, (o) soil temperature.

Rajasthan (northwestern end of India) of the onset monsoon isochrones for several recent years. That difference is proportional to the mean divergent between Kerala and Northern India. It is interesting to note that the mean northward-directed steering (as measured by this mean 200 hPa level, south to north divergent wind) is increasing in recent years.

In swift, it is noted that the environment ahead of isochrones is being modulated by the clouds, circulation and light rains from the anvils of the tall clouds of the onset isochrone. The slowly enhanced local Hadley circulation has an important role to steer the newly formed cloud elements ahead of isochrones. We examined local Hadley circulation from observations and model experiments for 2 to 16 June 2013. Local Hadley circulation (zonally averaged over 73–90°E) is displayed for 8 and 11 June, which is very active over 10–22°N (Figure 6(a) and (b)). Model predicted local Hadley circulation shows the vertically rising branch from 10 to 15°N on 8 June and 17 to 22°N on 11 June (Figure 6(c) and (d)). Local Hadley circulation for year 2014 is not as strong and well organize as compared to 2013, from WRF experiment SM25 (Figure 6(e) and (f)) for 8 and 11 June. For 2013 model experiments, local Hadley circulation show vertical intensification over several latitudes as is evidenced by stronger divergent winds of the

upper troposphere (the return branch of the local Hadley cell), which steers these clouds ahead of isochrones. Eventually newer buoyant elements are initiated ahead of isochrones. Generally buoyant elements are of the order of few kilometers and were difficult to visualize in present experiments with 25 km of resolution. As a result, model experiments were carried out at a horizontal resolution of 3 km in order to examine the time history of Buoyancy. These buoyant elements grew from 8 to 11 June over entire Indian landmass (Figure 6(g) and (h)). The growth of buoyant elements is indicative of the spread of convection over a wide area and a faster northward motion of the isochrones.

5. Concluding remarks and discussions

This research work dialogues about the importance of soil parameters and clouds in WRF simulations for the movement of monsoonal isochrones over Indian region. Two cases of unlike monsoonal isochrones movement were considered to compare and conclude. The modeling shows that the experiments SMSR25 and SR25 are the promising among the other experiments where we modified soil moisture, nonconvective clouds and soil temperature respectively. The modifications help in the

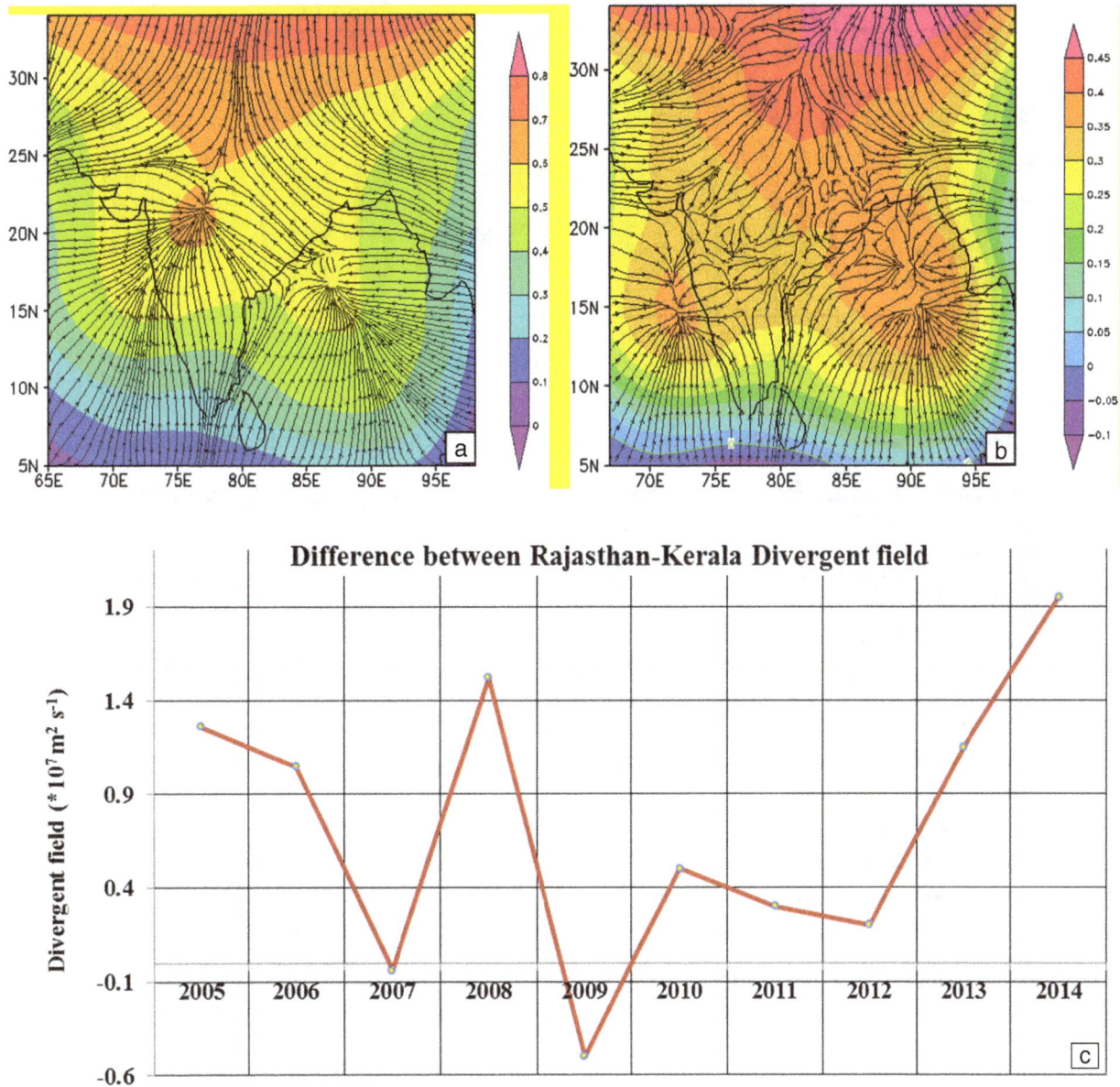

Figure 5. Mean of velocity potential filed from June 2 to 16 (*10⁷ m² s⁻¹) at 200 mb from (a) observation, (b) model experiment (SM25), (c) velocity potential difference between (*10⁷ m² s⁻¹ at 200 mb) Jaisalmer [Rajasthan, (71°E, 27°N)] and Trivandrum [Kerala, (77°E, 8.5°N)].

formation of a new line of convection that becomes the new position of the isochrone. That enhancement of soil moisture for the 2013 must have grown from spring season rains and also from the anvil rains of the parent isochrone. This moistening of the soil immediately ahead of the parent isochrone favored the formation of new clouds that grew in a conditionally unstable environment.

The central issue on fast versus slow meridional motion of the monsoonal isochrone appears to be strongly related to the upper branch of the local Hadley Cell over India. These divergent winds are directed from south to north and play an important role in the northward steering of the onset isochrone. The enhancement of the local Hadley cell appears to be related to this enhanced convection along the onset

isochrone that contributed to a fast steering northward and the establishment of the monsoon onset over most of India in a much shorter time compared to the climatological progress. The message that emerges from this study is that for real-time forecasts of this major societal scientific issue, monitoring of soil moisture and incorporation of nonconvective rain within a real-time high resolution nonhydrostatic cloud-resolving model may be necessary to address the march of the isochrones. Afterward, we planned to carry out a large number of sensitivity experiments using pixel-level soil moisture data from Navy satellite and asymmetry in the cloud cover across the isochrones. New study will be possible by using HRLDAS and 4DVAR components of WRF model. We suggest following steps, in the context of possible real-time application of this finding: (1) For

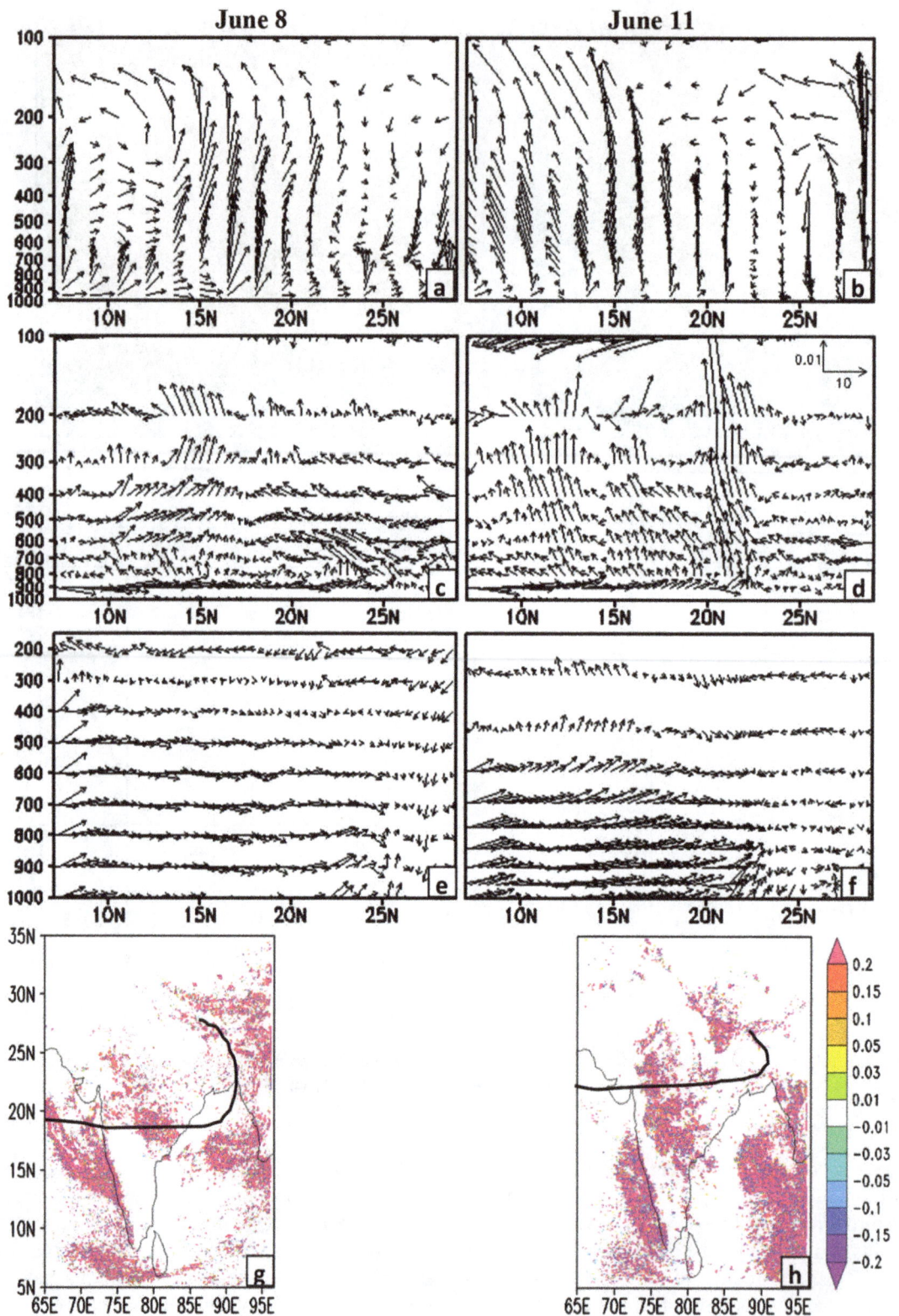

Figure 6. Local Hadley circulation (zonally averaged over 73–90°E) is displayed for 8 and 11 June 2013 over Indian region. (a) 8 June and (b) 11 June from observation; (c) 8 June and (d) 11 June from model; (e) 8 June and (f) 11 June from model for 2014; sum of the buoyancy (*10^{-2} m s^{-2}, 800–400 mb) from model experiment at 3 km horizontal resolution (g) 8 June and (h) 11 June.

the accurate prediction of monsoonal isochrones an acquisition of high-resolution soil moisture datasets on real time from ISRO facilities [e.g. Agro Metrological (AGROMET) Towers] is required; (2) Real-time availability of GPM precipitation and radar-based

vertical structure of hydrometeors (for definition of anvils and nonconvective rains) is important; and (3) There may be a need for real-time postprocessing of the model output to map the local Hadley cell and the divergent winds that steer the isochrones.

Acknowledgements

We wish to acknowledge Project MM/SERP/FSU-USA/2013/INT-8-002, entitled 'Sensitivity Studies for Indian Summer Monsoon Forecast Modeling', MoES-India.

Supporting information

The following supporting information is available:

Figure S1. Daily accumulated rainfall (mm) from for 2, 5, 8, 11, 14 and 16 June 2013 from TRMM3B42 and WRF-ARW model sensitivity experiments (control, enhancing soil moisture by 15%, enhancing stratiform rain by 15%, enhancing soil moisture and stratiform rain by 15% and reducing soil temperature by 7.5% ahead of the 1 June 2013 onset isochrone).

Figure S2. Daily accumulated rainfall (mm day^{-1}) from for 2, 5, 8, 11, 14 and 16 June 2013 from TRMM3B42 and WRF-ARW model sensitivity experiments (control, enhancing soil moisture by 20%, enhancing stratiform rain by 20% ahead of the 1 June 2013 onset isochrone).

Figure S3. ERA Interim volumetric soil moisture layer-1 for 2013: (a) 2–5 June, (b) 2–10 June, (c) 2–16 June, (d) to (f) except for low-level clouds, (g) to (i) except for soil temperature layer-1.

Figure S4. ARW-WRF sensitivity experiment of enhancing soil moisture by 25% ahead of the 1 June 2013 onset isochrone: (a) total soil moisture layer-1, 2–5 June; (b) total soil moisture layer-1, 2–10 June; (c) total soil moisture layer-1, 2–16 June; (d) total cloud fraction 2–5 June; (e) total cloud fraction 2–10 June; (f) total cloud fraction 2 1–16 June from 1000 to 100 mb; (g) average soil temperature layer-1, 2–5 June; (h) average soil temperature layer-1, 2–10 June; (i) average soil temperature layer-1, 2–16 June.

Figure S5. The IMD onset isochrones for year June 2014, a year of slower monsoonal isochrones.

References

Alfieri L, Claps P, D'Odorico P, Laio F, Over TM. 2008. An analysis of the soil moisture feedback on convective and stratiform precipitation. *Journal of Hydrometeorology* **9**: 280–291.

Ananthakrishnan R, Srinivasan V, Ramakrishnan AR, Jambunathan R. 1968. Synoptic features associated with onset of southwest monsoon over Kerala. Forecasting Manual Report IV-18.2, India Meteorological Department, Pune, 17 pp.

Chang CP, Harr PA, McBride J, Hsu HH. 2004. Maritime continent monsoon. In *East Asian Monsoon*, Chang CP (ed). World Scientific Singapore; 107–150.

Dee DP, Uppala SM, Simmons AJ, Berrisford P, Poli P, Kobayashi S, Andrae U, Balmaseda MA, Balsamo G, Bauer P, Bechtold P, Beljaars ACM, van de Berg L, Bidlot J, Bormann N, Delsol C, Dragani R, Fuentes M, Geer AJ, Haimberger L, Healy SB, Hersbach H, Hólm EV, Isaksen L, Kållberg P, Köhler M, Matricardi M, McNally AP, Monge-Sanz BM, Morcrette J-J, Park B-K, Peubey C, de Rosnay P, Tavolato C, Thépaut J-N, Vitart F. 2011. The ERA-Interim reanalysis: configuration and performance of the data assimilation system. *Quarterly Journal of the Royal Meteorological Society* **137**: 553–597.

Gill AE. 1980. Some simple solutions for heat-induced tropical circulation. *Quarterly Journal of the Royal Meteorological Society* **106**: 447–462.

Koster RD, Dirmeyer PA, Guo ZC, Bonan GB, Chan E, Cox P, Gordon CT, Kanae S, Kowalczyk E, Lawrence D, Liu P, Luo CH, Malyshev S, McAvaney B, Mitchell K, Mocko D, Oki T, Oleson K, Pitman A, Sud YC, Taylor CM, Verseghy D, Vasic R, Xue YK, Yamada T. 2004. Regions of strong coupling between soil moisture and precipitation. *Science* **305**: 1138–1140.

Krishnamurti TN, Ramanathan R. 1982. Sensitivity of the monsoon onset to differential heating. *Journal of Atmospheric Science* **39**: 1290–1306.

Krishnamurti TN, Simon A, Thomas A, Mishra AK, Sikka D, Niyogi D, Chakraborty A, Li L. 2012. Modeling of forecast sensitivity on the March of monsoon isochrones from Kerala to New Delhi: the first 25 days. *Journal of Atmospheric Science* **69**: 2465–2487.

Krishnan R, Ayantika DC, Kumar V, Samir P. 2011. The long-lived monsoon depressions of 2006 and their linkage with the Indian Ocean Dipole. *International Journal of Climatology* **31**: 1334–1352.

Rao YP. 1976. Southwest monsoon. IMD Meteorological Monograph No. 1/1976, India Meteorological Department, New Delhi, 107–185.

Samir P, Sikka DR. 2013. Variability of the TRMM-PR total and convective and stratiform rain fractions over the Indian region during the summer monsoon. *Climate Dynamics* **41**: 21–44.

Wang B. 2005. *Intraseasonal Variability of the Atmosphere–Ocean Climate System*, Lau KM, Waliser DE (eds). Springer-Verlag Berlin Heidelberg; 315–317.

Yanai M, Li C, Song Z. 1992. Seasonal heating of the Tibetan Plateau and its effects on the evolution of the Asian summer monsoon. *Journal of the Meteorological Society of Japan* **70**: 319–351.

Yeh TC, Wetherald RI, Manabe S. 1984. The effect of soil moisture on the short-term climate and hydrology change: a numerical experiment. *Monthly Weather Review* **112**: 474–490.

5

What controls the interannual variation of tropical cyclone genesis frequency over Bay of Bengal in the post-monsoon peak season?

Zhi Li,[1] Tim Li,[2,3]* Weidong Yu,[1] Kuiping Li[1] and Yanliang Liu[1]

[1]Center for Ocean and Climate Research, First Institute of Oceanography, SOA, Qingdao, China
[2]IPRC and Department of Meteorology, University of Hawaii, Honolulu, HI, USA
[3]International Laboratory on Climate and Environment Change and Key Laboratory of Meteorological Disaster, Nanjing University of Information Science and Technology, China

*Correspondence to:
 T. Li, IPRC and Department of Meteorology, University of Hawaii, No. 874, Dillingham Boulevard, Honolulu, HI 96822, USA.
E-mail: timli@hawaii.edu

Abstract

Tropical cyclone (TC) over Bay of Bengal (BoB) during its climatologic maximum peak season (October–November, post-monsoon season) exhibits a significant interannual variation between a negative Indian Ocean dipole (NIOD) and a positive IOD (PIOD) phase but not between El Niño and La Niña phase. Diagnosis of observational data reveals that the most important parameter that determines the interannual variation of BoB TC is the interaction between the mid-tropospheric relative humidity and the long-term mean states of absolute vorticity, vertical wind shear, and potential intensity. The change of mid-tropospheric moisture is primarily determined by vertical advection associated with low-level vorticity anomalies during IOD.

Keywords: tropical cyclone; Bay of Bengal; IOD; interannual variation

I. Introduction

The North Indian Ocean (NIO) is one of the main tropical cyclone (TC) genesis ocean basins. TCs over the NIO impose threats to a billion people each year in Arabia, India, Myanmar, Bangladesh, and Southeast Asian region (SEAR) (Webster, 2008; Lin *et al.*, 2013). Among the NIO TCs, almost 3/4 of those occur in Bay of Bengal (BoB). The climatological annual cycle of TC over BoB exhibits a marked bimodal character (Camargo *et al.*, 2009; Yanase *et al.*, 2012). The two peaks of BoB TC annual cycle take place in before and post-monsoon periods, April–May and October–November, respectively. The dynamic and thermodynamic factors that cause the bimodal characteristic were studied by Li *et al.* (2013b) based on the diagnosis of environmental parameters associated with the TC genesis potential index (GPI, Emanuel and Nolan, 2004). It was noted that due to the effect of mid-tropospheric relative humidity (RH), the overall TC genesis number during the post-monsoon peak is greater than that during the pre-monsoon peak (Li *et al.*, 2013b).

Besides the distinctive annual cycle, the BoB TC also undergoes a significant interannual variation. As dominant interannual modes in the tropical oceans, the Indian Ocean Dipole (IOD) and El Niño Southern Oscillation (ENSO) may exert a great influence on the frequency, intensity, and track of TCs over the Indian Ocean through induced large-scale atmosphere circulation (Wang and Chan, 2002; William and Young, 2007; Wing *et al.*, 2007; Eric and Chan, 2012; Clifford

et al., 2013; Sumesh and Kumar, 2013). Unlike ENSO whose peak phase typically occurs in boreal winter, the peak phase of IOD occurs in October–November and it almost overlaps the maximum peak of BoB TC frequency.

BoB TC frequency exhibits significant difference during October–November between positive IOD (PIOD) and negative IOD (NIOD) rather than between El Niño and La Niña. It may be attributed to the phase-locking relationship between IOD and BoB TC maximum peak phase. IOD, a tightly coupled air–sea interaction phenomenon, may exert a great impact on the interannual variation of October–November BoB TC (Singh *et al.*, 2008; Yuan and Cao, 2013). It was shown that more frequent BoB TC genesis in October–November occurred when there was a NIOD, and vice versa. While these observational studies pointed out the close relationship between IOD and BoB TC, it is not clear what large-scale controlling parameters associated with IOD influence the BoB TC frequency in the post-monsoon peak season. Yuan and Cao (2013) speculated that local warm sea surface temperature anomaly (SSTA) and low-level cyclonic vorticity associated with IOD might affect TC genesis. In this study we will use a quantitative analysis approach to reveal the relative roles of various environmental factors such as vertical wind shear (VWS), RH, vorticity, and sea surface temperature (SST) in controlling the interannual variation of TC frequency over BoB. Because the GPI is widely used in TC community and it well reflects to the characters of TC frequency at each ocean basin (Emanuel and Nolan, 2004; Camargo *et al.*, 2009; Yanase *et al.*,

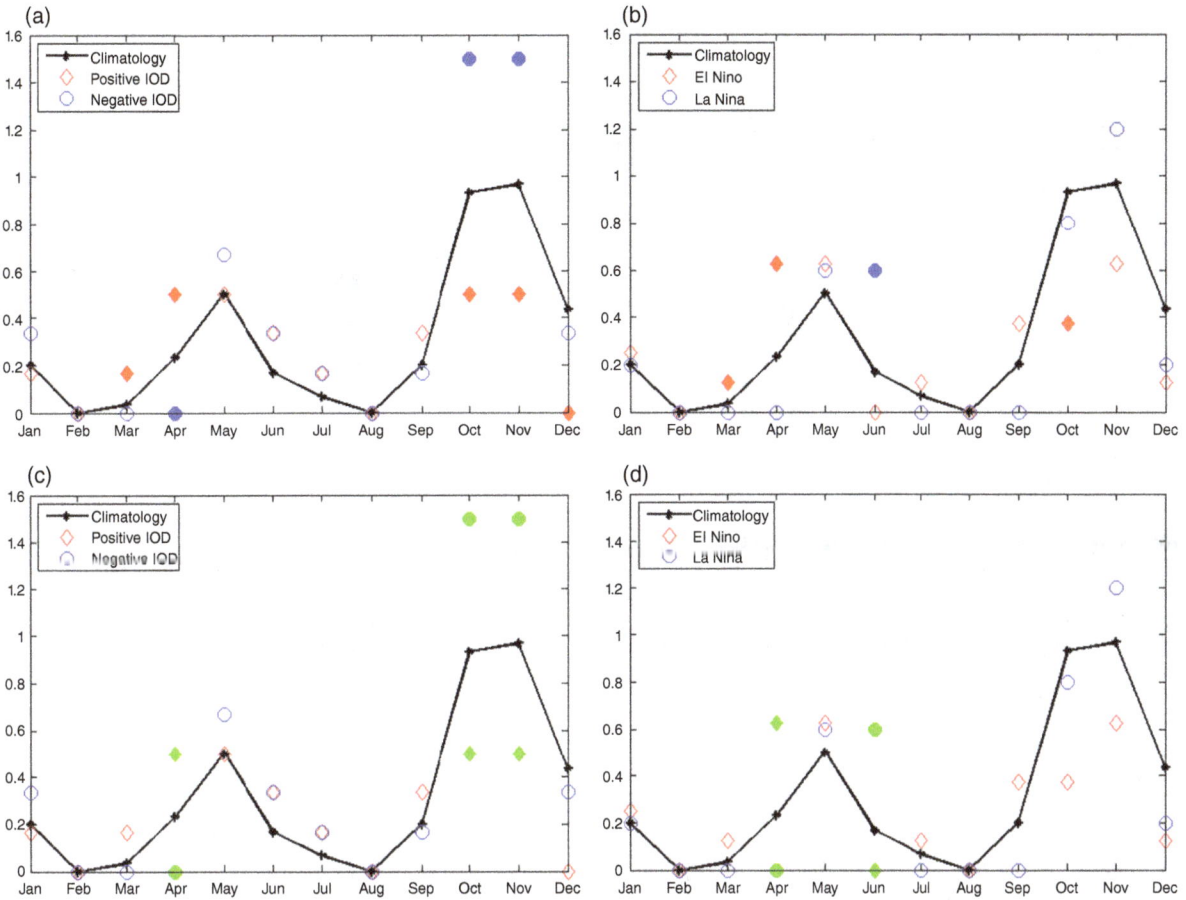

Figure I. Seasonal evolution of observed monthly averaged TC number over Bay of Bengal (BoB). In left panel, the solid line with asterisk represents climatology (1981–2010), whereas the open diamonds (circles) represent composite of positive (negative) IOD phase. In (a), values where t-test is statistically significant at the 90% and above confidence level are shown as closed diamonds (circles). In (c), the green closed diamonds and circles mean there is significant difference with more than 95% confidence in PIOD and NIOD. (b) and (d) same as in (a) and (c) but for El Niño (La Niña).

2012; Li *et al.*, 2013b), we will also use this index to quantitatively evaluate the relative effect of each environmental parameter in determining the difference of BoB TC frequency between NIOD and PIOD. In the following, we first introduce the data and methods to be used. Then we present the analysis results and conclusions.

2. Data and methods

TC best-track data from the Joint Typhoon Warning Center (JTWC) is utilized for determining TC number. Daily and monthly wind, air temperature, air-specific humidity, relative humidity from National Centers for Environmental Prediction (NCEP)–National Center for Atmospheric Research (NCAR) reanalysis and monthly SST from National Oceanic and Atmospheric Administration (NOAA) observed interpolated (OI) datasets are used to describe the large-scale environmental processes. Except for the SST data that have a horizontal resolution of 2° latitude by 2° longitude, the rest data sets, including wind, specific humidity, relative humidity, and air temperature, have a resolution of 2.5° latitude by 2.5° longitude.

It is well-known that TC genesis depends on several environmental factors including vorticity, VWS, SST, and water vapor content (Gray, 1968, 1979). Emanuel and Nolan (2004) further refined the TC genesis condition and proposed a GPI:

$$\text{GPI} = \text{Term1} \times \text{Term2} \times \text{Term3} \times \text{Term4} \quad (1)$$

where, $\text{Term1} = |10^5 \eta|^{3/2}$, $\text{Term2} = (1 + 0.1 V_{\text{shear}})^{-2}$, $\text{Term3} = \left(\dfrac{H}{50}\right)^3$, $\text{Term4} = \left(\dfrac{V_{\text{pot}}}{70}\right)^3$, η is absolute vorticity at 850 hPa (hereafter, vorticity), V_{shear} is the magnitude of VWS between 200 and 850 hPa, H is RH at 600 hPa , and V_{pot} is the maximum TC potential intensity (PI) defined by Emanuel and Nolan (2004):

$$V_{\text{pot}}^2 = C_{\text{p}} \left(T_{\text{s}} - T_{\text{o}}\right) \frac{T_{\text{s}}}{T_{\text{o}}} \frac{C_{\text{k}}}{C_{\text{D}}} \left(\ln \theta_{\text{e}}^* - \ln \theta_{\text{e}}\right) \quad (2)$$

In the PI formula, C_{p} is the heat capacity at constant pressure, T_{s} is the ocean temperature, T_{o} means outflow temperature, C_{k} is the exchange coefficient for enthalpy, C_{D} is the drag coefficient, θ_{e}^* is the saturation equivalent potential temperature at ocean surface, and θ_{e} is the boundary layer equivalent potential temperature. In order to quantitatively assess the relative contribution of

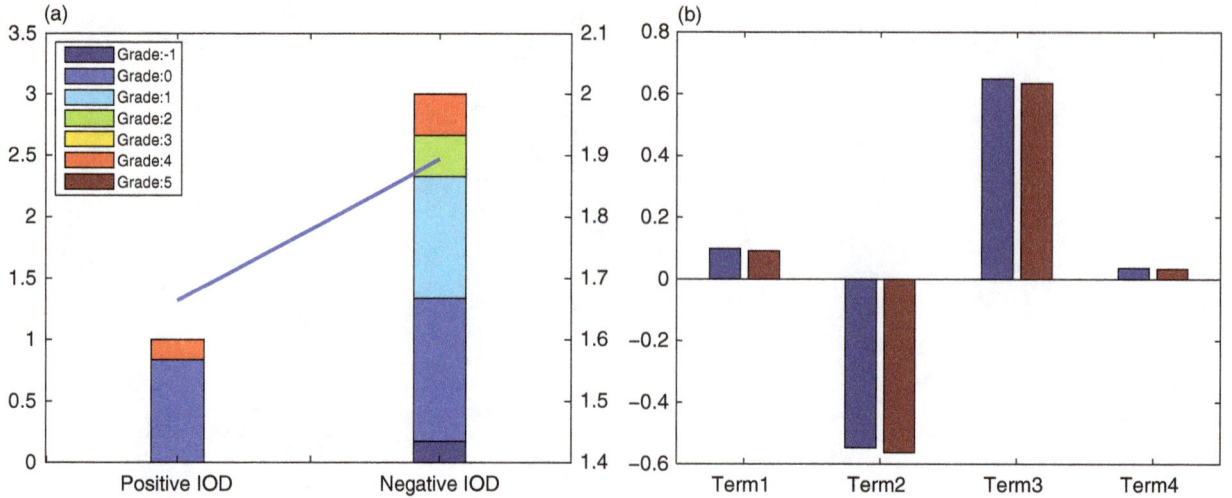

Figure 2. (a) Averaged BoB TC number in October–November during PIOD and NIOD. According to Saffir–Simpson scale, grades −1, 0, 1, 2, 3, 4, and 5 refer to tropical depression, tropical storm and 1 to 5 grade typhoons, respectively. The overlaid blue line represents the averaged GPI value in October–November during PIOD and NIOD. The left vertical-axis is for TC number and the right vertical-axis is for the GPI value. (b) The contributions of absolute vorticity, vertical wind shear, relative humidity, and potential intensity to the GPI difference between NIOD and PIOD. The sum of the four terms is approximately equal to a positive GPI difference value between NIOD and PIOD shown in (a). The blue bars are calculated using method 1 and the red brown bars are calculated using method 2.

each term, Li *et al.* (2013b) developed an analysis strategy by taking a natural logarithm at both sides of the GPI formula first and then applying a total differential to both sides. Thus the change of GPI may be separated into four terms as below:

$$\delta\,\mathrm{GPI} = \alpha_1\,\delta\mathrm{Term1} + \alpha_2\,\delta\,\mathrm{Term2} + \alpha_3\,\delta\mathrm{Term3}$$
$$+ \alpha_4\,\delta\,\mathrm{Term4} \qquad (3)$$

where

$$\alpha_1 = \begin{cases} \overline{\mathrm{Term2}}\;\overline{\mathrm{Term3}}\;\overline{\mathrm{Term4}} & \text{method 1} \\ \overline{\mathrm{Term2}}\;\overline{\mathrm{Term3}}\;\overline{\mathrm{Term4}} & \text{method 2} \end{cases} \qquad (4)$$

$$\alpha_2 = \begin{cases} \overline{\mathrm{Term1}}\;\overline{\mathrm{Term3}}\;\overline{\mathrm{Term4}} & \text{method 1} \\ \overline{\mathrm{Term1}}\;\overline{\mathrm{Term3}}\;\overline{\mathrm{Term4}} & \text{method 2} \end{cases} \qquad (5)$$

$$\alpha_3 = \begin{cases} \overline{\mathrm{Term1}}\;\overline{\mathrm{Term3}}\;\overline{\mathrm{Term4}} & \text{method 1} \\ \overline{\mathrm{Term1}}\;\overline{\mathrm{Term3}}\;\overline{\mathrm{Term4}} & \text{method 2} \end{cases} \qquad (6)$$

$$\alpha_4 = \begin{cases} \overline{\mathrm{Term1}}\;\overline{\mathrm{Term2}}\;\overline{\mathrm{Term3}} & \text{method 1} \\ \overline{\mathrm{Term1}}\;\overline{\mathrm{Term2}}\;\overline{\mathrm{Term3}} & \text{method 2} \end{cases} \qquad (7)$$

Through the above diagnosis method, one may quantitatively estimate the impacts of each environmental factor on TC frequency change between the negative and positive IOD phases.

A moisture budget analysis is further conducted to examine specific processes that give rise to the specific humidity anomaly (q') in PIOD and NIOD. Following Yanai *et al.* (1973), Hsu and Li (2012), and Li *et al.* (2013a), a tendency equation for interannual specific humidity anomaly (q') may be written as:

$$\frac{\partial q'}{\partial t} = -(V\,\nabla q)' - \left(\omega\,\frac{\partial q}{\partial p}\right)' - (Q_2/L)' \qquad (8)$$

where V is the horizontal velocity, ω is the p-vertical velocity, Q_2 is apparent moisture sink, and L is latent heat constant. In the equation above, $-(V\,\nabla q)'$ denotes anomalous horizontal moisture advection, $-\left(\omega\,\frac{\partial q}{\partial p}\right)'$ indicates anomalous vertical moisture advection, and $-(Q_2/L)'$ represents the anomalous moisture source or sink (Q_2 is primarily determined by surface evaporation and atmospheric condensation).

3. Analysis results

Six PIOD cases (1982, 1986, 1994, 1997, 2006, and 2007) and six NIOD cases (1984, 1992, 1996, 1998, 2005, and 2010) are selected during the period of 1981–2010 according to the original definition of IOD (Saji *et al.*, 1999; Webster *et al.*, 1999). In the same time, 1982, 1986, 1991, 1994, 1997, 2002, 2006, and 2009 are confirmed as eight El Niño cases and 1988, 1998, 1999, 2007, and 2010 also are considered as La Niña cases. Figure 1(a) presents BoB TC frequency number is significantly different with its climatology in October–November of IOD and the confidence level is above 90%. On the other hand, the difference of TC number exceeds 90% confidence level only in October of El Niño (Figure 1(b)). Furthermore, the difference of TC number between NIOD and PIOD exceeds 95% confidence level, respectively in October and November (Figure 1(c)). However, the difference significance between El Niño and La Niña is less than 75% (Figure 1(d)). In October–November, the difference of TC number between NIOD and PIOD is statistically significant at a confidence level exceeding 99%. Hence, it may be concluded that IOD mostly modulates the

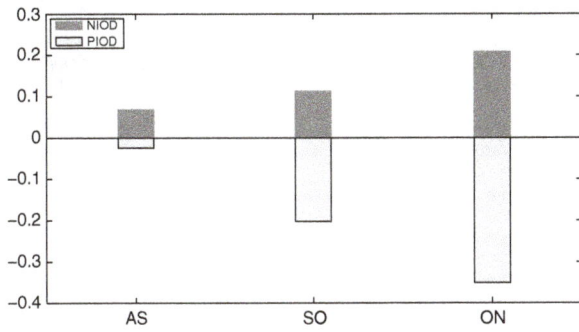

Figure 3. Area-average 600 hPa SH anomaly over the major BoB TC genesis region (5°–20°N, 80°–100°E) during August–September (AS), September–October (SO), and October–November (ON) of NIOD (gray bar) and PIOD (white bar).

interannual variation of BoB TC in post-monsoon peak season.

The paper examines the spatial distribution of GPI and TC genesis locations during October–November of both IOD phases over BoB, respectively. The horizontal distribution of the GPI anomaly well fit the TCs' feature (figure not shown). Moreover, according to BoB TC genesis region, a box (5°–20°N, 80°–100°E) is defined as the TC formation area. We calculated the area averaged GPI values during October–November of NIOD and PIOD and found that these values match quite well the averaged October–November TC number (Figure 2(a)).

Because the GPI well reflects TC genesis frequency difference between NIOD and PIOD, we further use GPI to diagnose the relative contribution of each environmental factor. Figure 2(b) shows the calculation result (i.e. difference between NIOD and PIOD). As one can see, more frequent TC frequency during NIOD is primarily attributed to the interaction between the mid-tropospheric RH and the long-term mean states of absolute vorticity, VWS, and potential intensity. Because mean states of absolute vorticity, VWS, and potential intensity are constant and independent of year-to-year changes, we may approximately state the contribution is attributed to RH term. In the same way, the vorticity term has a weak positive contribution. The VWS term has a negative contribution. The effect of SST or PI is negligible.

The result indicates that IOD primarily modulates BoB TC frequency through the change of RH field in middle troposphere (600 hPa). Given that RH is a function of specific humidity (SH) and air temperature (AT), a natural question is whether RH change is primarily controlled by SH or AT changes. By the diagnosis, we confirmed the change of the 600 hPa RH associated with IOD is primarily attributed to the change of SH, not AT.

The examination of time evolution of 600 hPa SH shows that it has a maximum tendency in September–October and reaches its peak in Octobe–November for both the PIOD and NIOD composites (Figure 3). Therefore, a question of why GPI has a maximum positive (minimum negative) value in October–November during NIOD (PIOD) may be converted to a question of why local SH at 600 hPa increases (decreases) rapidly during September–October. This motivates us to further

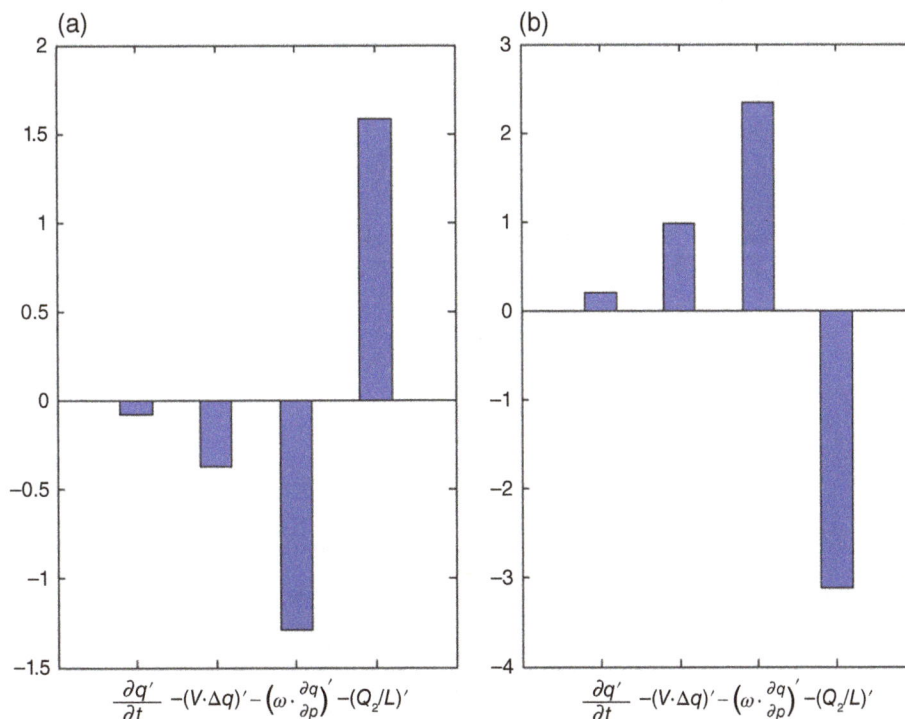

Figure 4. Composite 600 hPa-level SH anomaly tendency terms in September–October for the PIOD (a) and NIOD (b) groups, respectively.

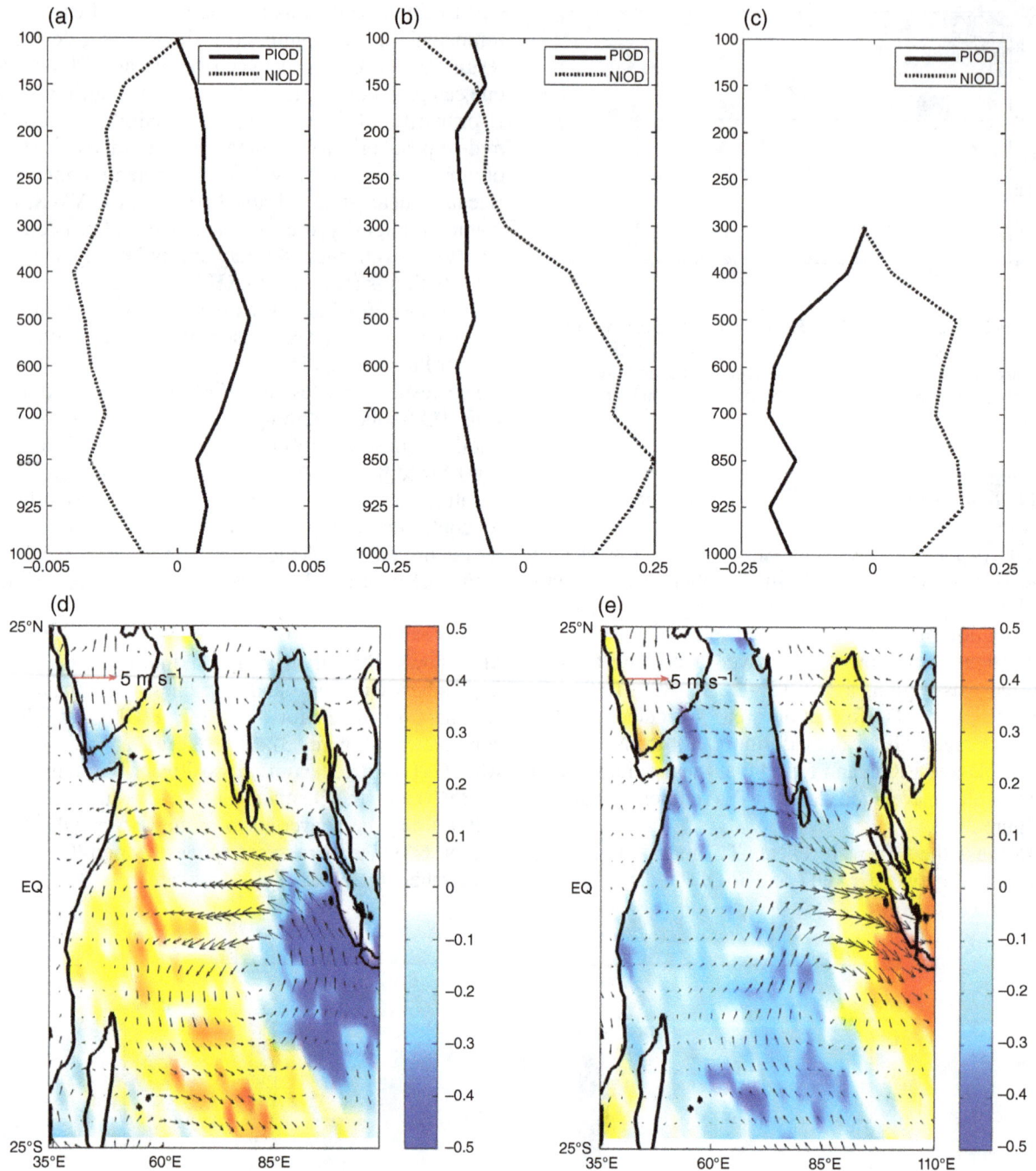

Figure 5. (Top) Vertical profiles of area-average anomalous vertical (a) p-velocity (ω), (b) vorticity and (c) SH over (5°–20°N, 80°–100°E) in September–October for NIOD (dashed) and PIOD (solid). (Bottom) Composite 850 hPa wind anomaly (vector) and SST anomaly (shading) in September–October during (d) PIOD and (e) NIOD.

diagnose the moisture budget at 600 hPa for PIOD and NIOD cases.

Figure 4 shows the SH budget analysis result. Because the PIOD and NIOD results are approximately a mirror image, we discuss in the following only the NIOD result (Figure 4(b)). The increase of SH at 600 hPa is primarily attributed to the vertical advection, while the apparent moisture sink term has a negative contribution. Physical interpretation is given below. Note that SST anomalies associated with NIOD are quite strong in September–October. Although the SSTA is asymmetric about the Equator in the eastern IO during IOD

events and its center generally occurs at nearby 8°S in the vicinity of Sumatra, the wind curl anomaly responding to the SSTA is quite symmetric to the Equator (Li *et al.*, 2003; Yu *et al.*, 2005; Schott *et al.*, 2009). In response to a warm (cold) SSTA in the eastern (western) equatorial IO during NIOD, there are pronounced westerly anomalies at the equator and two cyclonic gyre circulations at both sides of the equator (Figure 5(e)). The cyclonic flow at top of atmospheric boundary layer induces anomalous ascending motion through the Ekman pumping effect. The anomalous ascending motion advects the mean moisture upward and causes the increase of SH in the lower and middle troposphere (Figure 5(c)). This promotes a convectively unstable stratification and favors the deepening of ascending motion and moist layer (Figure 5(a) and (b)). The increase of mid-tropospheric moisture favors more frequent TC genesis in October–November during NIOD.

In addition to the moisture effect, the low-level cyclonic (anticyclonic) wind anomaly at 850 hPa also contributes to TC genesis frequency change during NIOD (PIOD). The cyclonic and anticyclonic wind anomaly makes the vorticity term to provide secondary positive contribution for TC number difference between NIOD and PIOD.

Because climatologic wind vertical shear in BoB is easterly shear, an increase (decrease) of such a shear in the BoB TC genesis region during NIOD (PIOD) prohibits (favor) cyclogenesis. This is why VWS has a negative impact on BoB TC on the interannual timescale. On the other hand, according to previous studies (e.g. Li, 2006; Li, 2012), compared with same amount of westerly shear, easterly vertical shear is more favorable for the growth of synoptic-scale perturbations. From this point, NIOD may provide a favorable environmental shear condition for TC formation. SST plays a key role in PI term. The effect of the PI on the GPI during October–November of IOD is very weak since BoB region averaged SST is similar to its climatology during this period.

4. Summary

Through quantitative diagnosis of various environmental parameters associated with the GPI, we found that the most important parameter that determines the interannual variation of TC genesis frequency in BoB during the post-monsoon season is interaction between the mid-tropospheric RH and long-term mean states of absolute vorticity, VWS, and PI. The contribution is attributed to RH term because the mean states of the rest terms are constant and independent of interannual variation. In the same way, for BoB TC-number difference in October–November between NIOD and PIOD, we can state low-level environmental vorticity has a weaker positive contribution, VWS has a negative contribution, and PI's contribution is almost negligible.

Due to the critical role of mid-tropospheric water vapor, a moisture budget analysis is further carried out to understand key processer responsible for the mid-tropospheric moisture change. It was found that the major process affecting the moisture change is anomalous vertical advection.

Based on the results above, a scenario through which a negative or positive IOD influences BoB TC genesis frequency in October–November is proposed. Because NIOD and PIOD scenarios are generally a mirror image, in the following we focus on the discussion of the NIOD case.

1. Although the SSTA is asymmetric about the equator, its impact on atmospheric circulation in the eastern IO is approximately symmetric. In response to a NIOD, there are pronounced low-level westerly anomalies at the equator and an anomalous cyclone over BoB. The cyclonic flow at top of boundary layer induces anomalous ascending motion, which transports mean moisture upward and moistens the lower troposphere. The so-induced convective instability further strengthens large-scale ascending motion. As a result, the moist layer deepens. Our moisture budget analysis confirms that the anomalous vertical advection, instead of anomalous horizontal advection, plays a critical role in increasing the specific humidity anomaly at 600 hPa. The change of SH mainly contributes to the change of RH in 600 hPa.

2. Meanwhile the low-level cyclonic vorticity anomaly at 850 hPa makes a weaker positive contribution to the GPI difference between NIOD and PIOD.

3. The VWS associated with IOD has a negative contribution to the GPI change. This is because the wind anomaly associated with NIOD has nearly same direction as the climatological wind over BoB TC genesis region.

4. The PI term does not significantly contribute to the GPI change, simply because the SSTA associated with IOD over BoB is very weak and the most significant SSTA appears south of the equator.

Acknowledgements

We thank the JTWC, NCEP/NCAR, ECMWF, and NOAA for their free datasets. This study is sponsored by China National 973 Program 2015CB453200, NRL grant N00173-13-1-G902, Basic Scientific Fund for National Public Research Institutes of China GY0214G04 and NSFC grant 41406030. The International Pacific Research Center is partially sponsored by the Japan Agency for Marine-Earth Science and Technology (JAMSTEC). This is SOEST contribution number 9568, IPRC contribution number 1159, and ESMC contribution number 068.

References

Camargo SJ, Wheeler MC, Sobel AH. 2009. Diagnosis of the MJO modulation of tropical cyclogenesis using an empirical index. *Journal of the Atmospheric Sciences* **66**: 3061–3074.

Clifford SF, Subrahmanyam B, Murty VSN. 2013. ENSO-modulated cyclogenesis over Bay of Bengal. *Journal of Climate* **26**: 9806–9818.

Emanuel KA, Nolan DS. 2004. Tropical cyclone activity and global climate. Preprints. In 26th Conference on Hurricanes and Tropical Meteorology, Miami, Florida, 2–7 May 2004. American Meteorological Society. https://ams.confex.com/ams/26HURR/techprogram/paper_75463.htm (accessed 5 May 2004).

Eric KW, Chan JCL. 2012. Interannual variations of tropical cyclone activity over north Indian Ocean. *International Journal of Climatology* **32**: 819–830.

Gray WM. 1968. Global view of the origin of tropical disturbances and storms. *Monthly Weather Review* **96**: 669–700.

Gray WM. 1979. Hurricanes: their formation, structure and likely role in the general circulation. In *Meteorology over the Tropical Ocean*, Shaw DB (ed). Royal Meteorological Society: Bracknell, UK; 155–218.

Hsu PC, Li T. 2012. Role of the boundary layer moisture asymmetry in causing the eastward propagation of the Madden-Julian Oscillation. *Journal of Climate* **25**: 4914–4931, doi: 10.1175/JCLI-D-11-00310.1.

Li T. 2006. Origin of the summertime synoptic-scale wave train in the western North Pacific. *Journal of the Atmospheric Sciences* **63**: 1093–1102.

Li T. 2012. *Synoptic and climatic aspects of tropical cyclogenesis in Western North Pacific*, Oouchi K, Fudeyasu H (eds). Nova Science Publishers, Inc.: New York, NY; 61–94.

Li T, Wang B, Chang CP, Zhang Y. 2003. A theory for the Indian Ocean dipole-zonal mode. *Journal of the Atmospheric Sciences* **60**: 2119–2135.

Li K, Yu W, Li T, Murty VSN, Khokiattiwong S, Adi TR, Budi S. 2013a. Structures and mechanisms of the first-branch northward-propagation intraseasonal oscillation over tropical Indian Ocean. *Climate Dynamics* **40**: 1707–1720.

Li Z, Yu W, Li T, Murty VSN, Tangang F. 2013b. Bimodal character of cyclone climatology in Bay of Bengal modulated by monsoon seasonal cycle. *Journal of Climate* **26**: 1033–1046.

Lin II, Black P, Price JF, Yang CY, Chen SS, Lien CC. 2013. An ocean coupling potential intensity index for tropical cyclones. *Geophysical Research Letters* **40**: 1878–1882.

Saji NH, Goswami BN, Vinayachandran PN, Yamagata T. 1999. A dipole mode in the tropical Indian Ocean. *Nature* **401**: 360–363.

Schott AF, Xie SP, McCreary JP. 2009. Indian Ocean circulation and climate variability. *Reviews of Geophysics* **47**: RG1002, doi: 10.1029/2007RG000245.

Singh OP, Gupta M, Santha K, Saikia D, Khanuja S. 2008. Indian Ocean dipole mode and tropical cyclone frequency. *Currentence* **94**: 29–31.

Sumesh KG, Kumar MRR. 2013. Tropical cyclones over north Indian Ocean during El Nino modoki years. *Natural Hazards* **68**: 1057–1074.

Wang B, Chan JCL. 2002. How strong ENSO events affect tropical storm activity over the western North Pacific. *Journal of Climate* **15**: 1643–1658.

Webster PJ. 2008. Myanmar's deadly daffodil. *Nature Geoscience* **1**: 488–490.

Webster PJ, Moore AM, Loschnigg JP, Leben RR. 1999. Coupled ocean–atmosphere dynamics in the Indian Ocean during 1997–98. *Nature* **401**: 356–360.

William MF, Young SG. 2007. The interannual variability of tropical cyclones. *Monthly Weather Review* **135**: 3587–3598.

Wing AA, Sobel AH, Camargo SJ. 2007. Realtionship between the potential and actual intensities of tropical cyclones on inter-annual time scales. *Geophysical Research Letters* **34**: 402–420, doi: 10.1029/2006GL028581.

Yanai M, Esbensen S, Chu JH. 1973. Determination of bulk properties of tropical cloud clusters from large-scale heat and moisture budgets. *Journal of the Atmospheric Sciences* **30**: 611–627.

Yanase W, Satoh M, Taniguchi H, Fujinami H. 2012. Seasonal and intraseasonal modulation of tropical cyclogenesis environment over the Bay of Bengal during the extended summer monsoon. *Journal of Climate* **25**: 2914–2930.

Yu W, Xiang B, Liu L, Liu N. 2005. Understanding the origins of interannual thermocline variations in the tropical Indian Ocean. *Geophysical Research Letters* **32**: 348–362, doi: 10.1029/2005GL024327.

Yuan JP, Cao J. 2013. North Indian Ocean tropical cyclone activities influenced by the Indian Ocean Dipole mode. *Science China Earth Sciences* **56**: 855–865.

6

Observational and dynamic downscaling analysis of a heavy rainfall event in Beijing, China during the 2008 Olympic Games

Huiqi Li,[1,2] Xiaopeng Cui,[1,3]* Wenlong Zhang[4] and Lin Qiao[5]

[1] Key Laboratory of Cloud-Precipitation Physics and Severe Storms (LACS), Institute of Atmospheric Physics, Chinese Academy of Sciences, Beijing, China
[2] College of Earth Science, University of Chinese Academy of Sciences, Beijing, China
[3] Collaborative Innovation Center on Forecast and Evaluation of Meteorological Disasters, Nanjing University of Information Science & Technology, Nanjing, China
[4] Institute of Urban Meteorology, China Meteorological Administration, Beijing, China
[5] Beijing Municipal Weather Forecast Center, Beijing Meteorological Service, Beijing, China

*Correspondence to:
X Cui, Key Laboratory of
Cloud-Precipitation Physics and
Severe Storms (LACS), Institute of
Atmospheric Physics, Chinese
Academy of Sciences, Beijing
100029, China.
E-mail: xpcui@mail.iap.ac.cn

Abstract

A local precipitation event with several dispersedly distributed heavy rainfall centers exceeding 50 mm occurred in Beijing, China on 14 August 2008 during the Beijing Olympic Games. The heavy rainfall event was produced by a few scattered convective storms. Detailed observational analysis with data from automatic weather stations (AWSs) as well as the meteorological radar in Beijing and a dynamic downscaling analysis with a diagnostic model, California Meteorological Model (CALMET), showed that convergence zones caused by small-scale topography and colliding outflow boundaries were key influencing factors in the initiation and development of the convective storms. Convergence helped to induce upward vertical motion as well as concentrate moisture to reduce the convective inhibition (CIN). Horizontal wind speed may modulate the effectiveness of convergence. Downscaled wind fields by CALMET not only retain the overall features of the original fields, but also present more detailed structures, especially near complex terrain, which is much helpful in analyzing and predicting the development of the storms.

Keywords: local heavy rainfall; convergence; outflows; topography; observational analysis; dynamic downscaling analysis

1. Introduction

The initiation and development of local convective storms and consequent heavy precipitation are determined by a combination of multiple-scale factors. Under favorable synoptic environment, local convergence is one of the significant factors influencing the formation and organization of convective storms and precipitation (Byers and Rodebush, 1948; Wilson and Schreiber, 1986; Wilson and Megenhardt, 1997). Outflows produced by precipitation systems can contribute to the development of convergence zones (Wilhelmson and Chen, 1982). Previous studies on outflows have given more attention to those produced by highly organized convective systems, while the outflows produced by localized, scattered convective storms, and their roles in the formation of new storms still lack detailed analysis. In the vicinity of complex terrain, the formation of convective storms and precipitation could also be modulated by local topography, making the situation more complicated. Houze (2012) summarized four possible orographic effects in convective precipitation.

Beijing is characterized by local complex terrain with the Yan Mountains to the north and the Taihang Mountains to the west with heights varying from 200 to 1600 m (Figure 1(a)). Relatively smooth terrain with low elevations dominates the central and southeast parts of Beijing where 'unexpected' (in the current observational network) local convective storms and consequent short-term local heavy rainfall often occur, being a big challenge to the forecasters. An unusual local heavy precipitation event produced by some scattered convective storms on 14 August 2008 during the Beijing Olympic Games provides a perfect case to investigate the local conditions related to the formation and organization of local convective storms and heavy rainfall centers.

An overview of this heavy rainfall event is given in Section 2. In Section 3, observational analysis using data from the automatic weather stations (AWSs) as well as the meteorological radar in Beijing is performed, followed by a dynamic downscaling analysis with a diagnostic model, California Meteorological Model (CALMET), to show the formation and organization of the local convective storms and consequent rainfall centers. A summary is given in Section 4.

2. Case overview

The local heavy rainfall event produced by several dispersedly distributed convective storms started from

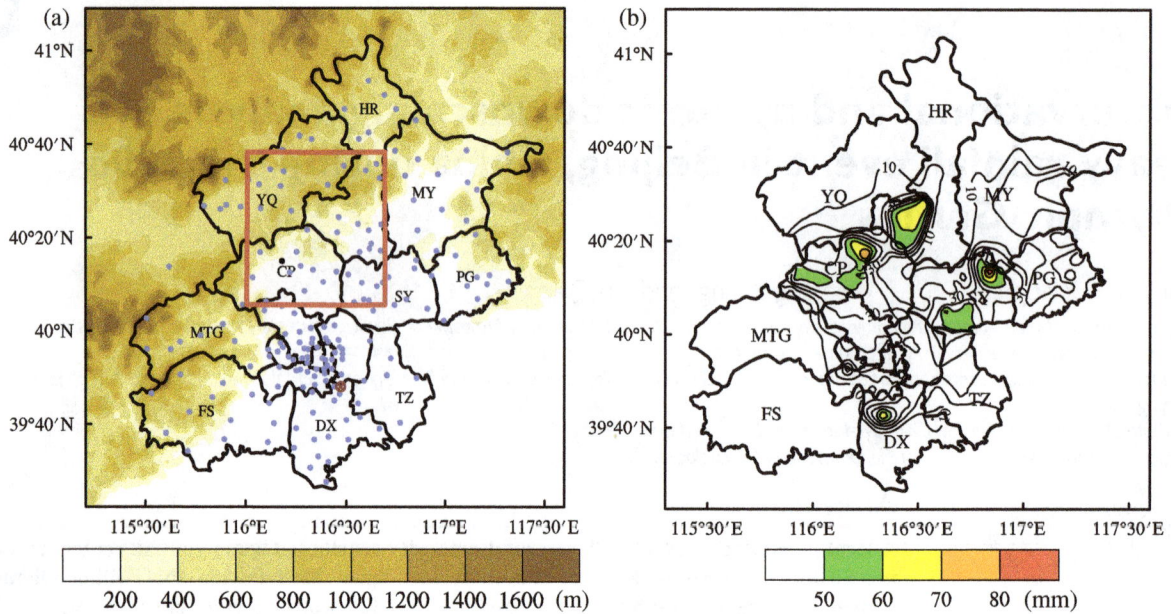

Figure 1. (a) Terrain in Beijing (shaded, m). Dots represent automatic weather stations. The red dot denotes the location of the Taipingzhuang station used in Figure 5. Red open circle with cross inside denotes the position of Beijing radar. The box is the computation domain for CALMET. Abbreviations in the figure are the names of corresponding districts: HR for Huairou, YQ for Yanqing, MY for Miyun, CP for Changping, SY for Shunyi, PG for Pinggu, MTG for Mentougou, FS for Fangshan, DX for Daxing, and TZ for Tongzhou. Those central parts without names are the urban areas of Beijing. (b) 12-h accumulated precipitation (mm) from 0000 UTC on 14 August 2008.

0300 UTC on 14 August 2008 and lasted for only about 5 h (Zhang *et al.*, 2014). Six local heavy rainfall centers with accumulated rainfall amounts exceeding 50 mm distributed irregularly in Beijing (Figure 1(b)), most of which last for only 1–2 h. Nine weather stations recorded an accumulated rainfall amount of more than 50 mm. Several matches of the ongoing 2008 Beijing Olympic Games were forced to be postponed (http://2008.sohu.com/20080814/n258856711.shtml and http://2008.qq.com/a/20080815/004363.htm).

At 0000 UTC on 14 August 2008, a low pressure located to the southwest of Beijing and a trough was present over the northeast of China at 500 hPa, leaving Beijing between them in a quasi-deformation field. The low pressure was related to the formation of convective cloud cluster to the southwest of Beijing (Figure 2(a)). At 850 hPa, Beijing was under the control of weak easterlies with no distinct shear line. At surface, there was no well-marked frontal system affecting Beijing. Local higher convective available potential energy (CAPE) and lower convective inhibition (CIN) in Beijing (Zhang *et al.*, 2014) were favorable for the development of local convective storms. It can be hypothesized that local conditions, such as temperature and moisture variation, small-scale topography could be important for the triggering and organization of those scattered convective storms and associated rainfall centers in this event.

3. Observational analysis

Zhang *et al.* (2014) analyzed the evolution of the formation location of storms in this event, and emphasized the importance of the boundary-layer convergence which existed more than half an hour before the emergence of several storms. They also pointed out that local topography and outflows from pre-existing thunderstorms played a crucial role in the event. However, the following questions remain unsolved. (1) How was the boundary-layer convergence generated in detail under the impact of local topography and the outflows? (2) What was the detailed role of convergence in the formation of convective storms? (l3) There were some convergence zones without deep convections in this event in Beijing. What were the possible reasons?

With the assistance of data from the AWSs and a diagnostic model, CALMET, we try to address the above problems in this section.

3.1. Observational analysis using data from AWSs and radar

Several isolated storms were important origins of the heavy rainfall (see storms with labels in Figure 2). At 0254 UTC on 14 August 2008, besides the relatively extensive radar reflectivity to the southwest of Beijing and some scattered cells in Mentougou, a separate storm cell ('A' in Figure 2(a)) with reflectivity over 45 dBZ emerged in Pinggu. Upslope flows converging in the north of Pinggu contribute to the initiation of cell A (Figure 2(a) and 3(b)). The outflows from the southwest precipitation system enhanced the local convergence in Daxing. Another significant contributor to the convergence in Daxing is the southwestwards-running flows through the river valley of Miyun, which were partly induced by the convective activity to the northeast of

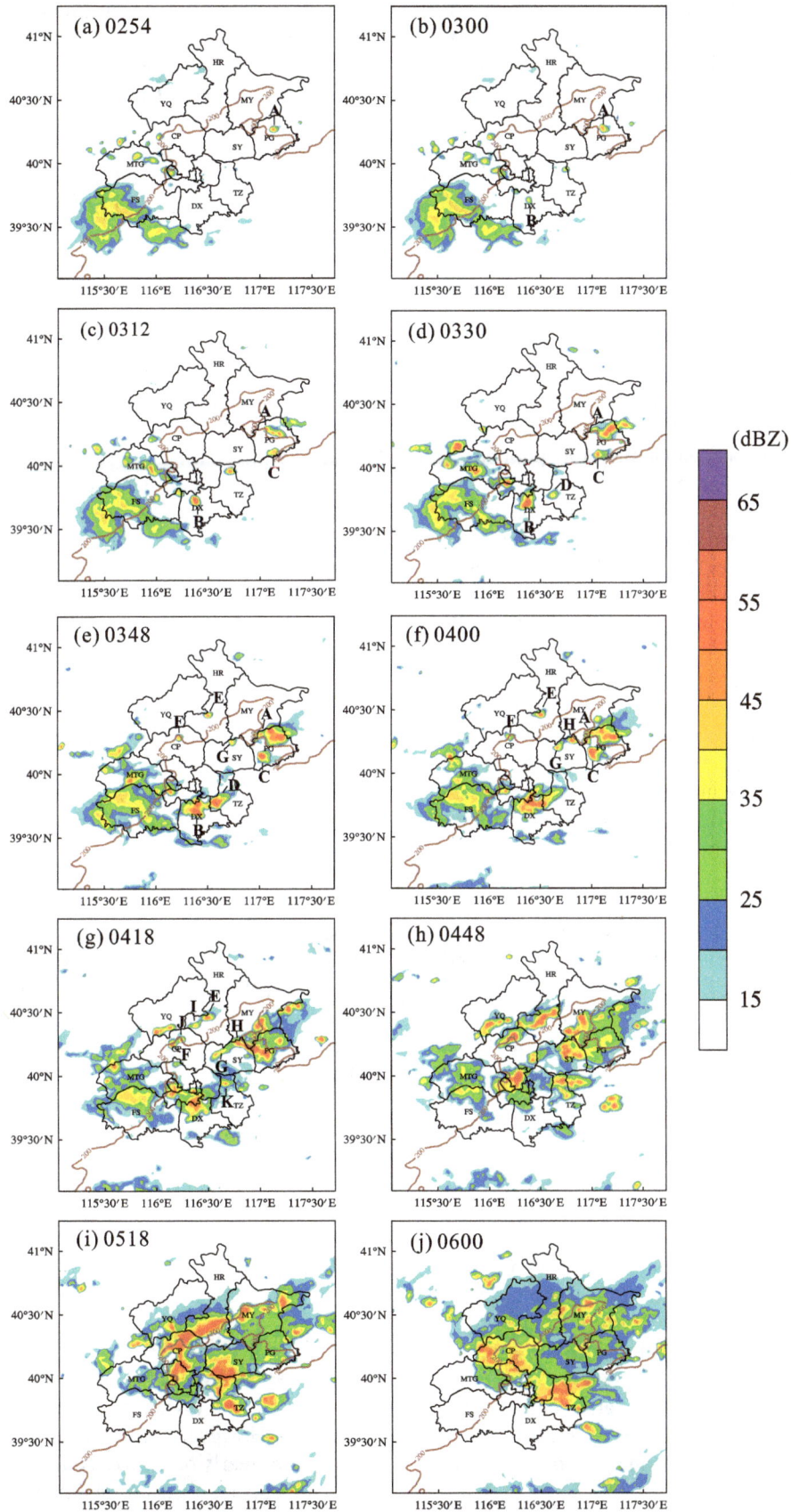

Figure 2. Vertical maximum reflectivity (dBZ) observed by Beijing radar at (a) 0254, 2 (b) 0300, (c) 0312, (d) 0330, (e) 0348, (f) 0400, (g) 0418, and (h) 0448 UTC. Black capital letters give numbers to the cells described in the text. Brown lines are isohypses of 200 m; similarly for the rest of figures.

Beijing under the favorable large-scale environment of the 500-hPa low (not shown). Accordingly, a new separate cell ('B' in Figure 2(b)) emerged in Daxing at 0300 UTC and developed fast in the next 10 min (Figure 2(c)). Distinct local heavy precipitation started to occur in Daxing after the formation of cell B and lasted for less than 2 h (Figure 4(a) and (b)). Accumulated 1-h rainfall amount of 51.1 mm was recorded in the Daxing station from 0300 to 0400 UTC. Another separate cell ('C' in Figure 2(c)) with reflectivity over 40 dBZ ensued in Pinggu near the 200-m isohypse at 0312 UTC, which was triggered by the local convergence in the south of Pinggu (Figure 3(d)). With the development of the local precipitation system in Daxing, an apparent cold pool with relatively drier air (indicated by purple circles in Figure 3(e)–(h)) was generated and corresponding outflows were formed. With the further expansion of the cold pool in Daxing, the local convergence zone was pushed to the Daxing–Tongzhou border (Figure 3(e)–(h)), contributing to the following formation and development of local convective storms there. At 0330 UTC, a new separate cell with reflectivity over 40 dBZ was observed in the west of Tongzhou ('D' in Figure 2(d)). As the cells in Pinggu continued to develop, interaction between the outflows near Pinggu and the northeasterly along the river valley of Miyun formed a local confluence zone near the 200-m isohypse near the border of Shunyi and Miyun, which triggered the cell G and the cell H (Figure 2(e) and (f)). With the development and merger of cells A, C, and H, a distinct rainfall center of hourly precipitation over 20 mm was observed near the border of Shunyi, Pinggu, and Miyun from 0400 to 0500 UTC (Figure 4(b)).

It is noticeable that distinct convergence was found in the center of Shunyi (as indicated by the confluence of streamlines in Figure 3(b), (d), (f), and (h)), but no convective storm developed. Horizontal wind speed was relatively large in the center of Shunyi (not shown). It may make the updraft associated with the convergence more tilted and more vulnerable to entrainment (Markowski *et al.*, 2006). Sounding at the southern observatory shows that it was drier at the upper boundary layer than at the surface, thus entrainment was unfavorable for the development of deep convective clouds.

At 0348 UTC, two separate storm cells with reflectivity over 40 dBZ were also seen in the south of Huairou ('E' in Figure 2(e)) and in the north of Changping ('F' in Figure 2(e)), which were related to the small-scale convergence zone under local complex terrain and the surface small-scale cyclonic circulation near local fork horn shaped topography, respectively (Figure 3(h), also can be seen in the following analysis by CALMET). Convergence could be seen in the north of Changping around 0300 UTC, and convection was detected by the radar at 0348 UTC. Convergence could induce upward vertical motion to lift parcels to reach their level of free convection (LFC). Were there any other factors influencing the formation of the convection?

As shown in Figure 5, the 2-m mixing ratio observed in the Taipingzhuang station (denoted by the red dot in Figure 1(a)) near the convergence area presented an increasing trend before the development of the cell F in Changping. It illustrates that convergence in the north of Changping also assisted in moisture pooling, which reduced CIN there. The combined effect of upward vertical motion and increasing moisture led to the formation of the cell F (Figure 2(e)) at 0348 UTC in Changping.

At 0415 UTC, cold pools and outflows near Daxing and Pinggu (indicated by purple circles and dark blue circles respectively in Figure 3(i) and (j)) grew rapidly as the storms and related local precipitation systems developed and decayed. In the meantime, the southeasterly winds between Daxing and Pinggu enhanced significantly, and the previous flow along the river valley of Miyun was almost totally cut off. Two new cells, I and J, began to form between E and F along the slantwise terrain in the northwest of Beijing. The combination of the southeasterly winds and the slantwise terrain, as well as convergence resulting from the colliding boundary between E and F contributed to the formation of cell I and J (Figure 2(g)). These convective storms induced the local precipitation in Changping from 0400 UTC (Figure 4b). With the enhancement of the outflows in Daxing and Pinggu, a new cell ('K' in Figure 2(g)) in the north of Tongzhou formed at the convergence zone between the two outflows. Later at 0445 UTC (Figure 3(k) and (l)), the cold pools near Pinggu and Daxing continue to enlarge. The outflows from Shunyi, Miyun, Pinggu, and from Daxing converged with the outflows from the four storm cells (E, F, I, and J) forming in Huairou and Changping previously. Thus, a local intense convergence zone developed along the 200-m isohypse in the northwest of Beijing (Figure 3(l)). The above cells (E, F, I, and J) began to merge and develop along the convergence zone, and a small-scale line convection was to be organized in the northwest of Beijing (Figure 2(h)). With the development and southwestwards movement of these convective systems (Figure 2(i) and (j)), precipitation increased sharply in Huairou and Changping from 0500 UTC and formed the most notable rainfall centers along the 200-m isohypse (Figure 4(c)). The 47.9-mm rainfall amount in Bohai station and 46.8-mm rainfall amount in Changping station were recorded from 0500 to 0600 UTC.

3.2. Downscaling analysis of surface wind fields

As analyzed in Section 3.1, the formations of the convective cells E and F were closely related to small-scale topography. However, the detailed formation processes could not be clearly recognized due to the spatial resolution of the current observational network (Figure 1(a)). Thus, a dynamic downscaling analysis using a diagnostic model, CALMET, was conducted below by using the AWS observations.

CALMET is a component of the California Puff (CALPUFF) Modeling System which is an advanced

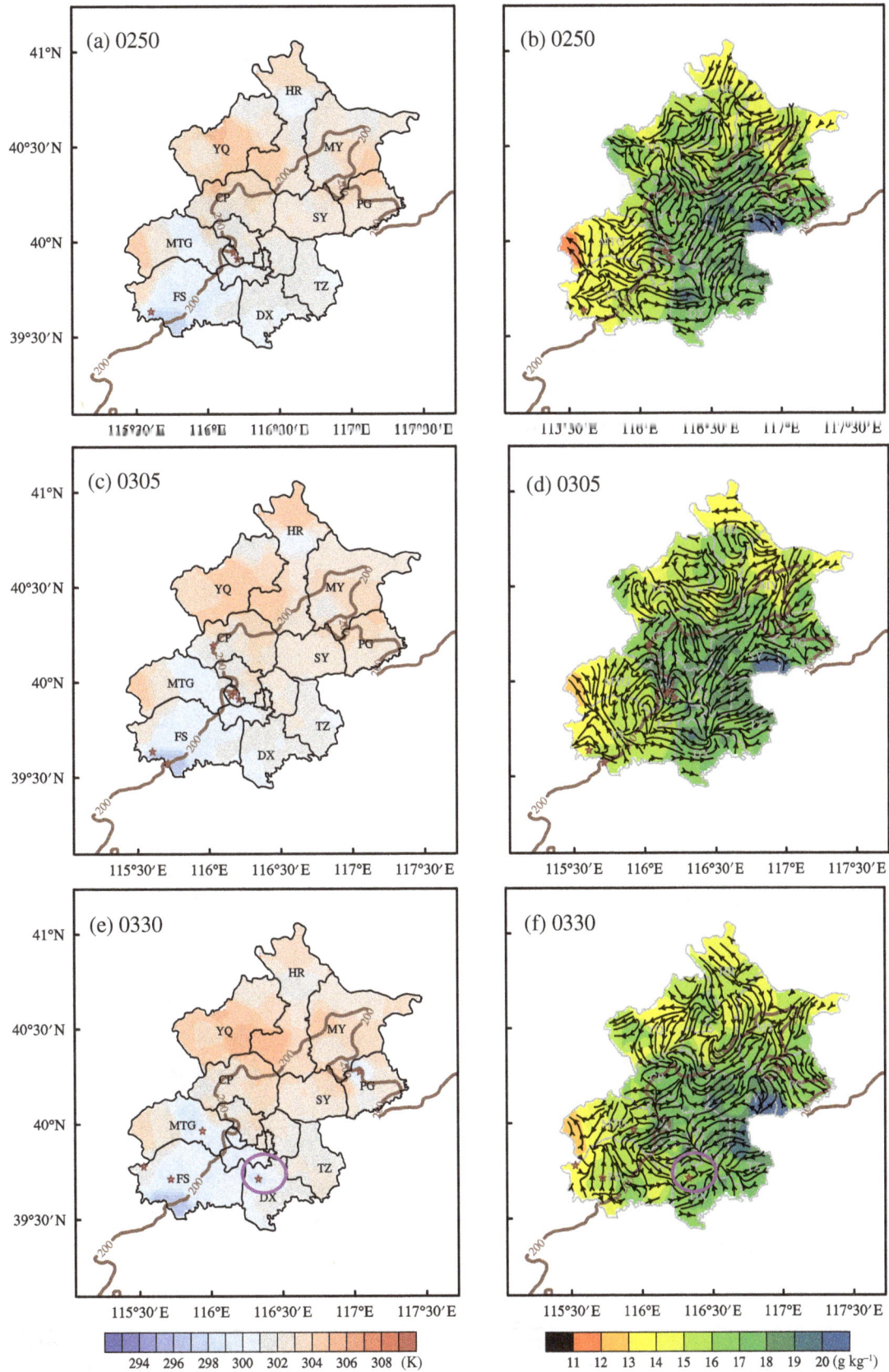

Figure 3. Distributions of potential temperature (K) (left), and mixing ratio (g kg^{-1}) superimposed by surface streamlines (right), observed by the AWSs at (a) and (b) 0250, (c) and (d) 0305, (e) and (f) 0330, (g) and (h) 0345, (i) and (j) 0415, (k) and (l) 0445 UTC, respectively. Red stars denote stations where rainfall was recorded during the past 5 min. Purple circles indicate the cold pools in Daxing, and dark blue circles indicate the cold pools in Pinggu.

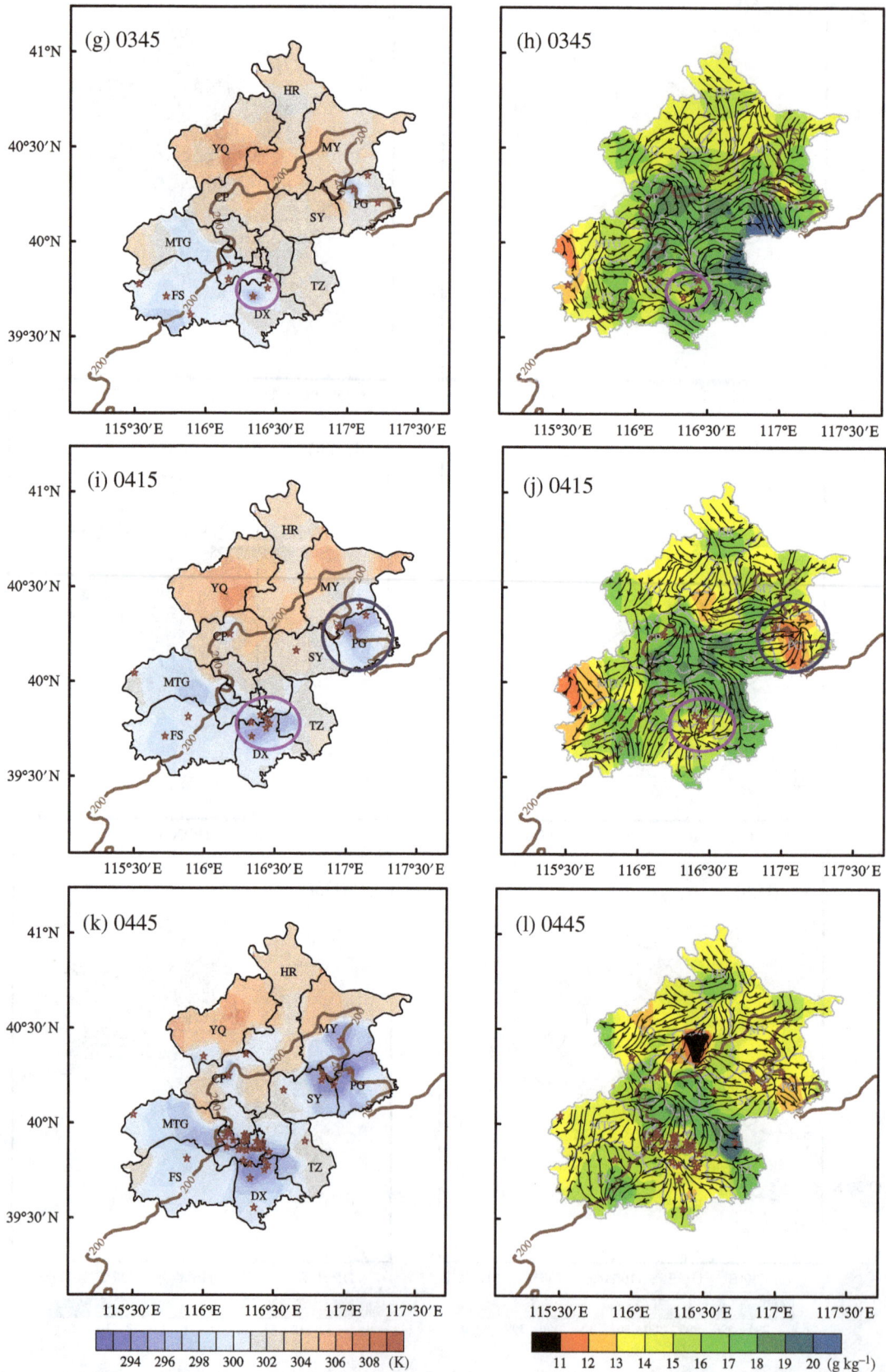

Figure 3. Continued.

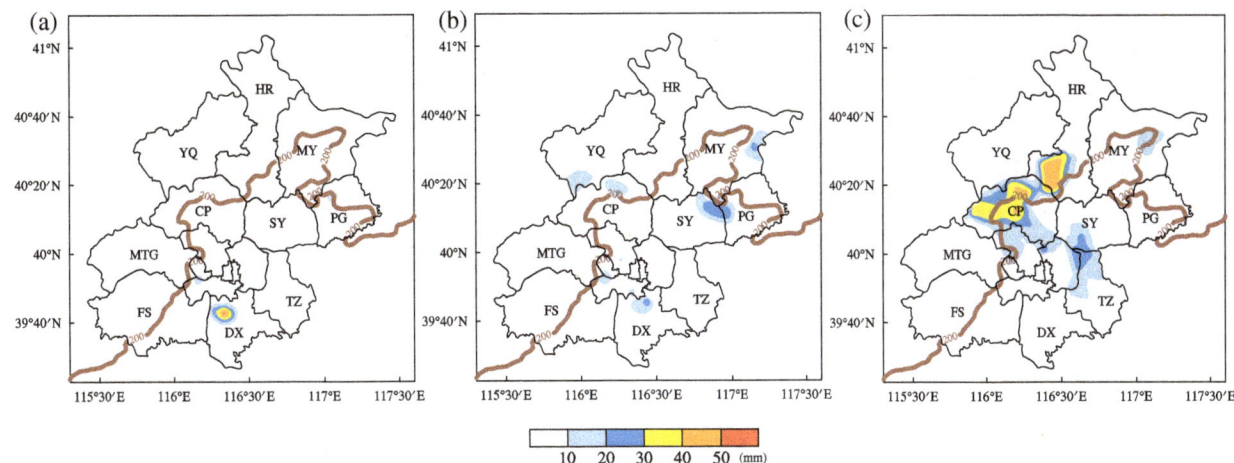

Figure 4. Observed hourly precipitation (shaded, mm) during the periods of (a) 0300–0400, (b) 0400–0500, (c) 0500–0600 UTC on 14 August 2008.

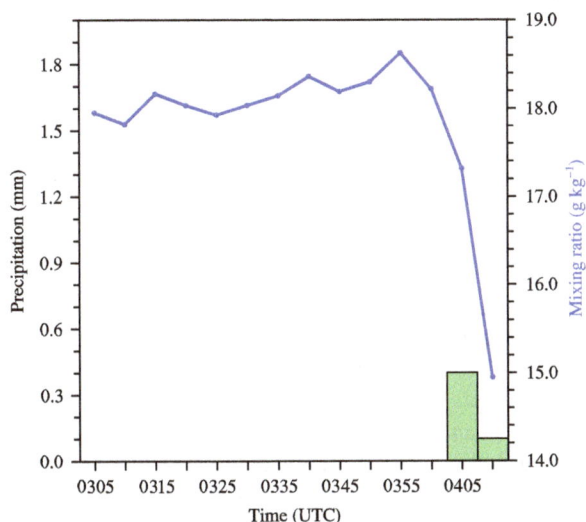

Figure 5. Time series of 5-min accumulated precipitation (green bars, mm), and 2-m mixing ratio (blue line, g kg^{-1}) observed in the Taipingzhuang station (denoted by the red dot in Figure 1(a)) in Changping.

nonsteady-state meteorological and air quality modeling system. CALMET has been widely applied to wind resource assessment (Yim *et al.*, 2007; Mari *et al.*, 2011), which is composed of a diagnostic wind field module and micrometeorological modules. It takes two steps for CALMET to approach the downscaling of wind fields. First, an initial-guess wind field from observations or mesoscale model output is adjusted according to kinematic effects of terrain, slope flows, and terrain blocking effects. Second, an objective analysis procedure is performed. Additionally, smoothing, O'Brien procedure and divergence minimization are optional (Scire *et al.*, 2000). Evaluations showed that when there were plentiful observational stations, CALMET could generate reasonable high-resolution wind fields, retaining original features of observations (Cox *et al.*, 2005; Wang and Shaw, 2009). A diagnostic domain of 60 km × 60 km with a resolution of

100 m in CALMET was set up in this work (The red box in Figure 1(a)). 1 arc-second (∼ 30 m) ASTER Global Dem (ASTER GDEM) terrain data [A product of the Ministry of Economy, Trade and Industry of Japan (METI) and the National Aeronautics and Space Administration (NASA)], and 1-km US Geological Survey (USGS) Global Land Use and Land Cover Data were used as input to help reveal the detailed effect of topography in the northwest of Beijing. The initial-guess wind fields were produced using hourly observations from the AWSs in Beijing.

More details of interactions between fine-scale complex terrain and local flows were showed in the downscaled wind fields (Figure 6(b) and (d)) than in the original fields (Figure 6(a) and (c)). The cell F in Changping (Figure 2(g)) was triggered and developed under complex terrain (Figure 1(a)). In the background of 1-arc-second (∼ 30 m) resolution terrain, a cyclonic circulation (the red rectangle in the lower-left corner of Figure 6(d)), in which fine-scale convergence zones related to fine-scale terrain variation were embedded, was more clearly showed to develop in the vicinity of inverted V-shaped terrain in Changping. Cell F formed and grew near the surface cyclonic circulation. And in the meantime, intense directional convergence of wind vectors (the wind vectors on either side nearly blew toward each other) could be much clearly seen in the southwest of Huairou after downscaling (The red rectangle in the upper-right corner of Figure 6(b) and (d)), which sustained the formation and development of the cell E in Huairou. It is notable that the convergence of local upslope flows related with cell E could not be recognized from the original AWS observational network shown in Figure 6(a) and (c), in which the wind vectors were more uniform.

CALMET not only retains general features of the original wind fields, but also presents more detailed structures and helps to understand the role of local topography in the formation and development of local convective storms.

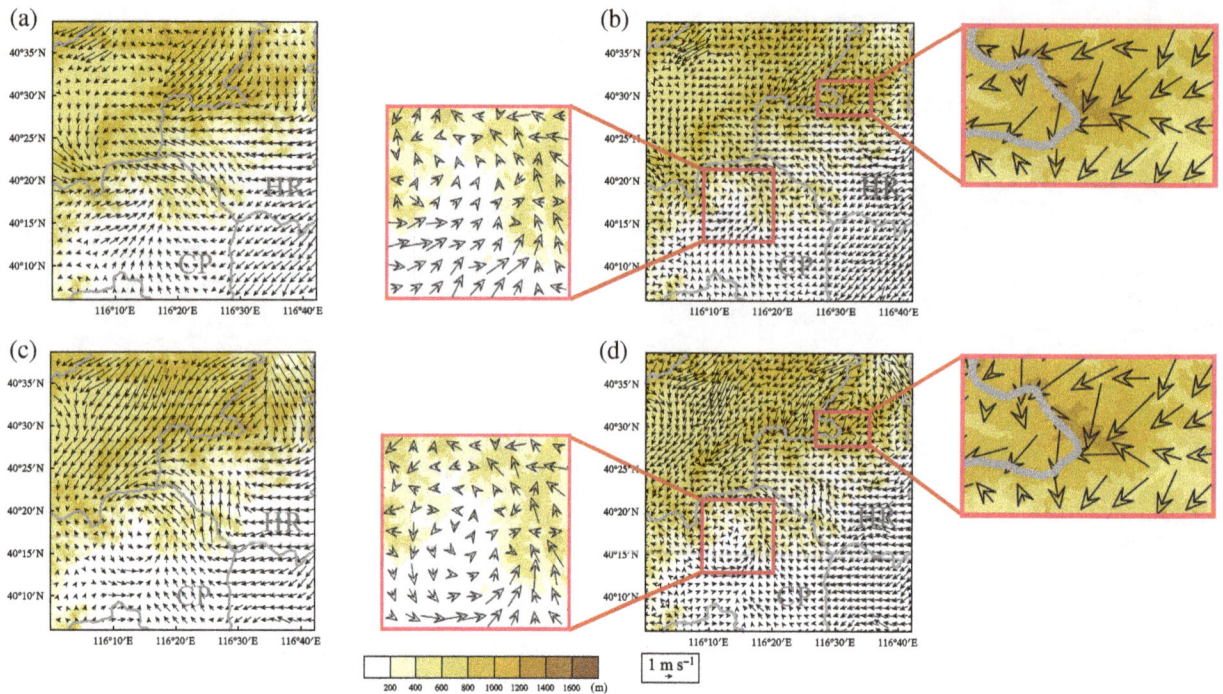

Figure 6. Surface wind vectors fields observed by AWSs (left), and obtained by CALMET (right) (a) and (b) at 0300 UTC, and (c) and (d) at 0400 UTC on 14 August 2008. Enlarged figures of red rectangles in (b) and (d) are shown on either side. Terrain shown in this figure is from 1 arc-second ASTER GDEM data (a product of METI and NASA).

4. Summary

A local heavy precipitation event with several dispersedly distributed heavy rainfall centers exceeding 50 mm occurred in Beijing, China on 14 August 2008 during the ongoing Olympic Games. The heavy rainfall event was produced by a few scattered convective storms. In this article, detailed observational analysis with data from the AWSs as well as the Doppler radar in Beijing and a dynamic downscaling analysis with the diagnostic model, CALMET, were performed to examine the initiation and development of the local convective storms and consequent precipitation. Some results were obtained.

1. Detailed observational analysis showed that convergence zones attributed to small-scale topography and colliding outflow boundaries were of vital importance in the development of those initial convective storms in this event. Convergence helped to induce upward vertical motion as well as concentrate moisture to reduce CIN. Horizontal wind speed may modulate the effectiveness of convergence and associated updraft, thus affecting the formation of convective storms.

2. The downscaled wind fields by CALMET not only retain the overall features of the original flow fields, but also present more detailed structures, especially in complex terrain. And some of the details could not be clearly recognized in the original fields.

The present study implies that it is important for forecasters to notice the development of convergence, especially near the mountains and CALMET may

serve as an auxiliary tool for analyzing the formation and evolution of storms. However, lacking the three-dimensional observations with high resolution, problems including the dynamic as well as thermodynamic effects of cold pool outflows on the convective initiation, and the interaction between topography and outflows are still unclear. Convective initiation mechanisms in this case need to be further explored by using numerical simulation in the future work.

Acknowledgments

This work was supported by the National Basic Research Program of China (973 Program) under grant Nos. 2014CB441402 and 2015CB452804, and the Key Research Program of the Chinese Academy of Sciences under grant No. KZZD-EW-05-01. The authors are grateful to the Ministry of Economy, Trade and Industry of Japan (METI) and the National Aeronautics and Space Administration (NASA) for providing ASTER Global Digital Elevation Model (ASTER GDEM) data, US Geological Survey (USGS) for providing Global Land Use and Land Cover Data (http://earthexplorer.usgs.gov/), European Centre for Medium-Range Weather Forecasts (ECMWF) for providing ERA-Interim data (http://apps.ecmwf.int/datasets/data/interim-full-daily/) and Beijing Meteorological Service for providing the observational data in Beijing.

References

Byers HR, Rodebush HR. 1948. Causes of thunderstorms of the Florida Peninsula. *Journal of Meteorology* **5**: 275–280.

Cox RM, Sontowski J, Dougherty CM. 2005. An evaluation of three diagnostic wind models (CALMET, MCSCIPUF, and SWIFT) with wind data from the Dipole Pride 26 field experiments. *Meteorological applications* **12**: 329–341.

Houze RAJ. 2012. Orographic effects on precipitating clouds. *Reviews of Geophysics* **50**: RG1001, doi: 10.1029/2011RG0 00365.

Markowski P, Hannon C, Rasmussen E. 2006. Observations of convection initiation "failure" from the 12 June 2002 IHOP deployment. *Monthly Weather Review* **134**: 375–405.

Mari R, Bottai L, Busillo C, Calastrini F, Gozzini B, Gualtieri G. 2011. A GIS-based interactive web decision support system for planning wind farms in Tuscany (Italy). *Renewable Energy* **36**: 754–763.

Scire JS, Robe FR, Fernau ME, Yamartino RJ. 2000. *A User's Guide for the CALMET Meteorological Model (Version 5)*. Earth Tech, Inc.: Concord, MA; 332 pp.

Wang W, Shaw WJ. 2009. Evaluating wind fields from a diagnostic model over complex terrain in the Phoenix region and implications to dispersion calculations for regional emergency response. *Meteorological applications* **16**: 557–567.

Wilhelmson RB, Chen CS. 1982. A simulation of the development of successive cells along a cold outflow boundary. *Journal of the Atmospheric Sciences* **39**: 1466–1483.

Wilson JW, Megenhardt DL. 1997. Thunderstorm initiation, organization, and lifetime associated with Florida boundary layer convergence lines. *Monthly Weather Review* **125**: 1507–1525.

Wilson JW, Schreiber WE. 1986. Initiation of convective storms at radar-observed boundary-layer convergence lines. *Monthly Weather Review* **114**: 2516–2536.

Yim SHL, Fung JCH, Lau AKH, Kot SC. 2007. Developing a high-resolution wind map for a complex terrain with a coupled MM5/CALMET system. *Journal of Geophysical Research* **112**: D05106, doi: 10.1029/2006JD007752.

Zhang WL, Cui XP, Huang R. 2014. Intensive observational study on evolution of formation location of thunderstorms in Beijing under complex topographical conditions. *Chinese Journal of Atmospheric Sciences* **38**: 825–837.

7

Testing parameterization schemes for simulating depositional growth of ice crystal using Koenig and Takahashi parameters: A pre-summer rainfall case study over Southern China

Xiaofan Li,[1]* Peijun Zhu,[1] Guoqing Zhai,[1] Rui Liu,[1] Xinyong Shen,[2] Wei Huang[2] and Donghai Wang[3]

[1]Department of Earth Sciences, Zhejiang University, Hangzhou, China
[2]Collaborative Innovation Center on Forecast and Evaluation of Meteorological Disasters, Key Laboratory of Meteorological Disaster of Ministry of Education, Nanjing University of Information Science and Technology, China
[3]State Key Laboratory of Severe Weather, Chinese Academy of Meteorological Sciences, Beijing, China

*Correspondence to:
X. Li, Department of Earth Sciences, Zhejiang University, 38 Zheda Road, Hangzhou 310027, China.
E-mail: xiaofanli@zju.edu.cn

Abstract

In this study, the three (Hsie, Krueger and Zeng) schemes that parameterize depositional growth of ice crystal are tested using Koenig and Takahashi parameters in modeling China pre-summer torrential rainfall event. The Krueger scheme with the Takahashi parameter produces the closest simulations to the observations because it increases cloud ice and snow in the upper troposphere. The increased ice hydrometeor enhances and suppresses infrared radiative cooling above and below 12 km, respectively. The suppressed infrared cooling reduces net condensation while the enhanced infrared cooling suppresses hydrometeor loss through the reduction in melting of ice hydrometeor.

Keywords: depositional growth of ice crystal; parameterization schemes; cloud and rainfall response; infrared radiative cooling; cloud-resolving model simulation

1. Introduction

One of the greatest uncertainties in global and regional climate modeling comes from the presentation of clouds and associated microphysical processes (e.g. Cess *et al.*, 1997). One way to reduce such uncertainties is to use cloud-resolving models with fine spatial resolutions and explicit microphysical parameterization schemes. However, there are uncertainties associated with simulations of ice anvil clouds, the important part of cloud systems. Ice hydrometeors are the major players in producing precipitation through the melting of precipitation ice and in regulating radiative energy balance through the reflection of incoming solar radiation back to the space and prevention of emitted infrared radiative energy from out of the Earth. The one of uncertain processes in modeling ice clouds may be related to the simulation of the Bergeron–Findeisen (Bergeron, 1935; Findeisen, 1938) process in which water vapor from cloud droplets is transferred to ice crystals due to the fact that the saturation vapor pressure over ice is lower than one over liquid water.

Most of the modern bulk schemes explicitly simulate the Bergeron–Findeisen process directly by calculating the growth rate of ice particles (e.g. Walko *et al.*, 1995; Morrison *et al.*, 2005; Phillips *et al.*, 2007; Thompson *et al.*, 2008). But there are the uncertainties due to the scheme-dependent parameterizations of the Bergeron–Findeisen process such as the depositional growth of snow from cloud ice (P_{SFI}) and the growth

of cloud ice by the deposition of cloud water (P_{IDW}). Hsie *et al.* (1980) used the mass of a natural ice nucleus (1.05×10^{-15} g) and the 50 µm radius of ice crystal in the calculations of P_{IDW} and P_{SFI}, respectively, whereas Krueger *et al.* (1995) replaced a natural ice nucleus with an averaged mass of ice crystal in the calculation of P_{IDW} and increased the radius of ice crystal to 100 µm in the calculation of P_{SFI}. Li *et al.* (1999) showed that the mixing ratios of cloud ice and snow simulated by the Krueger scheme are significantly larger than those simulated by the Hsie scheme. Zeng *et al.* (2008) abandoned the assumption that the ice crystal density is independent of the size in above two schemes and proposed a linear relation between the number and mass of ice crystal in the small range of ice crystal radius (less than 50 µm) in the development of new schemes for P_{SFI} and P_{IDW}. The other uncertainties may come from the parameter of Koenig (1971)'s formula that are used in the schemes of P_{IDW} and P_{SFI}. Westbrook and Heymsfield (2011) conducted a comparison study between the parameters from Koenig (1971) and derived by laboratory experiment data from Takahashi and Fukuta (1988) and Takahashi *et al.* (1991). The significant differences are found in temperature-dependent parameters of depositional growth of a single ice crystal (Koenig, 1971).

The objective of this study is to examine the responses of rainfall, clouds and associated radiation to the changes in schemes that parameterize the Bergeron–Findeisen process. A series of

Table 1. Model setup.

Prognostic equations	Potential temperature, specific humidity, perturbation zonal wind and vertical velocity, and mixing ratios of five cloud species
Cloud microphysical parameterization schemes	Hsie *et al.* (1980); Lin *et al.* (1983); Rutledge and Hobbs (1983, 1984); Tao *et al.* (1989); Krueger *et al.* (1995) and Zeng *et al.* (2008)
Radiative parameterization schemes	Chou *et al.* (1991, 1998) and Chou and Suarez (1994)
Boundary conditions	Cyclic lateral boundaries; model top is 42 hPa
Basic parameter set	Model domain (768 km), horizontal grid (1.5 km), time step (1.5 s) and 33 vertical levels

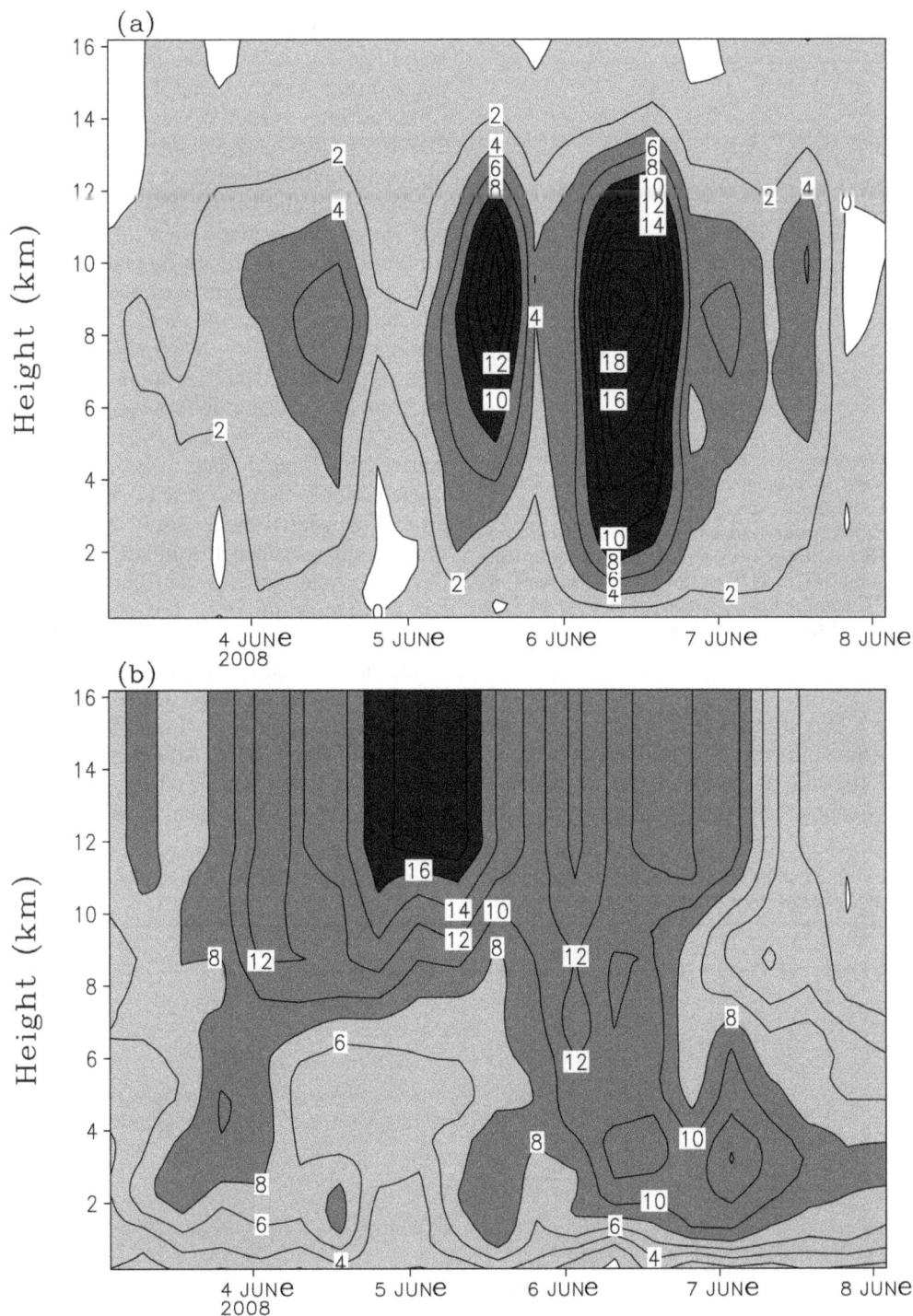

Figure 1. Temporal and vertical distribution of (a) vertical velocity (cm s^{-1}) and (b) zonal wind (m s^{-1}) imposed in the experiments. Ascending motion in (a) and westerly wind in (b) are shaded.

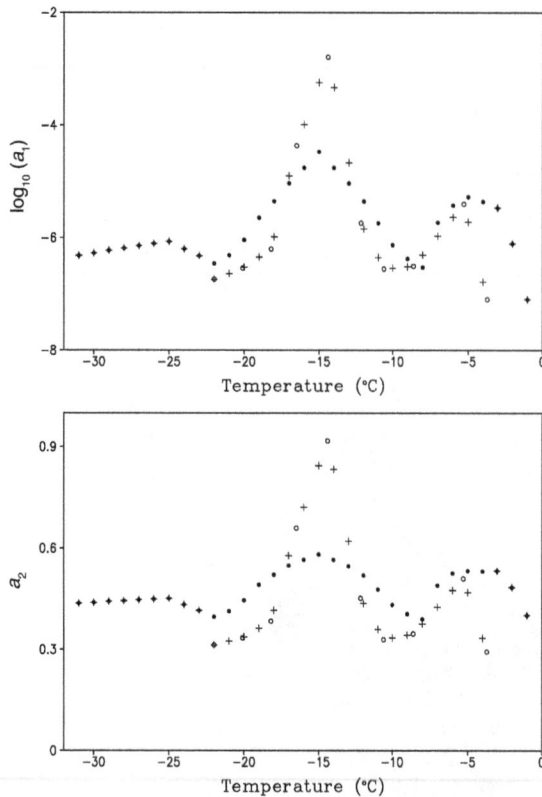

Figure 2. Parameters $\log_{10}(a_1)$ and a_2 for Koenig (1971)'s formula $(dm/dt = a_1 m^{a_2})$ as a function of air temperature. Closed and open circles denote the Koenig predicted parameters and the Takahashi laboratory-derived parameters by Westbrook and Heymsfield (2011), respectively. Cross denotes parameters interpolated from the Takahashi laboratory-derived parameters from -22 to $-3\,°C$ and the Koenig predicted parameters from -31 to $-23\,°C$ and -2 to $-1\,°C$, which are used in H80T, K80T and Z08T.

two-dimensional cloud-resolving sensitivity experiments of pre-summer torrential rainfall event occurred during June 2008 will be conducted using different bulk schemes of depositional growth of ice crystal with

Koenig to Takahashi parameter. In the next section, model, and control and sensitivity experiments are described. The results are presented in Section 3. A summary is given in Section 4.

2. Model and experiments

The model used here is the version of Goddard Cumulus Ensemble Model developed by Soong and Ogura (1980); Soong and Tao (1980) and Tao and Simpson (1993) and modified by Sui *et al.* (1994, 1998) and Li *et al.* (1999). The model was used to study pre-summer torrential rainfall event over southern China during June 2008 (Wang *et al.*, 2010; Shen *et al.*, 2011b). Before convection occurred, the subtropical high extended westward and southwesterly winds were strong. Warm and humid air transported by southwesterly winds encountered cold air from the north over southern China. A trough moved into southern China to trigger the development of a squall-line convection, which causes torrential rainfall. The maximum rain amount of 482.2 mm occurred in Yangjiang, Guangdong on 6 June. The maximum hourly rain rate was $51.3\,\mathrm{mm\,h^{-1}}$. The surface southerly winds increased to $15\,\mathrm{m\,s^{-1}}$. The model setup used in this study is summarized by Table 1. The two-dimensional framework is used in this study because of similarities in two- and three-dimensional model simulations in terms of thermodynamics, surface heat fluxes, rainfall, precipitation efficiency and vertical transports of mass, sensible heat and moisture (e.g. Tao and Soong, 1986; Tao *et al.*, 1987; Grabowski *et al.*, 1998; Tompkins, 2000; Khairoutdinov and Randall, 2003; Sui *et al.*, 2005). Detailed model dynamic framework and associated physical package can be found in Gao and Li (2008) and Li and Gao (2011). All experiments are integrated from 0200 local standard time (LST = UTC + 8h) 3 June to 0200 LST 8 June 2008 (a total of 5 days).

Table 2. Summary of schemes that parameterize the growth of cloud ice by the deposition of cloud water (P_{IDW}) and the depositional growth of snow from cloud ice (P_{SFI}).

Scheme	Description
Hsie et al. (1980)	$P_{IDW} = \frac{n_0 e^{\frac{1}{2}|T-T_0|}}{10^3 \rho} a_1 \left(m_n\right)^{a_2}$, $P_{SFI} = \frac{q_i}{\Delta t_i}$, where $n_0 = 10^{-8}m^{-3}$; $T_0 = 0°C$; $m_n = 1.05 \times 10^{-15}g$; ρ is the air density, which only is a function of height; a_1 and a_2 are the positive temperature-dependent coefficients. q_i is the mixing ratio of cloud ice; $\Delta t_i \left[= \left(m_{i50}^{1-b_2} - m_{i40}^{1-b_2}\right)/a_1\left(1-a_2\right)\right]$ is the timescale needed for a crystal to grow from radius r_{i40} to radius r_{i50}; $m_{i50}(=4.8 \times 10^{-10}kg)$ is the mass of an ice crystal r_{i50} (50 μm) and $m_{i40}(=2.46 \times 10^{-10}kg)$ is the mass of an ice crystal r_{i40} (40 μm).
Krueger et al. (1995)	$P_{IDW} = \frac{n_0 e^{\frac{1}{2}|T-T_0|}}{10^3 \rho} a_1 \left(\frac{\rho q_i}{n_0 e^{\frac{1}{2}|T-T_0|}}\right)^{a_2}$, $P_{SFI} = \frac{q_i}{\Delta t_i}$, where $\Delta t_i = \left(m_{i100}^{1-a_2} - m_{i40}^{1-a_2}\right)/a_1\left(1-a_2\right)m_{i100}(=3.84 \times 10^{-9}kg)$ is the mass and of an ice crystal r_{i100} (10² μm).
Zeng et al. (2008)	$P_{IDW} = \frac{2}{(a_2+1)(b_2+2)}\left[3a_2q_i + \left(1-a_2\right)m_{i50}\mu\rho^{-1}N_i\right]a_1 m_{i50}^{a_2-1}$, $P_{SFI} = \max\left[2a_1\left(3q_i - m_{i50}\mu\rho^{-1}N_i\right)m_{i50}^{a_2-1},0\right]$, where $N_i = n_0 e^{\beta(T_0-T)}$, β varies from 0.4 to 0.6, and n_0 varies from 10^{-9} to 10^{-6} cm^{-3} (Fletcher, 1962); μ (1.2) is the ice particle enhancement factor due to a riming–splintering mechanism (Hallet and Mossop, 1974).

Table 3. Summary of experiments with parameterization schemes of depositional growth of cloud ice (P_{SFI} and P_{IDW}) and their parameter choices.

Experiment	Description
H80K	Schemes from Hsie et al. (1980) and their parameters from Koenig (1971)
H80T	As in H80K except for parameter data are replaced with those from Westbrook and Heymsfield (2011)
K95K	Schemes from Krueger et al. (1995) and their parameters from Koenig (1971)
K95T	As in K95K except for parameter data are replaced with those from Westbrook and Heymsfield (2011)
Z08K	Schemes from Zeng et al. (2008) and their parameters from Koenig (1971); $\beta = 0.45$; $n_o = 5 \times 10^{-9}$ cm^{-3}
Z08T	As in Z08K except for parameter data are replaced with those from Westbrook and Heymsfield (2011)

The model is imposed with the large-scale vertical velocity and zonal wind (Figure 1) as well as horizontal advections (not shown) averaged over 108°–116°E, 21°–22°N. The experiment (K95K in this study) has been validated with rain gauge observation and temperature and specific humidity data from National Centers for Environmental Prediction/Global Data Assimilation System (NCEP/GDAS) (Wang et al., 2010; Shen et al., 2011b). The experiments have been conducted to study thermodynamic aspect of precipitation efficiency (Shen and Li, 2011) and the responses of pre-summer torrential rainfall to vertical wind shear (Shen et al., 2013) and cloud radiative processes (Shen et al., 2011b) and ice (Wang et al., 2010; Shen et al., 2011c) and water (Shen et al., 2011a) clouds and ice microphysical processes (Shen et al., 2012).

The temperature-dependent parameters (a_1 and a_2) of Koenig formula for simulating the Bergeron–Findeisen process,

$$\frac{dm}{dt} = a_1 m^{a_2} \qquad (1)$$

where m is mass of a single ice crystal. Figure 2 shows the two parameters as a function of air temperature. The differences between Koenig prediction and Takahashi laboratory data reach their peaks for both parameters at −14.4°C, where the Takahashi experiment-derived $a1$ is about two orders of magnitudes larger than the Koenig predicted $a1$ and the Takahashi experiment-derived $a2$ also is larger than the Koenig predicted $a2$. Takahashi parameter is smaller than the Koenig parameter when temperatures vary from −3.7 to −10.2°C and from −18.2 to −22.0°C.

Three pairs of sensitivity experiments that use schemes of Hsie et al. (1980; H80), Krueger et al. (1995; K95) and Zeng et al. (2008; Z08) will be carried out. The three parameterization schemes are briefly described in Table 2. Each pair of experiments use the Koenig parameter (Koenig, 1971) and Takahashi parameter derived by Westbrook and Heymsfield (2011) using the laboratory data from Takahashi and Fukuta (1988) and Takahashi et al. (1991). The experiment designs are summarized in Table 3.

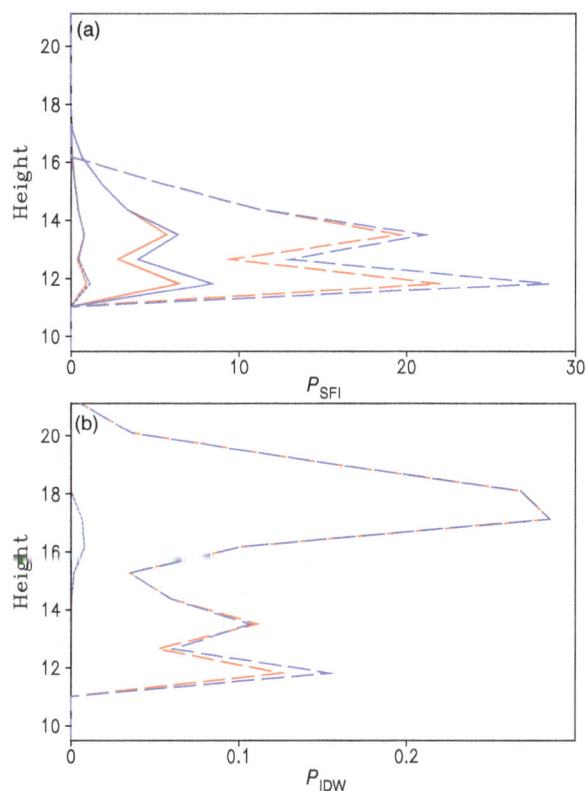

Figure 3. Vertical profiles of (a) P_{SFI} (g kg^{-1} h^{-1}) and (b) P_{IDW} (10^{-2} g kg^{-1} h^{-1}) calculated using vertical profiles of cloud ice mixing ratio and air temperature at a zonal grid point of 482 and 1000 LST 6 June 2008 in K95K. Solid, short dash and long dash lines denote H80, K95 and Z08, respectively. Red and blue lines denote calculations with the Koenig predicted parameters (K) and the Takahashi laboratory-derived parameters (T), respectively. The values of P_{IDW} in H80K, H80T and K95K are not shown in (b) because of negligibly small values.

Note that the Takahashi parameter is interpolated at 1°C interval from −22 to −3°C, which is used in H80T, K80T and Z08T. The Koenig parameter from −31 to −23°C and from −2 to −1°C is used in the experiments.

3. Results

Before the model responses to schemes and parameters of P_{SFI} and P_{IDW} are examined, the vertical profiles of P_{SFI} and P_{IDW} are calculated using the vertical profiles of cloud ice mixing ratio and air temperature (a zonal grid point of 482 at 1000 LST 6 June 2008 in K95K as cloud ice mixing ratio reaches a maximum), and the vertical profiles of P_{SFI} and P_{IDW} are shown in Figure 3. The P_{SFI} is larger in H80 than in K95 whereas it is less than in Z08. The P_{SFI} is smaller in the schemes with the Koenig parameter than in the schemes with the Takahashi parameter. The P_{IDW} is larger in Z08 than in H80 and K95. K95 is insensitive to the schemes with the Koenig and the Takahashi parameters, but the P_{IDW} is smaller in the schemes with the Koenig parameter than in the schemes with the Takahashi parameter at relatively warm air temperatures. The 5-day and model

Table 4. P_{SFI} and P_{IDW} averaged for 5 days over model domain in H80K, H80T, K95K, K95T, Z08K and Z08T.

	H80K	H80T	K95K	K95T	Z08K	Z08T
P_{SFI}	3.18	3.16	1.42	1.81	4.44	4.66
P_{IDW}	0.00	0.00	0.00	0.00	0.03	0.03

Unit is mm day^{-1}.

Table 5. Root mean squared differences in surface rain rate (RMSDP$_S$), temperature (RMSDT) and specific humidity (RMS-Dqv) between simulations and with rain gauge data and NCEP/GDAS data.

	H80K	H80T	K95K	K95T	Z08K	Z08T
RMSDP$_S$	25.27	26.91	25.41	24.59	25.19	27.01
RMSDT	0.65	0.69	0.65	0.61	0.69	0.72
RMSDqv	0.42	0.43	0.41	0.35	0.44	0.42

Units are mm day^{-1} for surface rain rate, °C for temperature and g kg^{-1} for specific humidity. RMSDP$_S$ is calculated using hourly data whereas RMSDT and RMSDqv are calculated using 6-h data.

domain mean analysis reveals that P_{SFI} is at least two orders of magnitudes larger than P_{IDW} (Table 4). The Takahashi parameter generally increases depositional growth of snow from cloud ice except that it barely changes P_{SFI} in the Hsie scheme.

The surface rain rate, temperature and specific humidity are compared with rain gauge data and NCEP/GDAS data, respectively. Their root mean squared differences (RMSD) between simulated model domain means and area means from gauge data and NCEP/GDAS data show that the RMSDs generally increase from H80K and Z08K to H80T and Z08T, respectively, whereas they decrease from K95K to K95T (Table 5). K95T has the smallest RMSDs among the six simulations for surface rain rate, temperature and specific humidity. This indicates that the Krueger scheme with the Takahashi parameter produces the best cloud-resolving model simulations.

The 5-day and model domain mean surface rain rates are lower in the simulations with the Takahashi parameter than those in the simulations with the Koenig parameter (Table 6). The largest decrease in the mean rain rate occurs from K95K to K95T, among the three pairs of simulations. Because all the mean simulated rain rates are higher than the mean observed rain rate, the mean rain rate simulated in K95T appears to be the closest to the mean observed rain rate, which is consistent with the analysis of the RMSD. The mean rain rates decrease from H80K and K95K to H80T and K95T, respectively, whereas Z08K and Z08T have similar mean rain rates (Table 6). To examine the change in the mean rain rate from the experiment with the Koenig parameter to the experiment with the Takahashi parameter, cloud budget is analyzed. Following Gao *et al.* (2005) and Cui and Li (2006), the model domain mean mass-integrated cloud budget is expressed by

$$P_S = Q_{NC} + Q_{CM} \qquad (2)$$

Table 6. (a) Cloud microphysical budgets (P_S, Q_{NC}, and Q_{CM}) averaged for 5 days over model domain in H80K, H80T, K95K, K95T, Z08K and Z08T and (b) their differences averaged over 5 days and model domain.

(a)

	H80K	H80T	K95K	K95T	Z08T	Z08K
P_S	32.86	32.33	32.74	31.96	33.08	32.96
Q_{NC}	32.16	31.70	31.99	31.67	31.54	32.45
Q_{CM}	0.69	0.63	0.75	0.28	1.54	0.52

(b)

	H80T–H80K	K95T–K95K	Z08T–Z08K
P_S	−0.53	−0.78	−0.12
Q_{NC}	−0.46	−0.32	0.91
Q_{CM}	−0.06	−0.47	−1.02

The observed 5-day mean rain rate is 29.23 mm day^{-1}. Unit is mm day^{-1}.

where

$$Q_{NC} = Sqv \qquad (2a)$$

$$Q_{CM} = -\frac{\partial q_5}{\partial t} \qquad (2b)$$

Here Q_{NC} is the net condensation ($Sqv > 0$) or evaporation ($Sqv < 0$) [see (1.2b) in Li and Gao (2011)]; Q_{CM} is the hydrometeor loss ($Q_{CM} > 0$) or gain ($Q_{CM} < 0$) since advection term is canceled due to the cyclic lateral boundary condition in the model used in this study; total hydrometeor mixing ratio q_5 is the sum of mixing ratios of cloud water (q_c), raindrops (q_r), cloud ice (q_i), snow (q_s) and graupel (q_g).

The decrease in the mean rain rate from H80K to H80T is mainly associated with the reduction in the mean net condensation (Table 6). The decrease in the mean rain rate from K95K to K95T is related to both the suppressions in the mean net condensation and the mean hydrometeor loss ($Q_{CM} > 0$). The increase in the mean net condensation from Z08K to Z08T is offset by the decrease in the mean hydrometeor loss, which leads to the similar rain rates in the two experiments. The decrease in the mean rain rate from K95K to K95T is larger than that from H80K to H80T because the reduction in the mean hydrometeor loss from K95K to K95T is larger than that from H80K to H80T. The decrease in the mean rain rate from K95K to K95T is larger than that from Z08K to Z08T because the net condensation is reduced from K95K to K95T but it is increased from Z08K to Z08T. To analyze the difference in the decrease of the mean hydrometeor loss between H80T–H80K and K95T–K95K, the mean hydrometeor change (Q_{CM}) can be further broken down to the mean hydrometeor change in cloud water (Q_{CMC}), raindrops (Q_{CMR}), cloud ice (Q_{CMI}), snow (Q_{CMS}) and graupel (Q_{CMG}).

$$Q_{CM} = Q_{CMC} + Q_{CMR} + Q_{CMI} + Q_{CMS} + Q_{CMG} \qquad (3)$$

where

$$Q_{CMC} = -\frac{\partial q_c}{\partial t} = -Sqc \qquad (3a)$$

Table 7. List of microphysical processes and their parameterization schemes.

Notation	Description	Scheme
P_{MLTG}	Growth of vapor by evaporation of liquid from graupel surface	RH84
P_{MLTS}	Growth of vapor by evaporation of melting snow	RH83
P_{REVP}	Growth of vapor by evaporation of raindrops	RH83
P_{IMLT}	Growth of cloud water by melting of cloud ice	RH83
P_{CND}	Growth of cloud water by condensation of supersaturated vapor	TSM
P_{GMLT}	Growth of raindrops by melting of graupel	RH84
P_{SMLT}	Growth of raindrops by melting of snow	RH83
P_{RACI}	Growth of raindrops by the accretion of cloud ice	RH84
P_{RACW}	Growth of raindrops by the collection of cloud water	RH83
P_{RACS}	Growth of raindrops by the accretion of snow	RH84
P_{RAUT}	Growth of raindrops by the autoconversion of cloud water	LFO
P_{IDW}	Growth of cloud ice by the deposition of cloud water	KFLC
P_{IACR}	Growth of cloud ice by the accretion of rain	RH84
P_{IHOM}	Growth of cloud ice by the homogeneous freezing of cloud water	
P_{DEP}	Growth of cloud ice by the deposition of supersaturated vapor	TSM
P_{SAUT}	Growth of snow by the conversion of cloud ice	RH83
P_{SACI}	Growth of snow by the collection of cloud ice	RH83
P_{SACW}	Growth of snow by the accretion of cloud water	RH83
P_{SFW}	Growth of snow by the deposition of cloud water	KFLC
P_{SFI}	Depositional growth of snow from cloud ice	KFLC
P_{SACR}	Growth of snow by the accretion of raindrops	LFO
P_{SDEP}	Growth of snow by the deposition of vapor	RH83
P_{GACI}	Growth of graupel by the collection of cloud ice	RH84
P_{GACR}	Growth of graupel by the accretion of raindrops	RH84
P_{GACS}	Growth of graupel by the accretion of snow	RH84
P_{GACW}	Growth of graupel by the accretion of cloud water	RH84
P_{WACS}	Growth of graupel by the riming of snow	RH84
P_{GDEP}	Growth of graupel by the deposition of vapor	RH84
P_{GFR}	Growth of graupel by the freezing of raindrops	LFO

The schemes are Lin et al. (1983, LFO); Rutledge and Hobbs (1983; 1984; RH83, RH84); Tao et al. (1989, TSM) and Krueger et al. (1995, KFLC).

$$Q_{CMR} = -\frac{\partial q_r}{\partial t} = -Sqr + P_S \quad (3b)$$

$$Q_{CMI} = -\frac{\partial q_i}{\partial t} = -Sqi \quad (3c)$$

$$Q_{CMS} = -\frac{\partial q_s}{\partial t} = -Sqs \quad (3d)$$

$$Q_{CMG} = -\frac{\partial q_g}{\partial t} = -Sqg \quad (3e)$$

Here, Sqc, Sqr, Sqi, Sqs and Sqg are sources and sinks of cloud water, raindrops, cloud ice, snow and graupel, respectively [see (1.2c – 1.2g) in Li and Gao (2011)], and $Sqc + Sqr + Sqi + Sqs + Sqg = Sqv$. The mean cloud water is changed from the loss in H80K to the gain in H80T, whereas the mean cloud water gain is enhanced from K95K to K95T (Table 8). The mean raindrops loss is enhanced from H80K to H80T, whereas the mean raindrops gain in K95K is changed to the raindrops loss in K95T. The change in the mean cloud ice from the loss in K95K to the gain in K95T becomes larger than that from the loss in H80K to the gain in H80T. The mean snow loss is intensified from H80K to H80T, but the weak mean snow gain is barely changed from K95K to K95T. The mean graupel is changed from the gain in H80K to the loss in H80T, whereas the mean graupel loss is enhanced from K95K to K95T. Thus, the reduction brought by use of the Takahashi parameter

weakened from the Hsie scheme to the Krueger scheme is associated with the changes in Q_{CMI}, Q_{CMS} and Q_{CMG}. Following Li and Gao (2011), the sum of Q_{CMI}, Q_{CMS} and Q_{CMG} [(3c) + (3d) + (3e)] can be further written as

$$\begin{aligned}
Q_{CMI} + Q_{CMS} + Q_{CMG} &= -(Sqi + Sqs + Sqg) \\
&= -P_{IHOM}\left(T < T_{oo}\right) + P_{IMLT}\left(T > T_o\right) \\
&\quad - P_{IDW}\left(T_{oo} < T < T_o\right) - P_{DEP} - P_{SACW}\left(T < T_o\right) \\
&\quad - P_{SFW}\left(T < T_o\right) + P_{RACS}\left(T > T_o\right) + P_{SMLT}\left(T > T_o\right) \\
&\quad - P_{SDEP}\left(T < T_o\right) + P_{MLTS}\left(T > T_o\right) - P_{GACW}\left(T < T_o\right) \\
&\quad - P_{IACR}\left(T < T_o\right) - P_{GACR}\left(T < T_o\right) - P_{GFR}\left(T < T_o\right) \\
&\quad + P_{GMLT}\left(T > T_o\right) - P_{GDEP}\left(T < T_o\right) + P_{MLTG}\left(T > T_o\right) \\
&\quad - P_{SACR}\left(T < T_o\right)
\end{aligned} \quad (4)$$

Here, $T_o = 0\,°C$ and $T_{oo} = -35\,°C$. The cloud microphysical terms in (Equation (4)) are defined in Table 7.

The mean ice hydrometer (sum of cloud ice, snow and graupel) is changed from the gain in H80K to the loss in H80T, but its loss is suppressed from K95K to K95T (Table 9), which leads to the enhanced reduction in the mean hydrometeor loss (Table 8). Further breakdown of ice hydrometeor into ice microphysical terms in Table 9 reveals the melting of cloud ice to cloud water (P_{IMLT}) only occurs in K95K, which leads to reduction in P_{IMLT} from K95K to K95T. The melting of graupel to raindrops (P_{GMLT}) is enhanced from H80K to H80T,

Table 8. Breakdown of Q_{CM} in H80K, H80T, K95K and K95T and their differences averaged over 5 days and model domain.

	H80K	H80T	K95K	K95T	H80T–K80K	K95T–K95K
Q_{CM}	0.69	0.63	0.75	0.28	−0.06	−0.47
Q_{CMC}	0.34	−0.66	−0.03	−0.59	−1.00	−0.56
Q_{CMR}	0.58	0.86	−0.12	0.23	0.28	0.35
Q_{CMI}	0.09	−0.07	0.50	−0.10	−0.16	−0.60
Q_{CMS}	0.10	0.40	−0.10	−0.06	0.30	0.04
Q_{CMG}	−0.42	0.09	0.51	0.80	0.51	0.29

Unit is mm day^{-1}.

but it is suppressed from K95K to K95T. This indicates that the change in the melting of ice hydrometeor to water hydrometeor (sum of cloud water and raindrops) brought by use of the Takahashi parameter from the increase in the Hsie scheme to the decrease in the Krueger scheme is responsible for the enhanced reduction in the mean hydrometeor loss from the Hsie scheme to the Krueger scheme.

The Koenig and the Takahashi parameters affect vertical profile of radiation by changing vertical structures of cloud hydrometeors, which in turn changes the mean net condensation and hydrometeor loss. Thus, the vertical profiles of mixing ratios of cloud hydrometeor and radiation are averaged over 5 days and model domain and their differences between H80T and H80K, K95T and K95H and Z08T and Z08K are shown in Figures 4 and 5, respectively. The radiation here is calculated using Q_R/c_p, where Q_R is the radiative heating rate due to convergence of net flux of solar and infrared radiative fluxes and c_p is the specific heat of dry air at constant pressure. The increase in snow mixing ratio from H80K to H80T (Figure 4(a)) mainly weakens the mean infrared radiative cooling below 6 km (Figure 5(a)), while the reduction in graupel mixing ratio enhances the mean infrared radiative cooling above 6 km. The water mixing ratios decrease from H80K to H80T. The enhanced mean

melting of graupel to raindrops from H80K to H80T corresponds to the suppressed mean infrared radiative cooling below 6 km.

The enhanced cloud ice and snow mixing ratios from K95K to K95T (Figure 4(b)) weakens the mean infrared radiative cooling below 12 km (Figure 5(b)). The water mixing ratios generally decrease from H80K to H80T except for the increase in raindrops mixing ratio around 5 km. The decreases in the mean melting of cloud ice and graupel from K95K to K95T is associated with the strengthened mean infrared radiative cooling above 12 km. The reduction in mean net condensation from K95K to K95T corresponds to the suppressed mean infrared radiative cooling below 12 km.

The decrease in graupel mixing ratio from Z08K to Z08T is stronger than the increase in snow mixing ratio from 4 to 8 km and the reduction in snow mixing ratio occurs from 8 to 11 km (Figure 4(c)), which leads to the increase in the mean infrared radiative cooling above 4 km (Figure 5(c)). The water mixing ratios increases from Z08K to Z08T. The increase in the mean net condensation from Z08K to Z08T corresponds to the enhancement in the mean infrared radiative cooling.

4. Summary

The Bergeron–Findeisen process has been simulated using the parameterization scheme for the depositional growth of ice crystal with the temperature-dependent parameters from Koenig (1971) in the past decades. Recently, Westbrook and Heymsfield (2011) calculated these parameters using the laboratory data from Takahashi and Fukuta (1988) and Takahashi et al. (1991) and found significant differences between the two parameter sets. In this study, we conducted three pairs of sensitivity experiments using three parameterization schemes [Hsie et al., 1980 (H80); Krueger et al., 1995 (K95);

Table 9. Breakdown of $Q_{CMI} + Q_{CMS} + Q_{CMG}$ in H80K, H80T, K95K and K95T and their differences averaged over 5 days and model domain.

	H80K	H80T	K95K	K95T	H80T–K80K	K95T–K95K
$Q_{CMI} + Q_{CMS} + Q_{CMG}$	−0.23	0.42	0.91	0.64	0.65	−0.27
$-P_{IHOM}$	−0.02	−0.02	−0.07	−0.05	0.00	0.02
P_{IMLT}	0.00	0.00	0.84	0.00	0.00	−0.84
$-P_{DEP}$	−6.87	−7.02	−6.81	−6.78	−0.15	0.03
$-P_{SACW}$	−2.03	−1.97	−2.12	−2.05	0.06	0.07
$-P_{SFW}$	−0.28	−0.24	−0.02	−0.03	0.04	−0.01
P_{RACS}	0.32	0.47	0.13	0.12	0.15	−0.01
P_{SMLT}	0.55	0.57	0.64	0.62	0.02	−0.02
$-P_{SDEP}$	−1.11	−1.17	−1.14	−1.23	−0.06	−0.09
P_{MLTS}	0.04	0.04	0.04	0.05	0.00	0.01
$-P_{GACW}$	−9.50	−9.02	−9.21	−8.65	0.48	0.56
$-P_{IACR}$	−0.48	−0.42	−0.39	−0.29	0.06	0.10
$-P_{GACR}$	−0.17	−0.36	−0.37	−0.09	−0.19	0.28
P_{GMLT}	20.01	20.42	20.30	19.67	0.41	−0.63
$-P_{GDEP}$	−1.24	−1.25	−1.20	−1.23	−0.01	−0.03
P_{MLTG}	0.77	0.76	0.71	0.64	−0.01	−0.07
$-P_{SACR}$	−0.22	−0.37	−0.41	−0.05	−0.15	0.36

Unit is mm day^{-1}.

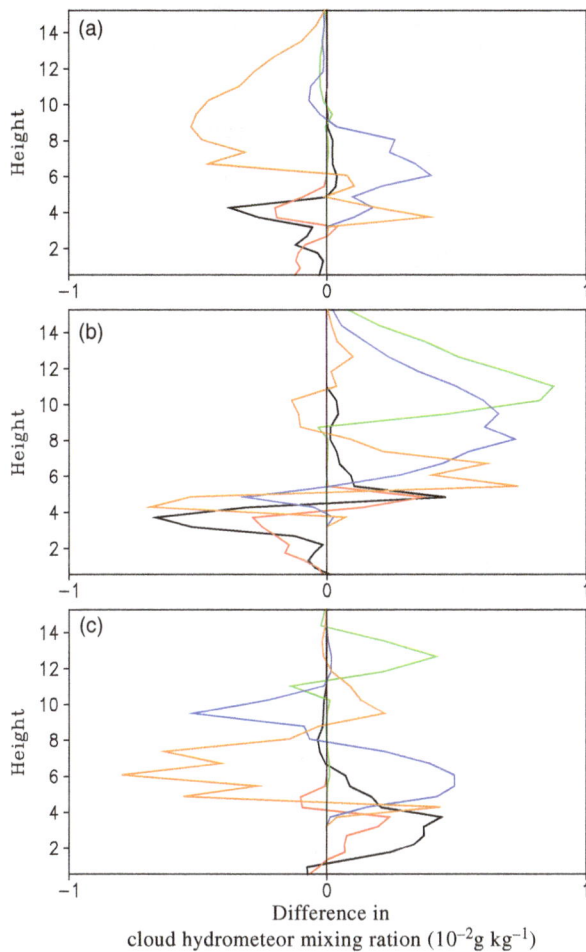

Figure 4. Difference in vertical profiles of 5-day and model domain mean cloud water (black), raindrops (red), cloud ice (green), snow (blue) and graupel (orange) between (a) H80T and H80K (H80T−H80K), (b) K95T and K95K (K95T−K95K) and (c) Z08T and Z08K (Z08T−Z08K). Unit is $10^{-2}\,g\,kg^{-1}$.

Figure 5. Difference in vertical profiles of 5-day and model domain mean radiation (orange) and its components of solar heating (red) and infrared cooling (blue) (a) H80T and H80K (H80T−H80K), (b) K95T and K95K (K95T−K95K) and (c) Z08T and Z08K (Z08T−Z08K). Unit is $°C\,day^{-1}$.

Zeng *et al.*, 2008 (Z08)] and the two parameter sets [Koenig (K) and Takahashi (T)]. The pre-summer torrential rainfall event occurred over southern China during June 2008 is chosen as the simulated rainfall case in this study. The major results include the followings:

• The analysis of RMSD in surface rain rate, temperature and specific humidity between the simulations and observations shows that the experiment K95T produces the best simulation.

• The calculations of 5-day and model domain mean rain rates reveal that the three schemes with the Takahashi parameter tend to reduce the mean rain rate compared to the three schemes with Koenig parameter. K95T generates the closest mean rain rate to the mean observational rain rate due to the largest decrease from K95K to K95T among the three pairs of experiments.

• The reduction in the rain rate enhanced by use of the Takahashi parameter from the Hsie scheme to the Krueger scheme corresponds to the strengthened slowdown in hydrometeor loss; it is related to the change in the melting of ice hydrometeor from the

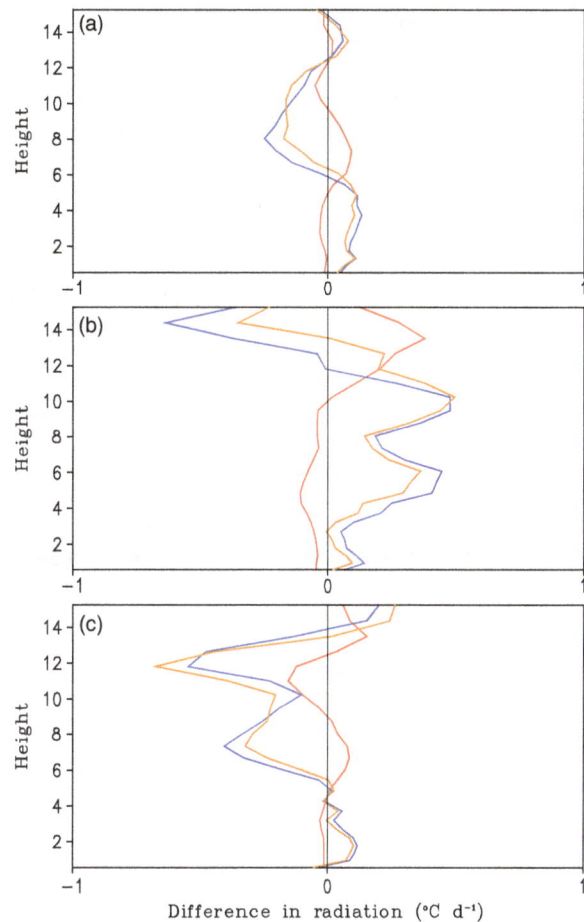

increase associated with the weakened infrared radiative cooling below 6 km brought by use of the Takahashi parameter in the Hsie scheme to the decrease associated with enhanced infrared radiative cooling above 12 km in the Krueger scheme.

• The reduction in the rain rate enhanced by use of the Takahashi parameter from the Zeng scheme to the Krueger scheme corresponds to the change in net condensation from the increase associated with the strengthened infrared radiative cooling in the Zeng scheme to the decrease associated with weakened infrared radiative cooling in the Krueger scheme.

The Takahashi parameter has been linearly interpolated into the model in this study, which may cause uncertainty. But the uncertainties may not be important compared to the uncertainties caused by different schemes listed in Table 2. The tests are carried out using the two-dimensional model only for one torrential rainfall event in this study. Therefore, further tests are required using three-dimensional cloud-resolving models for more torrential rainfall events under different environmental thermodynamic conditions to generalize the results from this study.

Acknowledgements

The authors thank W.-K. Tao at NASA/GSFC for his cloud-resolving model and the SCMREX science team and office sponsored by Chinese Academy of Meteorological Sciences, Chinese Meteorological Administration for providing the observational data used in this paper. The data are available through application at http://scmrex.cma.gov.cn. X. Li and R. Liu were supported by National Natural Science Foundation of China (41475039) and National Key Basic Research and Development Project of China (2015CB953601). P. Zhu and G. Zhai were supported by the National Natural Science Foundation of China (41175047) and the National Key Basic Research and Development Project of China (2013CB430100). X. Shen and W. Huang were supported by the National Key Basic Research and Development Project of China (2013CB430103 and 2015CB453201), the National Natural Science Foundation of China (41375058). D. Wang was supported by the China Meteorological Administration Special Public Welfare Research Fund (GYHY201006014 and GYHY201306005), National Natural Science Foundation of China (91437221, 41175064 and 41175047), National Basic Research and Development Project of China (2012CB417204) and Basic Research Fund of the Chinese Academy of Meteorological Sciences (2014R016 and 2014Z001).

References

Bergeron T. 1935. On the physics of clouds and precipitation. In *Proces Verbaux de l'Association de Météorologie*. International Union of Geodesy and Geophysics: Lisbon, Portugal; 156–178.

Cess RD, Zhang M, Potter G, Alekseev V, Barker H, Bony S, Colman R, Dazlich D, Del Genio A, Déqué M, Dix M, Dymnikov V, Esch M, Fowler L, Fraser J, Galin V, Gates W, Hack J, Ingram W, Kiehl J, Kim Y, Le Treut H, Liang X-Z, McAvaney B, Meleshko V, Morcrette J, Randall D, Roeckner E, Schlesinger M, Sporyshev P, Taylor K, Timbal B, Volodin E, Wang W, Wang W, Wetherald R. 1997. Comparison of the seasonal change in cloud-radiative forcing from atmospheric general circulation models and satellite observations. *Journal of Geophysical Research* 102(D14): 16593–16603.

Chou M-D, Suarez MJ. 1994. An efficient thermal infrared radiation parameterization for use in General Circulation Model. NASA Technical Memorandum 104606, Vol. 3, Code 913, 85 pp, NASA/Goddard Space Flight Center, Greenbelt, MD.

Chou M-D, Kratz DP, Ridgway W. 1991. Infrared radiation parameterization in numerical climate models. *Journal of Climate* 4: 424–437.

Chou M-D, Suarez MJ, Ho C-H, Yan MM-H, Lee K-T. 1998. Parameterizations for cloud overlapping and shortwave single scattering properties for use in general circulation and cloud ensemble models. *Journal of the Atmospheric Sciences* 55: 201–214.

Cui X, Li X. 2006. Role of surface evaporation in surface rainfall processes. *Journal of Geophysical Research* 111: D17112, doi: 10.1029/2005JD006876.

Findeisen W. 1938. Kolloid-meteorologische Vorgänge bei Neiderschlags-bildung. *Meteorologische Zeitschrift* 55: 121–133.

Fletcher HN. 1962. *The Physics of Rain Clouds*. Cambridge University Press: London, UK, 386 pp.

Gao S, Li X. 2008. *Cloud-Resolving Modeling of Convective Processes*. Springer: Dordrecht, The Netherlands, 206 pp.

Gao S, Cui X, Zhou Y, Li X. 2005. Surface rainfall processes as simulated in a cloud resolving model. *Journal of Geophysical Research* 110: D10202, doi: 10.1029/2004JD005467.

Grabowski WW, Wu X, Moncrieff MW, Hall WD. 1998. Cloud-resolving model of tropical cloud systems during phase III of GATE. Part II: effects of resolution and the third spatial dimension. *Journal of the Atmospheric Sciences* 55: 3264–3282.

Hallet J, Mossop SC. 1974. Production of secondary ice particles during the riming process. *Nature* 249: 26–28.

Hsie E-Y, Farley RD, Orville HD. 1980. Numerical simulation of ice-phase convective cloud seeding. *Journal of Applied Meteorology* 19: 950–977.

Khairoutdinov MF, Randall DA. 2003. Cloud-resolving modeling of the ARM summer 1997 IOP: model formulation, results, uncertainties, and sensitivities. *Journal of the Atmospheric Sciences* 60: 607–625.

Koenig LR. 1971. Numerical modeling of ice deposition. *Journal of the Atmospheric Sciences* 28: 226–237.

Krueger SK, Fu Q, Liou KN, Chin H-NS. 1995. Improvement of an ice-phase microphysics parameterization for use in numerical simulations of tropical convection. *Journal of Applied Meteorology* 34: 281–287.

Li X, Gao S. 2011. *Precipitation Modeling and Quantitative Analysis*. Springer: Dordrecht, The Netherlands, 240 pp.

Li X, Sui C-H, Lau K-M, Chou M-D. 1999. Large-scale forcing and cloud-radiation interaction in the tropical deep convective regime. *Journal of the Atmospheric Sciences* 56: 3028–3042.

Lin Y-L, Farley RD, Orville HD. 1983. Bulk parameterization of the snow field in a cloud model. *Journal of Climate and Applied Meteorology* 22: 1065–1092.

Morrison H, Curry JA, Khvorostyanov VI. 2005. A new double-moment microphysics parameterization for application in cloud and climate models. Part I: description. *Journal of the Atmospheric Sciences* 62: 1665–1677.

Phillips VT, Donner LJ, Garner ST. 2007. Nucleation processes in deep convection simulated by a cloud-system-resolving model with double-moment bulk microphysics. *Journal of the Atmospheric Sciences* 64: 738–761.

Rutledge SA, Hobbs PV. 1983. The mesoscale and microscale structure and organization of clouds and precipitation in midlatitude cyclones. Part VIII: a model for the "seeder-feeder" process in warm-frontal rainbands. *Journal of the Atmospheric Sciences* 40: 1185–1206.

Rutledge SA, Hobbs PV. 1984. The mesoscale and microscale structure and organization of clouds and precipitation in midlatitude cyclones. Part XII: a diagnostic modeling study of precipitation development in narrow cold-frontal rainbands. *Journal of the Atmospheric Sciences* 41: 2949–2972.

Shen X, Li X. 2011. Thermodynamic aspects of precipitation efficiency. In *Thermodynamics-Interaction Studies-Solids, Liquids and Gases*, Moreno Piraján JC (ed). InTech: Rijeka, Croatia; 73–94. ISBN: 978-953-307-563-1.

Shen X, Wang Y, Li X. 2011a. Radiative effects of water clouds on rainfall responses to the large-scale forcing during pre-summer heavy rainfall over southern China. *Atmospheric Research* 99: 120–128.

Shen X, Wang Y, Li X. 2011b. Effects of vertical wind shear and cloud radiative processes on responses of rainfall to the large-scale forcing during pre-summer heavy rainfall over southern China. *Quarterly Journal of the Royal Meteorological Society* 137: 236–249.

Shen X, Zhang N, Li X. 2011c. Effects of large-scale forcing and ice clouds on pre-summer heavy rainfall over southern China in June 2008: a partitioning analysis based on surface rainfall budget. *Atmospheric Research* 101: 155–163.

Shen X, Liu J, Li X. 2012. Torrential rainfall responses to ice microphysical processes during pre-summer heavy rainfall over southern China. *Advances in Atmospheric Sciences* 29: 493–500.

Shen X, Qing T, Huang W, Li X. 2013. Effects of vertical wind shear on pre-summer heavy rainfall budget: a cloud-resolving modeling study. *Atmospheric and Oceanic Science Letters* 6: 44–51.

Soong ST, Ogura Y. 1980. Response of tradewind cumuli to large-scale processes. *Journal of the Atmospheric Sciences* 37: 2035–2050.

Soong ST, Tao W-K. 1980. Response of deep tropical cumulus clouds to Mesoscale processes. *Journal of the Atmospheric Sciences* 37: 2016–2034.

Sui C-H, Lau K-M, Tao W-K, Simpson J. 1994. The tropical water and energy cycles in a cumulus ensemble model. Part I: equilibrium climate. *Journal of the Atmospheric Sciences* 51: 711–728.

Sui C-H, Li X, Lau K-M. 1998. Radiative-convective processes in simulated diurnal variations of tropical oceanic convection. *Journal of the Atmospheric Sciences* 55: 2345–2359.

Sui C-H, Li X, Yang M-J, Huang H-L. 2005. Estimation of oceanic precipitation efficiency in cloud models. *Journal of the Atmospheric Sciences* 62: 4358–4370.

Testing parameterization schemes for simulating depositional growth of ice crystal using Koenig...

63

Takahashi T, Fukuta N. 1988. Supercooled cloud tunnel studies on the growth of snow crystals between −4° and −20°C. *Journal of the Meteorological Society of Japan* **66**: 841–855.

Takahashi T, Endoh T, Wakahama G, Fukuta N. 1991. Vapor diffusional growth of free-falling snow crystals between −3° and −23°C. *Journal of the Meteorological Society of Japan* **69**: 15–30.

Tao W-K, Simpson J. 1993. The Goddard Cumulus Ensemble model. Part I: model description. *Terrestrial, Atmospheric and Oceanic Sciences* **4**: 35–72.

Tao W-K, Soong S-T. 1986. The study of the response of deep tropical clouds to mesoscale processes: three-dimensional numerical experiments. *Journal of the Atmospheric Sciences* **43**: 2653–2676.

Tao W-K, Simpson J, Soong S-T. 1987. Statistical properties of a cloud ensemble: a numerical study. *Journal of the Atmospheric Sciences* **44**: 3175–3187.

Tao W-K, Simpson J, McCumber M. 1989. An ice-water saturation adjustment. *Monthly Weather Review* **117**: 231–235.

Thompson G, Field PR, Rasmussen RM, Hall WD. 2008. Explicit forecasts of winter precipitation using an improved bulk microphysics scheme. Part II: implementation of a new snow parameterization. *Monthly Weather Review* **136**: 5095–5115.

Tompkins AM. 2000. The impact of dimensionality on long-term cloud-resolving model simulations. *Monthly Weather Review* **128**: 1521–1535.

Walko RL, Cotton WR, Meyers MP, Harrington JY. 1995. New RAMS cloud microphysics parameterization. Part I: the single-moment scheme. *Atmospheric Research* **38**: 29–62.

Wang Y, Shen X, Li X. 2010. Microphysical and radiative effects of ice clouds on responses of rainfall to the large-scale forcing during pre-summer heavy rainfall over southern China. *Atmospheric Research* **97**: 35–46.

Westbrook CD, Heymsfield AJ. 2011. Ice crystals growing from vapor in supercooled clouds between −2.5° and −22°C: testing current parameterization methods using laboratory data. *Journal of the Atmospheric Sciences* **68**: 2416–2429.

Zeng X, Tao W-K, Lang S, Hou AY, Zhang M, Simpson J. 2008. On the sensitivity of atmospheric ensembles to cloud microphysics in long-term cloud-resolving model simulations. *Journal of the Meteorological Society of Japan* **86**: 45–65.

8

Modulation of Pacific Decadal Oscillation on the relationship of El Niño with Southern China rainfall during early boreal winter

Gang Li,[1] Jiepeng Chen,[2]* Xin Wang,[2,3]* Yanke Tan[4] and Xiaohua Jiang[1]

[1] Xichang Satellite Launch Center, Xichang, China
[2] State Key Laboratory of Tropical Oceanography, South China Sea Institute of Oceanology, Chinese Academy of Sciences, Guangzhou, China
[3] Laboratory for Regional Oceanography and Numerical Modeling, Qingdao National Laboratory for Marine Science and Technology, Qingdao, China
[4] College of Meteorology and Oceanography, PLA University of Science and Technology, Nanjing, China

*Correspondence to:

X. Wang, State Key Laboratory of Tropical Oceanography, South China Sea Institute of Oceanology, Chinese Academy of Sciences, 164 West Xingang Road, Guangzhou 510301, China; Laboratory for Regional Oceanography and Numerical Modeling, Qingdao National Laboratory for Marine Science and Technology, Qingdao, China.
E-mail: wangxin@scsio.ac.cn
J. Chen, State Key Laboratory of Tropical Oceanography, South China Sea Institute of Oceanology, Chinese Academy of Sciences, 164 West Xingang Road, Guangzhou 510301, China.
E-mail: chenjiep@scsio.ac.cn

Abstract

The modulation of Pacific Decadal Oscillation (PDO) on the relationship of El Niño with southern China rainfall during early boreal winter (November–December) is investigated in this study. When El Niño occurs in positive phase of PDO, significantly positive rainfall anomalies only appears over the southwestern part of southern China. When El Niño occurs in negative phase of PDO, pronounced positive rainfall anomalies are observed over almost the whole southern China. Further analyses revealed that the anomalous Philippine Sea anticyclone (PSAC) and local meridional circulation play an important role in modulating the relationship of El Niño with southern China rainfall by PDO. During the negative PDO phase, under the combined influence of tropical Pacific and Indian Ocean dipole sea surface temperature anomalies pattern, the anomalous PSAC over the western North Pacific and rising motion associated with local meridional circulation over southern China provide sufficient water vapor and dynamical conditions for the formation of rainfall. During positive PDO phase, however, PSAC and local meridional circulation are only influenced by tropical Pacific and they become weak, not favoring the formation of southern China rainfall.

Keywords: Pacific Decadal Oscillation (PDO); El Niño; southern China rainfall; early boreal winter

1. Introduction

Rainfall variability over southern China has been investigated extensively (Chan and Zhou, 2005; Zhou and Chan, 2007; Gu et al., 2009; Feng et al., 2014; Yao et al., 2015; Zhang et al., 2015). Many above studies suggested that the boreal winter rainfall over southern China also shows significant variability on subseasonal, interannual and interdecadal scales. Moreover, the variability of boreal winter rainfall over southern China can induce disruptive weather and climate hazards, which cause seriously damage to agriculture and socioeconomic development (Zhou et al., 2011). Therefore, it is necessary to study the variability of boreal winter rainfall over southern China.

The variability of boreal winter rainfall over southern China is influenced by many factors (Zhou, 2011; Zhang et al., 2015; Zhang et al., 2017). In particular, the El Niño-Southern Oscillation (ENSO) may have the most significant relationship with boreal winter rainfall over southern China (Wang et al., 2000).

In fact, it should be mentioned that the relationship between El Niño and southern China climate is influenced by the Pacific Decadal Oscillation (PDO; Mantua et al., 1997; Mantua and Hare, 2002) (Chan and Zhou, 2005; Wang et al., 2008; Feng et al., 2014; Kim et al., 2014).

Although many above studies have focused on the modulation of PDO on the relationship between ENSO and East Asian winter monsoon (EAWM), the direct modulation of PDO on the relationship of El Niño with boreal winter rainfall over southern China has not been sufficiently studied. Especially, considering the significantly subseasonal variability of boreal winter rainfall over southern China (Yao et al., 2015), it is reasonable to hypothesize that the modulation of PDO on the relationship between El Niño and rainfall during early boreal winter (November–December) may be different from that during the boreal winter, which is usually defined as December–January–February or January–February–March. Therefore, the main purpose of this study is to investigate the direct modulation of PDO on the relationship of El Niño with southern China rainfall during early boreal winter.

The article is organized as follows. Section 2 describes the datasets and methods used in this study. In Section 3, we investigate the influence of PDO on the relationship of El Niño with southern China rainfall in during early boreal winter. The summary and discussion are provided in Section 4.

2. Datasets and methods

The monthly sea surface temperature (SST) data is provided by the Hadley Centre Sea Ice and Sea Surface Temperature (HadISST1.1) dataset (Rayner *et al.*, 2003). The monthly atmospheric data is taken from the National Oceanic and Atmospheric Administration Cooperative Institute for Research in Environmental Sciences (NOAA-CIRES) 20th Century Reanalysis V2c dataset (Compo *et al.*, 2011). The first dataset is the full data analysis (V7) rainfall dataset (Schneider *et al.*, 2015) and the Climatic Research Unit (CRU) Time Series (TS) 3.24.01 rainfall dataset (University of East Anglia Climatic Research Unit et al., 2017) are used as the rainfall datasets.

This study covers the period from 1901 to 2010 for the consistency of the above datasets. The average of November and December is defined as the early boreal winter. In addition, all datasets are detrended before analyses. The composite analysis is used in our study. The statistical significance of composite is examined by Student's *t*-test.

The PDO index from University of Washington is used to describe the interdecadal variability of Pacific (Mantua *et al.*, 1997) (figure not shown). To reveal the interdecadal variability of PDO, we apply the 11-year running mean to the time series of original PDO index during early boreal winter. Here, the positive (negative) phase of PDO is defined as the years that the 11-year running mean PDO index is above (below) 0. Therefore, the periods of 1903–1943 and 1978–1992 are referred to as the positive phase of PDO, while the periods of 1944–1977 and 1993–2010 are referred to as the negative phase of PDO.

The Niño-3.4 index is defined as area-averaged SSTA in the tropical central-eastern Pacific (5°S–5°N, 170°–120°W) (figure not shown). El Niño years are selected when the normalized Niño-3.4 index is greater than 0.5. Following this definition, 16 (16)

Table I. Classification of El Niño years in concurrent with positive and negative phase of the PDO during early boreal winter for the period 1901–2010.

	Positive PDO phase	Negative PDO phase
El Niño years	1904, 1905, 1911, 1913, 1914, 1918, 1923, 1925, 1930, 1940, 1941, 1979, 1982, 1986, 1987, 1991	1902, 1951, 1957, 1963, 1965, 1968, 1969, 1972, 1976, 1977, 1994, 1997, 2002, 2004, 2006, 2009

El Niño years occurred during the positive (negative) phase of the PDO (Table 1). For convenience, El Niño that occurs with the positive (negative) PDO phase is referred to as El Niño + positive PDO (El Niño + negative PDO).

3. Results

3.1. Rainfall anomalies

Figure 1 shows the composite of rainfall anomalies over southern China for El Niño + positive PDO and El Niño + negative PDO, respectively. During El Niño + positive PDO (Figure 1(a)), the northeast-southwest-oriented positive rainfall anomalies appear over the whole southern China with a centre around (25°N, 110°E). Significant rainfall anomalies mainly appear over the southwestern part of southern China and rainfall anomalies over the eastern part of southern China are not significant. During El Niño + negative PDO (Figure 1(b)), the magnitude of the northeast-southwest-oriented rainfall belt increases significantly. Almost the whole southern China except for the coast of southeast China is dominated by significantly positive rainfall anomalies. Besides, it should be mentioned that the whole rainfall belt moves northeastward to some extent. Figures 2(c) and (d) show that composite results based on GPCC V7 dataset clearly reproduce the results based on CRU TS 3.24.01 dataset.

The above results suggest that the influence of El Niño on southern China rainfall during early boreal winter is different during the different PDO phase. El Niño can exert significant influence on southern China rainfall for negative phase of PDO. However, the influence is very weak for the positive phase of PDO.

3.2. SST

Figure 2 shows the composite of early boreal winter SST anomalies (SSTA) for El Niño + positive PDO and El Niño + negative PDO, respectively. During El Niño + positive PDO (Figure 2(a)), positive SSTA dominates in the tropical central-eastern Pacific, extratropical South Pacific, Indian Ocean and South China Sea (SCS). Negative SSTA dominates in the North and South Pacific and the tropical western Pacific. Comparing SSTA between El Niño + positive PDO and El Niño + negative PDO (Figure 2(b)), three significant features can be emphasized. First, although spatial distribution of positive SSTA in the tropical central-eastern Pacific is similar for different PDO phase, the extent of maximum SSTA (>1.5 °C) for El Niño + negative PDO is larger and extends more westward than that for El Niño + positive PDO. Second, the magnitude of negative SSTA in the tropical western Pacific is larger for El Niño + negative PDO than that for El Niño + positive PDO. Third, a basin-wide positive SSTA pattern dominates in the tropical Indian Ocean for El Niño + positive PDO; however, a Indian

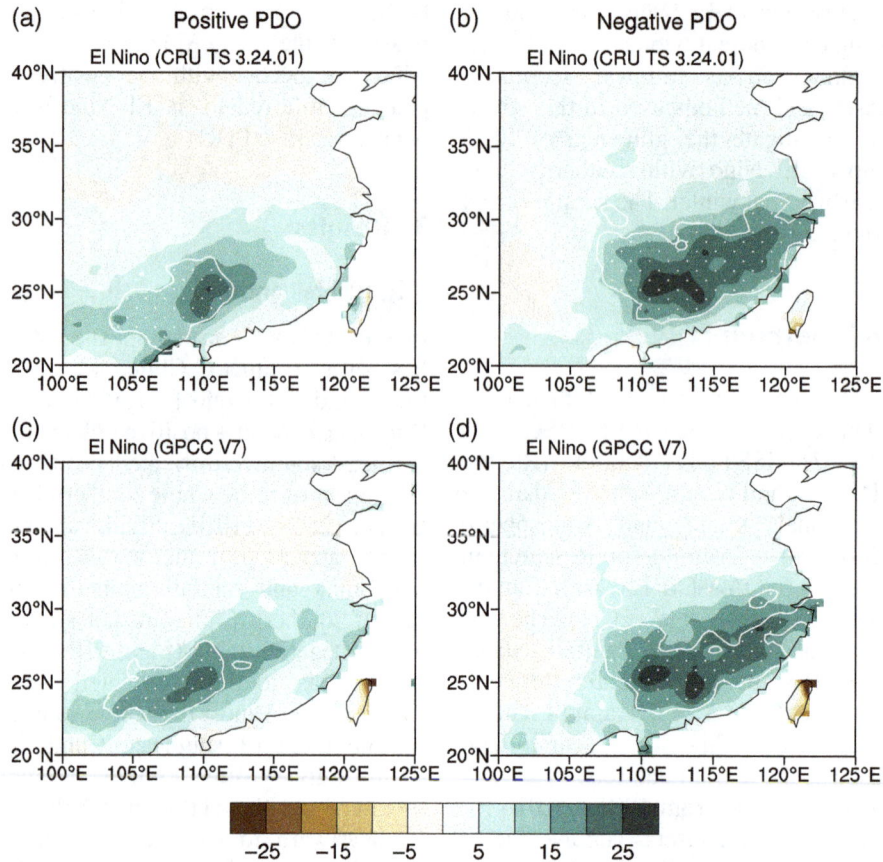

Figure 1. Composite of the early boreal winter rainfall anomalies (unit: mm) during (a) El Niño + positive PDO and (b) El Niño + negative PDO based on CRU TS 3.24.01 dataset, respectively. (c) and (d) As in (a) and (b), but based on GPCC V7 dataset. The white contours filled with white dots represents significance exceeding the 95% confidence level based on the Student's t-test.

Ocean dipole (IOD) SSTA pattern, with positive SSTA in the western Indian Ocean and negative SSTA in the eastern Indian Ocean, dominates in the tropical Indian Ocean for El Niño + negative PDO.

3.3. Possible mechanism

Figure 3 shows the composite of 850 hPa wind anomalies during early boreal winter for El Niño + positive PDO and El Niño + negative PDO, respectively. Previous studies have suggested that the anomalous Philippine Sea anticyclone (PSAC) can convey the influence of El Niño to southern China during boreal winter (Wang *et al.*, 2000). During El Niño + positive PDO (Figure 3(a)), PSAC appears over western North Pacific (WNP) with a centre at (17.5°N, 132°E). Significant southwesterly anomalies to the northwestern side of the PSAC dominate over southern China, favoring moisture transport from SCS to this region. The western part of southern China is associated with significantly anomalous moisture convergence (figure not shown), which induces the related rainfall anomalies in Figure 1(a). During El Niño + negative PDO (Figure 3(b)), PSAC strengthens and moves southwestward with a centre at (16°N, 128°E). Southwesterly anomalies over southern China are much stronger. Besides, moisture can be transported from not only SCS but also the Bay

of Bengal (BOB) to southern China and significantly anomalous moisture convergence dominates the whole southern China (figure not shown). Therefore, significantly positive rainfall anomalies are observed over southern China (Figure 1(b)).

During El Niño + positive PDO, significant westerly anomalies control the tropical western-central Pacific and easterly anomalies prevail over the Maritime Continent and tropical eastern Indian Ocean (Figure 3(a)). The anomalous divergence over the tropical western Pacific results in anomalous sinking motion (Figure 4(a)), which excites an anomalous local Hadley circulation over East Asia (110°–120°E) (Figure 4(c)). However, the anomalous rising motion associated with this local Hadley circulation over southern China (25°–30°N) is insignificant, thus not providing sufficiently dynamical condition for southern China rainfall. During El Niño + negative PDO (Figure 3(b)), significant westerly anomalies over the tropical western-central Pacific move westward slightly. Besides, easterly anomalies over the Maritime Continent and tropical eastern Indian Ocean intensify significantly and extend more westward. The anomalous divergence over the tropical western Pacific can induce significant sinking motion (Figure 4(b)), which results in an anomalous local Hadley circulation over East Asia (Figure 4(d)). The anomalous rising motion

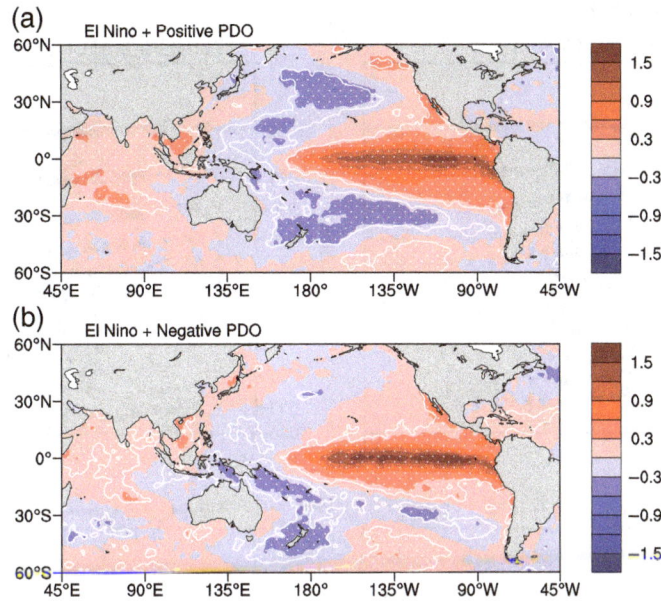

Figure 2. Composite of the early boreal winter SSTA (unit: °C) during (a) El Niño + positive PDO and (b) El Niño + negative PDO, respectively. The white contours filled with white dots represents significance exceeding the 95% confidence level based on the Student's *t*-test.

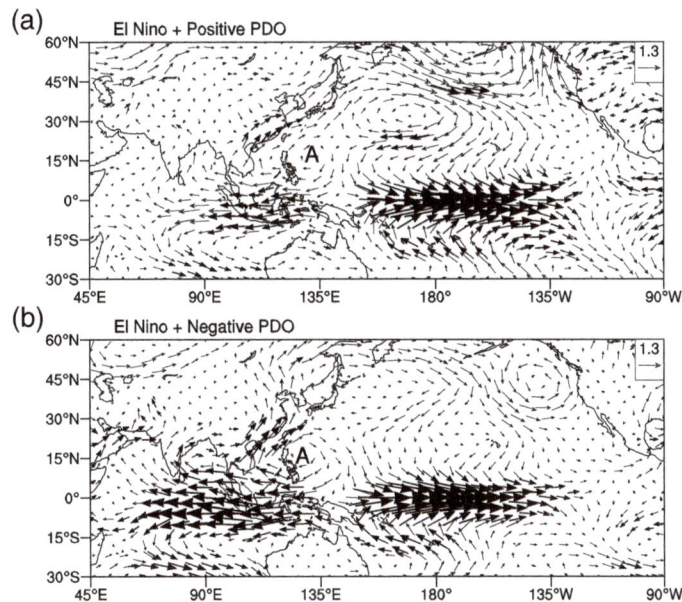

Figure 3. Composite of the early boreal winter wind anomalies (unit: m s⁻¹) at 850 hPa during (a) El Niño + positive PDO and (b) El Niño + negative PDO, respectively. The red vectors indicate zonal winds exceeding the 95% confidence level based on the Student's *t*-test.

associated with local Hadley circulation over southern China becomes significant compared to that during El Niño + positive PDO, and thus providing a dynamical condition for southern China rainfall.

4. Summary and discussion

The modulation of PDO on the influence of El Niño and southern China rainfall during early boreal winter from 1901 to 2010 is investigated in this study. Results show that, significantly rainfall anomalies only appear over the southwestern part of southern China during El Niño + positive PDO. However, during El Niño + negative PDO, almost the whole southern China is dominated by pronounced positive rainfall anomalies.

During El Niño + positive PDO, an anomalous PSAC can transport anomalous water vapor from SCS to southern China. Meanwhile, the anomalous rising motion associated with this local meridional circulation over southern China is insignificant. The above atmospheric circulation is unfavorable for the formation of rainfall over the eastern part of southern China. During El Niño + negative PDO, the anomalous PSAC

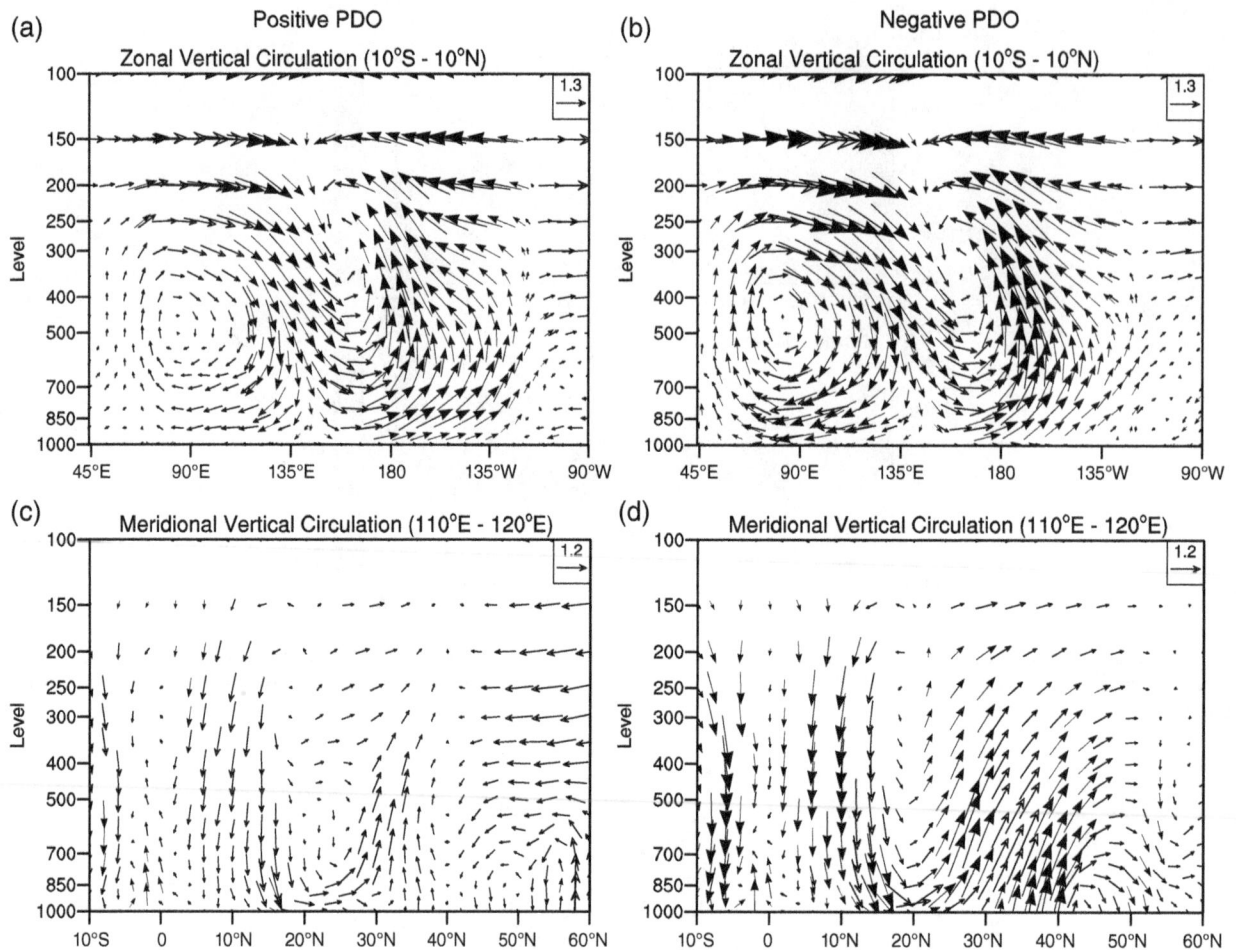

Figure 4. Composite of vertical circulation anomalies (unit: m s^{-1}) during early boreal winter in cases of El Niño + positive PDO (left panel) and El Niño + negative PDO (right panel). The top row is zonal vertical circulation anomalies by averaging zonal wind and vertical velocity (scaled by −100) between 10°S and 10°N. The bottom row is meridional vertical circulation anomalies by averaging meridional wind and vertical velocity (scaled by −100) between 110° and 120°E. The red vectors indicate vertical velocity exceeding the 95% confidence level based on the Student's t-test.

intensifies significantly and moves southwestward. Besides, moisture can be transported from not only SCS but also BOB to southern China. Anomalous sinking motion over the tropical western Pacific excites a stronger rising motion over southern China. Therefore, more rainfall appears over southern China.

The distinct features of PSAC over WNP and local meridional circulation over East Asian can be attributed to the different SSTA in the tropical Pacific and Indian Ocean. First, the location and intensity of PSAC depend on the maximum warm SSTA in tropical Pacific (Wang and Zhang, 2002). The extent of maximum SSTA in the tropical Pacific during El Niño + negative PDO is larger and extends more westward, and induces a stronger PSAC. Besides, negative SSTA in the tropical western Pacific can induce a stronger local meridional circulation over East Asia than that during El Niño-positive PDO.

Second, during El Niño + positive PDO, Indian Ocean shows a basin-wide warming. Previous studies have demonstrated that the basin-wide warming in Indian Ocean lags the mature phase El Niño and it is an important 'capacitor' during the decaying phase of

El Niño (Yuan *et al.*, 2008). It may increase the intensity and persistence of PSAC during El Niño decaying years based on both observational and modeling analyses (Annamalai *et al.*, 2005). However, during El Niño + negative PDO, Indian Ocean shows a positive IOD, which usually appears in the developing phase of El Niño. The boreal autumn positive IOD can induce a strengthened PSAC with a westward-extending position during the following boreal winter (Xie *et al.*, 2009; Zhang *et al.*, 2017).

However, the above mechanism should be further investigated by using a coupled atmospheric and oceanic general circulation model. Besides, it should be mentioned that there exists two types of El Niño (Ashok *et al.*, 2007; Kao and Yu, 2009). Recent studies have suggested that the tropical central (CP) El Niño occurs more frequently and its intensity increases significantly (Lee and McPhaden, 2010). The influence of two types of El Niño on China rainfall is widely investigated (Feng *et al.*, 2011; Li *et al.*, 2013; Li *et al.*, 2014). For example, the El Niño-induced more southern China rainfall only appears during the tropical eastern (EP) El Niño, not during the CP El Niño (Feng et al., 2010). Therefore, further studies are needed to understand the modulation of PDO on the relationship between two types of El Niño and southern China rainfall.

Acknowledgements

The authors thank three anonymous reviewers for their useful comments on the manuscript. This work is jointly supported by the Strategic Priority Research Program of the Chinese Academy of Sciences (Grant No. XDA11010403), National Department Public Benefit Research Foundation (Grant No. GYHY201406003), the CAS/SAFEA International Partnership Program for Creative Research Teams, National Natural Science Foundation of China (Grant Nos. 41422601, 41521005, 41475070, 41490642 and 41575097), and the National Basin Research Program of China (Grant No. 2013CB956203).

References

Annamalai IIP, Liu P, Xie SP. 2005. Southwest Indian Ocean SST variability: its local effect and remote influence on Asian monsoons. *Journal of Climate* **18**: 4150–4167.

Ashok K, Behera SK, Rao SA, Weng H, Yamagata T. 2007. El Niño Modoki and its possible teleconnection. *Journal of Geophysical Research* **112**: C11007.

Chan JCL, Zhou W. 2005. PDO, ENSO and the early summer monsoon rainfall over south China. *Geophysical Research Letters* **32**: L08810.

Compo GP, Whitaker JS, Sardeshmukh PD, Matsui N, Allan RJ, Yin X, Jr Gleason BE, Vose RS, Rutledge G, Bessemoulin P, Bronnimann S, Brunet M, Crouthamel RI, Grant AN, Groisman PY, Jones PD, Kruk MC, Kruger AC, Marshall GJ, Maugeri M, Mok HY, Nordli Ø, Ross TF, Trigo RM, Wang XL, Woodruff SD, Worley SJ. 2011. The twentieth century reanalysis project. *Quarterly Journal of the Royal Meteorological Society* **137**: 1–28.

Feng J, Wang L, Chen W, Fong SK, Leong KC. 2010. Different impacts of two types of Pacific Ocean warming on southeast Asian rainfall during boreal winter. *Journal of Geophysical Research* **115**: D24122.

Feng J, Chen W, Tam CY, Zhou W. 2011. Different impacts of El Niño and El Niño Modoki on China rainfall in the decaying phases. *International Journal of Climatology* **31**: 2091–2101.

Feng J, Wang L, Chen W. 2014. How does the east Asian summer monsoon behave in the decaying phase of El Niño during different PDO phases? *Journal of Climate* **27**: 2682–2698.

Gu W, Li CY, Li WJ, Zhou W, Chan JCL. 2009. Interdecadal unstationary relationship between NAO and east China's summer precipitation patterns. *Geophysical Research Letters* **36**: L13702.

Kao HY, Yu JY. 2009. Contrasting eastern Pacific and central Pacific types of ENSO. *Journal of Climate* **22**: 615–632.

Kim JW, Yeh SW, Chang EC. 2014. Combined effect of El Niño-Southern Oscillation and Pacific Decadal Oscillation on the east Asian winter monsoon. *Climate Dynamics* **42**: 957–971.

Lee T, McPhaden MJ. 2010. Increasing intensity of El Niño in the central-quatorial Pacific. *Geophysical Research Letters* **37**: L14603.

Li XZ, Zhou W, Li CY, Song J. 2013. Comparison of the annual cycles of moisture supply over southwest and southeast China. *Journal of Climate* **26**: 10139–10158.

Li XZ, Zhou W, Chen DL, Li CY, Song J. 2014. Water vapor transport and moisture budget over eastern China: remote forcing from the two types of El Niño. *Journal of Climate* **27**: 8778–8792.

Mantua NJ, Hare SR. 2002. The Pacific Decadal Oscillation. *Journal of Oceanography* **58**: 35–44.

Mantua NJ, Hare SR, Zhang Y, Wallace JM, Francis RC. 1997. A Pacific interdecadal climate oscillation with impacts on salmon production. *Bulletin of the American Meteorological Society* **78**: 1069–1079.

Rayner NA, Parker DE, Horton EB, Folland CK, Alexander LV, Rowell DP, Kent EC, Kaplan A. 2003. Global analyses of sea surface temperature, sea ice, and night marine air temperature since the late nineteenth century. *Journal of Geophysical Research* **108**: 4407.

Schneider U, A Becker, P Finger, A Meyer-Christoffer, B Rudolf, M Ziese. 2015. *GPCC Full Data Reanalysis Version 7.0 at 0.5°: Monthly Land-Surface Precipitation from Rain-Gauges Built on GTS-Based and Historic Data.* Deutscher Wetterdienst, Global Precipitation Climatology Centre. https://doi.org/10.5676/DWD_GPCC/FD_M_V7_050

University of East Anglia Climatic Research Unit, Harris IC, Jones PD. 2017. *CRU TS3.24.01: Climatic Research Unit (CRU) Time-Series (TS) Version 3.24.01 of High Resolution Gridded Data of Month-by-Month Variation in Climate (Jan. 1901-Dec. 2015).* Rutherford Appleton Laboratory, Oxon, Harwell: Centre for Environmental Data Analysis.

Wang B, Zhang Q. 2002. Pacific-east Asian teleconnection. Part II: how the Philippine Sea anomalous anticyclone is established during El Niño development? *Journal of Climate* **15**: 3252–3265.

Wang B, Wu RG, Fu X. 2000. Pacific-east Asian teleconnection: how does ENSO affect the east Asian climate? *Journal of Climate* **13**: 1517–1536.

Wang L, Chen W, Huang RH. 2008. Interdecadal modulation of PDO on the impact of ENSO on the east Asian winter monsoon. *Geophysical Research Letters* **35**: L20702.

Xie SP, Hu KM, Hafner J, Tokinaga H, Du Y, Huang G, Sampe T. 2009. Indian Ocean capacitor effect on indo-western Pacific climate during the summer following El Niño. *Journal of Climate* **22**: 730–747.

Yao YH, Lin H, Wu QG. 2015. Subseasonal variability of precipitation in China during boreal winter. *Journal of Climate* **28**: 6548–6559.

Yuan Y, Zhou W, Yang H, Li CY. 2008. Warming in the northwestern Indian Ocean associated with the El Niño event. *Advances in Atmospheric Sciences* **25**: 246–252.

Zhang L, Fraedrich K, Zhu XH, Sielmann F, Zhi XF. 2015. Interannual variability of winter precipitation in Southeast China. *Theoretical and Applied Climatology* **119**: 229–238.

Zhang L, Sielmann F, Fraedrich K, Zhi XF. 2017. Atmospheric response to Indian Ocean dipole forcing: changes of Southeast China winter precipitation under global warming. *Climate Dynamics* **48**: 1467–1482.

Zhou LT. 2011. Interdecadal change in sea surface temperature anomalies associated with winter rainfall over South China. *Journal of Geophysical Research* **116**: D11101.

Zhou W, Chan JCL. 2007. ENSO and South China Sea summer monsoon onset. *International Journal of Climatology* **27**: 157–167.

Zhou BZ, Gu LH, Ding YH, Shao L, Wu ZM, Yang XS, Li CZ, Li ZC, Wang XM, Cao YH, Zeng BS, Yu MK, Wang MY, Wang SK, Sun HG, Duan AG, An YF, Wang X, Kong WJ. 2011. The Great 2008 Chinese ice storm: its socioeconomic-ecological impact and sustainability lessons learned. *Bulletin of the American Meteorological Society* **92**: 47–60.

9

Impact of summer rainfall over Southern-central Europe on circumglobal teleconnection

Zhongda Lin* and Riyu Lu

State Key Laboratory of Numerical Modelling for Atmospheric Sciences and Geophysical Fluid Dynamics, Institute of Atmospheric Physics, Chinese Academy of Sciences, Beijing, China

*Correspondence to:
Z. Lin, Institute of Atmospheric Physics, Chinese Academy of Sciences, P.O. Box 9804, Beijing 100029, China.
E-mail: zdlin@mail.iap.ac.cn

Abstract

In this study, a significant link is found between summer rainfall over southern-central Europe (SCE) and the circumglobal teleconnection (CGT). It is proposed that the SCE rainfall drives the CGT. The downstream-extended CGT-like wave train is reproduced with a linearized barotropic vorticity equation model forced by an upper-tropospheric divergence over SCE. This study provides an alternative mechanism for the formation and maintenance of the CGT.

Keywords: circumglobal teleconnection; southern-central European summer rainfall; Mediterranean Sea heating; rainfall feedback

1. Introduction

The summertime circumglobal teleconnection (CGT) is characterized by a wave train along the upper-tropospheric westerly jet in the mid-latitudes of the Northern Hemisphere (Ding and Wang, 2005). It significantly affects mid-latitude climate on interannual timescales (Lu *et al.*, 2002; Ding and Wang, 2005; Huang *et al.*, 2011; Saeed *et al.*, 2011; Schubert *et al.*, 2011; Kosaka *et al.*, 2012; Wang *et al.*, 2013; Lin, 2014; Zhang and Zhou, 2015). Many studies, therefore, have been devoted to understanding mechanisms for the formation and maintenance of the CGT (Ding and Wang, 2005; Yasui and Watanabe, 2010; Ding *et al.*, 2011; Chen and Huang, 2012; Chen *et al.*, 2013; Hall *et al.*, 2013; Lee *et al.*, 2014). In particular, it would greatly enhance the understanding of and help predict the CGT if the heating sources responsible for triggering the CGT could be identified (Yasui and Watanabe, 2010).

The CGT could be induced by heat forcing related to tropical summer monsoon rainfall (Ding and Wang, 2005; Yasui and Watanabe, 2010; Yim *et al.*, 2014). Several studies have highlighted the role of the Indian summer monsoon rainfall (Ding and Wang, 2005; Lin, 2009; Ding *et al.*, 2011). They have proposed that Indian heat forcing could trigger a westward-propagating Gill-type Rossby wave response and form an anticyclonic anomaly at the entrance of the Asian westerly jet in the upper troposphere (Rodwell and Hoskins, 1996), initiating an eastward-propagating CGT-like wave train trapped within the strong westerly jet (Enomoto *et al.*, 2003). Similarly, this anticyclonic disturbance could also be excited by divergent flow-induced vorticity advection, because of heat forcing over the tropical western Indian Ocean (Chen and Huang, 2012). Yasui and Watanabe (2010) investigated dynamical forcings of the CGT in a dry atmospheric general circulation model. They proposed that the mid-latitude diabatic heat forcing around the Mediterranean Sea (MS) region, rather than those over tropical monsoon regions, can form the CGT pattern most efficiently.

However, rainfall exhibits a strong annual cycle over the MS region, with a wet winter season and a dry summer season. The absence of summer rainfall, because of the effect of the monsoon-desert mechanism (Rodwell and Hoskins, 1996), may hinder the development of a downstream-extended wave train similar to that proposed in winter by Watanabe (2004). Instead, intense summer rainfall occurs to the north over continental southern-central Europe (SCE) (Figure 2). This gives rise to the following question: Can the diabatic heating related to the SCE rainfall modulate the CGT? This question is investigated in this study.

2. Data and methods

The monthly atmospheric data were obtained from the European Centre for Medium-Range Weather Forecasts (ECMWF) 40-year Re-Analysis (ERA-40) from 1958 to 2002 (Uppala *et al.*, 2005). The monthly precipitation data are obtained from the Climatic Research Unit (CRU), University of East Anglia, UK, with a horizontal resolution of $0.5° \times 0.5°$ (Mitchell and Jones, 2005). Summer is defined as June, July, and August.

A barotropic model is used to investigate the effect of SCE rainfall on the CGT. The linearized barotropic vorticity equation is

$$\frac{\partial \zeta'}{\partial t} = -\overline{\mathbf{V}}_\psi \cdot \nabla \zeta' - \mathbf{V}'_\psi \cdot \nabla \left(f + \overline{\zeta}\right)$$

$$- \mathbf{V}'_\chi \cdot \nabla \left(f + \overline{\zeta}\right) - \left(f + \overline{\zeta}\right) \nabla$$

$$\cdot \mathbf{V}'_\chi - \kappa \zeta' - \varepsilon \nabla^4 \zeta',$$

where the over bar represents the zonal-mean variables and the prime is the deviation from the zonal-mean state. $\mathbf{V}_\psi$ and $\mathbf{V}_\chi$ are the rotational and divergent wind components, respectively, f is the Coriolis parameter, and ζ is the relative vorticity. The zonal-mean flow is set to the summertime climatological values at 200 hPa calculated from the ERA-40 data averaged over 1958–2002. The biharmonic diffusion coefficient, ε, is set to 2.34×10^{16} m^4 s^{-1} and the damping coefficient, $\kappa = 10$ day^{-1}, is used in this model. The vorticity equation is solved using the spectrum transform technique with a triangular truncation at wavenumber 21. The steady response is calculated as the 30-day mean averaged for 31–60 days.

3. Results

Figure 1(a) shows the spatial pattern of the negative phase of the summertime CGT, which is depicted by the anomalies of geopotential height at 200 hPa (H200) in the mid-latitudes of the Northern Hemisphere regressed against the minus CGT index (CGTI) for the period 1958–2002. The CGTI is defined as the normalized H200 anomalies averaged over west-central Asia (35°–40°N, 60°–70°E), in the same manner as Ding and Wang (2005), and its time series is shown in Figure 1(b). The CGT pattern is characterized by a mid-latitude wave train along the strong subtropical westerly jet in the upper troposphere with a zonal wavenumber of five. The negative H200 centers are significant over southwest Asia, east Asia, central North Pacific, and northeast North America, but weak over western Europe, consistent with the result of one-point correlation between the CGTI and the H200 anomalies identified by Ding and Wang (2005) in their Figure 1(b).

Related to the negative phase of the CGT, rainfall is significantly enhanced over SCE (Figure 2(a)). To better reveal their relationship, a rainfall index is calculated from the normalized summer-mean rainfall averaged over the SCE region (45°–50°N, 0°–30°E), whose time series is presented in Figure 1(b). The correlation coefficient between the CGTI and the SCE rainfall index (SCERI) is −0.43 for the period of 1958–2002 and is significant at the 99% confidence level based on Student's t-test, indicating a strong link between the summer rainfall over SCE and the CGT. In addition, no significant rainfall anomalies were found in the MS land region.

Previous studies have highlighted the role of diabatic heat forcing around the MS region in the CGT (Enomoto et al., 2003; Yasui and Watanabe, 2010). As shown in Figure 2(b), intense summer rainfall occurs over the SCE region. The climatology and standard deviation of summer-mean rainfall are approximately 3 and 0.7 mm per day, respectively, over the SCE region, compared with 0.7 and 0.3 mm per day, respectively, in the MS land region (30°–45°N, 0°–40°E) from 1958 to 2002. A similar result is obtained using the Climate Prediction Center (CPC) Merged Analysis of

Precipitation (CMAP) global precipitation data (Xie and Arkin, 1997) from 1979 to 2002, with the climatology and standard deviation of 2.6 and 0.6 mm per day, respectively, over the SCE region versus 0.6 and 0.3 mm per day, respectively, over the MS region. Rainfall is much stronger over the SCE region than the MS region in both climatology and interannual variability.

To investigate the possible feedbacks of the SCE rainfall, we calculated H200 anomalies related to the SCERI (Figure 3(a)). The SCE rainfall is linked to a mid-latitude wave train along the westerly jet around the entire Northern Hemisphere, similar to the CGT pattern (Figure 1(a)). To reveal characteristics of the associated Rossby wave propagation, the zonal and meridional components of wave-activity flux for stationary Rossby waves (W) were employed following Takaya and Nakamura (2001):

$$W = \frac{1}{2|\overline{V}|} \begin{pmatrix} \overline{u}\left(\psi_x'^2 - \psi'\psi_{xx}'\right) + \overline{v}\left(\psi_x'\psi_y' - \psi'\psi_{xy}'\right) \\ \overline{u}\left(\psi_x'\psi_y' - \psi'\psi_{xy}'\right) + \overline{v}\left(\psi_y'^2 - \psi'\psi_{yy}'\right) \end{pmatrix}$$

where $|V|$ is the magnitude of the horizontal wind vector (u, v), and ψ is the stream function; the over bar indicates the climatological summer mean averaged over 1958–2002, and the subscript and prime notations signify the partial derivatives and anomalies related to the SCERI, respectively.

Associated with the mid-latitude wave train, wave-activity flux propagates southeastward from eastern Europe into the Asian westerly jet at its entrance over west-central Asia, and then extends downstream along the westerly jet (Figure 3(b)). In theory, Rossby waves propagate eastward along strong westerlies (Hoskins and Ambrizzi, 1993). The propagation of the wave-activity flux downstream of the SCE rainfall suggests a possible role of the SCE rainfall in triggering the mid-latitude wave train or the CGT pattern.

The role of the SCE rainfall in the CGT is confirmed by the response to a divergence forced over the SCE region in the barotropic model (Figure 3(c)). The two-dimensional sinusoidal forcing of the divergence in the region (45°–50°N, 0°–30°E) was prescribed in the model. We took the maximum upper-tropospheric divergence to be 3.0×10^{-7} s^{-1}, which corresponds to an outflow associated with approximately one standard deviation of the SCE rainfall of 0.7 mm per day according to Hoskins and Karoly (1981). Details of the barotropic model are described in Section 2. The steady responses of the geopotential height (Figure 3(c)) resemble the H200 anomalies related to SCERI (Figure 3(a)). The geopotential height responses at mid-latitudes were calculated from stream function responses to the SCE divergence forcing based on the geostrophic balance. The resemblance includes positive anomalies over eastern Europe and a mid-latitude wave train along the Asian westerly jet. The negative geopotential height response over west-central Asia corresponds to a negative value of the CGTI, which agrees with a significant negative correlation of the

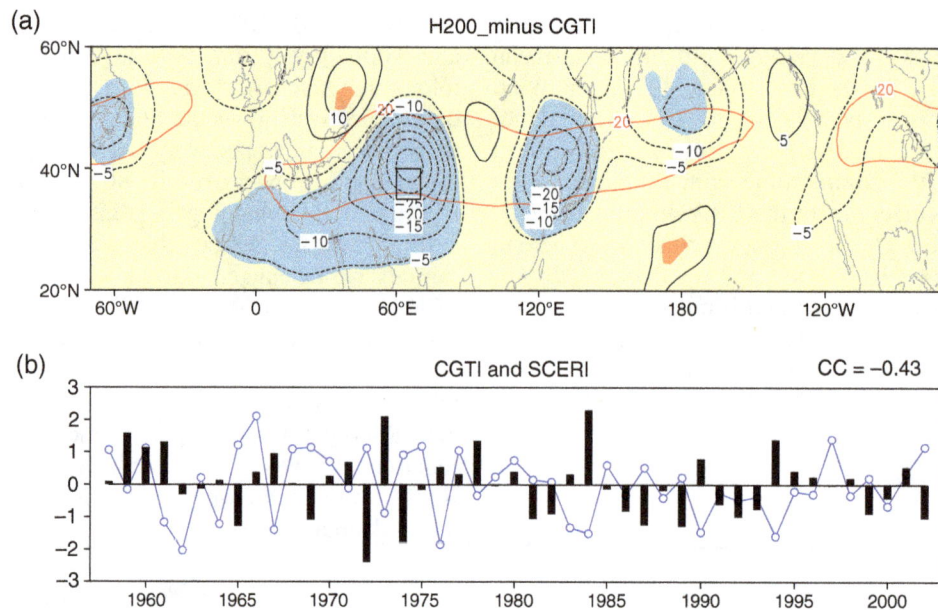

Figure 1. (a) Anomalies of the geopotential height at 200 hPa (H200, contour) regressed against the minus CGTI. The CGTI is defined as the 200-hPa geopotential height averaged over west-central Asia (35°–40°N, 60°–70°E), the region depicted by the black box, following Ding and Wang (2005). Shading indicates significance at the 95% confidence level, and the contour interval is 5 gpm. The red contour depicts the climatological westerly jet at 200 hPa with zonal winds exceeding 20 m s⁻¹. (b) The normalized time series of the CGTI (black bar) and the SCERI (blue solid line). The SCERI is calculated as the summer-mean rainfall averaged over the region (45°–50°N, 0°–30°E) surrounded by the blue box in Figure 2.

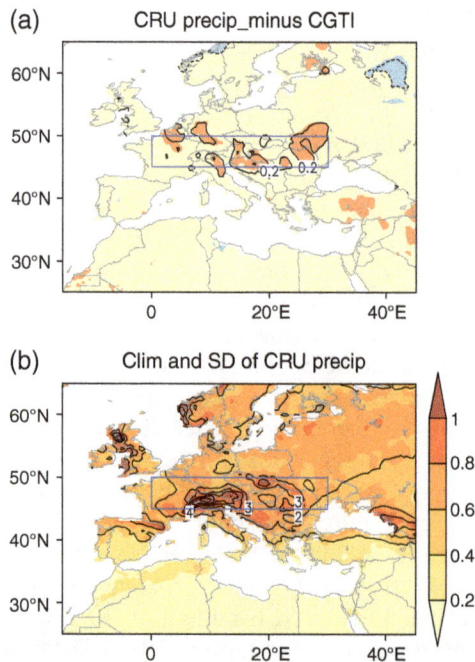

Figure 2. (a) Anomalies of the CRU precipitation (contour) regressed against the minus CGTI over Europe. Shading indicates significance at the 95% confidence level, and the contour interval is 0.2 mm day⁻¹. (b) Climatology (contour) and standard deviation (shading) of summer CRU precipitation. The contour interval is 1 mm day⁻¹ and the shading interval is 0.2 mm day⁻¹.

the SCERI-related anomalies (Figure 3(a)) and the barotropic model response to the SCE divergence forcing (Figure 3(c)). These differences suggest other factors or feedbacks may still be quite important. For example, Zuo *et al.* (2013) found that synoptic eddy-vorticity forcing over the North Atlantic plays an important role in triggering a downstream-extended wave train that links the North Atlantic sea surface temperature (SST) anomalies to East Asian summer monsoon. In addition, nonlinear effect is excluded in the current barotropic vorticity equation model. All these factors and feedbacks need to be considered to comprehensively understand the CGT–SCE rainfall connection.

4. Conclusion and discussion

Strong rainfall occurs over SCE in boreal summer. SCE rainfall can trigger the CGT pattern. Related to the SCE rainfall, the upper-tropospheric divergence forcing induces a southeastward-propagating wave flux into the Asian westerly jet at its entrance, corresponding to an anticyclonic anomaly over eastern Europe and a cyclonic anomaly over west-central Asia. Subsequently, the wave flux further extends eastward along the westerly jet because of the waveguide effect, forming a CGT-like wave train.

This study highlights the role of SCE rainfall in the CGT. However, the CGT may also exert some influence on the SCE rainfall. The stationary Rossby wave related to the CGT propagates southeastward from the exit of the North Atlantic upper-tropospheric westerly jet via western and eastern Europe toward southwest

CGTI with the SCE rainfall (Figure 1(b)). The result implies that SCE summer rainfall can trigger the CGT.

In addition to the weaker model response, a different H200 pattern over Europe exists between

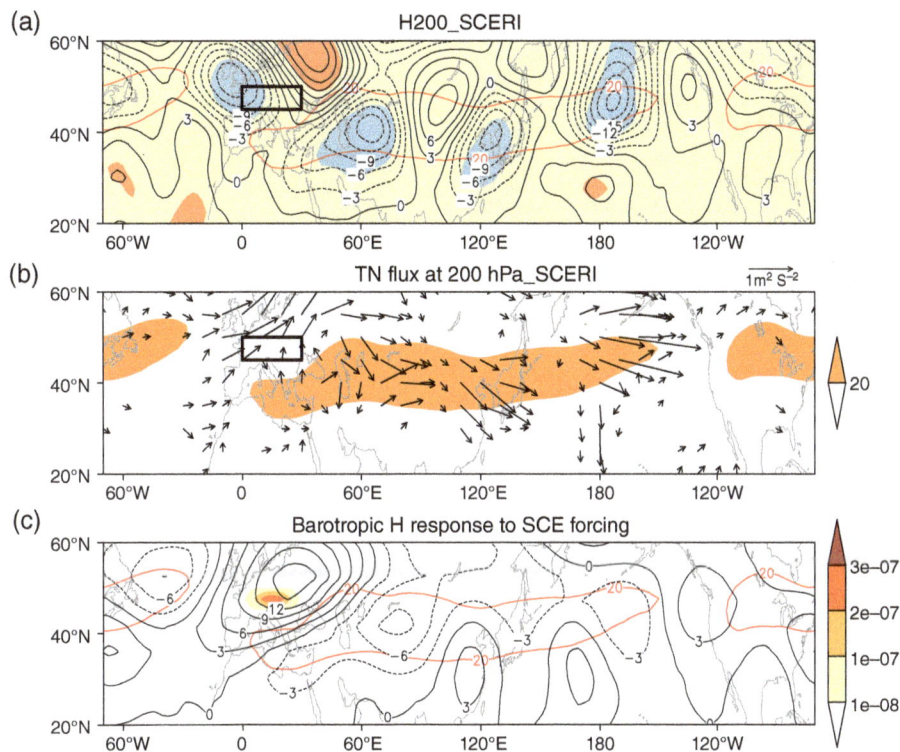

Figure 3. (a) Anomalies of H200 regressed against the SCERI and (b) the associated Takaya and Nakamura (TN) flux (vector). (c) Barotropic geopotential height responses (contour) to divergence forcing (shading) over the SCE region ($45°-50°N$, $0°-30°E$) under 200-hPa climatological summer-mean winds. The geopotential height responses in mid-latitudes are calculated from stream function responses based on the geostrophic balance. Shading depicts significance at the 95% confidence level in (a) and the divergence forcing with a maximum value of the half-period sinusoidal divergence forcing of $3.0 \times 10^{-7} s^{-1}$ in (c). The contour interval is 3 gpm in (a) and (c). The red contour in (a) and (c) and the shading in (b) depict the climatological westerly jet at 200 hPa with zonal winds exceeding $20 m s^{-1}$.

Asia along a great-circle route (Figure 1(a)). The eastward and northward moisture transport associated with the cyclonic anomaly over western Europe may contribute to enhanced rainfall over SCE (Figure 2(a)). The interaction between the SCE rainfall and the CGT leads to their significant linkage.

This study proposes that the SCE summer rainfall induces a CGT-like wave train (Figure 3), which is different from previous studies that highlighted the important role of summer diabatic heating over the eastern MS region (Enomoto *et al.*, 2003; Yasui and Watanabe, 2010). The fact that rainfall is much stronger over SCE than over the MS in the interannual variability suggests a more important contribution of the summer rainfall over SCE than over the MS to the interannual variation of the CGT. Their relative contribution is also indicated by the CGTI-related summer rainfall anomalies (Figure 2(a)), with a significant relationship of the CGT with the summer rainfall in SCE but not with that in the MS land region. The correlation coefficient of -0.43 between the CGTI and the SCERI suggests that approximately 20% of the variance of the CGT is associated with the SCE summer rainfall.

Acknowledgements

The authors thank the two reviewers and the editor for their valuable comments that improved the manuscript. This research was supported by the National Natural Science Foundation of China (grant nos. 41375086 and 41320104007).

References

Chen G, Huang R. 2012. Excitation mechanisms of the teleconnection patterns affecting the July precipitation in Northwest China. *Journal of Climate* **25**: 7834–7851, doi: 10.1175/jcli-d-11-00684.1.

Chen G, Huang R, Zhou L. 2013. Baroclinic instability of the silk road pattern induced by thermal damping. *Journal of the Atmospheric Sciences* **70**: 2875–2893, doi: 10.1175/jas-d-12-0326.1.

Ding Q, Wang B. 2005. Circumglobal teleconnection in the Northern Hemisphere summer. *Journal of Climate* **18**: 3483–3505.

Ding Q, Wang B, Wallace JM, Branstator G. 2011. Tropical-extratropical teleconnections in boreal summer: observed interannual variability. *Journal of Climate* **24**: 1878–1896, doi: 10.1175/2011jcli3621.1.

Enomoto T, Hoskins BJ, Matsuda Y. 2003. The formation mechanism of the Bonin high in August. *Quarterly Journal of the Royal Meteorological Society* **129**: 157–178, doi: 10.1256/qj.01.211.

Hall NMJ, Douville H, Li L. 2013. Extratropical summertime response to tropical interannual variability in an idealized GCM. *Journal of Climate* **26**: 7060–7079, doi: 10.1175/jcli-d-12-00461.1.

Hoskins BJ, Ambrizzi T. 1993. Rossby wave propagation on a realistic longitudinally varying flow. *Journal of the Atmospheric Sciences* **50**: 1661–1671.

Hoskins BJ, Karoly DJ. 1981. The steady linear response of a spherical atmosphere to thermal and orographic forcing. *Journal of the Atmospheric Sciences* **38**: 1179–1196.

Huang G, Liu Y, Huang R. 2011. The interannual variability of summer rainfall in the arid and semiarid regions of Northern China and its association with the Northern Hemisphere circumglobal teleconnection. *Advances in Atmospheric Sciences* **28**: 257–268, doi: 10.1007/s00376-010-9225-x.

Kosaka Y, Chowdary JS, Xie S-P, Min Y-M, Lee J-Y. 2012. Limitations of seasonal predictability for summer climate over East Asia and the Northwestern Pacific. *Journal of Climate* **25**: 7574–7589, doi: 10.1175/jcli-d-12-00009.1.

Lee JY, Wang B, Seo KH, Kug J-S, Choi Y-S, Kosaka Y, Ha KJ. 2014. Future change of Northern Hemisphere summer tropical–extratropical teleconnection in CMIP5 models. *Journal of Climate* **27**: 3643–3664.

Lin H. 2009. Global extratropical response to diabatic heating variability of the Asian summer monsoon. *Journal of the Atmospheric Sciences* **66**: 2697–2713, doi: 10.1175/2009jas3008.1.

Lin Z. 2014. Intercomparison of impacts of four summer teleconnections over Eurasia on East Asian rainfall. *Advances in Atmospheric Sciences* **31**: 1366–1376, doi: 10.1007/s00376-014-3171-y.

Lu R-Y, Oh J-H, Kim B-J. 2002. A teleconnection pattern in upper-level meridional wind over the North African and Eurasian continent in summer. *Tellus A* **54**: 44–55.

Mitchell TD, Jones PD. 2005. An improved method of constructing a database of monthly climate observations and associated high-resolution grids. *International Journal of Climatology* **25**: 693–712, doi: 10.1002/joc.1181.

Rodwell MJ, Hoskins BJ. 1996. Monsoons and the dynamics of deserts. *Quarterly Journal of the Royal Meteorological Society* **122**: 1385–1404.

Saeed S, Müller W, Hagemann S, Jacob D. 2011. Circumglobal wave train and the summer monsoon over northwestern India and Pakistan: the explicit role of the surface heat low. *Climate Dynamics* **37**: 1045–1060, doi: 10.1007/s00382-010-0888-x.

Schubert S, Wang H, Suarez M. 2011. Warm season subseasonal variability and climate extremes in the Northern Hemisphere: the role of stationary Rossby waves. *Journal of Climate* **24**: 4773–4792.

Takaya K, Nakamura H. 2001. A formulation of a phase-independent wave-activity flux for stationary and migratory quasigeostrophic eddies on a zonally varying basic flow. *Journal of the Atmospheric Sciences* **58**: 608–627.

Uppala SM, Kållberg P, Simmons A, Andrae U, Bechtold V, Fiorino M, Gibson J, Haseler J, Hernandez A, Kelly G. 2005. The ERA-40 re-analysis. *Quarterly Journal of the Royal Meteorological Society* **131**: 2961–3012.

Wang W, Zhou W, Wang X, Fong SK, Leong KC. 2013. Summer high temperature extremes in Southeast China associated with the East Asian jet stream and circumglobal teleconnection. *Journal of Geophysical Research [Atmospheres]* **118**: 8306–8319, doi: 10.1002/jgrd.50633.

Watanabe M. 2004. Asian jet waveguide and a downstream extension of the North Atlantic Oscillation. *Journal of Climate* **17**: 4674–4691.

Xie PP, Arkin PA. 1997. Global precipitation: a 17-year monthly analysis based on gauge observations, satellite estimates, and numerical model outputs. *Bulletin of the American Meteorological Society* **78**: 2539–2558.

Yasui S, Watanabe M. 2010. Forcing processes of the summertime circumglobal teleconnection pattern in a dry AGCM. *Journal of Climate* **23**: 2093–2114, doi: 10.1175/2009jcli3323.1.

Yim S-Y, Wang B, Liu J, Wu Z. 2014. A comparison of regional monsoon variability using monsoon indices. *Climate Dynamics* **43**: 1423–1437, doi: 10.1007/s00382-013-1956-9.

Zhang L, Zhou T. 2015. Drought over East Asia: a review. *Journal of Climate* **28**: 3375–3399, doi: 10.1175/jcli-d-14-00259.1.

Zuo J, Li W, Sun C, Xu L, Ren H-L. 2013. Impact of the North Atlantic sea surface temperature tripole on the East Asian summer monsoon. *Advances in Atmospheric Sciences* **30**: 1173–1186, doi: 10.1007/s00376-012-2125-5.

10

Empirical and modeling analyses of the circulation influences on California precipitation deficits

Yen-Heng Lin,[1]* Lawrence E. Hipps,[1] S.-Y. Simon Wang[1,2] and Jin-Ho Yoon[3]

[1]*Department of Plants, Soils and Climate, Utah State University, Logan, UT, USA*
[2]*Utah Climate Center, Utah State University, Logan, UT, USA*
[3]*School of Earth Sciences and Environmental Engineering, Gwangju Institute of Science and Technology, South Korea*

Correspondence to:
Y.-H. Lin, Department of Plants,
Soils and Climate, Utah State
University, 4820 Old Main Hill,
Logan, UT, USA.
E-mail: yheng@pie.com.tw

Abstract

Amplified and persistent ridges in western North America are recurring features associated with drought conditions in California. The recent drought event (2012–2016) lasted through both La Niña and El Niño episodes, suggesting additional climate drivers are important in addition to the commonly perceived El Niño-Southern Oscillation. Diagnostic analyses presented here suggest that, while the Pacific North American (PNA) and North Pacific Oscillation (NPO) do not directly cause drought in California, the relationships between them and with the upper air circulation pattern do modulate the spatial drought pattern. The positive PNA relative circulation leads drier northern California, and (-NPO) relative circulation leads southern California to be drier. The types of drought in this region emerge mostly from the combination of two PNA and NPO relative oceanic and atmospheric oscillations. At present, climate model projections do not indicate any significant change in these particular drought-modulating processes.

Keywords: California drought; circulation type; drought variation

1. Introduction

During the winters from 2011–2012 to 2014–2015, a persistent upper tropospheric ridge developed over the Northeastern Pacific, and this anomalous ridge prohibited much of the rain-producing weather disturbances from reaching California. The reduction in the rainy-season precipitation and warmer temperature led to a major drought with declined snowpack and subsequently less water during the dry seasons. While the occurrence of drought is not uncommon in California (Department of Water Resources, 1978, 1993, 2015), the fact that this recent drought episode has lasted four consecutive years was unprecedented in a 1200-year reconstructed history (Robeson, 2015).

Various climate modes impact the winter precipitation in California, such as the North Pacific Oscillation (NPO) or West Pacific (Linkin and Nigam, 2008), the Pacific North American (PNA; Renwick and Wallace, 1996), and the El Niño-Southern Oscillation (ENSO) patterns. Even though California droughts are closely associated with an amplified and stagnant ridge over the western United States, the formation mechanism of the ridge itself is elusive (Wang *et al.*, 2014), and the interpretation of the causes of historical droughts in California has been inconsistent. The notable 1976–1977 California drought winter was reported to associate with the El Niño (Namias, 1978). However, this was contradicted by the argument of Seager *et al.* (2015) that the 2012 California drought was initiated by the 2011 La Niña. Apart from the classic view that ENSO and the

Pacific Decadal Oscillation collectively contribute to California's dry winters (McCabe and Dettinger, 1999; Kam *et al.*, 2014), recent studies that focused on the post-2012 drought identified other unique atmospheric and oceanic features. For example, Wang and Schubert (2014) suggested that the 2013 sea surface temperature (SST) anomalies in the North Pacific produced a predilection for the ensuing California drought. Swain *et al.* (2014) pointed out the record-setting ridge in the upper troposphere as the cause of drought, while Wang *et al.* (2014) reported the associated geopotential height dipole (with a trough counterpart over the Great Lakes) was linked to an ENSO precursor. Subsequent studies (Hartmann, 2015; Lee *et al.*, 2015) also linked the 2013–2014 SST pattern to the North Pacific Mode (Deser and Blackmon, 1995) with an amplitude modulation from reduced sea ice content in the Arctic.

How do we reconcile these different observations and interpretations about which climatic features and variability influence drought conditions in California? In addressing this question, we shifted our attention on the atmospheric circulation and SST settings that affect *the pattern of drought*. The hypothesis addressed is that there are different dynamical processes that govern events leading to several distinct patterns of drought. This differs from previous studies searching for simply the causes of drought, implying that all droughts are the same. Additional understanding of the different types of atmospheric patterns associated with patterns of dry episodes could help society anticipate or mitigate the next drought.

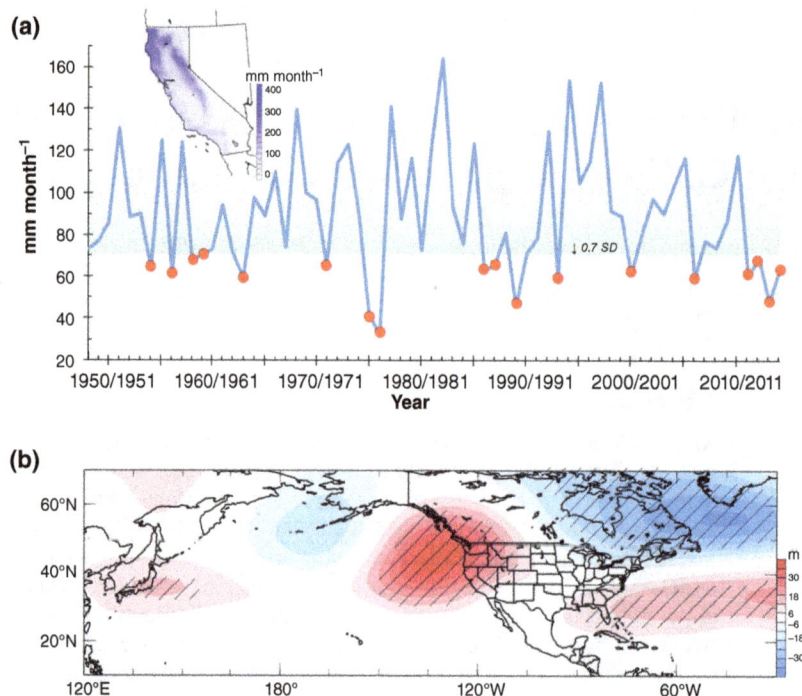

Figure 1. (a) Winter season (November–March) precipitation time series over California; the inset map shows the domain and winter mean precipitation in California. The red dots indicate the 18 dry winters in which precipitation was <0.7 standard deviation below average as shaded. (b) The composite anomaly of the 250-hPa geopotential height of the 18 drought winters. Hatches indicate significant level at $p < 0.05$ for the anomaly.

2. Data

To depict the winter season from November to March precipitation deficit over California, we utilized the parameter-elevation regressions on independent slopes model (PRISM) (Daly *et al.*, 2008) precipitation at 4 km horizontal resolution developed by Oregon State University (http://prism.oregonstate.edu). For the atmospheric circulation, we analyzed the monthly National Centers for Environmental Prediction (NCEP)/National Center for Atmospheric Research (NCAR) reanalysis data with a spatial resolution of 2.5° × 2.5° from 1948 to 2015 (Kalnay *et al.*, 1996). To explore the SST pattern, we analyzed the monthly NOAA extended reconstructed SST version 3 with a 2° × 2° spatial resolution (Smith *et al.*, 2008). For the purpose of documenting the drought connection to climate patterns, we also examined existing climate indices archived by the National Oceanic and Atmospheric Administration (NOAA) earth system research laboratory: http://www.esrl.noaa.gov/psd/data/climateindices/list/. Owing to the maximum extent of NCEP datasets, the analysis period of observational data is from 1948 to 2015. To reveal the drought distribution, the PRISM-derived Palmer Drought Severity Index (PDSI) (Palmer, 1965) data with 4 km horizontal resolution were also utilized. The PDSI is constructed by taking into account water supply, water demand, and soil moisture information influenced by surface temperature, and is used commonly to show the global and regional drought features (Heim, 2002; Dai *et al.*, 2004).

To further examine the variations of climate variability under external climate forcing and understand long-term changes, the historical and future simulations of Community Earth System Model version 1 (CESM1) (Hurrell *et al.*, 2013) with Community Atmosphere Model version 5.0 (CAM5) for the atmospheric component were used. The model version and setting follow Yoon *et al.* (2015a). Thirty ensemble members of CESM1 with 0.9° × 1.25° horizontal resolution through the large-ensemble project (Kay *et al.*, 2015) were utilized. The historical forcing scenario (HIS run), including greenhouse gases, aerosols, volcanic activity, ozone, land use change, and solar, covers 1920–2005 period, and the future Representative Concentration Pathway (RCP) 8.5 forcing scenario (RCP run) that represents a high-emission pathway (Taylor *et al.*, 2012) covers 2006–2080 period. The climate indices, such as PNA and NPO, are computed by correlating the 250-hPa geopotential height onto corresponding PNA (0°E–360°E, 20°N–90°N) and NPO (165°E–90°W, 10°N–70°N) loading patterns during the winter season in each ensemble. The loading patterns are generated by regressing NCEP's 250-hPa geopotential height onto observational PNA and NPO indices. The ensemble spread of initial conditions is generated by the 'round-off differences' method (Kay *et al.*, 2015).

3. Results

To understand the circulation variations in drought years in California, we first identify precipitation deficit

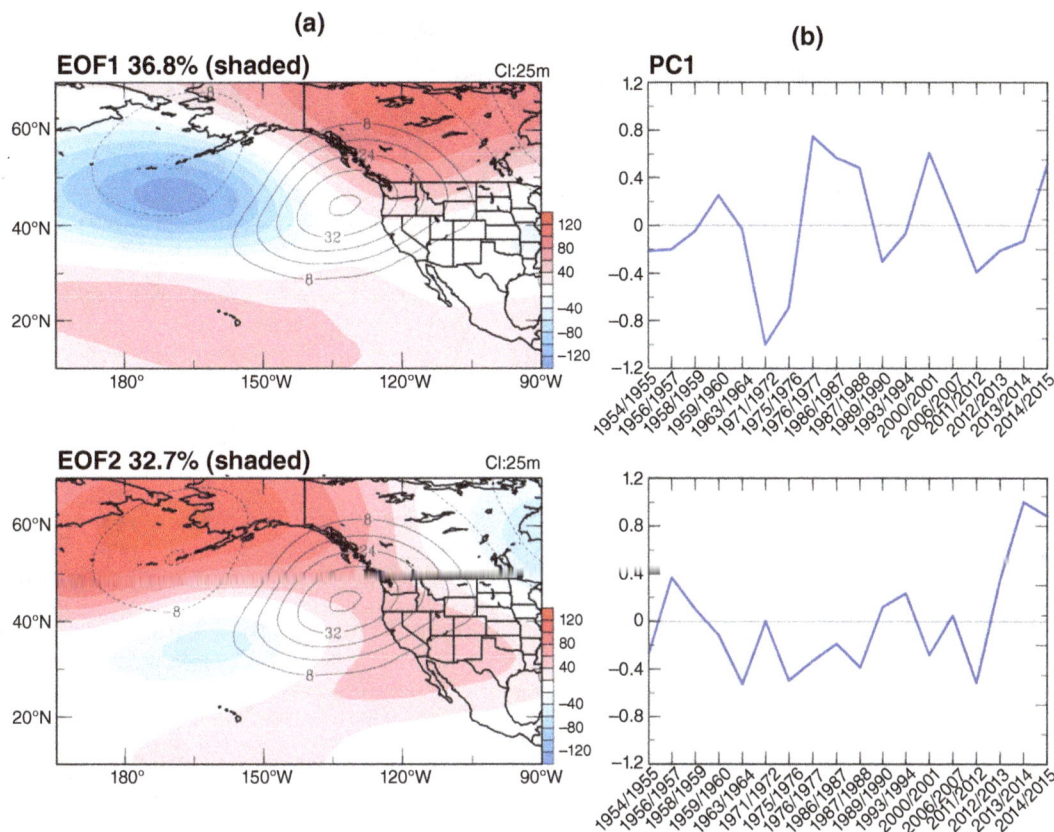

Figure 2. (a) The first two leading EOFs (shaded) of winter season (November–March) Z_{250} within the 18 dry winters, superimposed with their composite Z_{250} anomalies (contour). (b) The corresponding PCs in relation to each of the 18 dry winters.

events. Based upon the winter (November–March) precipitation in California (P_{CA}), which is shown in Figure 1(a) from 1948–1949 to 2014–2015, we defined the occurrence of drought to be when P_{CA} exceeded 0.7 standard deviation below the 57 years winter mean, a threshold to balance water deficit intensity with a sufficient number of cases. This definition of drought led to the inclusion of recent severe droughts of 1976–1977, 2011–2012, and 2014–2015, as well as other major low-precipitation years, isolating 18 drought events as indicated by red dots. Figure 1(b) shows the composite 250-hPa geopotential height anomalies of these 18 drought winters (as a departure from the 1948–2015 mean), depicting a prominent high-pressure anomaly centered off Northwestern United States and western Canada, but covering much of the western United States including California. Such a ridge will prohibit the occurrence of winter storms that produce rainfall in California (Swain *et al.*, 2014). A discernible yet weak wave train emerges in the upstream region over the North Pacific. Given its upstream source near the western North Pacific, this wave train appears to be different from the ENSO-induced teleconnection forced by the central/eastern tropical Pacific heating anomalies (Schonher and Nicholson, 1989). Another notable feature in the downstream side over northeastern North America is a robust anomalous trough that, together with the western ridge, forms the so-called North American dipole (Wang *et al.*, 2015).

To analyze the extent to which these 18 droughts differ case-to-case, we first applied the empirical orthogonal function (EOF) to depict the variation of the November–March 250-hPa geopotential height (Z_{250}) within these 18 drought events. The analysis here was focused on the region encompassing the composite ridge anomaly (165°E–90°W, 10°–70°N). The loading patterns of the first two leading modes (EOF 1 and 2) are shown in Figure 2(a) as shadings, which are superimposed on composite Z_{250} contours for comparison. These two EOFs constitute collectively about 70% of the total variance, meaning that the first two leading modes are the major circulation variations. It is noteworthy that they are also about the same fraction of variance, and thus importance. EOF1 depicts a northeastern extension of the anomalous ridge in the positive phase (according to the principal component or PC; Figure 2(b)) and a westward extension over the Gulf of Alaska in the negative phase. In EOF2, the anomalous ridge would expand mostly toward the northwest of the Bering Sea and into the southwestern United States as well. This pattern has shown prominence during the recent (2013–2014) drought as noted by Wang *et al.* (2014). Similar results are revealed in the upper troposphere at 200 and 500 hPa, as shown in Figure S1. These results are empirical evidence of the existence of two distinct climate circulation schemes that affect the pattern of drought in this region.

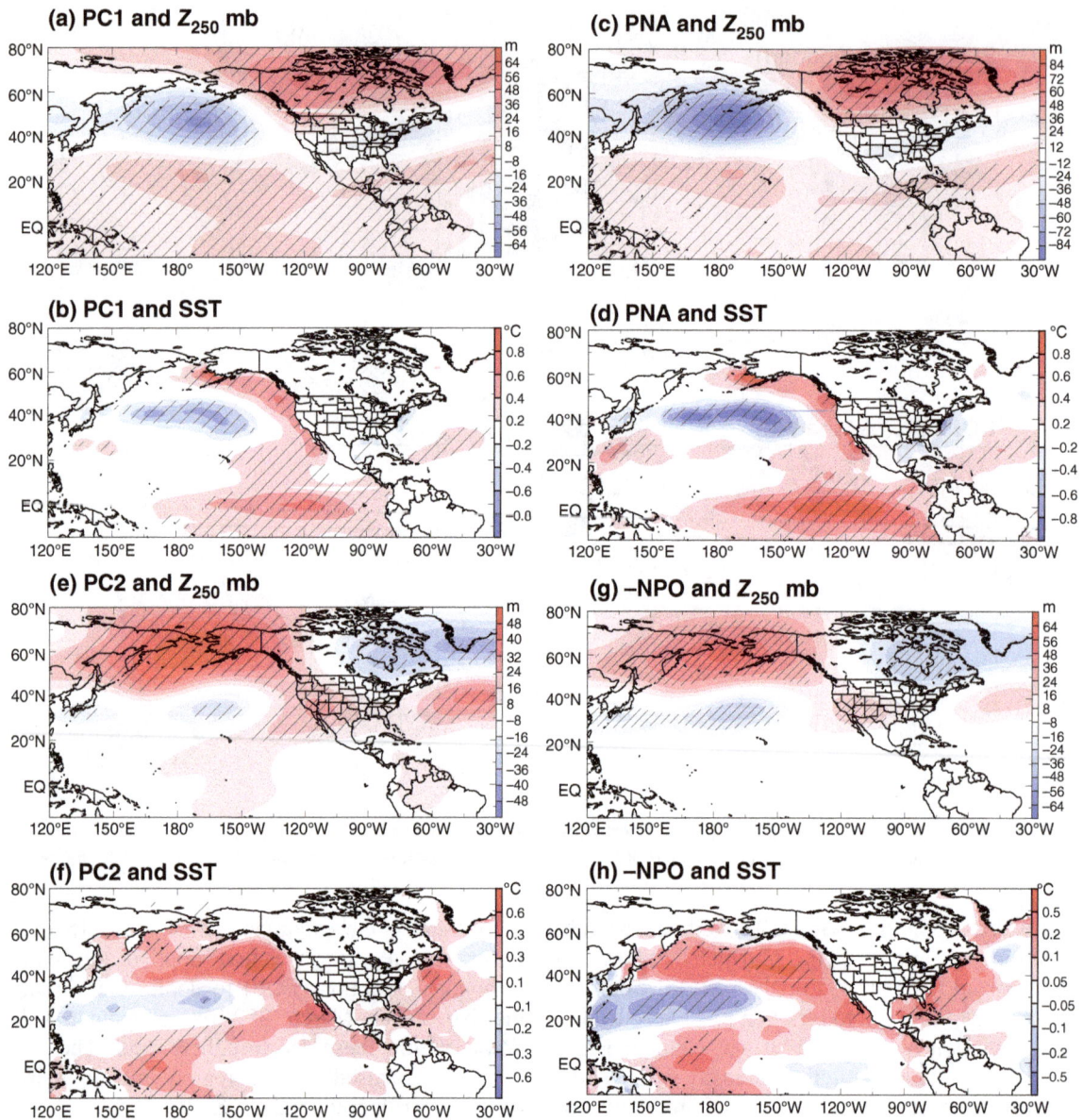

Figure 3. The Z_{250} and SST patterns from the 18 California dry winters regressed upon PC1 (a and b), PNA index (c and d), PC2 (e and f), and −NPO index (g and h). Hatches indicate significant level at $p < 0.05$ for the regression.

The intensity of these two EOFs was proportional to the amplitude of the PC values. Therefore, we correlated the PC of 18 drought winters with different climate indices of the same 18 drought winters; these correlation coefficients are summarized in Table S1, Supporting information. The results illustrate that PC1 has a high positive correlation ($r = 0.90$) with the PNA index (Barnston and Livezey, 1987), while PC2 has a high negative correlation ($r = -0.83$) with the NPO. The NPO is a leading atmospheric variation mode determined as the second PC of the November–March 1000-hPa height anomalies over North Pacific (Rogers, 1981). Similarly, the linear regression patterns of Z_{250} with PC1 (Figure 3(a)) and PNA (Figure 3(c)) for the 18 dry winters appear to be similar, while the corresponding SST regressions (Figures 3(b) and (d)) reveal an ENSO-like pattern. This is not surprising since ENSO is the prime forcing of the PNA teleconnection (Yu

and Zwiers, 2007), even though ENSO does not connect as prominently to the PC series of Z_{250} (Table S1). Meanwhile, the height and SST regression patterns with PC2 (Figure 3(e)) and −NPO (Figure 3(g)) both present a distinct high-latitude seesaw from the Tropics to the Bering Sea and its associated 'triband' SST anomalies (Figures 3(f) and (h)) as noted by Linkin and Nigam (2008). We note that the NPO's associated SST pattern (Figure 3(h)) is also similar to the 'North Pacific Mode' of SST identified by Deser and Blackmon (1995), which depicts the SST counterpart of the sea level pressure-based NPO. Given that PNA and NPO represent atmospheric modes, there may be intra-seasonal variability that is overlooked in this seasonal mean analysis.

The results shown in Figure 3 suggest that different climate forcing sources may influence the distribution and intensity of droughts by modulating

(a) PC1 and PR (18 dry winters)

(b) PC2 and PR (18 dry winters)

(c) PNA and PR (18 dry winters)

(d) −NPO and PR (18 dry winters)

**(e) PNA and PR
(1948/1949−2014/2015)**

**(f) −NPO and PR
(1948/1949−2014/2015)**

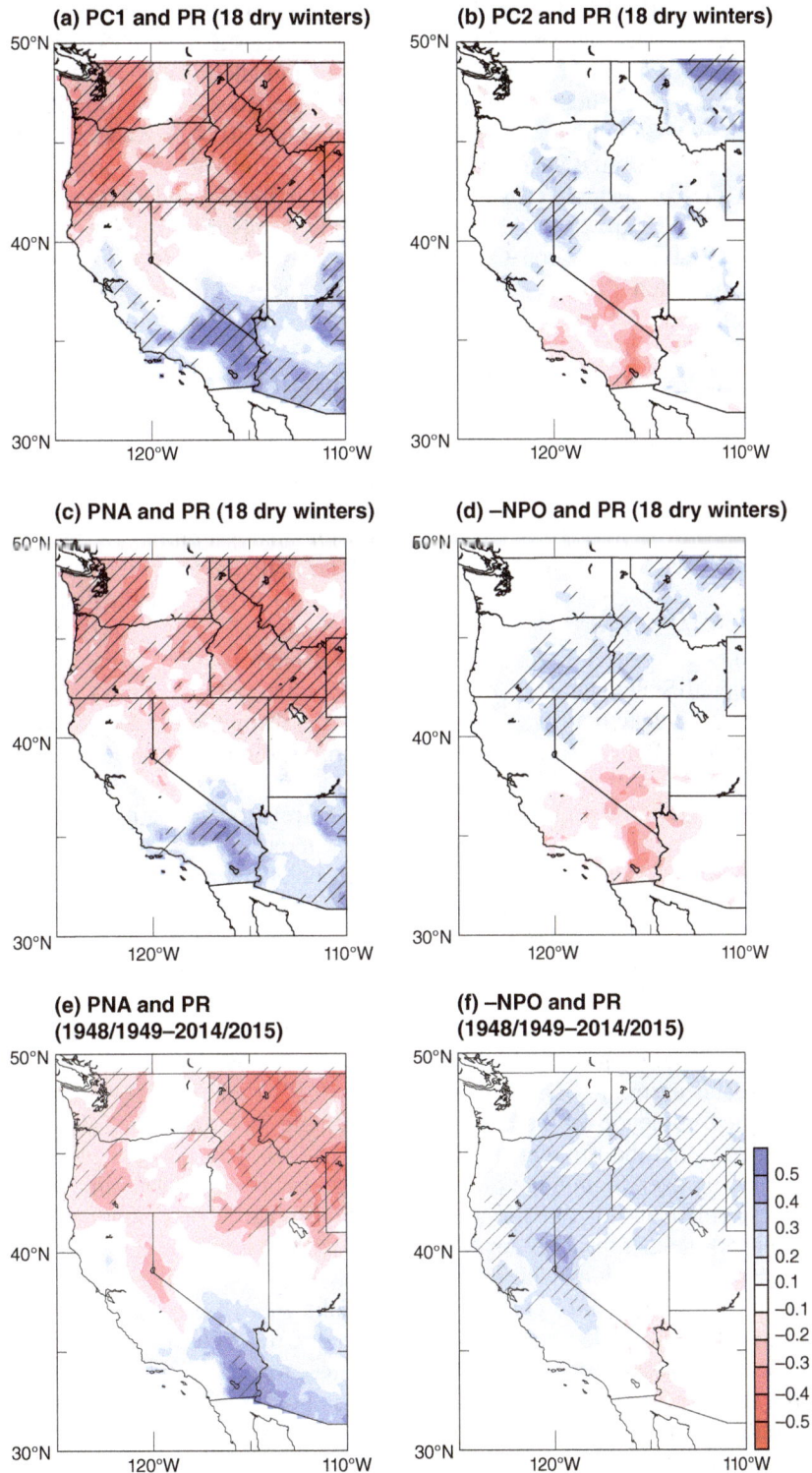

Figure 4. The influence ratio from the regression of 18 drought years precipitation onto (a) PC1, (b) PC2, (c) PNA, and (d) −NPO decided by the mean precipitation anomaly during drought years. (e) and (f) are the same as (c) and (d) but for 1948/1949−2014/2015 wintertime precipitation.

the drought-inducing ridge. To examine these tele-connection impacts on the drought pattern, we show in Figures 4(a) and (c) the association of PC1 and PNA with precipitation in the drought winters. Here, the values represent influence ratio on precipitation that is calculated from the regressions of precipi-tation onto each normalized index divided by the

mean precipitation anomaly within the 18 dry winters, ranging between −1 and 1. Positive cases of PC1 and PNA are associated with drier conditions in the Pacific Northwest, leaking into northern California, and less dry conditions in southern California and western/southwest coast of California. This shift of drought pattern is caused by the anomalous ridge being extended further northeastward (EOF1; Figure 2(a)). For PC2 and the −NPO, impacts on precipitation (Figures 4(b) and (d)), southern California and some of the Southwest United States experience more severe drought conditions, and correspondingly northern California and some of Northwest United States exhibit less intense drought with the southward extension of the high-pressure ridge (EOF2; Figure 2(a)). Although the influence fractions of PNA and NPO on California drought winters' precipitation do not show big differences (about 10–20%) in Sierra Nevada, the connections are larger in the coastal and agriculture intensive (valley) regions (Figures 4(b) and (d)). Moreover, it is instructive to consider that drought may not be defined only by precipitation. The regression results of PDSI in drought winter onto PC1/PNA and PC2/(−NPO) (Figures S3(a)–(d)) show the PNA and NPO are related to a measure of drought intensity, especially in southern California, with changes of over 1 point of PDSI value.

It appears that both the PNA and NPO, in addition to ENSO, modulate the drought pattern in California, as well as the western United States, but they do not directly cause the drought (i.e. due to their weak direct relationship with precipitation in California). This latter point is demonstrated in Figures 4(e) and (f) showing the influence ratio on precipitation by regressing wintertime precipitation onto normalized PNA and NPO indices during the entire 1948–2015 period divided by the winter mean precipitation. It is clear that the patterns for the entire period (Figures 4(e) and (f)) resemble in general terms those of the 18 drought years (Figures 4(c) and (d)). The record low snowpack in the Sierra Nevada in 2013–2014 that was identified as the lowest in the past 500 years (Belmecheri et al., 2016) coincided with a −PNA and a strong −NPO phases (Figure S2). This combined influence on the extreme low snowpack further intensified the drought making it harder to recover. However, even though the precipitation pattern bears resemblance with the variations within the 18 drought winters, neither the PNA nor NPO shows statistically significant correlations with precipitation within California. A similar lack of correlation was also found between the PNA/NPO and the winter PDSI (Figure S3). Altogether, these results suggest that the PNA and NPO play more of a modulating role of drought in California rather than causing it.

Further validation was carried out by analyzing the CESM1 large-ensemble simulations of HIS and RCP runs. By applying the same analysis as in the observational data, the simulated geopotential height anomalies in each of the 30 members reveal similar PNA and NPO features, both in the leading EOFs and regression patterns. In Figure S4, we show the ensemble mean of PNA-like and NPO-like EOFs of each scenario. The results show that the HIS and RCP runs have similar variances in the first two leading modes, suggesting that the CESM1 simulations agree that PNA and NPO are key circulation features that modulate the drought pattern in California. Figures S5 and S6 show

the averaged regression patterns of each ensemble's PC onto CESM1's Z_{250} and SST, suggesting that the influence of modeled PNA (NPO) on California droughts slightly increase (decrease) in RCP run. Recent studies (Zhou et al., 2014; Yoon et al., 2015a) have indicated that the anthropogenic warming would change North Pacific circulation and, in turn, would influence climate conditions in North America. The result from CESM1's RCP run suggests that this influence would be realized through the modulations of PNA and NPO.

4. Discussions

The PNA and NPO relative circulations are associated with the spatial distribution of precipitation during drought in California. Meanwhile, when applying the EOF analysis on the 18 drought winters' standardized normalized precipitation over California region (Figures S7(a) and (b)), the first two leading modes also have about 70% of total variance, which is similar with the EOF analysis of Z_{250} hPa. However, the first mode of EOF for precipitation shows no distinct spatial pattern, while the second mode does show a strong north–south pattern. The correlation patterns of Z_{250} hPa onto two leading PCs for precipitation do not show any significant correlation coefficients (at $p < 0.05$ level) (Figure S7(c)). It means these two orthonormal eigenvectors are not associated with a specific circulation pattern.

If one looks more closely, the first mode of EOF for the 18-drought winters precipitation is associated with the first and second EOFs of Z_{250} hPa being opposite in sign (Figure 2). Recall the value of the first mode of Z_{250} hPa is associated with the PNA, and the value of the second is associated with −NPO. Since the regressions of PNA and −NPO with precipitation have opposite dipole patterns near California (Figures 4(c) and (d)), the result is that the two regressions effectively counter each other and no dipole pattern is observed. So the first mode of precipitation is associated with PNA and −NPO having the same sign, it displays no spatial pattern. In contrast, the second leading mode of EOF for California 18-drought winters precipitation appears to result from a constructive effect of PNA and −NPO (Figures 4(c) and (d)) to enhance the dipole pattern. This is because, in this case the PC1 and PC2 of the Z_{250} hPa, associated with PNA and −NPO, have opposite signs, so the respective regressions with precipitation have similar spatial patterns, that re-enforce each other. Therefore, it is the combination of the signs of the circulation EOFs, related to those of the PNA and −NPO that relate to spatial patterns of drought year precipitation, and several types of droughts in California.

The forcing sources of the PNA are manifold, and previous studies have indicated that the intensity of the PNA is associated with the eastern tropical Pacific SST (Straus and Shukla, 2002; Yu and Zwiers, 2007) and the East Asian Jet Stream (Wallace and Gutzler, 1981; Leathers and Palecki, 1992). This explains the second highest correlation coefficient of the Niño indices with

(a) Frequency distribution

(b) Drought years correlation

(c) P_CA and PNA

(d) P_CA and −NPO

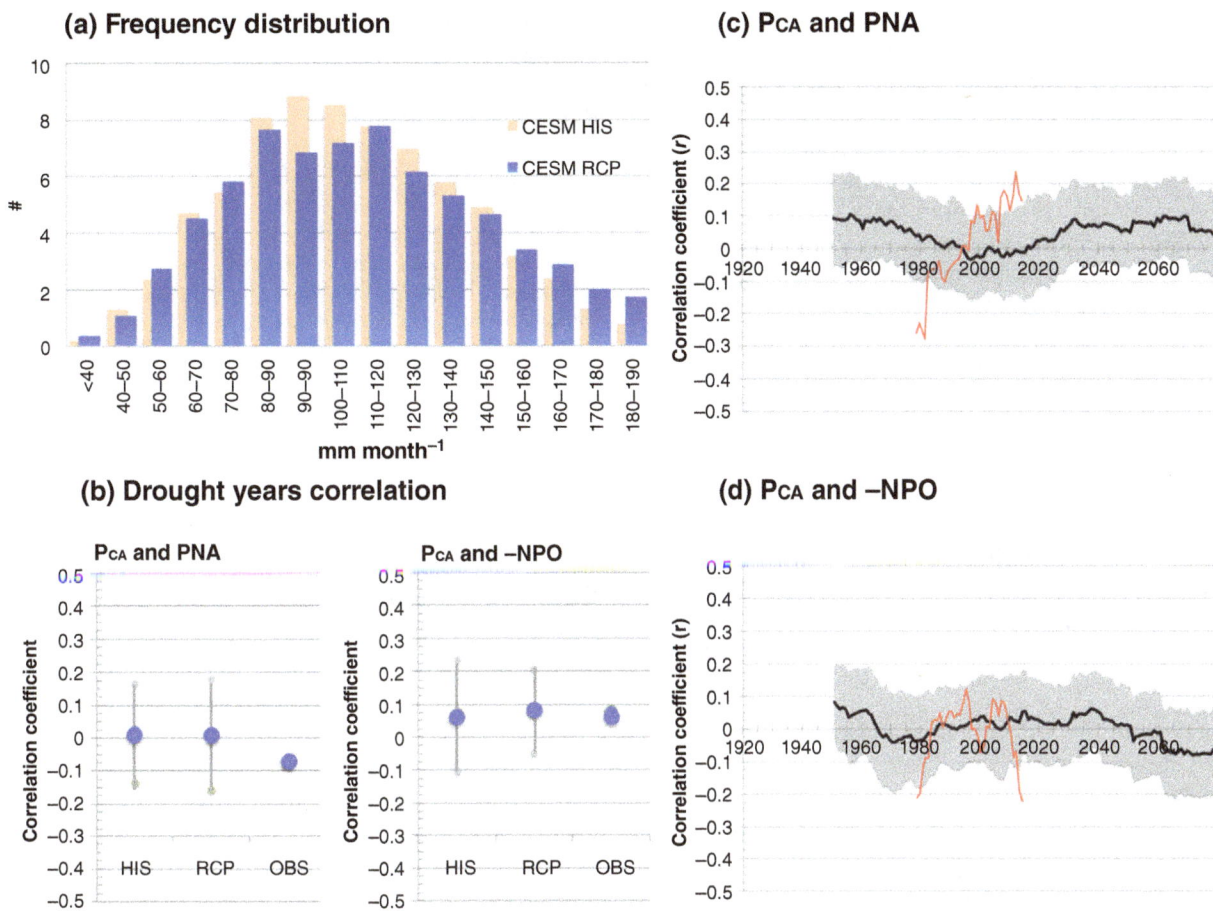

Figure 5. The precipitation changes in California simulated by CESM1 with 30 ensembles. (a) The ensemble mean of frequency distribution of November–March precipitation in California (P_{CA}). The historical scenario (HIS run) includes 74 years before 2005, and the RCP 8.5 scenario (RCP run) includes 74 years after 2005. (b) The correlation coefficient between P_{CA} and PNA (left) and between P_{CA} and −NPO (right) within drought years on ensembles' HIS/RCP runs and observational data (OBS). The blue circles show the correlation coefficient of ensemble mean or observation, and the gray bar indicates 50% of ensemble spread. (c) The 30-year window sliding correlation between PNA and P_{CA} over simulation period. The black solid line is the ensemble mean, the shaded areas indicate 50% of ensemble spread, and the red line is the observational data. (d) The same as (c) but for P_{CA} and −NPO.

PC1 as shown in Table S1. The NPO's role in the modulation of the California drought does connect to ENSO, since the NPO acts as an ENSO precursor (i.e. no direct correlation) through interactions with tropical SST and wind anomalies across the equatorial Pacific. The NPO's role in triggering ENSO occurs under the so-called 'Seasonal Footprinting Mechanism', from which the NPO imparts a surface wind stress to change the surface heat fluxes and underlying SST (Vimont *et al.*, 2003; Alexander *et al.*, 2010). These features supplement the common perception that ENSO and its different phases are responsible for California drought. Recent studies (Yoon *et al.*, 2015a, 2015b) projected that both intense drought and excessive flooding in California may increase by 50% toward the end of the 21st century, and this projection is based upon a strengthened relation to the ENSO cycle, not only through its warm and cold phases but also through its precursor (transition) patterns.

The long-term precipitation regression results with PNA and NPO indices from 1948–1949 to 2014–2015 show much less significance than the relationship of variations within California (Figures 4(e) and (f)). For example, the winters of 1975–1976 and 1976–1977 were associated with a distinct opposite phase of the PNA pattern and El Niño SST anomalies, while in 2011–2012 a La Niña SST pattern prevailed. The recent record droughts in 2013–2014 and 2014–2015 were associated with an amplified NPO (Figure 2(b)) without the presence of a mature-phase El Niño. The lack of association between the California drought and the PNA and the NPO is evident in Figure S2 indicating a mixture of phases in either index during the 18 dry winters. These reported features and the association with the NPO make simulation and/or prediction for California's winter climate difficult, since the majority of models do not simulate the ENSO precursors (i.e. the NPO) so well (Wang *et al.*, 2015; Yoon *et al.*, 2015a). Further research is needed in identifying the source of variability and predictability in the drought-producing ridge off the Northwest United States region as revealed in Figure 1(b).

Analysis of the CESM1 large-ensemble simulations with historical and RCP forcing scenarios is supportive of the respective roles of PNA and NPO in modulating the drought pattern. This finding led us to question the extent to which the relations between PNA/NPO and California's precipitation may change in the future climate. To answer this question, Figure 5(a) shows the frequency distribution of California winter precipitation superimposed with the CESM1's simulations. While the HIS run presents a normal distribution, the RCP run shifts the wet tail substantially and the dry tail slightly (far left). These results are consistent with the finding of the increased water cycle extremes in California projected by Yoon et al. (2015a), although the annual precipitation may not change with human-induced climate change (Pierce et al., 2013). Furthermore, the correlation coefficients of PNA/NPO and California precipitation within the drought winters (Figure 5(b)) and the sliding correlation coefficients of PNA/NPO and California precipitation with a 30-year window over all simulation period (Figures 5(c) and (d)) both show that, despite of the projected change in the frequency distribution of California precipitation, its association with either PNA or NPO remains very weakly and insignificantly correlated.

5. Conclusions

The upper-atmospheric high-pressure (ridge) anomaly that accompanies drought in California exhibits several patterns, which modulates the spatial distribution of precipitation during drought in this region. These findings suggest that the variations of the ridge are collectively *modulated* – but *not directly caused* – by a combination of geopotential heights, and the synchronization of the signs of the PNA and NPO. Neither the PNA nor the NPO appears to directly contribute to the formation of drought, at least for the winter season. Rather, they alter the pattern of drought. There are several different combinations of forcing factors (Z_{250} hPa, PNA, and NPO) that are associated with drought in California.

The analysis of CESM1 simulations indicates that these modulations of PNA and NPO will not change significantly in the future, although there is a projected increase in the extreme wet/dry anomalies in California.

In terms of future research, a couple of unsolved questions are worth pursuing: (1) investigating whether the appearance of drought-producing ridge anomaly is actually forced or caused by any prominent mode of climate variability, or is purely due to random changes in atmospheric states, and (2) examining the impacts of constructive and destructive superposition of the PNA and NPO and how well they can be simulated in seasonal predictions. Lastly, even though this study is focused on precipitation, the effect of temperature on exacerbating drought cannot be discounted; this temperature effect was recently demonstrated using paleoclimate record for the California drought (Griffin and Anchukaitis, 2014). The effect of anthropogenic warming as suggested by recent studies on the increasing chances of low-precipitation years in California (AghaKouchak et al., 2014; Diffenbaugh et al., 2015; Yoon et al., 2015a) and the associated dynamic processes warrants further investigation.

Acknowledgements

This research was supported by the Utah Agricultural Experiment Station, Utah State University (journal paper number 8951), and the Utah State University Libraries Open Access Fund. Yen-Heng Lin is supported by the Utah State University Presidential Doctoral Research Fellows (PDRF) program. JHYoon is supported by the Korea Meteorological Administration Research and Development Program under Grant KMIPA 2016-6030. The reanalysis data is provided by the NOAA/OAR/ESRL PSD, Boulder, Colorado, USA, from their Web site at http://www.esrl.noaa.gov/psd/.

Supporting information

The following supporting information is available:

Table S1. The correlation results between PCs and winter mean (November–March) of climate index over 18 California dry winters. The highest corollary/anticorollary index with PC1 or PC2 is shaded.

Figure S1. The same as Figure 2, the first two leading EOFs (shaded) and PCs of winter (November–March) geopotential height (Z) within the 18 dry winters, superimposed with dry winters' Z anomalies (contour), but for (a) Z_{200} hPa and (b) Z_{500} hPa.

Figure S2. Winter season (November–March) precipitation in California (blue line) overlaid with (a) the PNA index (orange line) and (b) the inverted NPO index (pink line) from 1948–1949 to 2014–2015, superimposed with the 18 dry winters as vertically shaded.

Figure S3. (a)–(d) Same as Figure 3 but for the regression patterns of the PDSI with the PNA and NPO indices. (e) and (f) Long-term regressions of the PDSI with the PNA and the inverted NPO over the 1948–2015 period. Hatches indicate significant level at $p < 0.05$ for the regression.

Figure S4. The 30 ensembles mean of the EOF analysis of 250-hPa HGT from each ensemble's drought years. (a) The PNA-like EOFs of HIS run, (b) the −NPO-like EOFs of HIS run, (c) the PNA-like EOFs of RCP run, and (d) the −NPO-like EOFs of RCP run. The contour shows the drought years' anomaly in HIS run (a and b) and RCP run (c and d).

Figure S5. Averaged Z_{250} hPa regression patterns from 30 ensembles in drought years by regressing with the PCs of (a) PNA-like EOFs from HIS run, (b) −NPO-like EOFs from HIS run, (c) PNA-like EOFs from RCP run, and (d) −NPO-like EOFs from RCP run.

Figure S6. Averaged SST regression patterns from 30 ensembles in drought years by regressing with the PCs of (a) PNA-like EOFs from HIS run, (b) −NPO-like EOFs from HIS run, (c) PNA-like EOFs from RCP run, and (d) −NPO-like EOFs from RCP run.

Figure S7. (a) The first two leading EOFs of standardized normalized winter season (November–March) precipitation within the 18 dry winters over California, (b) the corresponding PCs in relation to each of the 18 dry winters, and (c) the Z_{250} patterns from the 18 California dry winters correlated upon PC1 and PC2 of California precipitation.

References

AghaKouchak A, Cheng L, Mazdiyasni O, Farahmand A. 2014. Global warming and changes in risk of concurrent climate extremes: insights from the 2014 California drought. *Geophysical Research Letters* **41**: 8847–8852, doi: 10.1002/2014GL062308.

Alexander MA, Vimont DJ, Chang P, Scott JD. 2010. The impact of extratropical atmospheric variability on ENSO: testing the seasonal footprinting mechanism using coupled model experiments. *Journal of Climate* **23**: 2885–2901, doi: 10.1175/2010JCLI3205.1.

Barnston AG, Livezey RE. 1987. Classification, seasonality and persistence of low-frequency atmospheric circulation patterns. *Monthly Weather Review* **115**: 1083–1126.

Belmecheri S, Babst F, Wahl ER, Stahle DW, Trouet V. 2016. Multi-century evaluation of Sierra Nevada snowpack. *Nature Climate Change*, **6**: 2–3, doi: 10.1038/nclimate2809.

Dai A, Trenberth KE, Qian T. 2004. A global dataset of Palmer Drought Severity Index for 1870–2002: relationship with soil moisture and effects of surface warming. *Journal of Hydrometeorology* **5**: 1117–1130.

Daly C, Halbleib M, Smith JI, Gibson WP, Doggett MK, Taylor GH, Curtis J, Pasteris PP. 2008. Physiographically sensitive mapping of climatological temperature and precipitation across the conterminous United States. *International Journal of Climatology* **28**: 2031–2064, doi: 10.1002/joc.1688.

Department of Water Resources. 1978. *The 1976-1977 California Drought: A Review*. California Department of Water Resources, Sacramento, CA.

Department of Water Resources. 1993. *California's 1987-98 Drought: A Summary of Six Years of Drought*. California Department of Water Resources, Sacramento, CA.

Department of Water Resources. 2015. *California's Most Significant Droughts: Comparing Historical and Recent Conditions*. California Department of Water Resources, Sacramento, CA.

Deser C, Blackmon ML. 1995. On the relationship between tropical and North Pacific sea surface temperature variations. *Journal of Climate* **8**: 1677–1680, doi: 10.1175/1520-0442(1995)008, 1677:OTRBTA.2.0.CO;2.

Diffenbaugh NS, Swain DL, Touma D. 2015. Anthropogenic warming has increased drought risk in California. *Proceedings of the National Academy of Sciences of the United States of America* **112**: 3931–3936, doi: 10.1073/pnas.1422385112.

Griffin D, Anchukaitis KJ. 2014. How unusual is the 2012–2014 California drought? *Geophysical Research Letters* **41**: 9017–9023, doi: 10.1002/2014GL062433.

Hartmann DL. 2015. Pacific sea surface temperature and the winter of 2014. *Geophysical Research Letters* **42**: 1894–1902, doi: 10.1002/2015GL063083.

Heim RR Jr. 2002. A review of twentieth-century drought indices used in the United States. *Bulletin of the American Meteorological Society* **83**: 1149–1165.

Hurrell JW, Holland MM, Gent PR, Ghan S, Kay JE, Kushner PJ, Lamarque JF, Large WG, Lawrence D, Lindsay K, Lipscomb WH, Long MC, Mahowald N, Marsh DR, Neale RB, Rasch P, Vavrus S, Vertenstein M, Bader D, Collins WD, Hack JJ, Kiehl J, Marshall S. 2013. The community earth system model: a framework for collaborative research. *Bulletin of the American Meteorological Society* **94**(9): 1339–1360.

Kalnay E, Kanamitsu M, Kistler R, Collins W, Deaven D, Gandin L, Iredell M, Saha S, White G, Woollen J, Zhu Y, Leetmaa A, Reynolds R, Chelliah M, Ebisuzaki W, Higgins W, Janowiak I, Mo KC, Ropelewski C, Wang J, Jenne R, Joseph D. 1996. The NCEP/NCAR 40-year reanalysis project. *Bulletin of the American Meteorological Society* **77**(3): 437–471.

Kam J, Sheffield J, Wood EF. 2014. Changes in drought risk over the contiguous United States (1901-2012): the influence of the Pacific and Atlantic Oceans. *Geophysical Research Letters* **41**: 5897–5903, doi: 10.1002/2014GL060973.

Kay JE, Deser C, Phillips AS, Mai A, Hannay C, Strand G, Arblaster J, Bates S, Danabasoglu G, Edwards J, Holland M, Kushner P, Lamarque J-F, Lawrence D, Lindsay K, Middleton A, Munoz E, Neale R, Oleson K, Polvani L, Vertenstein M. 2015. The Community Earth System Model (CESM) large ensemble project: a community resource for studying climate change in the presence of internal climate variability. *Bulletin of the American Meteorological Society* **96**: 1333–1349, doi: 10.1175/BAMS-D-13-00255.1.

Leathers DJ, Palecki MA. 1992. The Pacific/North American teleconnection pattern and United States climate. Part II: temporal characteristics and index specification. *Journal of Climate* **5**: 707–716.

Lee M-Y, Hong C-C, Hsu H-H. 2015. Compounding effects of warm sea surface temperature and reduced sea ice on the extreme circulation over the extratropical North Pacific and North America during the 2013–2014 boreal winter. *Geophysical Research Letters* **42**: 1612–1618, doi: 10.1002/2014GL062956.

Linkin ME, Nigam S. 2008. The North Pacific Oscillation–West Pacific teleconnection pattern: mature-phase structure and winter impacts. *Journal of Climate* **21**: 1979–1997.

McCabe GJ, Dettinger MD. 1999. Decadal variations in the strength of ENSO teleconnections with precipitation in the western United States. *International Journal of Climatology* **19**(13): 1399–1410.

Namias J. 1978. Multiple causes of the North American abnormal winter 1976–77. *Monthly Weather Review* **106**: 279–295.

Palmer WC. 1965. Meteorological drought. Weather Bureau Research Paper 45, U.S. Department of Commerce, Washington, DC.

Pierce DW, Cayan DR, Das T, Maurer EP, Miller NL, Bao Y, Kanamitsu M, Yoshimura K, Snyder MA, Sloan LC, Franco G, Tyree M. 2013. The key role of heavy precipitation events in climate model disagreements of future annual precipitation changes in California. *Journal of Climate* **26**: 5879–5896, doi: 10.1175/JCLI-D-12-00766.1.

Renwick JA, Wallace JM. 1996. Relationships between North Pacific wintertime blocking, El Niño, and the PNA pattern. *Monthly Weather Review* **124**: 2071–2076.

Robeson SM. 2015. Revisiting the recent California drought as an extreme value. *Geophysical Research Letters* **42**: 6771–6779, doi: 10.1002/2015GL064593.

Rogers JC. 1981. The North Pacific Oscillation. *International Journal of Climatology* **1**: 39–57.

Schonher T, Nicholson SE. 1989. The relationship between California rainfall and ENSO events. *Journal of Climate* **2**: 1258–1269.

Seager R, Hoerling M, Schubert S, Wang H, Lyon B, Kumar A, Nakamura J, Henderson N. 2015. Causes of the 2011–14 California drought. *Journal of Climate* **28**: 6997–7024.

Smith TM, Reynolds RW, Peterson TC, Lawrimore J. 2008. Improvements to NOAA's historical merged land–ocean surface temperature analysis (1880–2006). *Journal of Climate* **21**: 2283–2296.

Straus DM, Shukla J. 2002. Does ENSO force the PNA? *Journal of Climate* **15**: 2340–2358.

Swain D, Tsiang M, Haughen M, Singh D, Charland A, Rajarthan B, Diffenbaugh NS. 2014. The extraordinary California drought of 2013/14: character, context and the role of climate change [in "Explaining Extremes of 2013 from a Climate Perspective"]. *Bulletin of the American Meteorological Society* **95**(9): S3–S6, doi: 10.1175/1520-0477-95.9.S1.1.

Taylor KE, Stouffer RJ, Meehl GA. 2012. An overview of CMIP5 and the experiment design. *Bulletin of the American Meteorological Society* **93**: 485–498, doi: 10.1175/BAMS-D-11-00094.1.

Vimont DJ, Wallace JM, Battisti DS. 2003. The seasonal footprinting mechanism in the Pacific: implications for ENSO. *Journal of Climate* **16**(16): 2668–2675.

Wallace JM, Gutzler DS. 1981. Teleconnections in the geopotential height field during the Northern Hemisphere winter. *Monthly Weather Review* **109**: 784–812.

Wang H, Schubert S. 2014. Causes of the extreme dry conditions over California during early 2013. *Bulletin of the American Meteorological Society* **95**(9): S7–S11.

Wang S-Y, Hipps L, Gillies RR, Yoon J-H. 2014. Probable causes of the abnormal ridge accompanying the 2013–2014 California drought: ENSO precursor and anthropogenic warming footprint. *Geophysical Research Letters* **41**: 3220–3226, doi: 10.1002/2014GL059748.

Wang S-YS, Huang W-R, Yoon J-H. 2015. The North American winter 'dipole' and extremes activity: a CMIP5 assessment. *Atmospheric Science Letters* **16**: 338–345, doi: 10.1002/asl2.565.

Yoon J-H, Wang S-Y, Gillies RR, Kravitz B, Hipps L, Rasch P. 2015a. Increasing water cycle extremes in California and in relation to ENSO cycle under global warming. *Nature Communications* **6**: 8657, doi: 10.1038/ncomms9657.

Yoon JH, Wang S-YS, Gillies RR, Hipps L, Kravitz B, Rasch PJ. 2015b. Extreme fire season in California: a glimpse into the future? *Bulletin of the American Meteorological Society* **96**: S5–S9, doi: 10.1175/BAMS-EEE_2014_ch2.1.

Yu B, Zwiers FW. 2007. The impact of combined ENSO and PDO on the PNA climate: a 1000-year climate modeling study. *Climate Dynamics* **29**: 837–851.

Zhou Z-Q, Xie S-P, Zheng X-T, Liu Q, Wang H. 2014. Global warming–induced changes in El Niño teleconnections over the North Pacific and North America. *Journal of Climate* **27**: 9050–9064, doi: 10.1175/JCLI-D-14-00254.1.

11

Multi-scale response of runoff to climate fluctuation in the headwater region of the Kaidu River in Xinjiang of China

Zuhan Liu,[1,2,3]* Lili Wang,[4] Xiang Yu,[1,2] Shengqian Wang,[2] Chengzhi Deng,[1,2] Jianhua Xu,[5] Zhongsheng Chen[5,6] and Ling Bai[5]

[1]Jiangxi Province Key Laboratory for Water Information Cooperative Sensing and Intelligent Processing, Nanchang Institute of Technology, Nanchang, China
[2]School of Information Engineering, Nanchang Institute of Technology, Nanchang, China
[3]Key Laboratory of the Education Ministry for Poyang Lake Wetland and Watershed Research, Jiangxi Normal University, Nanchang, China
[4]School of Science, Nanchang Institute of Technology, China
[5]The Research Center for East-west Cooperation in China, East China Normal University, Shanghai, China
[6]College of Land and Resources, China West Normal University, Nanchong, China

*Correspondence to:
Z. H. Liu, School of Information Engineering, Nanchang Institute of Technology, 289 Tianxiang Road, Nanchang, Jiangxi 330099, China.
E-mail: lzh512@nit.edu.cn

Abstract

Based on the climatological-hydrological daily data recorded in the headwater region of the Kaidu River during 1972–2011, the multi-scale characteristics of runoff variability and four climatic factor fluctuations (i.e. temperature, precipitation, relative humidity and evaporation) are analyzed using detrended fluctuation analysis. Furthermore, multi-scale response of runoff to climate fluctuation is investigated using detrended cross-correlation analysis. Main findings are as follows: (1) The temporal scaling behaviors of runoff and four climate factor series all exhibit two different power laws. In shorter temporal scaling, all the series indicate the similar persistence corresponding to the annual cycle. However, in longer temporal scaling, their different trends reflect the different inherent dynamic nature of various hydro-climatic change. (2) In the double logarithm curve $\log F^2(s) \sim \log s$, the long-range correlation of runoff and temperature, long-range correlation of runoff and precipitation and long-range correlation of runoff and relative humidity (hereafter referred to as Lrc-R-T, Lrc-R-P and Lrc-R-H, respectively) show two scaling regimes with two different scale indexes and a critical time scale of about 1 year; the long-range correlation of runoff and evaporation (hereafter referred to as Lrc-R-E) presents three scaling regime with three different scale indexes and with two critical time scales of about 1 and 10 years. These results reflect the multi-scale response characteristics of the runoff to climate change on different time scales.

Keywords: runoff; multi-scale; Kaidu River; power laws; long-range correlation

1. Introduction

Climate change is always a critical environmental issue. Global warming, which has garbed society's more attention, will possibly change the current regimes of precipitation, hydrological cycle and water resources (Chen *et al.*, 2010; Lan *et al.*, 2010). It is significant to study impacts of climate change on hydrology and water resources, which could provide scientific supports for sustainable utilization of water resources and healthy life maintenance of river. The study of multi-scale response of runoff to climate fluctuation can help to realize the forecast of runoff changing trend and reasonable development, configuration and use of runoff, ensure the sustainable use of water resources, which is of great theoretical and practical significance for planning and design, development, utilization and operation management of the water resources system (Gajbhiye *et al.*, 2016).

It is well known that the supply of water (rainfall) and demand (potential evaporation) are the main factors influencing the water balance for a long time (Budyko,

1974; Milly, 1994; Wang *et al.*, 2016). In addition, runoff and its components are controlled by main natural factors, such as meteorological factors and topographic condition (Yu *et al.*, 2005; Wang *et al.*, 2016). The response of runoff to climate fluctuation mainly embodies in two aspects: one is runoff change as climate change, the other is the runoff fluctuation with temperature, for which the main reason is that variations of temperature have simultaneous impacts on evaporation and snow melting. A lot of research generally believe that both runoff and precipitation change have the same trend of variability. While for different watershed, their characteristics and the water resources status of runoff were very diverse, so the influence of main climate factors on runoff changes is more different. As for the response of runoff to temperature, it is that highly significant intra-annual distribution of runoff is affected by temperature rising, especially in snow-based watershed.

Although the above studies revealed the change reason of runoff and the relationship between runoff and climate factor to a certain extent, the research works were limited to a temporal scaling and lack of multi-scale response of runoff to climate fluctuation. And because of the temporal scaling change, an important climate factor influencing on runoff on a certain scale is likely to be a rather little effect on the other one. Therefore, experimental research on the temporal scaling behaviors of runoff and climate fluctuation and the former may help to how we strengthen the management and control of water resources in making research on applicable countermeasures of water against climate change.

As for the expiration, it is an important link of the water balance and energy balance, and is an important process for contacting the hydrology dynamic changes with the variations of vegetation ecology. Namely, the expiration process is the most important link in coupling effect of hydrologic cycle of land atmosphere system (Miller and Russell, 1992; Chemel et al., 2015). In addition, it should be pointed out that river runoff has been evolving with the change of the earth and the world. And its changes are a complex process with multiple dynamic coupled factors such as nature events and human activities. Humid or arid weather conditions occurred in some previous period may impact now and future, accordingly neighboring observed values in the time series are correlated to some extent. Some traditional approaches such as power spectrum analysis, correlation analysis and statistical analysis are suitable for determining the correlation characteristics of stationary signals (Białous et al., 2016; Fong et al., 2016; Li et al., 2016; Rysak et al., 2016). However, the runoff time series usually is affected by noise or nonstationary signals, the mean, standard deviation, high order values and correlation function change as time goes on. In order to clearly understand the scaling behavior of inherent mechanism of runoff evolution and robustly analyze its long-range correlation, it is necessary to identify potential trend patterns caused by inherent long-range fluctuation in the data. The trend patterns caused by outside factors are usually smooth or oscillate slowly, hence if the potential trend patterns were not filtered before series analysis, the strong trend patterns remaining in the series would interfere with the long-range correlation analysis and as a result it would not be able to reveal the evolution process and laws of the runoff system in hydro-meteorological environments. The scaling index computation method proposed in the mechanism of deoxyribo nucleic acid, in other words, the detrended fluctuation analysis (DFA) method, can effectively solve this type of problem (Strychalski et al., 2008). Furthermore, multi-scale response of the runoff system to multiple climatic factor can be effectively detected through the detrended cross-correlation analysis (DCCA) method, which is an extension of the DFA method, and very suitable for the power-law long-range correlation analysis of two nonstationary time series (Podobnik and Stanley, 2007). Moreover, the two methods have been widely applied in climatological-hydrological process. For instance, Livina et al. (2007) applied the DFA method on the scaling properties of river runoff records of the Naab (26 years), the Regnitz (30 years) and the Vils (26 years); Kantelhardt et al. (2006) studied the temporal correlations and multifractal properties of long river discharge records from 41 hydrological stations around the globe using the DFA, multifractal DFA and wavelet analysis methods. Through calculating the DCCA cross-correlation coefficient ρ, Shen et al. (2015) found the cross-correlation between diurnal temperature ranges and the the daily air pollution index (API) was persistent at time scales, more specifically, the correlation with the API presented persistent cross-correlation at smaller time scales, and antipersistent cross-correlation at larger time scales.

The Kaidu River, located in the southern slope of the Tianshan Mountains, was honored as the 'Water Tower of Central Asia' and 'Solid Reservoir'. Its runoff supply is mainly from glacial meltwater and precipitation in the Tianshan Mountains, so both temperature and precipitation changes have significant impacts on the runoff from the mountain pass. These changes in runoff from mountain pass not only affect the water supply of downstream industry and agriculture, but also relate to the regional social and economic sustainable development and ecological safety maintenance. In addition, the global warming mitigation in recent years, the complexities and effects of climate change in the Tianshan Mountains have increased the uncertainty of future regional climatological-hydrological process, and the inflow from origin area shows overall increase while the amount of water into the mainstream of the Tarim River decreases, bringing worries for the future security of water resources; therefore, the management and control of water resources shall be strengthened in making research on applicable countermeasures of water against climate change. Thus, a typical catchment of the headwater region of the Kaidu River was selected as the study target region in this article. In this article, the hydrological and meteorological daily data in the region from 1972 to 2011 were selected as the research object. The multi-scale characteristics of runoff and climate factors were quantitatively evaluated by the DFA method. Moreover, the hydrological responses to climate change were analyzed using the DCCA method. These analyses are scientifically significant for improving our understanding of the inherent mechanism of hydrological process of the Kaidu River, its watershed hydrology and practically significant for improving our water resources management.

2. Study area, data and methodology

2.1. Study area

The Kaidu River is the largest river which flows into the Yanqi Basin, the river flows through Hejing, Yanqi and Bohu, originate from Eren Habirga covered by snow all the year around in the middle Tianshan Mountains, Xinjiang, North-west China, and is enclosed between latitudes $42°14'-43°21'$N and longitudes $82°58'-86°55'$E (Figure 1). The terrain of the Kaidu River Basin is

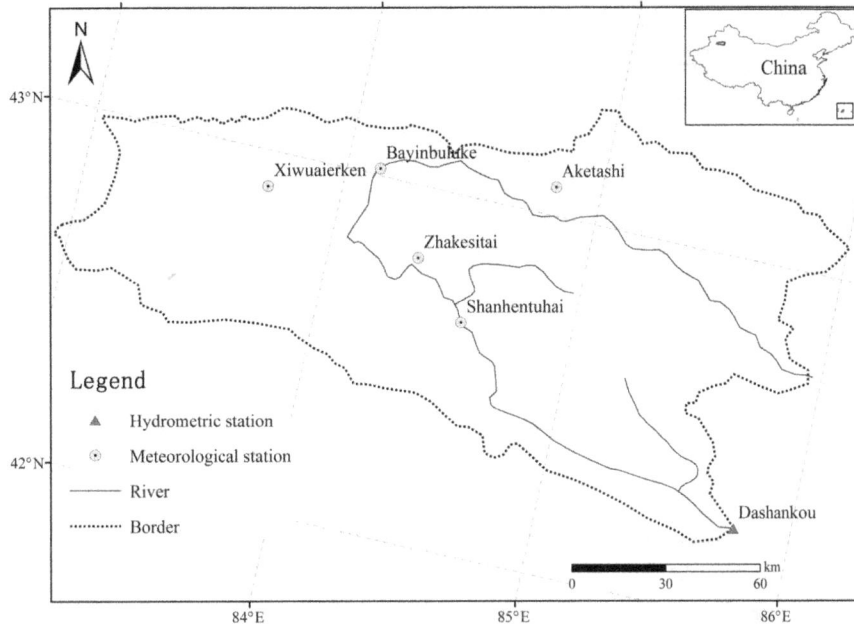

Figure 1. Location of the headwater region of the Kaidu River and the distribution of meteorological and hydrological stations.

higher in the north-west than in the south-east, so the whole basin is divided into three kinds of types. And the upstream segment length of about 200 km through Eren Wulu, Dayultuz Basin and the canyon, the total length of upper course is about 160 km. The gorge of the middle reaches of the Kaidu River, there is a great difference in the height of hypsography, and stream is rapid, and hydropower resource is mainly concentrated in this segment, where is recharge area of the meltwater from snow and ice, and is the main source of Kaidu River flood. It ends in the Bosten Lake which is located in Bohu County with segment length of the about 120 km (Li *et al.*, 2012; Chen *et al.*, 2013; Bai *et al.*, 2015).

2.2. Data

The Dashankou hydrological station and five meteorological stations located in the study area as shown in Figure 1. The records used in this article were obtained from a high-quality daily runoff and climate sequence (namely, temperature, precipitation, relative humidity and evaporation) data set spanning from 1 January 1972 to 31 December 2011, processed by the Xinjiang Tarim River Basin Management Bureau and National Meteorological Information Center, respectively. In this article, climate sequences are average daily temperature, average daily precipitation, average daily relative humidity and average daily evaporation from only five meteorological stations, respectively. To determine which data have higher quality, the data have been subjected to extremum, time consistency and other tests. In addition, to overcome the natural nonstationarity of the data due to season trends, we remove the annual cycle from the raw data e by computing the anomaly series $e' = e - <e>_d$ for five data series, where $< >_d$ denotes the long-time average value for the given calendar day (Chen *et al.*, 2002). In addition, we apply the

standard normal homogeneity test, Buishand and Pettit homogeneity test method to check these data (Pettit, 1979; Buishand, 1982; Alexandersson, 1986). The stepwise multiple linear regression method was employed to revise the inhomogeneity of time series.

2.3. Methodology

The DFA is an advanced method for determining the scaling behavior of data in the presence of possible trends without knowing their origin. For further detail computation, see Peng *et al.* (1994). In the method, the most important parameter is the root mean square fluctuation $F(n)$, which behaves as a power-law function of n then the data present scaling: $F(n) \propto n^\alpha$. The DFA exponent (α) is defined as the slope of the regression line for all points $[\log(n), \log[F(n)]]$. Specifically, $\alpha = 0.5$ indicates the series corresponds to a random walk (namely white noise); $0.5 \leq \alpha \leq 1$, indicates persistent long-range power-law correlations; $0 < \alpha \leq 0.5$, power-law anticorrelations are present; when $\alpha > 1$, correlations exist but cease to be of the power-law form; $\alpha = 1.5$ indicates brown noise, the integration of white noise.

In analogy to the DFA, which was proposed by Podobnik and Stanley (2007) for a single time series, DCCA was used for analyzing power-law long-range cross-correlations between different nonstationary time series (Pal *et al.*, 2016; Yin and Shang, 2016). If the detrended fluctuation covariance function $F(s)$ and scale s obey power-law cross-correlations in double logarithmic coordinates as shown $F^2(s) \sim s^\lambda$, where λ is the long-range cross-correlation scale index, there is long-range interrelation between two sequences (Ferreira, 2016). In particular, a value of $\lambda > 0.5$ indicates a positive long-range cross-correlation between two sequences. To be specific, if a sequence presents

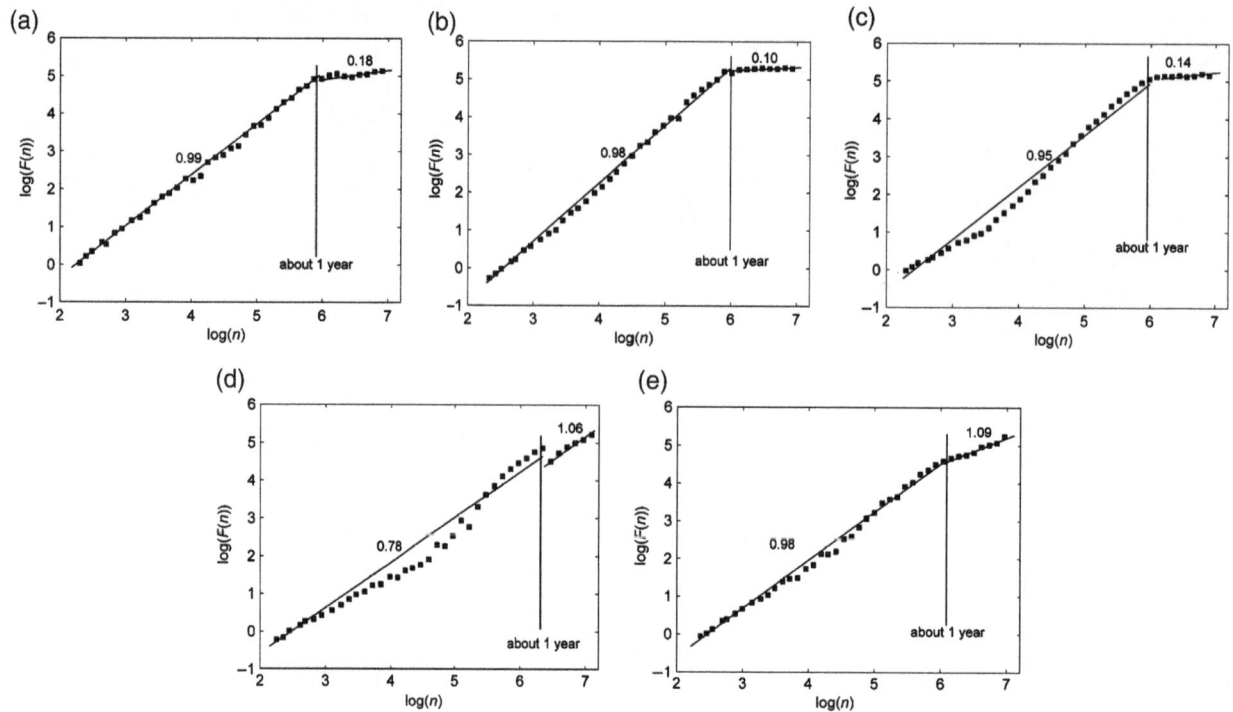

Figure 2. DFA of the daily runoff (a), average daily temperature (b), evaporation (c), precipitation (d) and relative humidity (e), respectively.

growth trend, the other sequence will also show growth trend. $\lambda < 0.5$ indicates that there is negative long-range cross-correlation. When $\lambda = 0.5$, there is nonlong-range cross-correlation between two sequences, that is, the change trend of a sequence exerts no effect on the change of another sequence (Ma *et al.*, 2016; Fan and Lin, 2017).

3. Results and discussion

3.1. The multi-scale characteristics of runoff and climate factors

Figure 2(a) shows the DFA analysis for the daily runoff. In this case, the plot exhibits curvature, showing obviously two different period regimes. n_c is the critical time scale where obvious dividing point occurs. And n_c is about 1 year, reflecting an influence of the annual cycle. For shorter time periods ($n < n_c$), the plot can be fitted to a straight line with a DFA exponent (α_1) of 0.99 which exhibits high persistence. Over longer time periods, $n > n_c$, a line with a decreased slop ($\alpha_2 \approx 0.18$) that the high persistence changes to antipersistence when the temporal scale is larger than 1 year. Those results indicate that high persistence or long-term memory of the runoff comes up to about 1 year. For time spans greater than 1 year, the runoff displays a high antipersistent behavior. This result is not very clear, which perhaps relates to long-term hydrological processes, the length of the data and the internal dynamics of runoff. It needs more researches and longer series to make an interpretation.

Moreover, the DFA method is applied to the climate factor series of average daily temperature, evaporation, precipitation and relative humidity in the headwater region of Kaidu River from five meteorological stations. The results are shown in Figures 2(b)–(e), respectively. These data fit not one but two visible lines, which are similar to $\log(F(n)) \sim \log n$ in Figure 2(a) and all n_c are about 1 year. As to precipitation and relative humidity, for $n < n_c$, α_1 are approximately 0.78 and 0.98, respectively; while fort $n > n_c$, α_1 are approximately 1.06 and 1.09, respectively. The relation $\alpha_2 > \alpha_1 > 0.5$ comes as a surprise. This shows that when the temporal scale is larger than 1 year, the persistence becomes higher. Hence, the trend dependence may persist more than 39 years for the two climate factors. Moreover, as to the temperature and evaporation time series, the result is similar to that of runoff.

3.2. The response of the runoff to climate change

Figure 3 shows the DCCA analysis results of the Lrc-R-T, Lrc-R-P, Lrc-R-H and Lrc-R-E. There obviously are two or even three scaling regions in the double logarithm curve $\log F^2(s) \sim \log s$ for the two types of four long-rang correlation within the same scaling regime. Namely, the Lrc-R-T, Lrc-R-P and Lrc-R-H show two scaling regimes with two different scale indexes (λ_1 and λ_2) and with a critical time scale (s_c) of about 1 year with the same meaning as n_c in Section 3.1, however, the Lrc-R-E presents three scaling regime with three different scale indexes (λ_1, λ_2 and λ_3) and with two critical time scales (s_c and s_c') of about 1 and 10 years. Linear fitting was respectively conducted on the

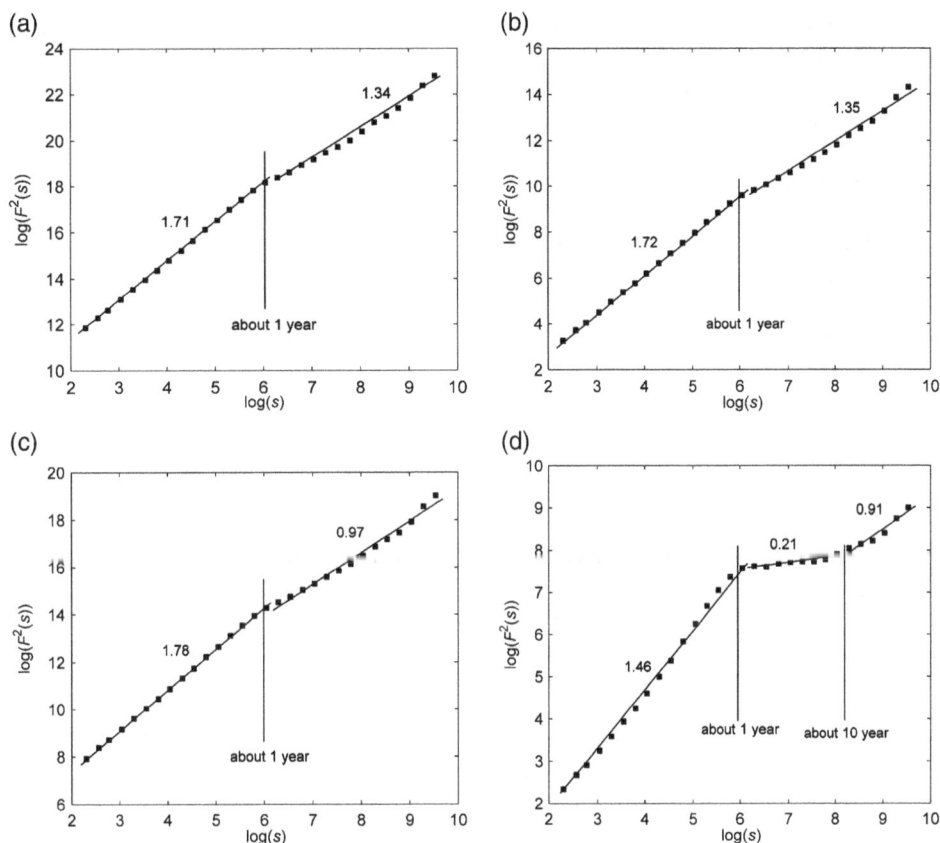

Figure 3. DCCA of the daily runoff and average daily temperature (a), precipitation (b), relative humidity (c) and evaporation (d), respectively.

scaling regimes and their scale indexes λ are obtained. By taking s_c as the cut-off point, the time that s_c position also corresponds to is exactly 1 year, reflecting the multi-scale response of the runoff to climate change in temporal scaling.

In the first scaling range, namely, over shorter time periods ($s < s_c$), the scale indexes λ_1 of the Lrc-R-T, Lrc-R-P, Lrc-R-H and Lrc-R-E are respectively 1.71, 1.78, 1.72 and 1.46, which all are >1, suggesting that there all are positive long-range correlation with nonpower law form. As an example of the Lrc-R-T, the runoff volume increased (decreased) with increasing (decreasing) temperature, both have stronger synchronicity.

In the second scaling range, namely, over longer time periods ($s > s_c$), the scale indexes λ_2 of the Lrc-R-T, Lrc-R-H and Lrc-R-P are 1.34, 1.35 and 0.97, respectively. This indicates that when the temporal scale is larger than 1 year, their positive correlation becomes weaker. Hence, the trend dependence may persist more than 39 years for the three correlations. As to the Lrc-R-T and Lrc-R-H, the analysis results are both similar to their trend in the first scaling range. That is, the two correlations also display positive correlation with nonpaw-law form; while for the Lrc-R-P, $\lambda_2 = 0.97$, indicates that the correlation of runoff and relative humidity manifested as $1/f$ noise behavior with power law at a large temporal scale. To point out, compared with the first scaling range, these positive correlations

become weaker. As to the correlation of the Lrc-R-E is contrary to that of above the three correlations. While for $s_c < s < s_c'$, $\lambda_2 = 0.21$. Namely, the runoff volume decreased (increased) with increasing (decreasing) evaporation in the corresponding time period scaling exponent is <0.5, this show that runoff and evaporation is power-law antilong-range correlation within the scaling regime of more than 1 year. The strong positive long-range correlation changes to antilong-range correlation as the time becomes longer. The runoff volume increased (decreased) with increasing (decreasing) temperature.

In the third scaling range, the correlation of runoff and evaporation is different from the former scaling range. The antilong-range correlation changes to strong positive long-range correlation as the time becomes much longer. That is to say, for $s > s_c'$, $\lambda_3 = 0.91$, over much longer time periods, the result fully restored transcription to shorter time periods ($s < s_c$) once more in the corresponding time period. This suggests that runoff and evaporation still show strong positive long-range correlation when the temporal scale is >10 years.

The multi-scale response of runoff to the four climate factors reflects mutual influence, dependence and change characteristics of runoff and climate change. In the shorter temporal scaling ($s < s_c$), the Lrc-R-T, Lrc-R-P, Lrc-R-H and Lrc-R-E respectively all show stronger positive long-range correlation. With the long-range correlation between the Lrc-R-T as an example, if runoff volume of the Kaidu River within 1 year increases (decreases), the temperature also will increase (decrease). However, the degree of the four long-range correlations is different. The strongest correlation is the Lrc-R-P, next are the Lrc-R-T and Lrc-R-H, and Lrc-R-E is the last. When it is more than a year ($s > s_c$), there are two kinds of correlation. The Lrc-R-T, Lrc-R-P and Lrc-R-H continue to indicate positive long-range correlation. However, the Lrc-R-E turns into negative correlation over time, becomes longer ($s_c < s < 10a$). For much longer time, ($s > 10a$), the Lrc-R-T, Lrc-R-P and Lrc-R-H still keep stronger positive correlation, while the Lrc-R-E turn into positive correlation from negative correlation in longer time ($s_c < s < 10a$).

The above research results show the long-range correlation characteristics and the temporal evolution properties of runoff between and temperature, precipitation, relative humidity and evaporation, respectively. The different response existed runoff to different climate factor in scale-invariant region. For instance, as all of you know, the effect of humidity on runoff mainly generated by precipitation and evaporation changes. The spatial pattern of the relative humidity change should also be a factor influencing the response of total runoff to the change in the mean relative humidity (Tang *et al.*, 2013). Therefore, a direct quantification research can make us more intuitively see the multi-scale response of runoff to relative humidity or the close relationship between them.

4. Conclusions

The DFA method is a modified root-mean-square analysis of random walk with advantages. It can avoid spurious detection of correlations that are artifacts of nonstationarity, which often affects the time series data. When the DFA is applied to the time series of hydro-climatic factors, which are runoff, temperature, precipitation, relative humidity and evaporation, respectively, the persistence or long-term memory of intrinsic time scales of hydro-climatic change can be extracted, it is helpful to determine their scaling behavior (Liu *et al.*, 2014). Moreover, the DCCA method is applied to detect the multi-scale response of runoff to four climate factors are studied, which are long-range correlation of runoff and temperature (Lrc-R-T), runoff and precipitation (Lrc-R-P), runoff and relative humidity (Lrc-R-H) and runoff and evaporation (Lrc-R-E), respectively. In this study, based on the hydrological and meteorological data in the Kaidu River Basin during 1972–2011, the multi-scale response of runoff to climate fluctuation were analyzed using the DFA and DCCA methods, the main findings are as follows:

1. The runoff variability and four climate factor fluctuation series follow two different power laws in shorter and longer temporal scaling regimes through the DFA method. In annual cycle, the DFA exponent

α_1, indicating some similar dynamic characteristics of various hydro-climatic change's temporal evolution. Meantime, in longer temporal scaling regimes, α_2 may reveal the inherently different dynamic nature of various pollutant series. The persistence duration may persist about 1 year for runoff and temperature series, while over 39 years for precipitation, relative humidity and evaporation series.

2. The temporal scaling behaviors of runoff and four climate factor series all possess different power laws, which is a significant finding in the article. Those results further validated the dynamic characteristics of temporal evolution of runoff and temperature, runoff and precipitation, and runoff and relative humidity are just the same, and further illustrates multi-scale response of runoff to climate change has consistency. This shows the dynamic characteristics of temporal evolution of runoff and evaporation inter-decade difference, but it returns to positive long-range correlation.

Climatological-hydrological system is composed by several subsystems, and by the interaction of multi-spheres, multi-factors and multi-scale. One or more ways can be internal or external interaction among subsystems, which result in interaction structure of more complex form not only in the time, but also in the space. The results form complex-huge system with oppression outside and nonlinear dissipative inside, and the whole shows its complexity. There are many influence factors of climatological-hydrological system, for example, human activities, geographical location, complex surface characteristics and so on. But these factors are not independent of each other, but there are nonlinear interactions at various spatial and temporal scales. These reasons result in which the time evolution process of climatological-hydrological process shows inherent nonlinear and external complex characteristics, which can be difficult to investigate scientifically and accurately their correlation and multiple-time-scale characteristics by way of statistics for stationary time series. DFA and DCCA methods are put forward in nonlinear scientific field, and they can more effectively investigate these features including multi-scale and response characteristics. The findings can help to develop effective warning strategies to reduce impacts on climatological-hydrological environment.

Acknowledgements

This work was supported by the Science and Technology Project of Jiangxi Provincial Department of Education (GJJ161097), Open Research Fund of Jiangxi Province Key Laboratory of Water Information Cooperative Sensing and Intelligent Processing (2016WICSIP012), the 13th Five-year Plan Project of Social Science of Jiangxi Province(16BJ36), the Key Project of Jiangxi Provincial Department of Science and Technology (20161BBF60061), China Postdoctoral Science Foundation (2016M600515) and the National Key Technology R & D Program (2015BAH50F00).

References

Alexandersson H. 1986. A homogeneity test applied to precipitation data. *International Journal of Climatology* **6**: 661–675.

Bai L, Chen ZS, Xu JH, Li WH. 2015. Multi-scale response of runoff to climate fluctuation in the headwater region of Kaidu River in Xinjiang of China. *Theoretical and Applied Climatology* **125**: 703–712.

Białous M, Yunko V, Bauch S, Ławniczak M, Dietz B, Sirko L. 2016. Power spectrum analysis and missing level statistics of microwave graphs with violated time reversal invariance. *Physical Review Letters* **117**: 144101.

Budyko MI. 1974. Climate and life. Analysis of non-stationary signals by recurrence dissimilarity. In: Miller DH, Dmowska R, Holton JR (eds). International geophysics series: a series of monographs and textbooks. Springer Proceedings in Physics, pp. 152–168, Academic Press, California.

Buishand TA. 1982. Some methods for testing the homogeneity of rainfall records. *Journal of Hydrology* **l58**: 11–27.

Chemel C, Russo MRJ, Hosking S, Telford PJ, Pyle JA. 2015. Sensitivity of tropical deep convection in global models: effects of horizontal resolution, surface constraints, and 3D atmospheric nudging. *Atmospheric Science Letters* **16**: 148–154

Chen Z, Ivanov PC, Hu K, Stanley HE. 2002. Effect of nonstationarities on detrended fluctuation analysis. *Physical Review E* **65**: 041107.

Chen YD, Zhang Q, Xu CY, Lu X, Zhang S. 2010. Multiscale streamflow variations of the Pearl River basin and possible implications for the water resource management within the Pearl River delta, China. *Quaternary International* **226**: 44–53.

Chen ZS, Chen YN, Li BF. 2013. Quantifying the effects of climate variability and human activities on runoff for Kaidu River Basin in arid region of northwest China. *Theoretical and Applied Climatology* **111**: 537–545.

Fan XX, Lin M. 2017. Multiscale multifractal detrended fluctuation analysis of earthquake magnitude series of Southern California. *Physica A* **479**: 225–235.

Ferreira P. 2016. Does the Euro crisis change the cross-correlation pattern between bank shares and national indexes?. *Physica A* **463**: 320–329.

Fong S, Cho K, Mohammed O, Fiaidhi J, Mohammed S. 2016. A time series pre-processing methodology with statistical and spectral analysis for classifying non-stationary stochastic biosignals. *Journal of Supercomputing* **72**: 3887–3908.

Gajbhiye S, Meshram C, Singh SK, Srivastava PK, Islam T. 2016. Precipitation trend analysis of Sindh River basin, India, from 102-year record (1901–2002). *Atmospheric Science Letters* **17**: 71–77.

Kantelhardt JW, Koscielny-Bunde E, Rybski D, Braun P, Bunde A, Havlin S. 2006. Long-term persistence and multifractality of precipitation and river runoff records. *Journal of Geophysical Research. Atmospheres* **111**: 93–108.

Lan YC, Zhao GH, Zhang YN, Wen J, Liu JQ, Hu XL. 2010. Response of runoff in the source region of the yellow river to climate warming. *Quaternary International* **226**: 60–65.

Li Q, Li LH, Bao AM. 2012. Snow cover change and impact on streamflow in the Kaidu River Basin. *Resources Science* **34**: 91–97 (in Chinese).

Li XF, Girin L, Gannot S, Horaud R. 2016. Non-stationary noise power spectral density estimation based on regional statistics. IEEE International Conference on Acoustics, Speech and Signal Processing; 181–185.

Liu ZH, Xu JH, Shi K. 2014. Self-organized criticality of climate change. *Theoretical and Applied Climatology* **115**: 685–691.

Livina V, Kizner Z, Braun P, Molnarc T, Bunded A, Havlinb S. 2007. Temporal scaling comparison of real hydrological data and model runoff records. *Journal of Hydrology* **336**: 186–198.

Ma PC, Li DY, Li S. 2016. Efficiency and cross-correlation in equity market during global financial crisis: Evidence from China. *Physica A* **444**: 163–176.

Miller JR, Russell GL. 1992. The impact of global warming on river runoff. *Journal of Geophysical Research. Atmospheres* **97**: 2757–2764.

Milly PCD. 1994. Climate, soil water storage, and the average annual water balance. *Water Resources Research* **30**: 2143–2156.

Pal M, Kiran VS, Rao PM, Manimaran P. 2016. Multifractal detrended cross-correlation analysis of genome sequences using chaos-game representation. *Physica A* **456**: 288–293.

Peng CK, Buldyrev SV, Havlin S, Simons M, Stanley HE, Goldberger AL. 1994. Mosaic organization of DNA nucleotides. *Physical Review E* **49**: 1685–1689.

Pettit AN. 1979. A non-parametric approach to the change-point detection. *Applied Statistics* **28**: 126–135.

Podobnik B, Stanley E. 2007. Detrended cross-correlation analysis: a new method for analyzing two nonstationary time series. *Physical Review Letters* **100**: 38–71.

Rysak A, Litak G, Mosdorf R. 2016. Analysis of non-stationary signals by recurrence dissimilarity. In: Webber, Jr. C., Ioana C., Marwan N (eds). Recurrence plots and their quantifications: Expanding Horizons. Springer Proceedings in Physics, **180**: 37–42, Springer, Cham.

Shen CH, Li CL, Si YL. 2015. A detrended cross-correlation analysis of meteorological and API data in Nanjing, China. *Physica A* **419**: 417–428.

Strychalski EA, Levy SL, Craighead HG. 2008. Diffusion of DNA in nanoslits. *Macromolecules* **41**: 7716–7721.

Tang Y, Tang Q, Tian F, Zhang Z, Liu G. 2013. Responses of natural runoff to recent climatic variations in the Yellow River basin, China. *Hydrology and Earth System Sciences Discussions* **10**: 4489–4514.

Wang LL, Chen DH, Bao HJ. 2016. The improved Noah land surface model based on storage capacity curve and Muskingum method and application in GRAPES model. *Atmospheric Science Letters* **17**: 190–198.

Yin Y, Shang P. 2016. Detection of multiscale properties of financial market dynamics based on an entropic segmentation method. *Nonlinear Dynamics* **83**: 1743–1756.

Yu B, Seed A, Pu L, Malone T. 2005. Integration of weather radar data into a raster GIS framework for improved flood estimation. *Atmospheric Science Letters* **6**: 66–70.

12

Heat content of the Arabian Sea Mini Warm Pool is increasing

P. V. Nagamani,[1] M. M. Ali,[1]* G. J. Goni,[2] T. V. S. Udaya Bhaskar,[3] J. P. McCreary,[4] R. A. Weller,[5] M. Rajeevan,[6] V. V. Gopala Krishna[7] and J. C. Pezzullo[8]

[1] National Remote Sensing Centre, ISRO, Hyderabad, India
[2] National Oceanic and Atmospheric Administration, AOML, Washington, DC, USA
[3] Indian National Center for Ocean Information Services, Hyderabad, India
[4] University of Hawaii, Honolulu, HI, USA
[5] Woods Hole Oceanographic Institution, Woods Hole, MA, USA
[6] Ministry of Earth Sciences, New Delhi, India
[7] National Institute of Oceanography, Goa, India
[8] Georgetown University, Washington, DC, USA

*Correspondence to:
M. M. Ali, Centre for
Ocean-Atmospheric Prediction
Studies, FSU, Tallahassee,
FL 32310, USA
E-mail: mmali110@gmail.com

Abstract

Sea surface temperature in the Arabian Sea Mini Warm Pool has been suggested to be one of the factors that affects the Indian summer monsoon. In this paper, we analyze the annual ocean heat content (OHC) of this region during 1993–2010, using *in situ* data, satellite observations, and a model simulation. We find that OHC increases significantly in the region during this period relative to the north Indian Ocean, and propose that this increase could have caused the decrease in Indian Summer Monsoon Rainfall that occurred at the same time.

Keywords: tropical cyclone heat potential; Arabian Sea Mini Warm Pool; satellite altimetry; ocean heat content; all India monsoon rainfall

1. Introduction

The region in the southeastern Arabian Sea from $4°–14°N$ to $68°–78°E$ is referred to as the Arabian Sea Mini Warm Pool (ASMWP; Shenoi *et al.*, 1999). There, sea surface temperature (SST) exceeds $30°C$, 2–3 months before the onset of the Indian Summer Monsoon (Deepa *et al.*, 2007; Vinaychandran *et al.*, 2007; Vinaychandran and Kurian, 2008). A number of studies have explored the connection between the summer monsoon and ASMWP variables. For example, Joseph (1990), Rao and Sivakumar (1999), Deepa *et al.* (2007), Vinaychandran *et al.* (2007), and Sanil Kumar *et al.* (2004) investigated the impact of ASMWP SST on the summer-monsoon onset vortex. Rao and Sivakumar (1999) observed a direct correspondence between upper-ocean heat content (OHC) in the region from the ocean surface to the depth of the $28°C$ isotherm during May–June and the genesis location of the onset vortex. Vinaychandran *et al.* (2007), however, concluded that, although the ASMWP impacts the onset vortex, it is not a sufficient condition for its formation. Rao *et al.* (2015) investigated the interannual variability of ASMWP variables in terms of their phase, amplitude, and spatial extent. Joseph (1990) suggested a need to study the ASMWP and its impact on the Indian Summer Monsoon Rainfall (ISMR).

In this study, we continue the effort to explore the relationship between the ASMWP and ISMR. Specifically, we show that ASMWP OHC has been increasing from 1993 to 2010, while ISMR has been decreasing at the same time. This connection points toward a possible linkage between the two variables, with the increased OHC leading to a southward shift of atmospheric convection thereby weakening rainfall over India (see Section 4).

2. Data and methodology

As a measure of OHC, we use tropical cyclone heat potential (TCHP), which is similar to other OHC proxies except that it integrates temperature from the surface to the depth of the $26°C$ isotherm instead of a fixed depth. We use three methods to calculate TCHP: a direct method using temperature profiles from *in situ* observations and model simulations through a numerical solution to version 3.1 of the Modular Ocean Model (MOM3.1), and an indirect one using satellite altimeter data.

In the first approach, we compute TCHP from the *in situ* and model profiles by,

$$\text{TCHP} = \rho C_p \int_0^{D_{26}} (T - 26) \, dz \qquad (1)$$

where ρ is the sea water density, C_p is the specific heat capacity at constant pressure, $T(z)$ is ocean temperature ($°C$), and D_{26} is the depth of the $26°C$ isotherm (Leipper and Volgenau, 1972). In this calculation, the product of ρC_p is taken as $4 \times 10^6 \, \text{J K}^{-1} \, \text{m}^{-3}$. Temperature profiles for Equation (1) are taken from observational data and MOM3.1 (courtesy: R. Sharma, 2013, pers. comm.).

The observed profiles are taken from all available temperature profiles during 1993–2010, namely, those collected during ship campaigns by conductivity temperature and depth (CTD) instruments, bathythermographs (BT), expendable BTs (XBT), and expendable CTDs (XCTD), as well as by Argo floats. All together 5761 *in situ* profiles (XBT – 2505, XCTD – 122, CTD – 356, Argo profiles – 2778) were considered during the study period. The model has been set up for the global domain (80°S–80°N) excluding polar regions, with a horizontal grid resolution varying from 0.58° × 0.58° in the Indian Ocean to 2.8° × 2.8° in the other oceans. There are 38 levels in the vertical, with 8 levels in the upper 40 m. The bottom topography is based on 1/12831/128 resolution data from the U.S. National Geophysical Data Center. Wind stress is computed from wind velocity using a wind-dependent, drag coefficient. Sharma *et al.* (2010) used this model to study the sea surface salinity variability. Temperature profiles from this model are used to compute TCHP as given in Equation (1).

If SST is less than 26 °C, TCHP for the layer is assumed to be zero. If the temperature observation is not available at a depth of 26 °C, D_{26} is determined by a linear interpolation. We retained only profiles with observations that begin at a depth of 5 m or less. As the Argo profiles do not have surface (0 m) observations, the shallowest observation (most profiles start from 4 m) is used for the surface value. We computed TCHP at all the locations within the region 4°–14°N and 68°–78°E, wherever *in situ* observations are available. The data used in plots are spatially and annually averaged TCHP values.

In the second approach, we use all the existing altimeter-derived, sea-surface-height anomaly (SSHA) data to determine TCHP (Goni *et al.*, 1996; Shay *et al.*, 2000). Accordingly, we assume, to first order, that the ocean can be approximated by a two-layer system with the upper-layer thickness (h_1) at latitude (x), longitude (y) and time (t). Then, provided that the mean upper-layer thickness ($\overline{h_1}$) and reduced gravity (g') fields are known from historical measurements, h_1 can be estimated from the altimeter-derived SSHA (η') field from

$$h_1(x,y,t) = \overline{h}_1(x,y) + \frac{g}{g'(x,y)}\eta'(x,y,t) \quad (2)$$

where $g' = \varepsilon g$, g is the acceleration of gravity and,

$$\varepsilon(x,y) = \frac{\rho_2(x,y) - \rho_1(x,y)}{\rho_2(x,y)} \quad (3)$$

where $\rho_1(x,y)$ and $\rho_2(x,y)$ represent upper- and lower-layer densities, respectively. Once the depth of the 26 °C isotherm (h_1) is estimated, and SSHA is obtained from satellite observations, the TCHP is the excess heat contained above the 26 °C isotherm. These TCHP values are obtained from Atlantic Oceanic and Meteorological Laboratory (AOML), National Oceanic and Atmospheric Administration (NOAA).

In addition, we analyzed SST trends from Tropical Rainfall Measuring Mission Microwave Imager during 1997–2010 and net heat flux from OAFlux (Yu *et al.*, 2007) during 1993–2009. All these observations are annually averaged. We also examined ISMR data from 1993 to 2010 obtained from India Meteorological Department (IMD). Total all India rainfall during the monsoon period, June–September, is used in this study.

3. Results

The comparison between *in situ* and satellite-derived TCHP shows a bias (*in situ* values being on the higher side) of 8.9 kJ cm^{-2} with a coefficient of determination, R^2, of 0.82 and root mean square error (RMSE) of 9.4 kJ cm^{-2} (Figure 1). Nagamani *et al.* (2012) compared the two estimations over the entire north Indian Ocean (NIO) and found an RMSE of 20.95 kJ cm^{-2} with an R^2 of 0.65 and bias of 11.27 kJ cm^{-2}.

During 1993–2010, the OHC of the region increased from 70.5 kJ cm^{-2} (61 kJ cm^{-2}) to 85 kJ cm^{-2} (78 kJ cm^{-2}), with an overall warming of 14.5 kJ cm^{-2} (17 kJ cm^{-2}) in the *in situ* (satellite) observations (Figure 2). These increasing OHC trends are statistically significant, with a p value of 0.002 (i.e., the probability p that there is no trend is less than 0.2%). Using the temperature profiles from MOM3.1, there is an overall increase of 20.5 kJ cm^{-2} during this period (from 42.5 kJ cm^{-2} in 1993 to 63 kJ cm^{-2} in 2010). Though the model results show higher warming compared to satellite and *in situ* observations, the slope is highly significant ($p < 0.001$) with R^2 of 0.565 between the year and OHC. Although the *in situ*, model and satellite observations have biases, all the three observations show an increasing trend. It is noteworthy that such large increasing trends are not observed in other parts of the NIO, although Roxy *et al.* (2014) observed an overall warming trend in the entire NIO. For example, satellite-derived OHC increased by only 3.5 kJ cm^{-2} in the Bay of Bengal and decreased by about 1 kJ cm^{-2} in the rest of the Arabian Sea (i.e. outside the ASMWP).

The increasing trend in ASMWP heat content is consistent with SST changes in the region. On the average, SST increased from 28.4 to 28.9 °C during 1993–2010. If the average temperature of this column were equal to SST, OHC of the column would increase by 22.4 kJ cm^{-2} over an 80-m thick upper layer (the average depth of the 26 °C isotherm). As the average temperature this column is generally less than SST owing to mixing processes, the estimated increase in OHC could be slightly less as well. At the same time, the net surface heat flux (NHF) in the ASMWP decreased by 24 W m^{-2} (Figure 3), a reduction that reduces the OHC of an 80-m water column by 34 kJ cm^{-2}. Thus, the net increase in OHC due to the changes in both SST and NHF is slightly less than 19.44 kJ cm^{-2}, which is closer to 17 kJ cm^{-2} of the estimation from satellite observation, if the ocean mean temperature is considered to be less than SST. Even these calculations suggest the

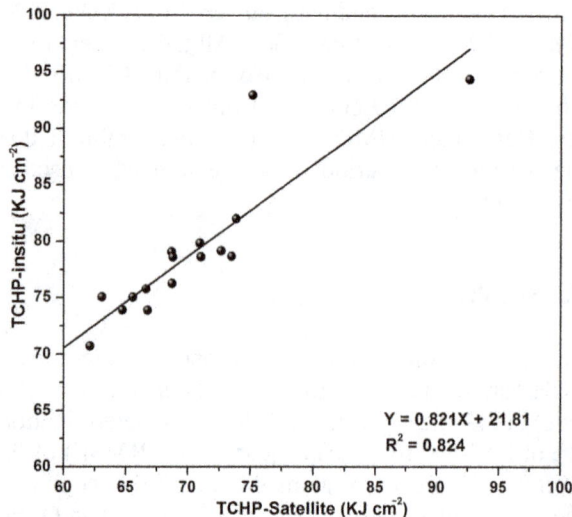

Figure 1. Comparison of *in situ* and satellite-derived OHC over the Arabian Sea Mini Warm Pool region during 1993–2010. The slope of the regression line is found to be significantly non-zero at the 95% level ($p < 0.001$).

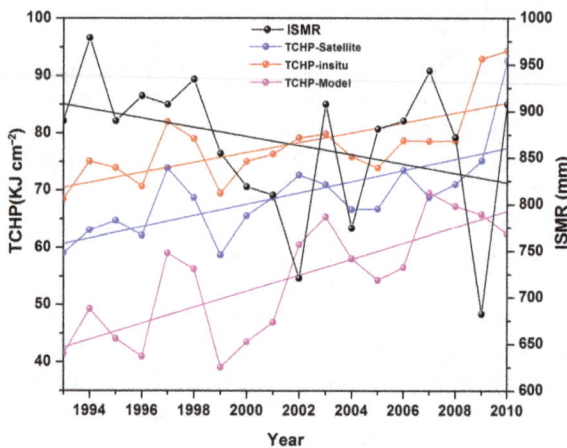

Figure 2. Comparison of Indian Summer Monsoon Rainfall (green dots and line) with tropical cyclone heat potential over the Arabian Sea Mini Warm Pool region from (a) *in situ* (red dots and line), (b) satellite (blue dots and line) and (c) model (pink dots and line) observations during 1993–2010.

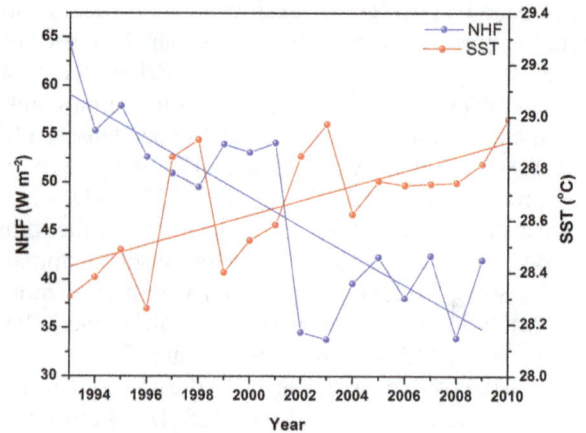

Figure 3. Trends in net heat flux (NHF, in blue) and sea surface temperature (SST, in red) during 1993–2009 for NHF (we did not have access beyond 2009) and 1993–2010 for SST.

the temperature difference averaged from 60°–100°E between the equator and the 25°N reduces monsoon rainfall. We suggest that the increase in OHC of the study region covering 4°–14°N and 68°–78°E was a significant factor in reducing the northward SST gradient. However, until a detailed modeling study, is carried out to confirm this idea, this relation is speculative.

4. Summary and discussion

The major emphasis of this paper is to show that the annual average TCHP of the study region is increasing. The ASMWP is known to impact the summer monsoon through its influence on the location of the monsoon onset vortex. Using *in situ* observations, satellite estimations, and model simulations, we show that OHC (and SST) in the ASMWP increased during 1993–2010 while ISMR decreased. Further, we suggest that the former leads to the latter by weakening the northward temperature gradient and, hence, allowing shifting convection southward. Here, we propose a possible linkage. More detailed modeling affords are required to prove this, which is beyond the scope of this study.

We recognize that other factors also affect ISMR, and therefore blur (and possibly overwhelm) any connection to ASMWP OHC. These factors include: increased aerosols, regional SST and OHC changes elsewhere in the Indian Ocean, and 'noise' due to climate variability such as El Nino Southern Oscillations (ENSO) and the Indian Ocean Dipole (IOD). Regarding the impact of aerosols, some studies suggest black carbon (Ramanathan *et al.*, 2005) or sulfate (Bollasina *et al.*, 2011) aerosols mask surface warming and the resulting cooler SST decreases monsoon rainfall, whereas, Wang (2004) draws the opposite conclusion. Regarding the impact of other regions of SST and OHC anomalies, Annamalai *et al.* (2013) noted the importance of regional heat sources over the Asian monsoon region, concluding that dynamical feedbacks among

increase in OHC of the region. However, the changes in OHC are due to a combination of surface fluxes and internal heat changes due to advection.

It is noteworthy that ISMR dropped from 910 to 820 mm during 1993–2010 (Figure 2), although the trend is not statistically significant ($p = 0.230$), suggesting that there is an inverse relationship between ISMR and ASMWP OHC. Such a linkage is possible as follows: Increased OHC and SST in the ASMWP lead to enhanced convection there, thereby shifting convection southward that would normally occur farther north and, hence, weakening the tropical easterly jet stream (Abish *et al.*, 2013) and ISMR. Earlier, Chung and Ramanathan (2006), using version 3 of the Community Climate Model, attributed a reduction in monsoon rainfall in their solution to an increase of equatorial SST; specifically, they report that the reduction in them complicate the hypothesis that increased ocean temperatures increase tropospheric moisture and hence rainfall. Further, Swapna *et al.* (2013) noted the importance of warming in the equatorial IO, concluding that the weakening of the summer monsoon circulation accelerated the warming of the equatorial IO, and that this warming in turn contributed to a further weakening of the monsoon with more monsoon breaks. When this small region can have a relation with the onset of the Southwest monsoon (which is proved through several publications), we suggest that OHC of the ASMWP could also be one of the factors that impact ISMR. Since we studied the overall annual trend in TCHP, we have not isolated the seasonal changes. This increase, as a part of the climatic trend, is possibly reducing the summer monsoon rainfall because the TCHP affects the atmosphere only at the time of the year when convection is possible at all, that is during the summer. In addition, the summer-averaged TCHP signal compares well with the annually averaged one. To confirm this property, we compared the satellite-derived TCHP during June to September with its 12-month average value. The comparison has a bias of $20 \, kJ \, cm^{-2}$ (June-September OHC being less than annual value) with a Pearson's correlation coefficient of 0.8, indicating that even the June-to-September TCHP is also increasing similar to the annual value, over the study period. (We are aware that TCHP during this season alone need not influence the Indian summer monsoon.)

In conclusion, our study suggests the importance of further investigation in to the influence of the ASMWP heat content on monsoon dynamics. It points toward the need for a sustained observational effort that allows the statistical relationship between OHC and Indian rainfall to be accurately determined. Further, it suggests the need for comprehensive modeling studies to demonstrate the dynamical processes involved in this linkage.

Acknowledgements

The authors acknowledge the TCHP data from AOML, NOAA, *in situ* temperature profiles from Indian National Centre for Ocean Information Services, rainfall data from India Meteorological Department, SST from remote sensing systems, and flux observations from Woods Hole Oceanographic Institution. The authors thank Dr Rashmi Sharma, Space Applications Centre for providing the TCHP values from the MOM-3.1 model. The support and encouragement provided at the respective organizations are gratefully acknowledged. The authors have no conflict of interest.

References

Abish B, Joseph PV, Johannessen OM. 2013. Weakening trend of the tropical easterly jet stream of the boreal summer monsoon season 1950–2009. *Journal of Climate* **26**: 9408–9414.

Annamalai H, Hfner J, Sooraj KP, Pillai P. 2013. Global warming shifts the monsoon circulation drying south Asia. *Journal of Climate* **26**: 2701–2718.

Bollasina MA, Yi M, Ramaswamy V. 2011. Anthropogenic aerosols and the weakening of the south Asian summer monsoon. *Science* **334**: 502–504.

Chung CE, Ramanathan V. 2006. Weakening of North Indian SST gradients and the monsoon rainfall in India and the Sahel. *Journal of Climate* **13**: 2036–2045.

Deepa R, Seetaramayya P, Nagar SG, Gnanaseelan C. 2007. On the plausible reasons for the formation of onset vortex in the presence of the Arabian Sea mini warm pool. *Current Science* **92**: 794–800.

Goni G, Kamholz S, Garzoli S, Olson D. 1996. Dynamics of the Brazil-Malvinas confluence based on inverted echo sounders and altimetry. *Journal of Geophysical Research* **101**(C7): 16273–16289.

Joseph PV. 1990. Warm pool over the Indian Ocean and monsoon onset. *Tropical Ocean Global Atmosphere Newsletters* **53**: 1–5.

Leipper D, Volgenau D. 1972. Hurricane heat potential of the Gulf of Mexico. *Journal of Physical Oceanography* **2**: 218–224.

Nagamani PV, Ali MM, Goni GJ, Dinezio PN, Pezzullo JC, Udaya Bhaskar TVS, Gopalakrishna VV, Kurian N. 2012. Validation of satellite-derived tropical cyclone heat potential with *in situ* observations in the north Indian Ocean. *Remote Sensing Letters* **3**(7): 615–620.

Ramanathan V, Chung C, Kim D, Bettge T, Buja L, Kiehl JT, Washington WM, Fu Q, Sikka DR, Wild M. 2005. Atmospheric brown clouds: impacts on South Asian climate and hydrological cycle. *Proceedings of the National Academy of Sciences of the United States of America* **102**: 5326–5333.

Rao RR, Sivakumar R. 1999. On the possible mechanisms of the evolution of a mini-warm pool during the pre-summer monsoon season and the onset vortex in the southeastern Arabian Sea. *Quarterly Journal of the Royal Meteorological Society* **125**: 787–809.

Rao RR, Jitendra V, GirishKumar MS, Ravichandran M, Ramakrishna SSVS. 2015. Interannual variability of the Arabian Sea Warm Pool: observations and governing mechanisms. *Climate Dynamics* **44**(7-8): 2119–2136.

Roxy M, Kapoor R, Terray P, Masson S. 2014. Curious case of Indian Ocean warming. *Journal of Climate* **27**: 8501–8509, doi: 10.1175/JCLI-D-14-00471.1.

Sanil Kumar KV, Hareesh Kumar PV, Joseph J, Panigrahi JK. 2004. Arabian Sea mini warm pool during May 2000. *Current Science* **86**: 180–184.

Sharma R, Agarwal N, Momin IM, Basu S, Agarwal VK. 2010. Simulated sea surface salinity variability in the tropical Indian Ocean. *Journal of Climate* **23**: 6542–6554.

Shay LK, Goni GJ, Black PG. 2000. Effect of a warm ocean ring on hurricane Opal. *Monthly Weather Review* **128**: 1366–1383.

Shenoi SSC, Shankar D, Shetye SR. 1999. On the sea surface temperature high in the Lakshadweep Sea before the onset of the southwest monsoon. *Journal of Geophysical Research* **104**: 15703–15712.

Swapna P, Krishnan R, Wallace JM. 2013. Indian Ocean and monsoon coupled interactions in a warming environment. *Climate Dynamics* **42**(9-10): 2439–2454, doi: 10.1007/s00382-013-1787-8.

Vinaychandran PN, Kurian J. 2008. Modeling Indian Ocean circulation: Bay of Bengal fresh plume and Arabian Sea min warm pool. In *Proceedings of the 12th Asian Congress of Fluid Mechanics*, Daejeon, Korea, 18–21 August 2008.

Vinaychandran PN, Shankar D, Kurian J, Durand F, Shenoi SSC. 2007. Arabian Sea mini warm pool and the monsoon onset vortex. *Current Science* **93**(2): 203–214.

Wang CA. 2004. A modeling study on the climate impacts of black carbon aerosols. *Journal of Geophysical Research* **109**: D03106, doi: 10.1029/2003jd004084.

Yu L, Jin X, Weller RA. 2007. Annual, seasonal, and inter annual variability of air–sea heat fluxes in the Indian Ocean. *Journal of Climate* **20**: 3190–3209.

13

Classification and forecast of heavy rainfall in Northern Kyushu during Baiu season using weather pattern recognition

Dzung Nguyen-Le,[1]* Tomohito J. Yamada[1] and Duc Tran-Anh[2]

[1] Faculty of Engineering, Hokkaido University, Japan
[2] National Center for Hydro-Meteorological Forecasting, Hanoi, Vietnam

*Correspondence to:
D. Nguyen-Le, Faculty of
Engineering, Hokkaido University,
N13-W8, Kita-ku, Sapporo,
Hokkaido 060-8628, Japan.
E-mail:
dzungnl@eng.hokudai.ac.jp

Abstract

In this study, the Self-Organizing Maps in combination with K-means clustering technique are used for classification of synoptic weather patterns inducing heavy rainfall exceeding 100 mm day^{-1} during the Baiu season (June–July) of 1979–2010 over northern Kyushu, southwestern Japan. It suggests that these local extreme rainfall events are attributed to four clustered patterns, which are primarily related to the Baiu front and the extratropical/tropical cyclone/depression activities and represented by the intrusion of warm and moist air accompanied by the low-level jet or cyclonic circulation. The classification results are then implemented with the analogue method to predict the occurrence (yes/no) of local heavy rainfall days in June–July of 2011–2016 by using the prognostic synoptic fields from the operational Japan Meteorological Agency (JMA) Global Spectral Model (GSM). In general, the predictability of our approach evaluated by the Equitable Threat Score up to 7-day lead times is significantly improved than that from the conventional method using only the predicted rainfall intensity from GSM. Although the false alarm ratio is still high, it is expected that the new method will provide a useful guidance, particularly for ranges longer than 2 days, for decision-making and preparation by weather forecasters or end-users engaging in disaster-proofing and water management activities.

Keywords: self-organizing map; heavy rainfall; clustering; forecast; Baiu; Kyushu

Introduction

During the Baiu season (June–July; JJ) in northern Kyushu, southwestern Japan (Figure 1), heavy rain frequently occur causing flooding and serious damages to life and properties. Thus, it is critical to provide an early prediction and warning for such rainfall events in the region. Generally, issuing an alert for the occurrence of locally intensive rain is mainly based on the predicted rainfall amount, e.g. from numerical models. Nowadays, owing to the progress in computing performance and atmospheric modelling, numerical models can run operationally with horizontal resolution down to a few kilometres. This fine resolution allows to explicitly resolving small-scale processes such as deep convection and orography effect, improves dramatically the predictability of rainfall. Nevertheless, it has always been challenging to accurately forecast rainfall in practical, especially for ranges longer than 2 days, owing to the imperfection of models and constraint of computational expenses. Meanwhile, medium-range (3–7 days) forecasts are crucial for reducing the impact of extreme rainfall by providing more time for decision-making and preparation. However, the coarse spatial resolution of numerical models used in medium-range forecast, typically larger than 10 km, limits their ability to predict the rainfall amount exactly.

Rainfall has nonlinear relationships with various meteorological factors and many local heavy rainfall events correspond to large-scale atmospheric condition (e.g. Baiu front). Classification and identification of such synoptic weather patterns (WP) are thus not only fruitful for understanding the genesis of local heavy rainfall but also potentially improving the forecasting capability. In this framework, an artificial neural network learning mechanism such as self-organizing map (SOM; Kohonen, 1982) can be an effective tool for analysing the atmospheric data (e.g. Nishiyama et al., 2007; Ohba et al., 2015, 2016). In this study, the SOM approach is used to objectively classify the anomalous WPs inducing heavy rainfall in Fukuoka-Saga prefectures, northern Kyushu, during the Baiu season. Further, the results are used with the analogue method to predict the occurrence (yes/no) of heavy rainfall days over the region. The analogue method (Lorenz, 1969) is based on the assumption that if the current WP is similar to those of historical observation, the local rainfall can be similar to that of the past.

The reminder of this paper is organized as follows. Section 2 briefly describes the data and our method. The main results are given in Section 3. Finally, Section 4 presents conclusions.

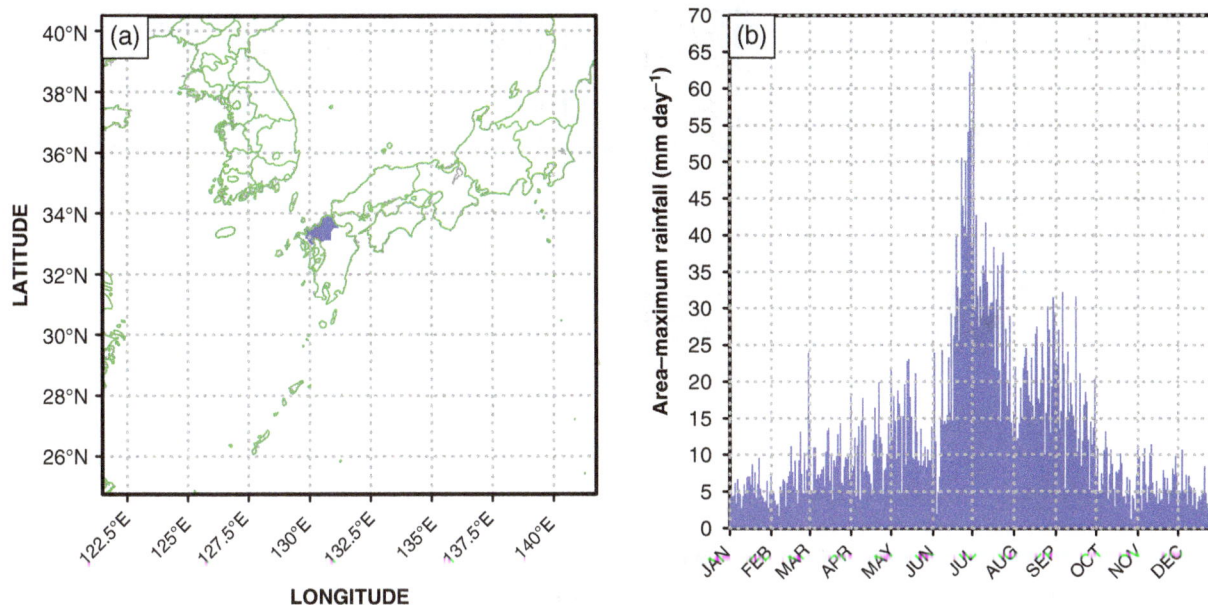

Figure 1. (a) Area of study used to define the synoptic patterns. The shading denotes the Fukuoka-Saga prefectures, northern Kyushu. (b) Climatological annual cycle of maximum rainfall intensity (mm day^{-1}) during 1979–2010 in Fukuoka-Saga prefectures.

Datasets and methodology

Datasets

The European Centre for Medium-Range Weather Forecasts ERA-Interim reanalysis (Dee *et al.*, 2011) on a 0.75° grid during JJ 1979–2011 is utilized for atmospheric variables. We also use the gridded dataset of APHRO_JP V1207 (Kamiguchi *et al.*, 2010) from 1979 to 2010 and the Japan Meteorological Agency (JMA) radar/rain gauge-analysed precipitation (R/A) (Nagata, 2011) from 2011 to 2016 as rainfall observation, in which rain gauge and radar estimation are compiled in a 0.05° and 1 km grid, respectively.

The scope of this study is to issue a warning of daily heavy rainfall using prognostic synoptic fields from JMA Global Spectral Model (GSM) and to evaluate the capability of our method up to 7-day lead times. We use atmospheric forecasts, specifically the operational TL959 (approximately 20-km resolution) and 60-vertical level version of GSM, which is executed for 96 h starting from 1200 UTC every day. Forecast is limited to JJ 2011–2016 and first 12-h is discarded. For comparison, predicted rainfall from GSM is also analysed.

Methodology

We applied SOM to daily-averaged atmospheric fields extracted from ERA-Interim, with a set of four variables was selected: 850-hPa zonal U and meridional wind V, 850-hPa and 500-hPa equivalent-potential temperature θ_e. Since θ_e is a thermodynamic parameter involving temperature and humidity, its distribution in the lower troposphere can be used to characterize the Baiu front activity (Tomita *et al.*, 2011). Other studies (e.g. Ninomiya, 2000; Ninomiya and Shibagaki, 2007)

also suggest that the differential advection of θ_e to the Japan Islands from the tropics that consists of a poleward moisture flux by the low-level jet (LLJ) is crucial to active the Baiu rainband. In addition, the importance of dry intrusion at the mid-level for heavy Baiu rainfall event is shown (e.g. Kato and Aranami, 2005; Kato, 2006). A quite similar selection of atmospheric variables was used by Ohba *et al.* (2015) to identify WPs that frequently provide heavy rainfall in Japan during the Baiu season. We also considered other candidates (e.g. 500-hPa geopotential high, mean sea-level pressure and vertically integrated moisture flux); however, selection of these variables was likely to degrade the performance of the classification and prediction.

Those four variables were obtained within a specific domain (24.75°–40.5°N, 121.5–141.75°E; Figure 1(a)) to form a total of 166 input vectors, corresponds to 166 heavy rainfall days were observed in JJ 1979–2010. Here, the heavy rainfall are defined when rainfall at least one grid box in Fukuoka-Saga prefectures exceeds 100 mm day^{-1}. By multiplying the dimension of the domain with the number of reanalysis fields, the size of each input vector is 4 (variables) × 28 (longitude points) × 22 (latitude points) = 2464. However, all elements are not at all independent due to spatial and inter-variable correlations among four fields, and the use of all elements in the training process is not an effective way in practice. To remove the redundant information contained in input vectors, principal component analysis (PCA) is applied for all four original reanalysis fields. Since both PCA and SOM are sensitive to the scales and thus dimensions of those fields, each variable was normalized prior to PCA. We only retain the empirical orthogonal functions that explain 90% of variance in the results. Then the corresponding principal components (PC) are digested into the SOM

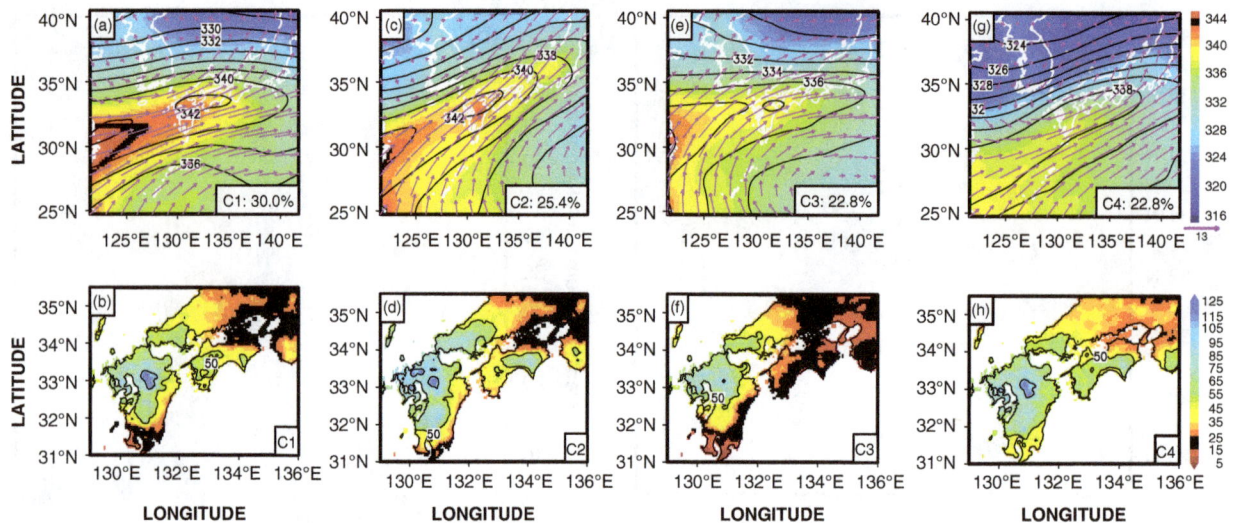

Figure 2. (a), (c), (e) and (g) Cluster-averaged (composited) daily-mean 850 hPa horizontal wind U, V (m s^{-1}: vector), equivalent-potential temperature θ_e (°K: shading), and 500 hPa equivalent-potential temperature θ_e (°K: contour) and (b), (d), (f) and (h) observed rainfall (mm day^{-1}) from Clusters 1–4 (C1–C4), respectively. Percentage denotes the occurrence frequency of each cluster.

instead of the original data. The number of retained PCs is $d = 31$, which means the size of input vectors is reduced by a factor of 80. This improved both the independency of elements and the computational time.

Note that the SOM training process depends on the size of SOM lattice and various training parameters (e.g. learning rate, radius, and training length). Trial and error is thus used to obtain the most suitable SOM size and parameters. The quantization error (QE) and topographic error (TE) are also used as performance indices. As a result, the SOM configuration of 7×7 hexagonal nodes (i.e. 49 WPs), a learning rate of 0.2 and a radius of 3 was selected. To keep the stability of SOM, the training length of two million steps was set. However, owing to concerns regarding the quality of SOM with a small number of samples, the bootstrap learning was incorporated into the SOM training process, in which the training sample was randomly drawn from the original training data. $B = 1000$ SOM maps generated by using the aforementioned configuration were compared. Accordingly, the map having the lowest QE and TE (Figure S1, Supporting information) and a relatively flat Sammon map (Figure S2), was defined as the master SOM.

Although SOM is powerful to project high-dimensionally nonlinear atmospheric features onto a visually understandable two-dimensional lattice, its drawbacks are unclear clustering boundaries between SOM nodes. To improve the clustering accuracy, the second stage is thus implemented, in which the SOM nodes are clustered again by the K-means method (Vesanto and Alhoniemi, 2000). However, the major disadvantage of K-means is the difficulty to decide an appropriate number of clusters K. Here, this problem can be avoided by predetermining K from the master SOM by using the U-matrix (unified distance matrix) method (Ultsch and Siemon, 1990), which is similar

to Nishiyama *et al.* (2007). The optimal clustering is determined based on the Davies-Bouldin index (DBI) proposed by Davies and Bouldin (1979). For more details on the SOM and U-matrix, one can refer to other studies (e.g. Nishiyama *et al.*, 2007).

In the forecast phase, daily-mean of the four variables 850-hPa U, V, θ_e and 500-hPa θ_e on each day in JJ 2011–2016 are extracted from GSM over the same region in the SOM training. The GSM forecast is then normalized using the same values and is projected onto the same PCA space as used for ERA-Interim. Next, we calculate the Mahalanobis distances md_j (Mahalanobis, 1936), which is essentially the distance between the new input vector $p = [p_1, p_2, \ldots, p_d]$ ($d = 31$) and the centroid $c_j = [c_{1j}, c_{2j}, \ldots, c_{dj}]$ of cluster j ($1 \leq j \leq K$), normalized by the standard deviation of the cluster in each dimension. Thus, it measures the number of standard deviations away the new vector is from the centroid of cluster j. In PCA space, the md_j is $\sqrt{\sum_{i=1}^{d} \left(\frac{p_i - c_{ij}}{\sigma_{ij}} \right)^2}$, where σ_{ij} is the standard deviation of cluster j in the ith dimension. Subsequently, if $\min_j md_j \leq 6$, the prognostic WP can be assigned to its best-match cluster c, where $c = \arg \min_j r_j$. Based on analogue method, the heavy rainfall will be predicted to occur in Fukuoka-Saga prefectures. To evaluate the predictability of our method, the probability of detection (POD), false alarm ratio (FAR) and equitable threat score (ETS) (see Appendix A2) were used.

Results

Classification

The clustering result (Figure S3) suggests four main different WPs inducing the heavy rainfall in Fukuoka-Saga

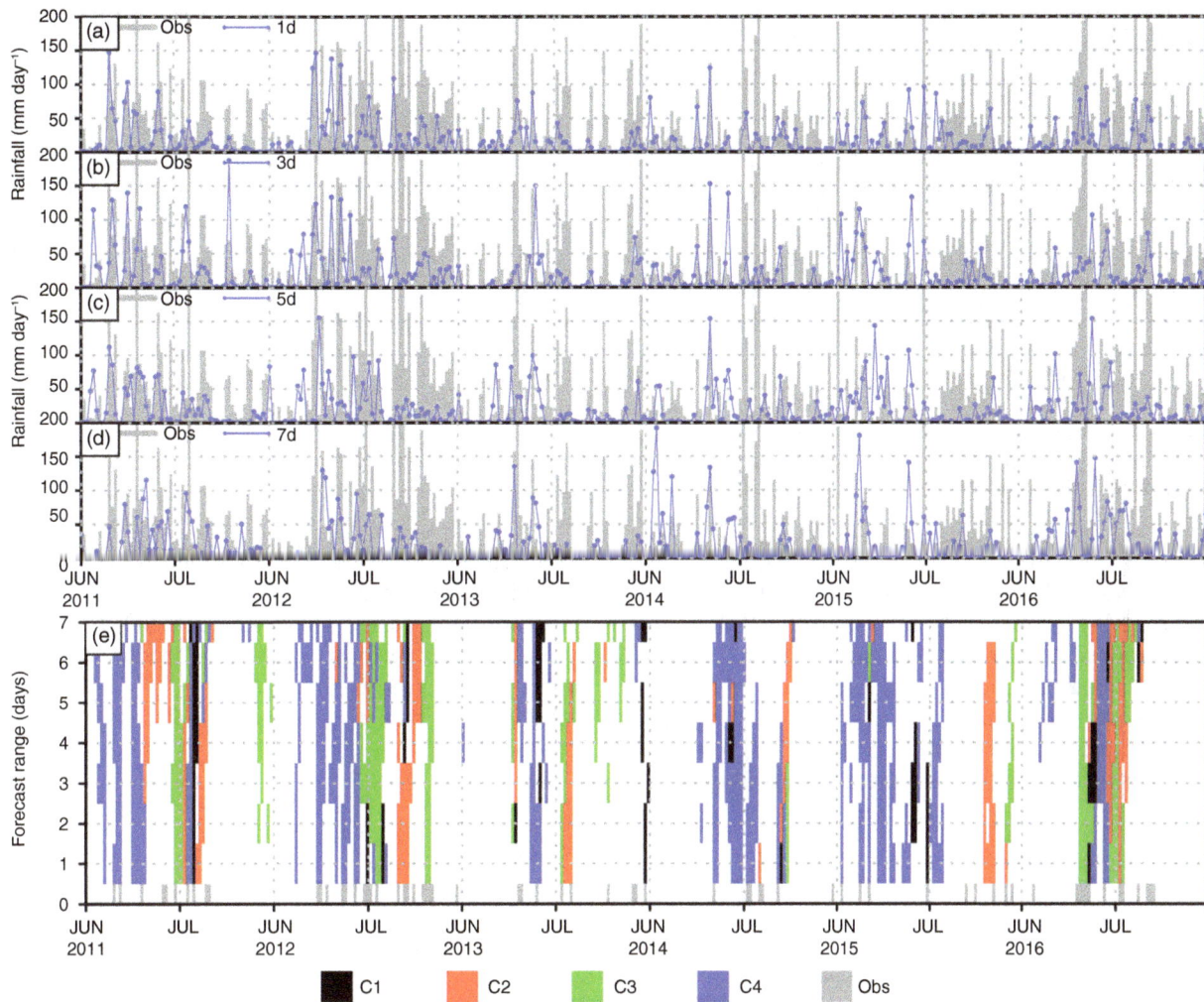

Figure 3. (a)–(d) Maximum rainfall intensity (mm day^{-1}) in Fukuoka-Saga prefectures, northern Kyushu during June–July from 2011–2016 derived from the observation (bars) and GSM forecast (lines) with 1-, 3-, 5- and 7-day lead times, respectively. (e) The occurrence (yes/no) of heavy rainfall day and causing clustered WP (solid boxes denote yes) predicted by using synoptic fields derived from the GSM forecasts with 1–7-day lead times (1–7 d). Here, forecast range of 0 day represents observation.

prefectures during the Baiu season, which are shown in Figure 2. The existence of a Baiu front may be observed in both the Clusters 1 and 2 (C1 and C2), in which at the low-level, the southwestern Japan region including Kyushu is covered by an intrusion of remarkably warm and humid air with the LLJ. Meanwhile, at the upper troposphere, the tongue of high θ_e air is separated west of northern Kyushu with the in-between lower θ_e air is drier but warmer than that just to the east (see Figure S4). This low θ_e air at 500 hPa is located over the high θ_e air at 850 hPa, leading to the strong convective instability around northern Kyushu and heavy rainfall is brought in the region (Kato, 2006). Although having higher low-level θ_e, the rain-producing influence of the LLJ and warm moist tongue of C1 on the Japan Islands is less than that of C2, possibly because their axis is more west–east. The area often experiences more than 50 mm day^{-1} of rainfall in C1 is relatively smaller than that in C2, including only mid- to northern Kyushu as well as the western edges of the Honshu and Shikoku Islands. Meanwhile, in C2, the high θ_e accompanying

the LLJ extends further from the southwest to the northeast elongated the Japan Islands; result in a larger rainband with moderate to heavy amount. Also note that in northern Kyushu, this cluster typically reproduces higher rainfall than C1.

Meanwhile, Cluster 3 (C3) is characterized by a relatively warm and moist air originating from the west to southwestern Japan and strong clockwise flow related to the Western North Pacific sub-tropical high. The lower θ_e around northern Kyushu, which is produced from drier but warmer air (Figure S4), is also observed at the mid-level, causing the convective instability there; although it is not as strong as that of C1 and C2. Nishiyama *et al.* (2007) suggest that this WP leads to the genesis of local afternoon shower due to a local atmospheric circulation enhanced by the sunshine. The authors also show that it cannot cause intensive rainfall since water vapour is insufficient, corresponds to the impacts of C3 are only limited to northern Kyushu and westernmost Honshu with only few spots exceeding 100 mm day^{-1}. On the other hand, C4 is linked to the low-level cyclonic circulation related to low pressure

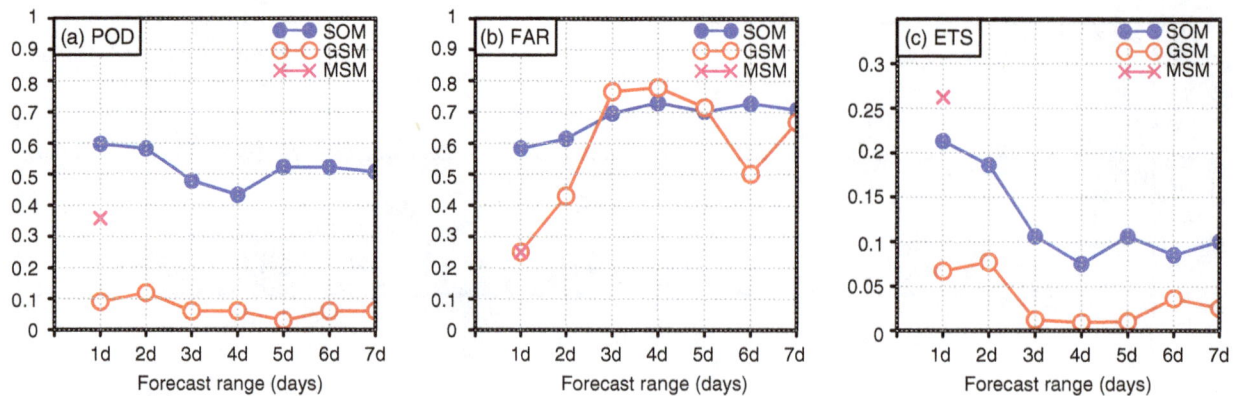

Figure 4. Forecast skill score, (a) probability of detection (POD), (b) false alarm ratio (FAR) and (c) equitable threat score (ETS), for the heavy rainfall occurrence over Fukuoka-Saga prefectures from the new method (SOM) and traditional method (GSM and MSM) with 1–7-day lead times (1–7 d).

system (i.e. extratropical/tropical cyclone/depression; ETCD) located around northern Kyushu. Although observed with the lowest θ_e, the WPs of this cluster reproduce an abundant amount of rainfall over the broadest area of southwestern Japan, as compared to other clusters. The characteristic features of the ETCD, however, are still not clearly recognized due to the relatively coarse resolution of current reanalysis data.

Predictability

Figure 3 shows the maximum daily rainfall in Fukuoka-Saga prefectures in JJ 2011–2016 from observation and the GSM forecast (Figures 3(a)–(d)) and the heavy rainfall occurrence predicted based on prognostic WP from GSM (Figure 3(e)). The forecast skill of this new method and the traditional method based on the predicted rainfall intensity from GSM, are shown in Figure 4. Basically, the operational GSM failed to detect the outbreak of local heavy rainfall with very poor POD since its predicted rainfall amounts are usually much smaller than observations. It also provides high FAR, especially with the forecast range from 3 days. As a result, the overall ETS with a threshold of 100 mm day^{-1} of GSM is 0.06 and 0.07 for 1 and 2 days forecast, respectively, and is generally less than 0.04 for ranges from 3 days.

Although both based on the GSM output, the predictability on the occurrence of local heavy rain by using the prognostic synoptic information is remarkably improved. The POD is generally higher than 0.4 for all forecast ranges, with the best of 0.6 are given from the two first lead times. However, many overestimated forecasts still exists, with FAR varying from 0.6 to 0.7. It may be due to the fact that although the synoptic atmospheric condition is highly favourable, extreme rainfall is not always generated inside the studied region. Another potential reason is the coarse resolution of reanalysis data used as training sample for the SOM puts strong limits on our current capability in obtaining an adequate classification of the heavy rainfall-inducing WP. Nevertheless, the ETS of

SOM reaches 0.22 and 0.19 for 1 and 2 days ahead, respectively, and then decreases to around 0.1 for the longer forecast ranges. This shows that SOM not only significantly outperforms GSM for the same forecast ranges but also its medium-ranges (3–7 days) forecast are better than short-ranges (1–2 days) forecast from GSM.

Conclusions

The SOM in combination with K-means cluster are conducted for classification of WP causing the heavy rainfall exceeding 100 mm day^{-1} during the Baiu season (JJ) of 1979–2010 over northern Kyushu, southwestern Japan. It results in four clustered patterns, which are primarily attributed to the Baiu front and the ETCD activities and characterized by high θ_e intrusion accompanying the LLJ or cyclonic circulation. These features are in good agreement with previous studies such as Nishiyama *et al.* (2007) and Ohba *et al.* (2015). The classification results are then implemented with the analogue method to forecast the occurrence (yes/no) of local heavy rainfall days in JJ of 2011–2016 using the prognostic synoptic fields from GSM. In general, the quantitative forecasting skill by the POD and ETS up to 7 days in advance under our approach is significantly improved than that from the conventional method based only on the predicted rainfall intensity from GSM, although the FAR is still high.

Note that our method is effective under the assumption that local rainfall is mostly controlled by the synoptic condition. It may be more difficult to apply to the region where rainfall process is mainly related to local factor such as the sea breeze circulation and local heating. Since the method can predict whether the heavy rainfall occurs but not being capable of providing the actual amount, questions may remain about its relative availability compared with other methods such as statistical downscaling by Ohba *et al.* (2016). Additionally, the forecast skill of SOM is still lower than that from the

JMA meso-scale model (MSM) that has a fine resolution of 5 km and the lead time up to 39 h (see Figure 4). Thus, there is still room for improvement. The spatial resolution of ERA-Interim is only 0.75°, which falls short for a precise description of heavy rainfall-inducing WPs, particularly for the ETCD activities. Given the weaknesses of the current dataset used as training sample for SOM, classification and predication results can be improved by using higher resolution and more accurate reanalysis and observations. Also further studies employed improved method should be considered. Nevertheless, in addition to previous studies, the present results encourage the idea of using weather pattern recognition for heavy rainfall prediction, by which it can provide a fruitful and first-order guidance, particularly for ranges longer than 2 days, for decision-making by weather forecasters or end-users engaging in disaster management activities.

Acknowledgements

The authors greatly appreciate three anonymous reviewers for their constructive comments. We also thank Dr. Le Duc of JAMSTEC for his critical suggestions. This research is supported by the Social Implementation of Climate change Adaptation Technology (SICAT) project of the Japanese Ministry of Education, Culture, Sports, Science and Technology (MEXT); and the Advancing Co-design of Integrated Strategies with Adaptation to Climate Change (ADAP-T) of the Japan International Collaboration Agency (JICA)/Japan Science and Technology Agency (JST) Science and Technology Research Partnership for Sustainable Development (SATREPS).

Supporting information

The following supporting information is available:

Appendix A1. Methodology

Figure S1. The Quantization Error (QE) and Topographic Error (TE) of 1000 SOM maps using bootstrap learning algorithm. The blue mark denotes the optimal master SOM.

Figure S2. Sammon map for the master SOM with the four corner nodes labelled.

Figure S3. Scaled U-matrix of the master SOM with the optimum clustering by K-means is shown.

Appendix A2. Forecast skill score

Appendix A3. Results of classification

Figure S4. Cluster-averaged (composited) of daily-mean (a), (c), (e) and (g) 500-hPa potential temperature θ (oK: shading) and (b), (d), (f) and (h) 500-hPa relative humidity (%: shading) from Clusters 1–4 (C1–C4), respectively. The contours denote each cluster-averaged of daily-mean 500-hPa equivalent-potential temperature θ_e.

References

Davies DL, Bouldin DW. 1979. A cluster separation measure. *IEEE Transactions on Pattern Analysis and Machine Intelligence* **1**: 224–227.

Dee DP, Uppala SM, Simmons AJ et al. 2011. The ERA-Interim reanalysis: configuration and performance of the data assimilation system. *Quarterly Journal of the Royal Meteorological Society* **137**: 553–597. https://doi.org/10.1002/qj.828.

Kamiguchi K, Arakawa O, Kitoh A, Yatagai A, Hamada A, Yasutomi N. 2010. Development of APHROJP, the first Japanese high-resolution daily precipitation product for more than 100 years. *Hydrological Research Letters* **4**: 60–64. https://doi.org/10.3178/HRL.4.60.

Kato T. 2006. Structure of the band-shaped precipitation system inducing the heavy rainfall observed over northern Kyushu, Japan on 29 June 1999. *Journal of the Meteorological Society of Japan* **84**: 129–153.

Kato T, Aranami K. 2005. Formation factors of 2004 Niigata-Fukushima and Fukui heavy rainfalls and problems in the predictions using a cloud-resolving model. *Scientific Online Letters on the Atmosphere* **1**: 1–4.

Kohonen T. 1982. Self-organized formation of topologically correct feature maps. *Biological Cybernetics* **43**: 59–69. DOI: https://doi.org/10.1007/ BF00337288.

Lorenz EN. 1969. Atmospheric predictability as revealed by naturally occurring analogues. *Journal of the Atmospheric Sciences* **26**: 639–646.

Mahalanobis PC. 1936. On the generalised distance in statistics. *Proceedings of the National Institute of Sciences of India* **2**: 49–55.

Nagata K. 2011. Quantitative precipitation estimation and quantitative precipitation forecasting by the Japan Meteorological Agency. *Technical review of RSMC Tokyo – Typhoon Center* **13**: 37–50http:// www.jma.go.jp/jma/jma-eng/jma-center/rsmc-hp-pub-eg/techrev/ text13-2.pdf.

Ninomiya K. 2000. Large- and meso-∝-scale characteristics of Meiyu–Baiu front associated with intense rainfalls in 1–10 July 1991. *Journal of the Meteorological Society of Japan* **78**: 141–157.

Ninomiya K, Shibagaki Y. 2007. Multi-scale features of the mei-yu–baiu front and associated precipitation systems. *Journal of the Meteorological Society of Japan* **85B**: 103–122. https://doi .org/10.2151/jmsj.85B.103.

Nishiyama K, Endo S, Jinno K, Uvo CB, Olsson J, Berndtsson R. 2007. Identification of typical synoptic patterns causing heavy rainfall in the rainy season in Japan by a self-organizing map. *Atmospheric Research* **83**: 185–200.

Ohba M, Nohara D, Yoshida Y, Kadokura S, Toyoda Y. 2015. Anomalous weather patterns in relation to heavy precipitation events in Japan during the Baiu season. *Journal of Hydrometeorology* **16**: 688–701. https://doi.org/10.1175/JHM-D-14-0124.1.

Ohba M, Kadokura S, Nohara D, Toyoda Y. 2016. Rainfall downscaling of weekly ensemble forecasts using self-organising maps. *Tellus A* **68**: 29293. https://doi.org/10.3402/tellusa.v68.29293.

Tomita T, Yamaura T, Hashimoto T. 2011. Interannual variability of the Baiu season near Japan evaluated from the equivalent potential temperature. *Journal of the Meteorological Society of Japan* **89**: 517–537.

Ultsch A, Siemon HP. 1990. Kohonen's self organizing feature maps for exploratory data analysis. In Proceedings of INNC'90, International Neural Network Conference. Kluwer Academic Publishers: Dordrecht; 305–308.

Vesanto J, Alhoniemi E. 2000. Clustering of the self-organizing map. *IEEE Transactions on Neural Networks* **11**: 586–600. https://doi.org/ 10.1109/72.846731.

14

Changes in the ENSO–rainfall relationship in the Mediterranean California border region

Edgar G. Pavia*

Centro de Investigación Científica y de Educación Superior de Ensenada (CICESE), México

*Correspondence to:
E. G. Pavia, CICESE, PO BOX
434844, San Diego, CA
92143-4844, USA.
E-mail: epavia@cicese.mx

Abstract

The El Niño-Southern Oscillation (ENSO)–rainfall relationship in the Mediterranean California border region (MCBR) changed recently: considering 30-year running periods (1951–1952 to 1980–1981, until 1986–1987 to 2015–2016) in four stations, ENSO–rainfall correlations (r), and mean precipitation (P) decreased almost simultaneously. Moreover, r and P appear related throughout the entire record: at the beginning and recently both seem lower than at an intermediate stage. Similar results are found for gridded precipitation data in the vicinity of the MCBR, suggesting that periods of recurrent dry seasons have been less related to ENSO than periods of numerous rainy seasons in this region during the last 65 years.

Keywords: ENSO; California rainfall; changing relationship

1. Introduction

The relationship between El Niño-Southern Oscillation (ENSO) and rainfall in the Mediterranean California border region (MCBR), plus its modulation by decadal variability have been discussed recently (Pavia *et al.*, 2016). In particular, the MCBR has experienced a drought in the last few years, while the ENSO-influence on local precipitation seems to have decreased (e.g. contrary to most cases, the 2010–2011 La Niña was a wet season, and the 2015–2016 El Niño was a dry season). Focusing solely in the ENSO–rainfall relationship, we wonder if the present situation (drought and apparently decreased ENSO-influence in the MCBR's rainfall) is just a coincidence or it corresponds to wider time- and space-patterns. For example, is the opposite true (wet periods corresponding to strong ENSO-influence on MCBR's rainfall)? Is this just a recent coincidence or similar situations have occurred before? Is all this exclusive of the MCBR or neighboring regions have experienced similar situations? The purpose of this work is to try to answer these questions, mainly because within the MCBR several major cities are located (Figure 1); but also because we think this study might help elucidate whether the recent drought may be explained in terms of natural variability (Seager *et al.*, 2015), anthropogenic warming (Diffenbaugh *et al.*, 2015), or a combination of both.

2. Data and methods

We use precipitation data from four stations within the MCBR: Los Angeles (LA), San Diego (SD), Tijuana (TJ), and Ensenada (EN) from the National Weather Service and the Mexican Water Authority, and the available gridded data of Livneh *et al.* (2015). For ENSO we use the El Niño 3.4 index (Nino3.4) from the Physical Science Division of NOAA's Earth System Research Laboratory. We begin by constructing Nino3.4 and rainfall annual indices (July to June) for the four stations from the 1951–1952 to the 2015–2016 season, as in Pavia *et al.* (2016). Similarly we construct Nino3.4 and annual indices for the gridded rainfall data from the 1950–1951 to the 2012–2013 season. We use correlation analysis for the entire records (r_o) plus for 30-year sliding periods (r), for the latter we also calculate annual rainfall mean values (P). We calculate the correlations of P and r [$r'(P,r)$] for the gridded and the four stations rainfall data, in the latter case we also fit a straight line to P versus r. Finally, we test the statistical significance of all correlation values in terms of their degrees of freedom (df) in order to calculate p-values; and also to easily identify, with the help of a list of the different periods used (Table 1), the time series used in the correlation analyses by their sample size (N), i.e. $N = df + 2$.

3. Results

3.1. Entire period

The map of Nino3.4 and rainfall baseline correlations, from 1950–1951 to 2012–2013, with $df = 61$, for the gridded data are shown in Figure 1(a), for $0.5 < r_o < 0.6$ (red areas) $p_o(<0.0001)$. Similarly, baseline correlations, from 1951–1952 to 2015–2016, with $df = 63$, for the four stations are: $r_o = 0.35$, $p_o(0.002)$, for LA; $r_o = 0.49$, $p_o(<0.0001)$, for SD; $r_o = 0.45$, $p_o(<0.0001)$, for TJ; and $r_o = 0.52$, $p_o(<0.0001)$, for EN.

3.2. The 30-year sliding periods

The map of r' from periods 1: 1950–1951 to 1979–1980, to 34: 1983–1984 to 2012–2013,

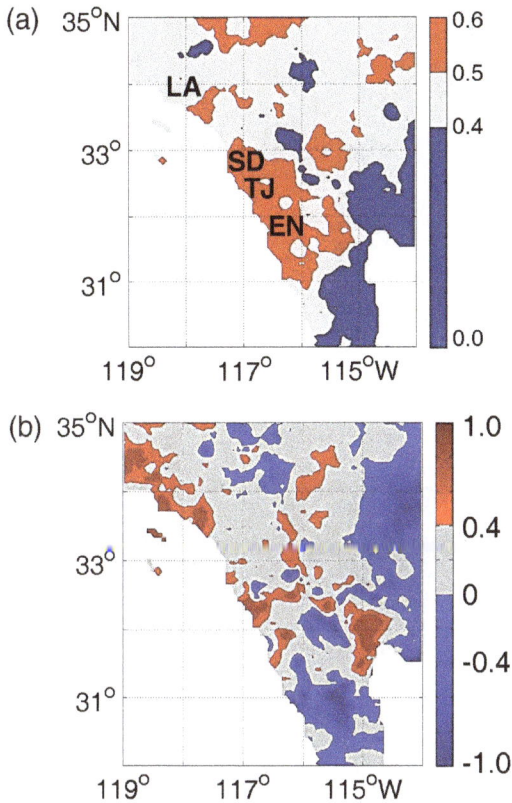

Figure 1. The region of study. (a) The color bar indicates the r_0 value, and the approximate location of stations considered (black acronyms). (b) The color bar indicates the r' value. For statistical significance, see text.

Table 1. The 30-year periods used in this study (periods 3–33 not shown).

Period number	30-year range	Gridded data	LA, SD, TJ, and EN
1	1950–1951 to 1979–1980	Yes	No
2	1951–1952 to 1980–1981	Yes	Yes
…	…	Yes	Yes
34	1983–1984 to 2012–2013	Yes	Yes
35	1984–1985 to 2013–2014	No	Yes
36	1985–1986 to 2014–2015	No	Yes
37	1986–1987 to 2015–2016	No	Yes

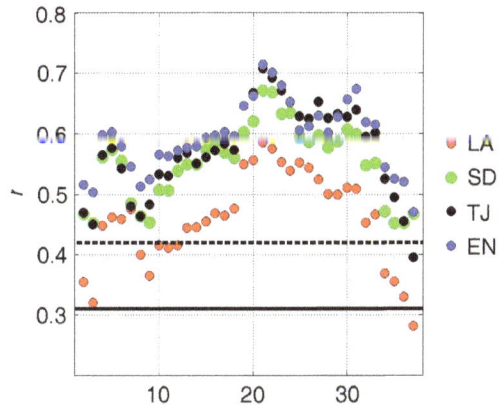

Figure 2. Time series of r for the four stations. The dashed line indicates the $p(0.01)$ significance level, and the heavy continuous line indicates the $p(0.05)$ significance level. The abscissas indicate the period number.

with $df = 32$, for the gridded data are presented in Figure 1(b), for $r'(P, r) > 0.4$ (red areas) $p(0.01)$ or better. Time series of r, from periods 2: 1951–1952 to 1980–1981, to 37: 1986–1987 to 2015–2016, and statistical significance levels, with $df = 28$, for the four stations are presented in Figure 2. For LA, SD, TJ, and EN plots of P versus r are presented in Figure 3. A linear fit to each of these four cases yield a positive slope, and $r'(P, r)$ and statistical significance, with $df = 34$, are as follows: $r' = 0.75$, for LA; $r' = 0.76$, for SD, $r' = 0.83$, for TJ; and $r' = 0.79$ for EN; all four cases at $p(<0.0001)$.

4. Discussion and conclusions

The relationship between ENSO and rainfall in the continental-coastal region approximately between 31°–35°N, 116°–119°W, labeled here as MCBR, has been recognized for some time (see, e.g. Schonher and Nicholson, 1989; Pavia and Badan, 1998; Fierro, 2014). However, this relationship seems to have weaken recently: correlations between Nino3.4 and gridded rainfall annual indices (from 1950–1951 to 2012–2013) are just $0.4 < r_0 < 0.6$ [gray and red areas in Figure 1(a)], and correlations between Nino3.4 and the four representative stations rainfall annual indices (from 1951–1952 to 2015–2016)

are just $0.35 < r_0 < 0.52$; when, for example, Pavia and Badan (1998) reported correlations above 0.64 (from 1951–1952 to 1996–1997) within this region. These four stations reached highest correlations of $0.67 < r < 0.72$ for the period 21: 1970–1971 to 1999–2000 (Figure 2) which have been declining since. Likewise, in SD, TJ, and EN precipitation mean values (P) reached a peak in the period 25: 1974–1975 to 2003–2004 (LA in period 24: 1973–1974 to 2002–2003) and also have been declined since. Indeed for gridded data r and P are positively correlated in and around the region [gray and red color in Figure 1(b)], although some of these correlations are low and non-statistically significant at $p(0.01)$ [gray color in Figure 1(b)]. Similarly, negative correlations are found mostly to the south and east of the MCBR [blue colors in Figure 1(b)]; gridded data is used only to get an idea of the spatial pattern of our analyses, and correlation values are used only qualitatively. Plots of P versus r for the four stations (Figure 3), all show similar features. In addition to yielding linear-fit positive slopes, the four cases show mostly bluish (early stage periods) and reddish (late stage periods) for lower P and r, and mostly greenish and yellowish (middle-stage periods) for higher P and r (Figure 3); these station data are considered more reliable for specific locations since they have not been spatially interpolated. Nevertheless the qualitatively good agreement between gridded and

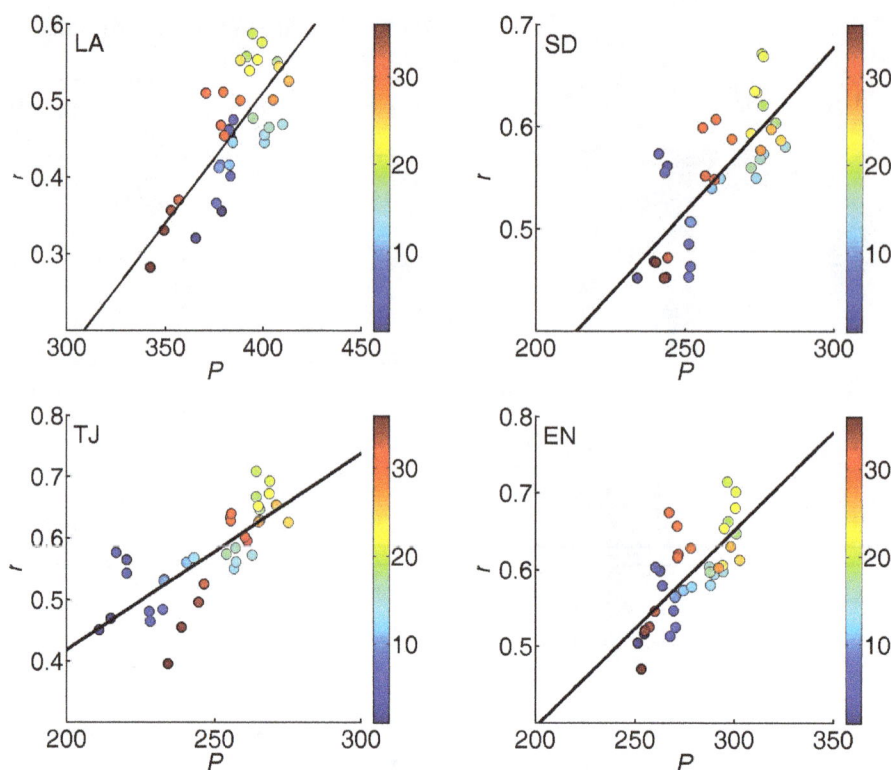

Figure 3. P (mm/year) versus r for the four stations. A straight line indicates their linear-fit (see text) and the color bar indicates the period number.

station data is a robust indication of a positive relationship between P and r in this region. Consequently, we conclude that in the MCBR, especially in the zones near the coast, periods of recurrent dry seasons have been less related to ENSO than periods of numerous rainy seasons, at least during the last 65 years. This conclusion (valid for the red areas of Figure 1(b), and the four representative stations in Figure 3) is more in line with the explanation of natural variability, rather than anthropogenic warming, as the cause of the recent MCBR drought, since the present conditions are similar to the conditions at the beginning of our record. The spatial patterns represent the 1950–2013 gridded data record, however the station data show that the present condition has been enhanced in the last couple years, since compared to previous results using 1951–2013 records (Pavia *et al.*, 2016), P and r have continued to decrease in all cases but the last two periods (36 and 37) in SD (Figure 2). Although the late 20th century increase in Central Pacific El Niño events (Yeh *et al.*, 2009) may help to explain the strengthening of ENSO–rainfall relationship during the middle-stage periods, it may not explain the early and late stage of the opposite condition. It is also worth noting that the northern-most LA station reports the less statistical significant correlations [e.g. $r = 0.28$, $p(0.07)$, for period 37]. Nonetheless, we believe the answer to why extended periods of below-average (above-average) rainfall concur with weak (strong) ENSO–rainfall relationship will be found only by examining sea surface patterns in the Pacific Ocean and elsewhere

(Seager *et al.*, 2015). The latter is a task which is obviously beyond the aim of this short contribution, whose goal is rather to incite further research on this topic. As the 2016–2017 La Niña becomes a wet season in the MCBR, ENSO–rainfall correlations would decrease while the drought ameliorates but continues. This result should strenghten our conclusions.

Acknowledgements

This work was funded by CONACYT (Mexico). I thank Federico Graef, Ramón Fuentes-Franco, and Ismael Villanueva for several brief, but fruitful discussions. The ENSO indices were obtained from NOAA, California precipitation data is from the NWS, and Baja California precipitation data is from the Mexican Water Authority.

References

Diffenbaugh N, Swain DL, Touma D. 2015. Anthropogenic warming has increased drought risk in California. *Proceedings of the National Academy of Sciences* **112**: 3931–3936.

Fierro AO. 2014. Relationships between California rainfall variability and large-scale climate drivers. *International Journal of Climatology* **34**: 3626–3640.

Livneh B, Bohn TJ, Pierce DW, Munoz-Arriola F, Nijssen B, Vose R, Cayan DR, Brekke L. 2015. A spatially comprehensive, meteorological data set for Mexico, the U.S., and southern Canada (NCEI Accession 0129374). Version 1.1. NOAA National Centers for Environmental Information. Dataset. doi: 10.7289/V5X34VF6 (accessed 26 September 2016).

Pavia EG, Badan A. 1998. ENSO modulates rainfall in the Mediterranean Californias. *Geophysical Research Letters* **25**: 3855–3858.

Pavia EG, Graef F, Fuentes-Franco R. 2016. Recent ENSO-PDO precipitation relationships in the Mediterranean California border region. *Atmospheric Science Letters* **17**: 280–285, doi: 10.1002/asl.656.

Schonher T, Nicholson SE. 1989. The relationship between California rainfall and ENSO events. *Journal of Climate* **2**: 1258–1269.

Seager R, Hoerling M, Schubert S, Wang H, Lyon B, Kumar A, Nakamura J, Henderson N. 2015. Causes and predictability of the 2011-14 California drought. *Journal of Climate* **28**: 6997–7024.

Yeh S-W, Kug J-S, Dewitte B, Kwon M-H, Kirman BP, Jin F-F. 2009. El Niño in a changing climate. *Nature* **461**: 511–514.

15

Effect of changing tropical easterly jet, low level jet and quasi-biennial oscillation phases on Indian summer monsoon

P. Rai and A. P. Dimri*

School of Environmental Sciences, Jawaharlal Nehru University, New Delhi, India

*Correspondence to:
A. P. Dimri, School of
Environmental Sciences,
Jawaharlal Nehru University,
New Delhi 110067, India.
E mail: apdimri@hotmail.com

Abstract

Using the National Center for Environmental Prediction and National Center for Atmospheric Research (NCEP/NCAR) reanalysis wind and homogeneous Indian summer monsoon (ISM) rainfall data during June, July, August and September (JJAS) for the period of 1953–2012, a long-term trend has been extracted in the tropical easterly jet (TEJ) and lower level jet (Somali jet/LLJ) along with the rainfall trend over Indian region using empirical mode decomposition (EMD) technique. The zonal wind speed at 100 hPa shows an increasing trend, while a decreasing trend is observed at 850 hPa. The quasi-biennial oscillation (QBO) phases, i.e. west and east phase, show a clear difference in wind anomalies at 100 and 850 hPa. A good (bad) monsoon year during the west (east) phase is confirmed on the basis of wind anomalies, velocity potential and divergent wind.

Keywords: Indian summer monsoon; tropical easterly jet; low level jet; quasi-biennial oscillation

1. Introduction

The Indian summer monsoon (ISM) rainfall, primarily caused by the differential heating between continent and ocean is the result of the general circulation features on global to local scale, affecting the Indian subcontinent. The precipitation due to significant amount of moisture incursion by southwesterly winds over the Indian subcontinent plays a vital role in various sectors viz. agriculture, water management, economy, etc., thus making it important to study its long-term trends and variability. The presence of low level jet (LLJ) at lower level (850 hPa) and tropical easterly jet (TEJ) at upper level (100 hPa) characterizes the ISM within the troposphere. The first detailed study on TEJ was carried out by Koteswaram (1958). TEJ having a wind speed of $40–50\,\mathrm{m\,s^{-1}}$ and strongest winds over the Arabian Sea during the Northern Hemispheric summer (Reiter, 1961) is a key feature of the ISM circulation and can be observed in the months of June, July, August and September (JJAS) at 100–150 hPa (Krishnamurti and Bhalme, 1976).

In the Tropics, the positive (negative) anomalies in rainfall are found to be associated with strong (weak) TEJ (Kobayashi, 1974). Owing to its synoptic nature, changes in strength of TEJ are considered as one of the indicators of long-term changes in atmospheric general circulation (Pielke *et al.*, 2001). A study by Kanamitsu *et al.* (1972) on the TEJ intensity during the summers of 1967 and 1972 found a weaker jet prevailing during 1972, which resulted to a drought condition over India. Various studies hint that the size of TEJ has reduced in

the recent years (Pattanaik and Satyan, 2000; Rao *et al.*, 2004). Decrease in strength of TEJ in the recent years has been reported by Sathiyamoorthy (2005) but an increase of $1\,\mathrm{m\,s^{-1}}\,\mathrm{year^{-1}}$ in strength of TEJ from 2000 onwards has been shown by Roja Raman *et al.* (2009) and Venkat Ratnam *et al.* (2013). Different physical mechanisms accountable for the maintenance of the TEJ are found in literature. Koteswaram (1958) and Flohn (1964) related the low level monsoonal flow and rainfall to the variation in intensity of the TEJ. Previous study by Krishnamurti and Bhalme (1976) observed that the cumulus convection results in latent heat release which leads to coupling of the lower level circulation with the upper level circulation.

In addition, quasi-biennial oscillation (QBO) reported in the 1960s, is a major oscillation in the equatorial stratospheric zonal wind (Veryard and Ebdon, 1961; Lindzen and Holton, 1968). It has a cycle of about 26 months in which there is periodic change in its zonally symmetric wind regimes, i.e. easterly and westerly regimes. The downward propagation with a speed of nearly $1\,\mathrm{km\,month^{-1}}$ through lower stratosphere is a key feature of this oscillation (Holton, 1968; Quiroz, 1981). Very few studies have been done previously on relationship between phases of QBO at 30 hPa and monsoon rainfall of India. A study by Mukherjee *et al.* (1985) found a significant relationship between the QBO phases at lower stratosphere and the percentage of departure of the ISM rainfall based on a short span of data for an equatorial station. Their main findings are on the association of weak (strong) ISM with the strong easterly (weak westerly) phase of the QBO. They also observed a variability of about 15% related to QBO phases in the ISM rainfall. Madhu (2014) also reported the presence of westerly anomaly in zonal wind during the deficit ISM years in the equatorial upper troposphere and presence of easterly anomaly during excess ISM years. Equatorial lower stratosphere on the other hand is marked by the opposite zonal wind anomaly during the deficit and excess years of ISM.

Climatic time series are often nonlinear and non-stationary. Therefore, the use of traditional techniques such as Fourier analysis may not be suitable. Although wavelet analysis can be used to deal with non-stationary series, but it cannot handle nonlinear series. To tackle the issue of both nonlinearity and non-stationarity of the series, Huang *et al.* (1998) developed a method named empirical mode decomposition (EMD). Since then, EMD has been widely used to analyze the nonlinear and non-stationary time series. Time series can be decomposed adaptively via EMD method into time–frequency space, thus enabling us to find both the dominant modes of variability in a time series and variation of these modes in time. Very few studies have been done previously using nonlinear EMD technique related to ISM and its components in Indian scenario. Iyengar and Kanth (2005) proposed a new forecasting strategy based on the combination of nonlinear and linear modeling for the ISM rainfall using EMD technique. The change in temperature under global warming scenario was studied using EMD technique by Molla *et al.* (2006). Dwivedi and Mittal (2007) used EMD technique to forecast the duration of active and break spells in intrinsic mode functions (IMFs) of intraseasonal oscillations.

In this study, we focus on the extraction of embedded nonlinear trend present in TEJ and LLJ along with the rainfall time series. In addition, the relationship of ISM rainfall with different phases of QBO and the physical mechanism responsible for the variability during the QBO period has also been analyzed.

2. Data and methodology

The 60-year (1953–2012) monthly mean wind data from National Center for Environmental Prediction and National Center for Atmospheric Research (NCEP/NCAR) reanalysis (http://www.esrl.noaa.gov/psd/) (Kalnay *et al.*, 1996) available at 2.5° resolution has been used in this study. Monthly homogeneous rainfall data (Parthasarathy *et al.*, 1995) for Indian region was obtained from Indian Institute of Tropical Meteorology, Pune, India (IITM, http://www.tropmet.res.in) from 1953 to 2012 over all the 316 rain gauge stations. East and west phase of QBO was extracted on the basis of wind speed at 30 hPa, i.e. wind speed $>5\,\mathrm{m\,s^{-1}}$ is considered as a west phase year and wind speed $<-5\,\mathrm{m\,s^{-1}}$ is considered as an east phase year. Based on this criterion, 20 west phase years and 33 east phase years were found and are presented in Table 1.

The rainfall and zonal wind data are decomposed using EMD method. Compared to other decomposition methods, EMD is empirical, intuitive, direct and adaptive (Wu and Huang, 2009). The main idea of EMD is to decompose the data locally into different oscillatory components called IMF. The lower IMFs represent fast oscillation components while the higher IMFs represent slow oscillations (Zhang *et al.*, 2010). The last component will not have any cyclic component and is a monotonic function representing the trend present in the time series data. To check whether the extracted IMFs are information rich or just contain noise, a statistical approach developed by Wu and Huang (2004) is used in which confidence limit level is defined and the spread lines (upper and lower) are determined. Based on the location of energy of IMFs, i.e. above the upper bound and below the lower bound, they are considered to contain information at the predefined confidence interval.

3. Results and discussion

EMD based analysis of rainfall and zonal wind time series data at 850 and 100 hPa [Figures 1(a)–(c)] extracts six embedded modes each and a residual trend in the data. The time scale characteristics have increased with the mode index, that is to say, frequency of each IMF has gradually reduced from IMF1 to IMF6. The last residual component represents the trend of the series. It is evident from the figures that the trend is devoid of any modulation and represents the timeline profile of the embedded trend in all the three-time series, which could not have been possible by conventional methods. In Figures 1(a)–(c), only zonal wind at 100 hPa shows a positive trend during the study period in accordance with the results of Sathiyamoorthy (2005), Sreekala *et al.* (2014) and Pattanaik and Satyan (2000). Zonal wind at 850 hPa and rainfall time series show a declining trend during the study period. The decreasing trend in the ISM rainfall is confirmed by the very recent study of Naidu *et al.* (2015). According to Naidu *et al.* (2015), the rainfall over Indian region during the last three decades has decreased because of the decreased circulation during monsoon season over India and the north–south SST gradient over the North Indian Ocean.

Spatial distribution of zonal wind at 100 hPa level has been shown in Figures 2(a)–(c) during the early (1953–1982) and the recent years (1983–2012). This division of early and recent years has been done on the

Table 1. West and east phase years of QBO.

QBO phases	Years
West phase years	1953, 1955, 1957, 1961, 1966, 1969, 1971, 1973, 1975, 1978, 1980, 1985, 1990, 1995, 1997, 1999, 2002, 2004, 2006, 2008
East phase years	1954, 1956, 1958, 1960, 1962, 1963, 1965, 1968, 1970, 1972, 1974, 1976, 1977, 1979, 1981, 1982, 1984, 1986, 1987, 1989, 1991, 1992, 1994, 1996, 1998, 2000, 2001, 2003, 2005, 2007, 2009, 2010, 2012

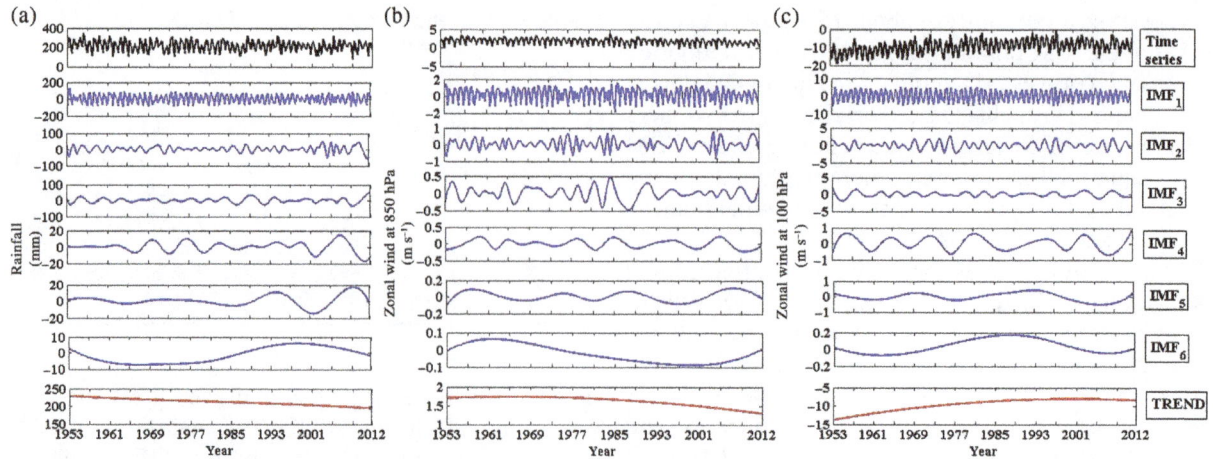

Figure 1. Extracted modes and trends using EEMD technique for, (a) rainfall (mm month^{-1}), (b) 850 hPa (m s^{-1}) zonal wind and (c) 100 hPa (m s^{-1}) zonal wind, averaged over (40°–120°E, 10°S–40°N) for the period 1953–2012. The top most panels show the original time series, while lower most panels show the corresponding trends and rest of the panels show IMF$_1$ to IMF$_6$.

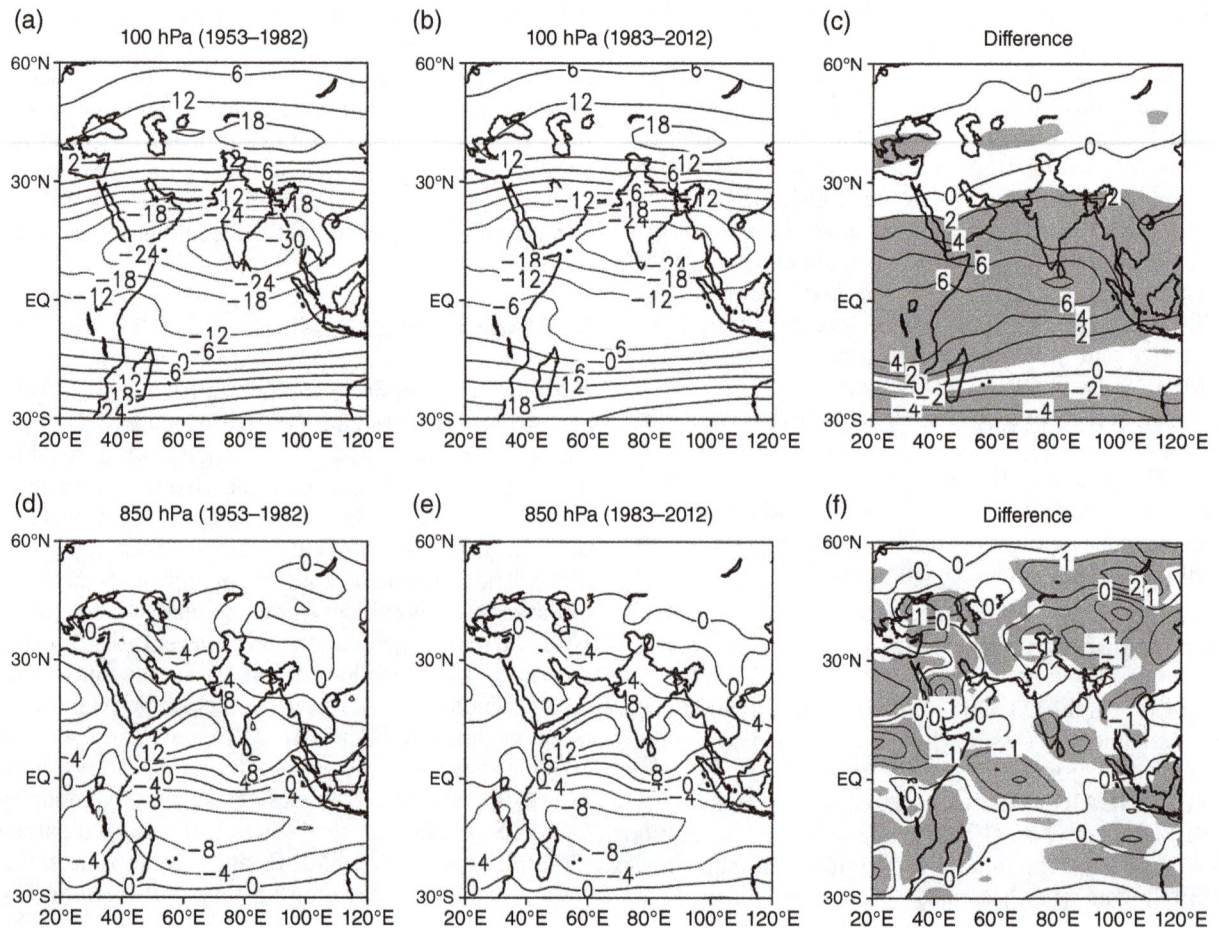

Figure 2. Spatial distribution of zonal wind (m s^{-1}) anomaly at 100 hPa during (a) early phase (1953–1982), (b) recent phase (1980–2011) and (c) their difference (recent phase – early phase) at 95% significant level. Panels (d)–(f) are same but for zonal wind (m s^{-1}) at 850 hPa.

basis of observed nonlinear trend in the time series in Figures 1(a)–(c) and the whole study period has been divided into two equal time spans of 30 years for further study of atmospheric features. There is a transition of westerlies to easterlies above 400 hPa level with the

core speed in easterlies at 150–100 hPa level. A current study by Sreekala *et al.* (2014) on decreasing trend in TEJ at 100, 150 and 200 hPa, shows a maximum decrease in zonal wind speed at 100 hPa and a minimum decrease at 200 hPa. In addition, studies by Rao *et al.* (2004), Sathiyamoorthy *et al.* (2007) and Bansod *et al.* (2012) also pointed out the decrease in intensity and distribution of TEJ. Naidu *et al.* (2011) observed a weak ISM period (1995–2005) for the Indian region, which is intimately linked with the weakening of TEJ stream in the recent decades. The reduction in core wind speed from 30 to 24 m s^{-1} with maximum decrease of 6 m s^{-1} is observed over the Arabian Sea region. Thus, both, core wind speed and its horizontal extent have reduced significantly at 100 hPa level which is in accordance with the earlier studies. The reduction in extent and strength of TEJ stream is mainly because of the weaker lower westerlies which causes decreased moisture advection, and thus affects the monsoonal rainfall. Joseph and Sijikumar (2004) reported the variation in LLJ during active and break monsoon period at intraseasonal time scale. In case of 850 hPa zonal wind, wind speed does not show much reduction as compared to zonal wind speed at 100 hPa level but during recent years, the core speed of LLJ shows a decreased value in its intensity. Maximum significant decrease of 1 m s^{-1} has been observed over western Indian Ocean.

It is noted that the anomalies in the zonal winds are associated with the QBO phases at 30 hPa during the summer monsoon in concordance with Mukherjee *et al.* (1985). In the monsoon season as shown in Figure 3[ii(a)], during west phase composites of QBO, the LLJ at 850 hPa shows a positive anomaly in zonal wind which represents the active monsoon year. During east phase composites as shown in Figure 3[ii(b)], the anomaly value is negative for the zonal wind which indicates weak monsoon year. The zonal wind anomaly in TEJ at 100 hPa shows a strong negative anomaly over the Indian region during the west phase composites, Figure 3[iii(a)]. Whereas, during the east phase composites, Figure 3[iii(b)], the anomaly in zonal wind shows weak negative values. Thus, zonal wind anomaly at 850 and 100 hPa shows strong connection of QBO phases with the monsoonal activity over the Indian region. It establishes the interaction of upper tropospheric circulation with lower stratosphere, especially over the monsoon region.

Wind anomaly pattern has been shown during the east and west phase composite years of QBO for LLJ at 850 hPa and TEJ at 100 hPa, Figures 3[iii(a)–(d)]. Wind anomaly at 850 hPa, Figure 3[iii(a)], during west phase composites clearly shows the moisture transport to the Indian subcontinent from the Arabian Sea through the stronger LLJ while wind anomaly shows a weaker westerly during the east phase, Figure 3[iii(b)]. The wind anomaly pattern at 100 hPa, Figure 3[iii(c)], during west phase composites shows a strong northeasterlies flow over the Indian region which develops at the upper level during the ISM. Whereas, in east phase it shows weakened easterlies over the Indian region. The wind anomaly plots during different phases of QBO also clearly indicate the strong monsoon during the west phase composites and a weak monsoon during the east phase composites of QBO.

To further confirm this relationship between ISM and QBO phases, anomalous velocity potential and divergent wind at sigma (σ) levels 0.8458 and 0.2106 along with anomalous outgoing long wave radiation (OLR) has been studied for west and east phase composite years, Figures 4[i(a)–(d)]. The distribution of velocity potential anomaly reflects the changes in strength of convection patterns during west and east phase composites. The flow of divergent wind takes place from region of negative anomalous velocity potential over Eastern Pacific to the region of positive anomalous region over Western Pacific. During west phase composites at lower level, extended strong inflow dominates over the equatorial Eastern Pacific (also over Indian region) and outflow dominates over equatorial Central Pacific, Figure 4[i(a)]. The OLR distribution, Figure 4[i(a)], shows decreased (increased) value over the equatorial Eastern (Western) Pacific leading to strong (weak) convection during west phase. In case of east phase composites, stronger convergence of divergent wind and decreased OLR over the equatorial Central Pacific is present. In addition, divergent source and increased OLR values over Indian region leads to reduced convection over the Indian subcontinent, Figure 4[i(b)]. This convergent source leads to sinking (rising) of air and decreased (increased) convection at upper (lower) level. During the west phase, the increased convective activity at upper level is mainly associated with the divergence of wind at 200 hPa over the Indian region and convergence over the equatorial Central Pacific, Figure 4[i(c)]. While the opposite pattern has been observed during east phase at upper level, Figure 4[i(d)]. The above-mentioned pattern clearly shows the signature of increased response of Walker circulation with downward (upward) motion over equatorial Western Pacific and equatorial Central Pacific (due to the brevity of the article it is not discussed in details).

Further, the time averaged difference in zonal velocity potential anomaly along 75°E at 850 and 200 hPa for west and east phase composites is presented in Figures 4[ii(a)–(d)]. During west phase (850 hPa), high value of velocity potential anomaly around ~25–35°N in Northern Hemisphere and low value around Equator to 5°S in Southern Hemisphere is clearly seen in Figure 4[ii(a)]. Somewhat opposite pattern has been observed during east phase composite, Figure 4[ii(b)]. During west phase, Figure 4[ii(a)], positive (negative) anomaly peaks at ~30°N (~5°S) indicating stronger rising (sinking) motion in Southern (Northern) Hemisphere. At 200 hPa, an opposite pattern at lower level (850 hPa) has been observed during both the phases. This signifies the more strengthening of Hadley circulation during west phase than east phase (due to the brevity of the article it is not discussed in details).

4. Conclusions

The main aim of this study is to explore the trends present in TEJ and LLJ and relationship between QBO

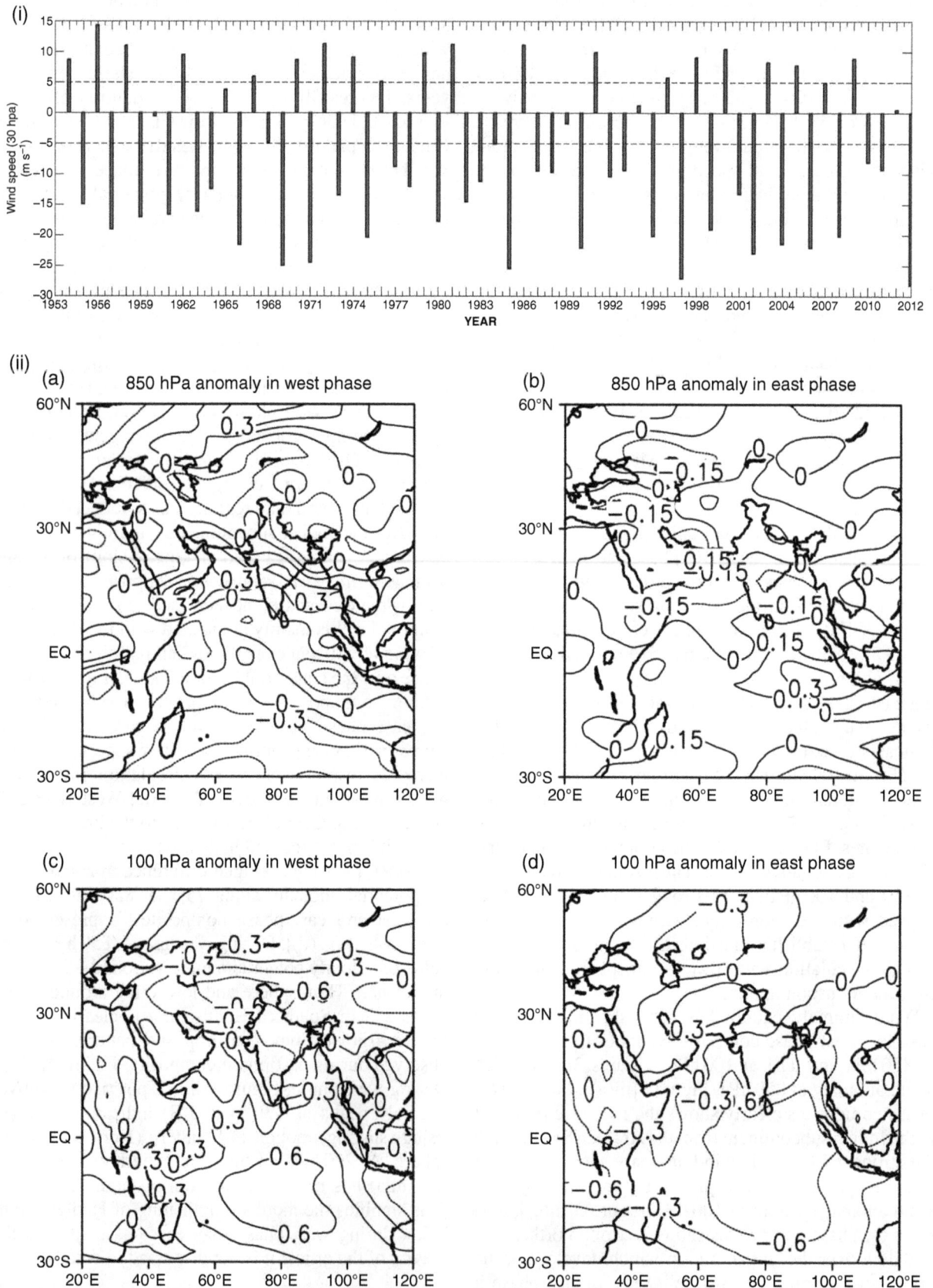

Figure 3. (i) Extraction of east and west phase of QBO based on wind speed. The dotted lines show the threshold, i.e. >+5 (west phase) and >−5 (east phase), (ii) zonal wind (m s^{-1}) anomaly distribution in QBO (a) west phase at 850 hPa, (b) east phase at 850 hPa, (c) west phase at 100 hPa and (d) east phase at 100 hPa. Panel (iii) is same as panel (ii) but for wind anomaly (m s^{-1}).

(iii)

(a)   850 hPa anomaly in west phase

(b)   850 hPa anomaly in east phase

(c)   100 hPa anomaly in west phase

(d)   100 hPa anomaly in east phase

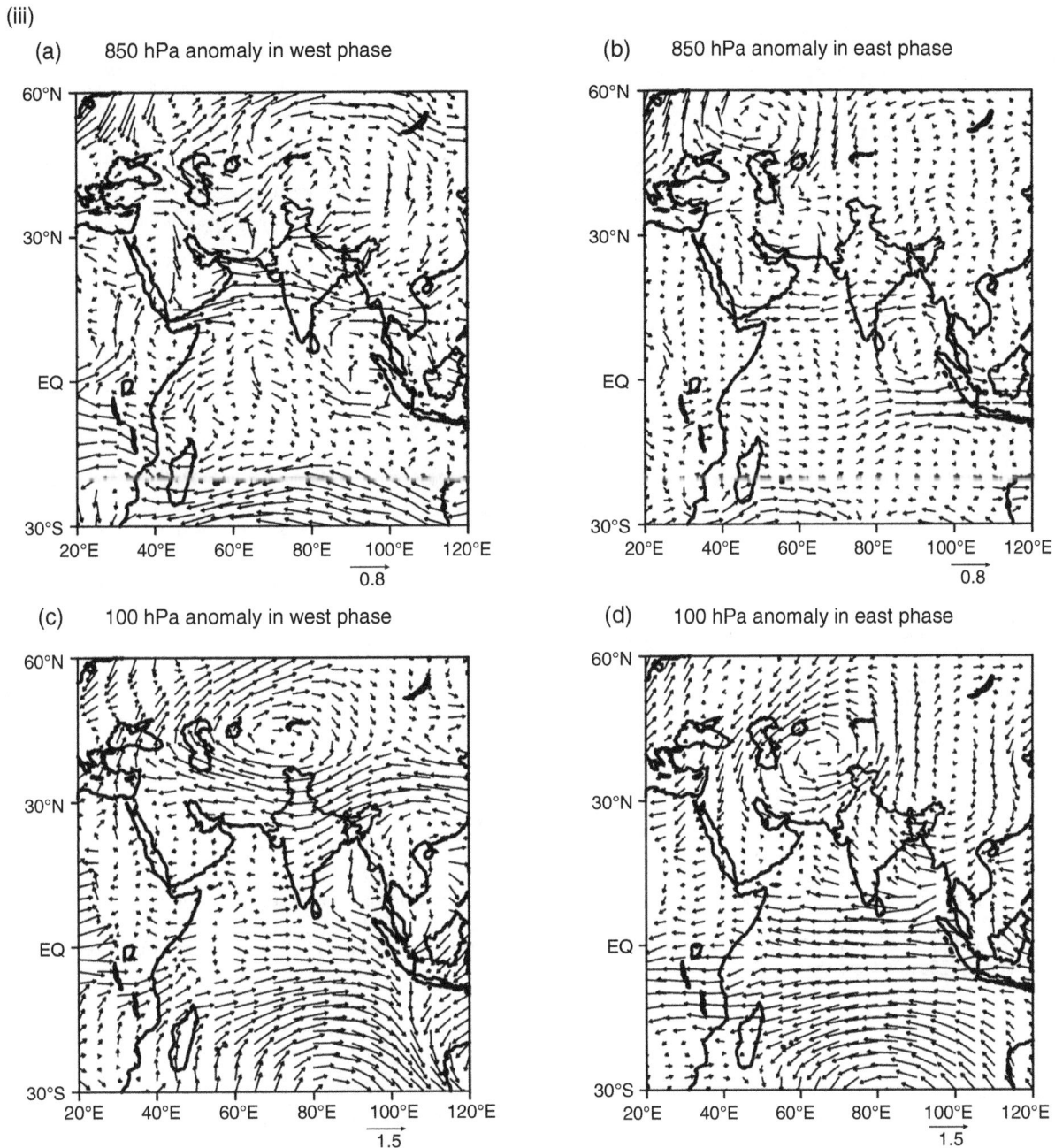

Figure 3. continued

in different phases (west and east phase) with the ISM and to understand the physical mechanism responsible for the ISM variability during the QBO period. Using the NCEP/NCAR reanalysis data, it is found that the intensity of the TEJ has decreased in the recent period (1983–2012) in comparison with the early period (1953–1982). The reduction in core wind speed from 30 to 24 m s^{-1}, with a maximum decrease of 6 m s^{-1}, is observed over the equatorial Indian Ocean region. This decrease in extent and intensity of the TEJ is mainly credited to the reduced moisture advection as related with weaker surface westerlies, which affects the ISM rainfall. A small reduction of nearly 1 m s^{-1} in the intensity of LLJ over the Arabian Sea has also been observed during the recent years.

The association between different phases of QBO and ISM rainfall has been established in this study. A total of 850 and 100 hPa zonal wind shows a strong westerly and easterly pattern respectively during west phase composites. While during the east phase composites, the intensity of zonal wind becomes weaker at both levels. This is indicative of a strong ISM rainfall during the west phase and weak ISM rainfall during the east phase. To further elaborate on their relationship, velocity potential and divergent wind along with the OLR at 850 and 200 hPa has been presented during the both phases. It shows the convergent source at lower levels (850 hPa) and a divergent source at upper levels (200 hPa) during the west phase and vice versa during the east phase over the Indian region.

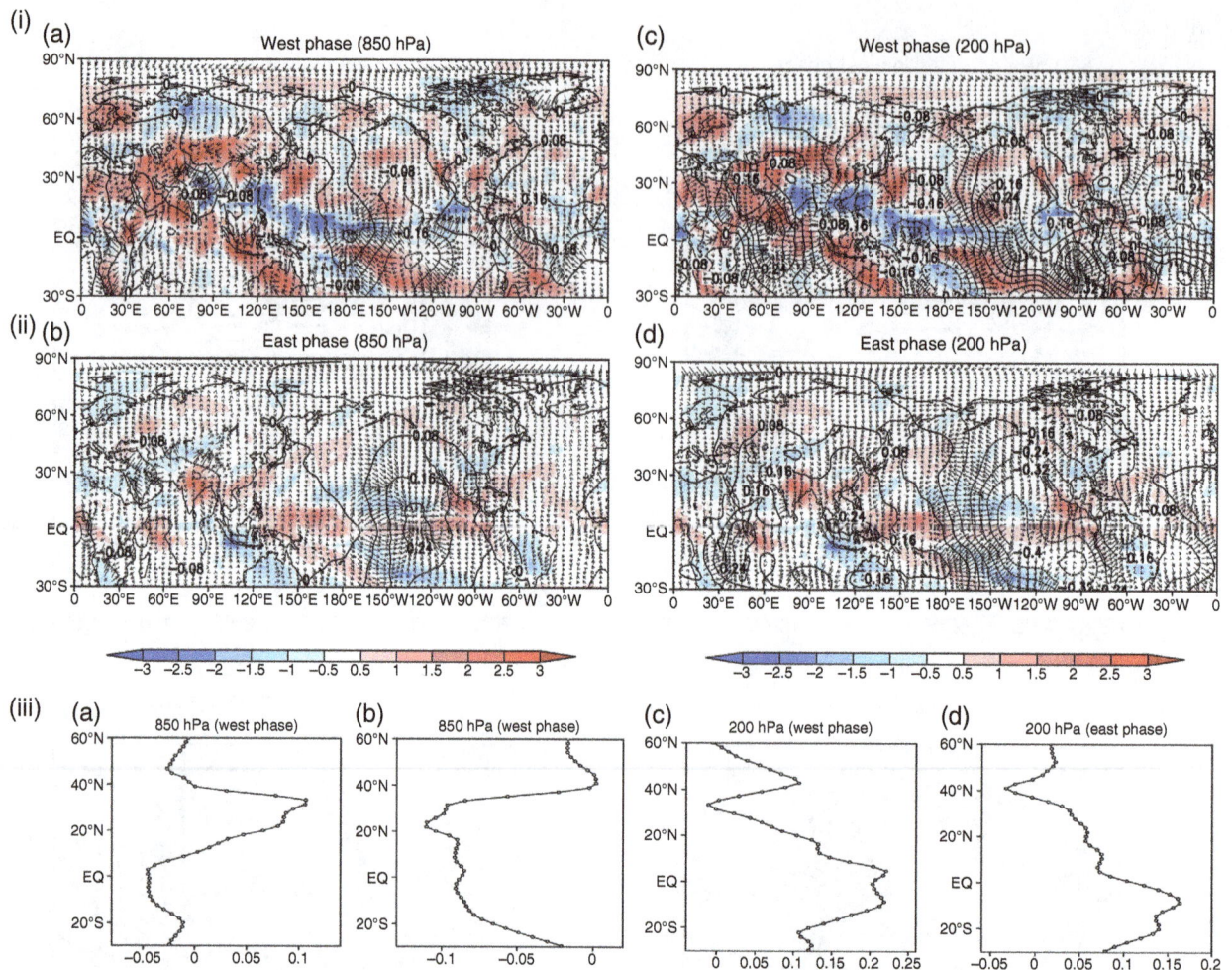

Figure 4. Time average anomalous velocity potential ($\times 1e^{-6}$ m^2 s^{-1}: contour) with corresponding anomalous divergent wind (m s^{-1}: arrow) and anomalous OLR at $\sigma = 0.8458$ and 0.2106 for (i) − (a) west phase at 850 hPa, (b) east phase at 850 hPa, (c) west phase at 200 hPa and (d) east phase at 200 hPa and (ii) along 75°E for (a) 850 hPa (west phase), (b) 850 hPa (east phase), (c) 200 hPa (west phase) and (d) 200 hPa (east phase).

Acknowledgements

P.R. acknowledges Council of Scientific and Industrial Research (CSIR) funding for Junior Research Fellowship. Data from NCEP/NCAR (http://www.esrl.noaa.gov/psd/), USA; NOAA, USA and IITM, Pune, India is used for the study and thus acknowledged. The authors also thank Jawaharlal Nehru University (JNU) for financial support.

References

Bansod SD, Singh HN, Patil SD, Singh N. 2012. Recent changes in the circulation parameters and their association with Indian summer monsoon rainfall. *Journal of Atmospheric and Solar-Terrestrial Physics* **77**: 248–253.

Dwivedi S, Mittal AK. 2007. Forecasting the duration of active and break spells in intrinsic mode functions of Indian monsoon intraseasonal oscillations. *Geophysical Research Letters* **34**(16): L16827, doi: 10.1029/2007GL030540.

Flohn H. 1964. *Investigations on the Tropical Easterly Jet*. Dümmlers Vlg. Contract No. DA-91-591-EUC-2781, Final Report No. l. Meteorological Institute, University of Bonn: Germany.

Holton JR. 1968. A note on the propagation of the biennial oscillation. *Journal of the Atmospheric Sciences* **25**(3): 519–521.

Huang NE, Shen Z, Long SR, Wu MC, Shih HH, Zheng Q, Yen N-C, Tung CC, Liu HH. 1998. The empirical mode decomposition and the Hilbert spectrum for nonlinear and non-stationary time series analysis. *Proceedings of the Royal Society of London A: Mathematical, Physical and Engineering Sciences* **454**(1971): 903–995.

Iyengar RN, Kanth SR. 2005. Intrinsic mode functions and a strategy for forecasting Indian monsoon rainfall. *Meteorology and Atmospheric Physics* **90**(1–2): 17–36.

Joseph PV, Sijikumar S. 2004. Intraseasonal variability of the low-level jet stream of the Asian summer monsoon. *Journal of Climate* **17**: 1449–1458.

Kalnay E, Kanamitsu M, Kistler R, Collins W, Deaven D, Gandin L, Iredell M, Saha S, White G, Woollen J, Zhu Y, Leetmaa A, Reynolds R, Chelliah M, Ebisuzaki W, Higgins W, Janowiak J, Mo KC, Ropelewski C, Wang J, Jenne R, Joseph D. 1996. The NCEP/NCAR 40-year reanalysis project. *Bulletin of the American Meteorological Society* **77**(3): 437–471.

Kanamitsu M, Krishnamurti TN, Depradine C. 1972. On scale interactions in the Tropics during northern summer. *Journal of the Atmospheric Sciences* **29**(4): 698–706.

Kobayashi N. 1974. Interannual variations of tropical easterly jet stream and rainfall in South Asia. *Geophysical Magazine* **37**: 123–134.

Koteswaram P. 1958. The easterly jet stream in the Tropics. *Tellus* **X**: 43–57.

Krishnamurti TN, Bhalme HN. 1976. Oscillations of a monsoon system. Part I. Observational aspects. *Journal of the Atmospheric Sciences* **33**(10): 1937–1954.

Lindzen RS, Holton JR. 1968. A theory of the quasi-biennial oscillation. *Journal of the Atmospheric Sciences* **25**(6): 1095–1107.

Madhu V. 2014. Variation of zonal winds in the upper troposphere and lower stratosphere in association with deficient and excess Indian summer monsoon scenario. *Atmospheric and Climate Sciences* **4**(04): 685.

Molla MKI, Rahman MS, Sumi A, Banik P. 2006. Empirical mode decomposition analysis of climate changes with special reference to rainfall data. *Discrete Dynamics in Nature and Society* **2006**: 1–17.

Mukherjee BK, Indira K, Reddy RS, Ramana Murty BV. 1985. Quasi-biennial oscillation in stratospheric zonal wind and Indian summer monsoon. *Monthly Weather Review* **113**(8): 1421–1424.

Naidu CV, Krishna KM, Rao SR, Kumar OB, Durgalakshmi K, Ramakrishna SSVS. 2011. Variations of Indian summer monsoon rainfall induce the weakening of easterly jet stream in the warming environment? *Global and Planetary Change* **75**(1): 21–30.

Naidu CV, Satyanarayana GC, Rao LM, Durgalakshmi K, Raju AD, Kumar PV, Mounika GJ. 2015. Anomalous behavior of Indian summer monsoon in the warming environment. *Earth-Science Reviews* **150**: 243–255.

Parthasarathy B, Munot AA, Kothawale DR. 1995. Monthly and seasonal rainfall series for all-India homogeneous regions and meteorological subdivisions. 1871–1994. Research Report No. RR-065, Indian Institute of Tropical Meteorology, Pune, 113 pp.

Pattanaik DR, Satyan V. 2000. Fluctuations of tropical easterly jet during contrasting monsoons over India: a GCM study. *Meteorology and Atmospheric Physics* **75**(1–2): 51–60.

Pielke RA, Chase TN, Kittel TGF, Knaff JA, Eastman J. 2001. Analysis of 200 mbar zonal wind for the period 1958–1997. *Journal of Geophysical Research: Atmospheres* **106**(D21): 27287–27290, doi: 10.1029/2000JD000299.

Quiroz RS. 1981. Period modulation of the stratospheric quasi-biennial oscillation. *Monthly Weather Review* **109**(3): 665–674.

Rao BRS, Rao DVB, Rao VB. 2004. Decreasing trend in the strength of tropical easterly jet during the Asian summer monsoon season and the number of tropical cyclonic systems over Bay of Bengal. *Geophysical Research Letters* **31**: L14103, doi: 10.1029/2004GL019817.

Reiter ER. 1961. *Jet Stream Meteorology*. University of Chicago Press: Chicago, IL; 515 pp.

Roja Raman M, Jagannadha Rao VVM, Venkat Ratnam M, Rajeevan M, Rao SVB, Narayana Rao D, Prabhakara Rao N. 2009. Characteristics of the tropical easterly jet: long-term trends and their features during active and break monsoon phases. *Journal of Geophysical Research: Atmospheres* **114**(D19): D19105, doi: 10.1029/2009JD012065.

Sathiyamoorthy V. 2005. Large scale reduction in the size of the tropical easterly jet. *Geophysical Research Letters* **32**: L14802, doi: 10.1029/2005GL022956.

Sathiyamoorthy V, Pal PK, Joshi PC. 2007. Intraseasonal variability of the tropical easterly jet. *Meteorology and Atmospheric Physics* **96**(3–4): 305–316.

Sreekala PP, Rao SB, Arunachalam MS, Harikiran C. 2014. A study on the decreasing trend in tropical easterly jet stream (TEJ) and its impact on Indian summer monsoon rainfall. *Theoretical and Applied Climatology* **118**(1–2): 107–114.

Venkat Ratnam M, Krishna Murthy BV, Jayaraman A. 2013. Is the trend in TEJ reversing over the Indian subcontinent? *Geophysical Research Letters* **40**(13): 3446–3449, doi: 10.1002/grl.50519.

Veryard RG, Ebdon RA. 1961. Fluctuations in tropical stratospheric winds. *Meteorology Magazine* **90**: 125–143.

Wu Z, Huang NE. 2004. A study of the characteristics of white noise using the empirical mode decomposition method. *Proceedings of the Royal Society of London A: Mathematical, Physical and Engineering Sciences* **460**(2046): 1597–1611.

Wu Z, Huang NE. 2009. Ensemble empirical mode decomposition: a noise assisted data analysis method. *Advances in Adaptive Data Analysis* **1**(01): 1–41.

Zhang J, Yan R, Gao RX, Feng Z. 2010. Performance enhancement of ensemble empirical mode decomposition. *Mechanical Systems and Signal Processing* **24**(7): 2104–2123.

16

Multidecadal convection permitting climate simulations over Belgium: Sensitivity of future precipitation extremes

Sajjad Saeed,[1,5]* Erwan Brisson,[2] Matthias Demuzere,[1] Hossein Tabari,[3] Patrick Willems[3,4] and Nicole P. M. van Lipzig[1]

[1] Department of Earth and Environmental Sciences, KU Leuven, Belgium
[2] Goethe University, Frankfurt, Germany
[3] Hydraulics Division, Department of Civil Engineering, KU Leuven, Belgium
[4] Department of Hydrology and Hydraulic Engineering, Vrije Universiteit Brussel, Belgium
[5] Center of Excellence for Climate Change Research (CECCR), King Abdulaziz University, Jeddah, Saudi Arabia

*Correspondence to:
S. Saeed, Department of Earth
and Environmental Sciences,
Celestijnenlaan 200e – box
2409, BE-3001 KU Leuven,
Leuven, Belgium.
E-mail:
sajjad.saeed@kuleuven.be

Abstract

We performed five high resolution (2.8 km) decadal convection permitting scale (CPS) climate simulations over Belgium using the COSMO-CLM regional climate model and examined the future changes in daily precipitation extremes compared to coarser resolution simulations. The CPS model underestimates the higher percentiles during both seasons, however, some improvements in the higher percentile values are noticed during the summer season. Analysis of three future climate simulations indicates that the CPS model modifies the future signals of daily precipitation extremes compared to their forcing non-CPS simulations during summer. During this season, the increase (decrease) in the daily precipitation extremes is stronger in the CPS compared to the non-CPS simulations. During winter, no significant changes between CPS and non-CPS were found.

Keywords: regional climate model; convection permitting simulations; extreme precipitation

1. Introduction

Extreme precipitation events largely influence society and ecosystems through floods, drought, infrastructure damage and even human causalities (Tabari et al., 2014). According to the 5th assessment report (AR5) of the International Panel on Climate Change (IPCC), the frequency and intensity of the precipitation extremes are likely to increase in the future warmer climate (IPCC, 2013). Understanding and quantifying the magnitude and frequency of such extremes for both the present-day climate and possible future climates is therefore relevant. IPCC's future climate projections are generally based on coarser resolution (e.g. 150–200 km or more) Global Climate Model simulations. Owing to their coarse resolution, not all processes, notably those occurring on mesoscales, are reasonably taken into account. These limitations result in important misrepresentations of extreme precipitation (Willems et al., 2012; Tabari et al., 2015).

To overcome this problem, Regional Climate Models (RCMs) are frequently used to downscale the coarser resolution global climate simulations to regional and local scales. The RCMs are capable of providing additional regional details such as an improved representation of topographical features (e.g. mountains and coastlines), land cover heterogeneity etc. (Christensen and Christensen, 2007; Prein et al., 2015). The recent internationally coordinated projects, e.g. PRUDENCE, ENSEMBLES and EURO-CORDEX employed RCMs with horizontal resolutions of 50, 25

and 12 km, respectively (Christensen and Christensen, 2007; Mearns et al., 2009; Kotlarski et al., 2014). Some recent studies performed climate simulations even at 7 km spatial resolution (Wagner et al., 2013). The increasing model resolution, however, does not guarantee a reduction of the model deficiencies and associated biases compared to the observations. Some studies (Clark et al., 2007; Walther et al., 2013) indicate that the increasing model resolution does not improve the representation of the precipitation diurnal cycle. This problem has been solved partly by resolving deep convection, a process that can be modeled in RCMs with grid mesh size as fine as at least 2–4 km (Weisman et al., 1997). Such high resolution RCM simulations where deep convection is explicitly resolved are generally referred as convective permitting scale (CPS) simulations. Previous studies (Kendon et al., 2012; Prein et al., 2013; Ban et al., 2014; Fosser et al., 2014; Brisson et al., 2016) showed that CPS models improve the representation of the diurnal cycle of precipitation. However, due to the high computational cost of CPS models, the number of climate change impact studies using such models is limited (Prein et al., 2015).

In most studies discussed above, the CPS model simulations are carried out to investigate their added value in the present-day climate. Only few studies examined future precipitation extremes in the CPS model simulations (Kendon et al., 2014; Ban et al., 2015). The latter two studies slightly disagree in their findings due to use of different statistical methods (Schär et al., 2016). Kendon et al. (2014) noticed a future intensification of short-duration rain in summer, with significantly more events exceeding the high thresholds. Whereas, Ban *et al.* (2015) showed that the extreme events are projected to become more frequent and more intense, but not as pronounced as in some previous studies. Additional simulations are therefore needed to investigate the reasons for diverging conclusions in these recent studies. This study aims to fill this gap by performing a set of present-day and future decadal CPS simulations over Belgium. Among other issues, an important question remains: how do high resolution CPS model integrations modify the representation of future precipitation extremes compared to the non-CPS simulations? The advantage of this study is that it relies on three separate 10-year future climate simulations, which makes it possible to study the robustness of the signals. The article is organized as follows: The next section gives an overview of the COSMO-CLM (CCLM) model and the configuration of the CPS simulations. A brief model evaluation and the simulated future changes in the precipitation extremes are described in the Section on Results and Discussions. The summary and conclusions are presented in the last section.

2. Data and methodology

2.1. Model

This study uses the CCLM RCM, a non-hydrostatic model based on the COSMO numerical weather prediction model (Steppeler *et al.*, 2003). Later on, the model was adapted by the climate limited-area modeling (CLM) community to perform both short- and long-term climate integrations by adding specific modules such as dynamic surface boundaries, a more complex soil model and the possibility to use various CO_2 concentrations (Böhm *et al.*, 2006; Rockel *et al.*, 2008).

Following Brisson *et al.* (2016), here we adopt the third order Runga-Kutta split-explicit time stepping scheme (Wicker and Skamarock, 2002), the lower boundary fluxes provided by the TERRA model (Doms *et al.*, 2011) and the radiative scheme after Ritter and Geleyn (1992). In this respect, it is noted that the precipitation change might be adversely affected by some deficiencies shortwave water vapor absorption in older radiative transfer schemes (DeAngelis *et al.*, 2015).

2.2. Experimental setup and methodology

The experimental setup in terms of model domains, physical parameterizations etc. generally follows that of Brisson *et al.* (2016). A three-step nesting strategy (shown in Figure S1) has been applied in this study. Five simulations are performed (Table S1). For simplicity, here LTS refers to the long-term simulation and the subscript represents the lateral boundary conditions. The ERA-Interim reanalysis data and the global EC-Earth model (Dee *et al.*, 2011; Hazeleger *et al.*, 2012) provides the necessary initial and boundary conditions to nest a 100×100 grid points domain with

Table 1. Definition of indices used in this study.

Index	Definition
Wet days	Days for which daily precipitation exceeds 90th percentile of reference period
Very wet days	Days for which daily precipitation exceeds 95th percentile of reference period
Extreme wet days	Days for which daily precipitation exceeds 99th percentile of reference period
Heavy rainfall	The mean of the upper 5% of daily precipitation intensities
Extreme event	Daily precipitation over a particular grid cell exceeding $30\,mm\,day^{-1}$
Very extreme event	Daily precipitation over a particular grid cell exceeding $60\,mm\,day^{-1}$

a 0.22° (approximately 25 km) grid mesh size. The resulting 3-h outputs are employed to nest a 0.0625° (approximately 7 km) domain. Finally, the hourly outputs of the latter nest, characterized by 150×150 grid points, are used as input for the 0.025° (approximately 2.8 km) simulation on a 192×175 grid points domain. The added value of CPS can best be assessed when the model is driven with ERA-Interim boundary conditions ($LTS_{ERA-Int}$), as deficiencies in the EC-Earth model might propagate into the regional model. The non-CPS model with horizontal resolutions of 25 and 7 km does not explicitly resolve deep convection and hence use the convection scheme after Tiedtke (1989). Such parameterization is unnecessary in the CPS (2.8 km) model setup where deep convection is dynamically resolved.

All simulations performed in this study employ 40 vertical levels. Observational data for daily accumulated precipitation for 199 stations covering the full simulation period (2001–2010) obtained from the Royal Meteorological Institute (RMI) of Belgium are employed to evaluate the model in present-day climate. In the case of comparison with station data, we extracted model information from the nearest grid point. This method is commonly used for model and station data comparison but it may introduce some uncertainties as the model data are grid averaged values whereas the station data present point values. Several indices (Chan *et al.*, 2013; Ban *et al.*, 2015) that are used here are summarized in Table 1. The precipitation percentiles are computed from continued time series, which also includes dry days. To compare data on different grids, the simulated datasets are first regridded to 0.22° regular grid using conservative first order regridding method. This regridding method is more desirable than the bilinear interpolation for discontinuous variables such as precipitation (Jones, 1999).

3. Results and discussions

3.1. Present-day analysis

Brisson *et al.* (2016) performed a detailed evaluation of CCLM over Belgium driven with ERA-Interim

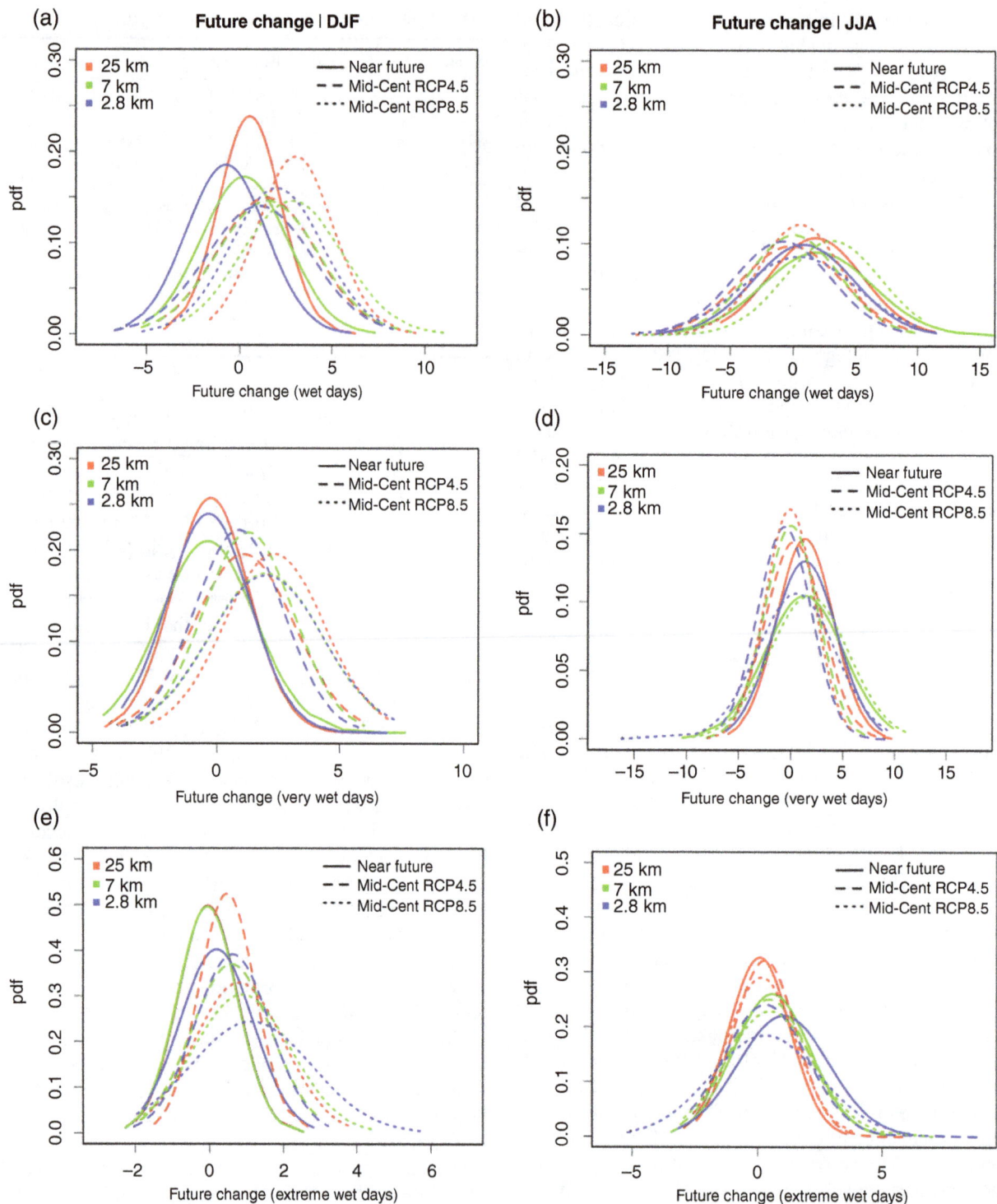

Figure 1. Future change in the wet, very wet and extreme wet days. The future changes are shown for Near Future (2026–2035) and Mid-Century (2060–2069) with respect to the present-day (2001–2010) period. In each case, the blue (red, green) lines show the CPS (non-CPS) model simulations.

boundary conditions. In this study we extend this evaluation for present-day simulation driven by EC-Earth boundary conditions ($LTS_{EC\text{-}Earth}$, Table S1). Table S2 gives the mean and higher percentiles for winter and summer for the set of present-day simulations ($LTS_{ERA\text{-}Int}$ and $LTS_{EC\text{-}Earth}$). Note that, due to the sparse network of station data it was not possible to aggregate the observations to a grid. The model data is

not area averaged, it just consists of all points and dates for which there exists a station in the RMI database. Compared to the observations the CPS model underestimates the higher percentiles during both seasons, however, some improvements in the higher percentile values are noticed during the summer season. This is especially evident for the $LTS_{ERA\text{-}Int}$ run. For the $LTS_{EC\text{-}Earth}$ simulation, the CPS model does not show

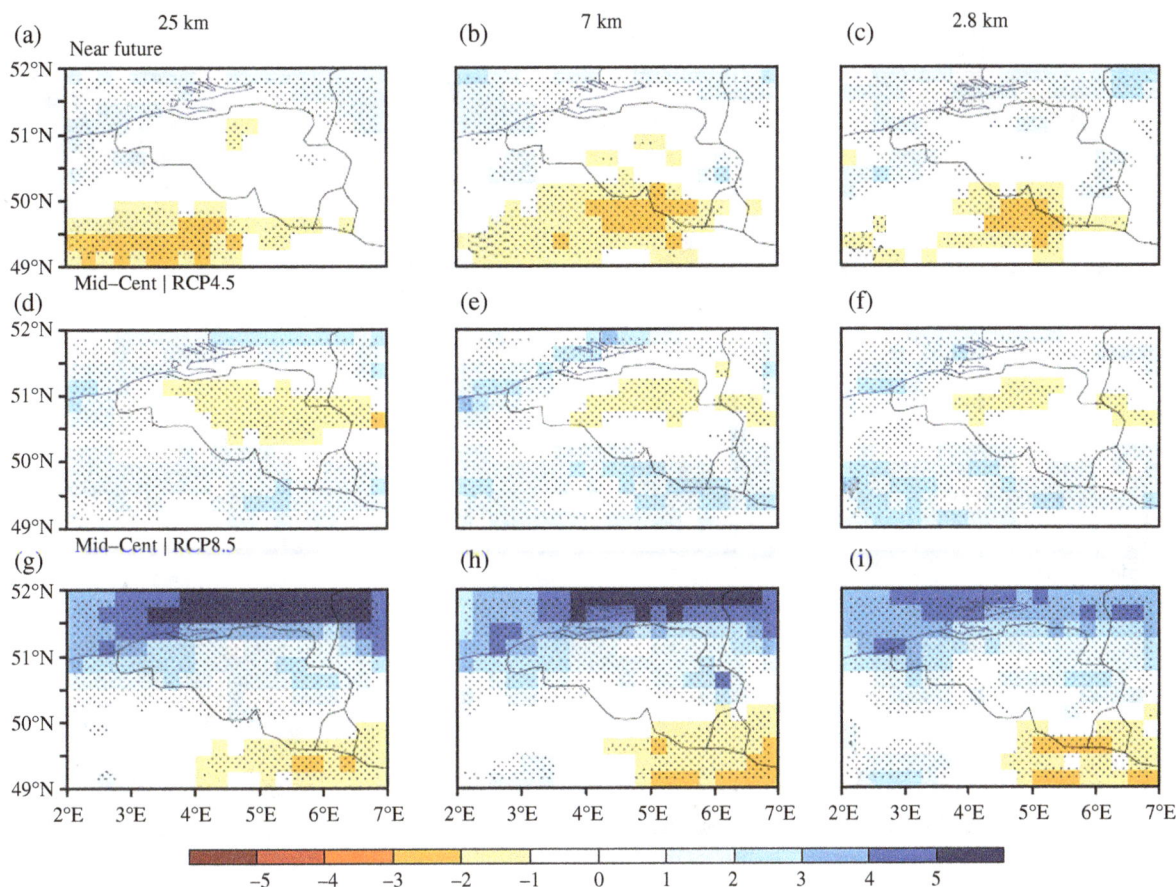

Figure 2. Future change in heavy precipitation (mm day^{-1}) during winter season. Similar to Figure 1, the future changes are computed for Near Future (2026–2035) and Mid-Century (2060–2069) with respect to the present-day (2001–2010) period. The left two panels show future changes for the non-CPS model simulations whereas the right panel show the future changes for CPS model simulations. The model resolutions are shown on the top of panel. The dotted areas indicate differences that are significant at 99% confidence level based on t-test.

marked improvements or deteriorations compared to the non-CPS simulations, at least not on a daily time scale.

3.2. Future simulation of extreme precipitation over Belgium

The main aim of this article is to investigate the CPS model response to increasing greenhouse gas forcing compared to non-CPS models. We therefore first examine the future change in number of simulated wet, very wet and extreme wet days as defined in Table 1. For this purpose, we computed for each grid cell the change in number of days for each category between the future and the present-day. We then constructed a sample consisting of these change values for each grid cell. In the final step, we examined the spatial distribution of future change in wet, very wet and extreme wet days (Figure 1). The most robust signal is found for extreme wet days where all three CPS simulations show an amplification of the future signals for both seasons compared to the non-CPS simulations (Figures 1(e) and (f)). During the summer season, the extreme wet days show a wider distribution in the CPS model simulations (Figure 1(f)). In this season, the enhancement

(reduction) in the future extreme wet days is more pronounced in all three CPS simulations compared to the non-CPS simulations. Note that the CPS model simulations also reveal a slight reduction in number of future wet and very wet days compared to the non-CPS simulations during both seasons (Figures 1(a)–(d)).

We further examine the intensity of future change in the heavy rainfall. During the winter season, both the CPS and non-CPS model simulations reveal a similar future change in the spatial distribution of heavy rainfall intensity over Belgium (Figure 2). Although, the area showing negative future change in heavy rainfall in the non-CPS model (Figures 2(a), (b), (d), (e), (g) and (h)) diminishes in the CPS model simulations (Figures 2(c), (f) and (i)), the difference between CPS and non-CPS is marginal during winter.

During summer, the CPS model simulations show amplification in the future signals of the heavy rainfall over Belgium compared to the non-CPS model simulations (Figures 3 and 4). Significant differences between CPS and non-CPS are found over some regions where precipitation increase (decrease) is stronger in the CPS compared to non-CPS simulations. However, some grid cells display minor changes in the simulated

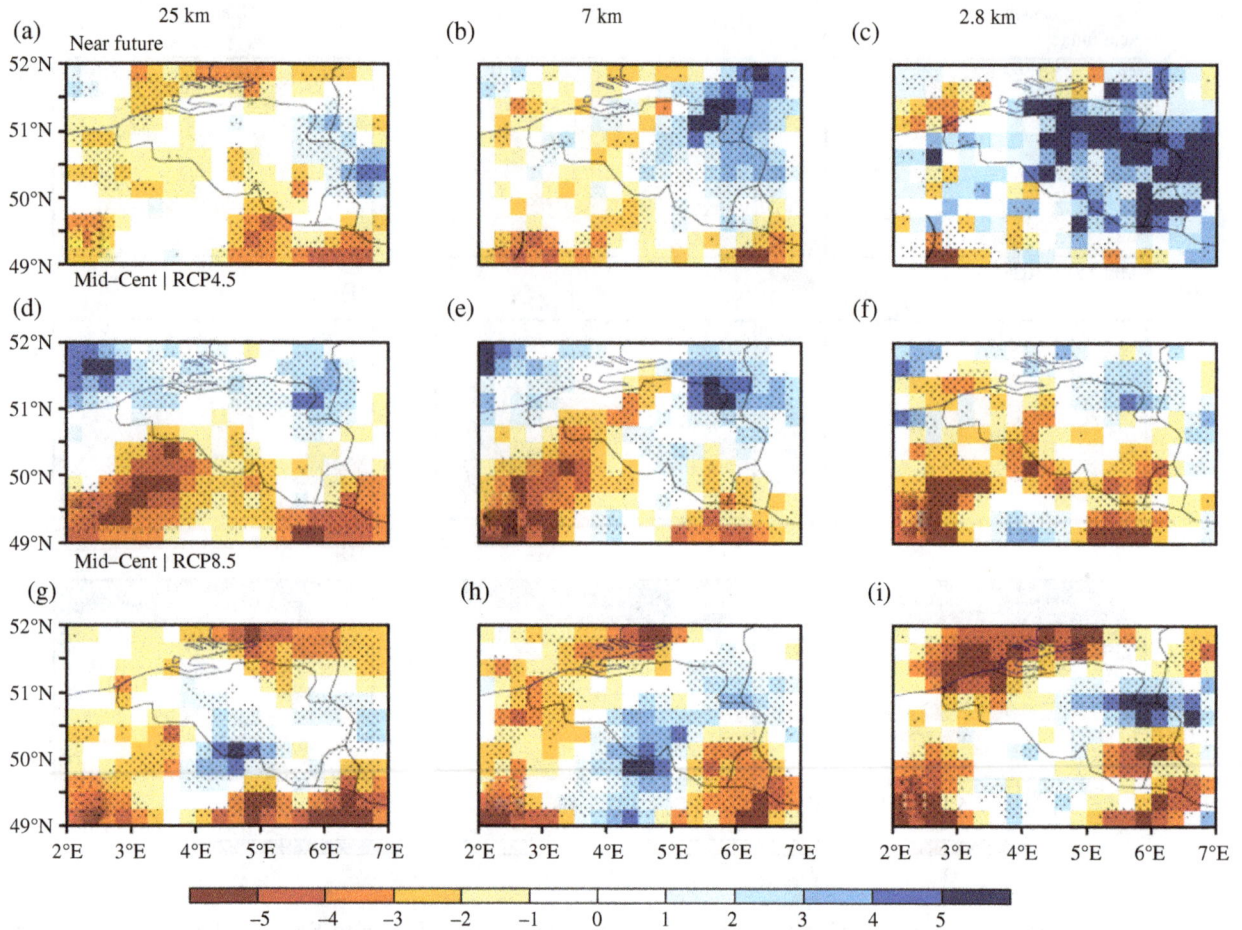

Figure 3. Same as Figure 2, but for summer season. The dotted areas indicate differences that are significant at 99% confidence level based on t-test.

patterns between CPS and non-CPS during the summer season (Figure 4). Based on the t-test, the regression coefficients for all these three simulations are significantly (99% level) different from one, hence rejecting the null hypothesis that both CPS and non-CPS have the same changes in heavy rainfall (Figure 4). Moreover, the slope of the regression lines between CPS (x-axis) and non-CPS (y-axis) is significantly (99% level) lower than 1 for all three simulations, indicates that the amplitude of the change, whether positive or negative, is significantly larger in CPS compared to the non-CPS (Figure 4).

We further examined the extreme and very extreme precipitation events (Table 1) by analyzing a cumulative distribution of the daily rainfall over Belgium (Figure 5). Each day and grid cell is treated as a sample of the distribution, so in total, the sample consists of the grid points in the analysis domain (198) multiplied with the number of days in the simulation (902 for winter and 920 for summer, yielding 178 596 values for winter and 182 160 values for summer). The cumulative distribution is plotted only for that part of the sample with daily rainfall rates above the threshold (30 and 60 mm day^{-1} respectively for very wet and extreme wet days). The upper limit of the precipitation imposed to define the daily extreme and

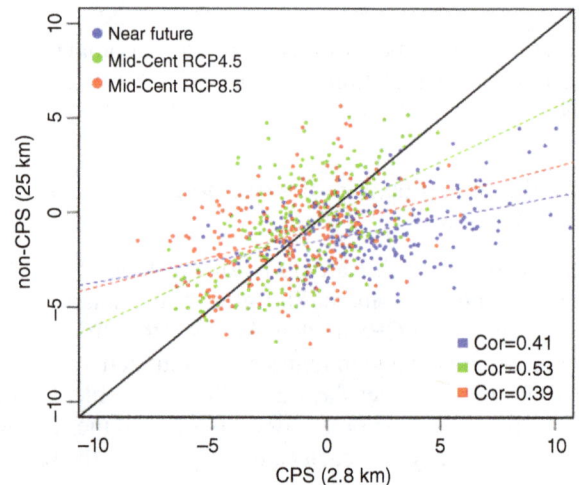

Figure 4. Scatter plot between CPS and non-CPS (25 km) simulated future change signal of heavy rainfall (mm day^{-1}) averaged over all summer seasons. Each dot in the scatter plot refers to a model grid cell. The blue, green and red dots show the change for Near Future, Mid-Cent$_{RCP4.5}$ and Mid-Cent$_{RCP8.5}$ simulations. The dashed blue, green and red lines indicate the regression lines for above three simulations respectively. The solid black line represents the (1 : 1) line. The correlation coefficients between CPS and non-CPS future change signals are significant to 95% level in all three cases and are shown in the lower left corner of the panel.

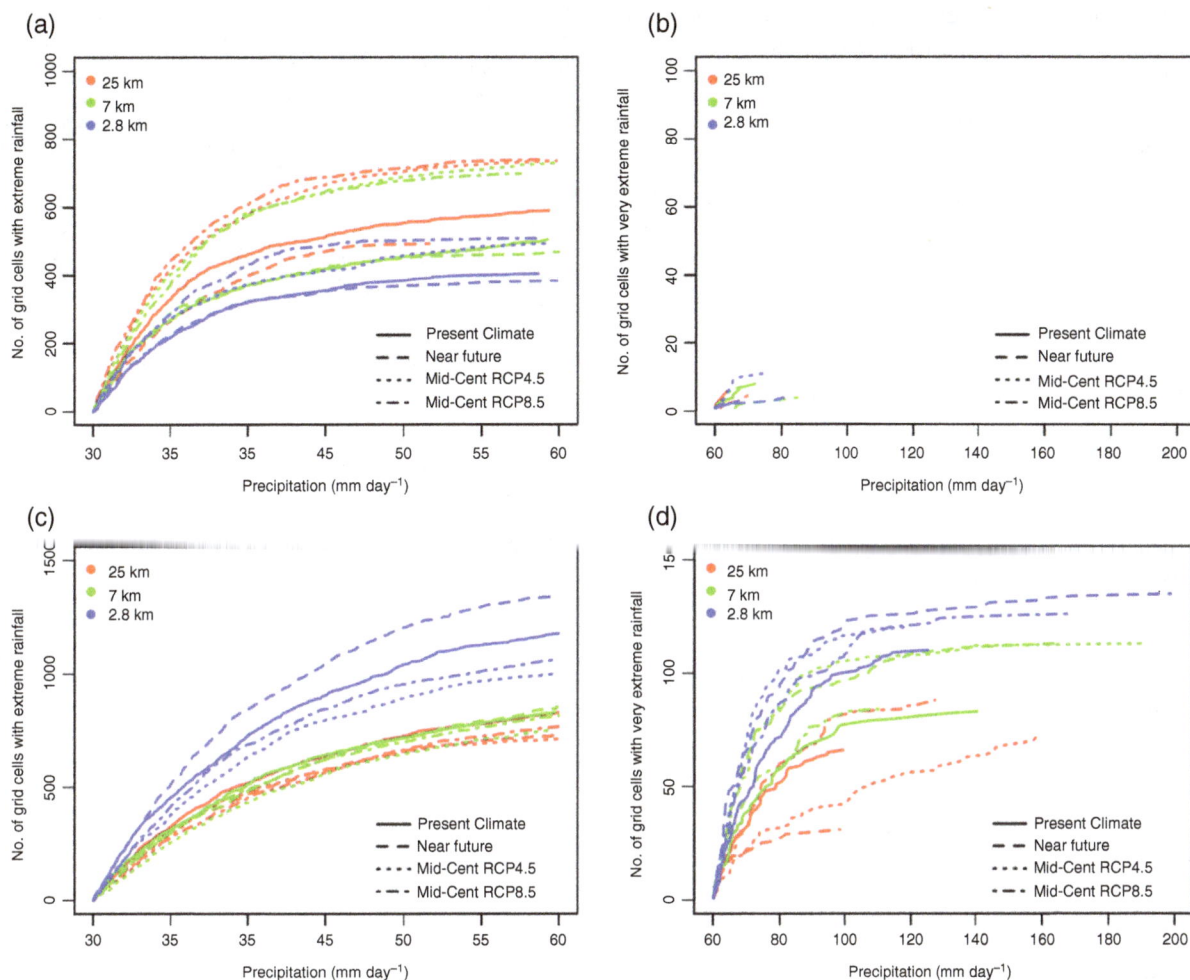

Figure 5. Cumulative distribution of the simulated daily precipitation extreme (a and c) and very extreme (b and d) events (mm day^{-1}) over Belgium during winter (a and b) and summer (c and d). The climate simulations performed in this study (Table S1) are represented by the solid, dashed, dotted and dot-dashed lines. In all cases, the blue (red, green) lines show the CPS (non-CPS) simulations.

very extreme events (in this case) is subjective and a small variation in this limit does not affect the overall results (not shown). During the winter season, neither CPS nor non-CPS simulations show robust signals between all three future simulations at grid cell scale (Figures 5(a) and (b)). However, during summer, both the frequency and intensity of the daily precipitation extremes covering a grid cell area also increase notably in the CPS simulations compared to the forcing non-CPS simulations (Figures 5(c) and (d)). Note that very high extremes are reached only in a very limited number of cases, while lower intensities are exceeded more frequently. The Near Future (Mid-Cent RCP4.5 and RCP8.5) CPS simulation has more (less) extreme compared to the present-day CPS (Figure 5(c)). We do not know the exact reason for this but since the last two simulations cover a different decade (2060–2069), there might be some decadal climate variability in the global model simulations that influence the RCM simulation results. However, for very extreme events (Figure 5(d)), there is clear amplification in the future signals in all cases compared to the present-day CPS simulation.

4. Summary and conclusions

The sensitivity of the future precipitation extremes on the daily time scale have been compared between the CPS simulations and their forcing non-CPS simulations. For this purpose, we performed five decadal high resolution (2.8 km) convection permitting scale (CPS) climate simulation over Belgium using the CCLM RCM. A detailed evaluation of the CCLM has been carried out by Brisson *et al.* (2016). This follow-up study assesses how the high resolution CPS model integrations may modify future signals of the daily precipitation extremes over Belgium compared to their forcing non-CPS simulations.

The analysis of three future simulations over Belgium with reference to the present-day climate reveals an amplification of the future daily precipitation extremes in the CPS simulations compared to the non-CPS simulations, both in frequency and intensity. This amplification is larger during the summer season. During winter, the difference between CPS and non-CPS is marginal. On the other hand, during the summer season, some regions where daily precipitation extremes increase (decrease) in the forcing non-CPS simulations, they increase (decrease) more in the CPS simulations.

Acknowledgements

The authors thank two anonymous reviewers for helpful and insightful comments. This research has been carried out under the MACCBET and CORDEX-BE projects. MACCBET and CORDEX-BE are funded by the Belgian Science Policy (BELSPO). M.D. is funded by the Flemish regional government through a contract as a FWO (Fund for Scientific Research) post-doctoral position. The computational resources and services used in this work were provided by the VSC (Flemish Supercomputer Center), funded by the Hercules Foundation and the Flemish Government department EWI.

Supporting information

The following supporting information is available:

Figure S1. Model domains used for 25, 7 and 2.8 km resolution simulations. The analysis domain is shown by the red rectangle.

Table S1. Simulations performed in this study with COSMO-CLM regional climate model.

Table S2. Daily precipitation intensity spatio-temporal average, 95th and 99th quantiles for the observation and the different simulations for the winter and the summer period. The model data was extracted from the COSMO-CLM grid cells, which encompasses the coordinates of the Royal Meteorological Institute's observations (RMI-OBS) locations.

References

Ban N, Schmidli J, Schär C. 2014. Evaluation of the convection-resolving regional climate modeling approach in decade-long simulations. *Journal of Geophysical Research. Atmospheres* **119**(13): 7889–7907.

Ban N, Schmidli J, Schär C. 2015. Heavy precipitation in a changing climate: does short-term summer precipitation increase faster? *Geophysical Research Letters* **42**: 1165–1172.

Böhm U, Kücken M, Ahrens W, Block A, Hauffe D, Keuler K, Rockel B, Will A. 2006. CLM – the climate version of LM: brief description and long-term applications. *COSMO Newsletter* **6**: 225–235.

Brisson E, Van Weverberg K, Demuzere M, Devis A, Saeed S, Stengel M, van Lipzig NPM. 2016. How well can a convection-permitting climate model reproduce decadal statistics of precipitation, temperature and cloud characteristics? *Climate Dynamics* **47**: 3043–3061, doi: 10.1007/s00382-016-3012-z.

Chan SC, Kendon EJ, Fowler HJ, Blenkinsop S, Ferro CAT, Stephenso DB. 2013. Does increasing the spatial resolution of a regional climate model improve the simulated daily precipitation? *Climate Dynamics* **41**: 1475–1495.

Christensen JH, Christensen OB. 2007. A summary of the PRUDENCE model projections of changes in European climate by the end of this century. *Climatic Change* **81**: 7–30.

Clark AJ, Gallus WA, Chen TC. 2007. Comparison of the diurnal precipitation cycle in convection-resolving and non-convection-resolving mesoscale models. *Monthly Weather Review* **135**(10): 3456–3473.

DeAngelis AM, Qu X, Zelinka MD, Alex Hall A. 2015. An observational radiative constraint on hydrologic cycle intensification. *Nature* **528**: 249–253, doi: 10.1038/nature15770.

Dee DP, Uppala SM, Simmons AJ, Berrisford P, Poli P, Kobayashi S, Andrae U, Balmaseda MA, Balsamo G, Bauer P, Bechtold P,

Beljaars ACM, van de Berg L, Bidlot J, Bormann N, Delsol C, Dragani R, Fuentesn M, Geer AJ, Haimberger L, Healy SB, Hersbach H, H'olm EV, Isaksen L, Kållberg P, Köhler M, Matricardi M, McNally AP, Monge-Sanz BM, Morcrette JJ, Park BK, Peubey C, de Rosnay P, Tavolato C, Thépaut JN, Vitart F. 2011. The ERA-Interim reanalysis: configuration and performance of the data assimilation system. *Quarterly Journal of the Royal Meteorological Society* **137**: 553–597.

Doms G, Forstner F, Heis E, Herzog HJ, Raschendorfer M, Reinhardt T, Ritter B, Schrodin R, Schulz JP, Vogel G. 2011. A Description of the Nonhydrostatic Regional COSMO Model Part II: Physical Parameterization. Technical Report September.

Fosser G, Khodayar S, Berg P. 2014. Benefit of convection permitting climate model simulations in the representation of convective precipitation. *Climate Dynamics* **44**: 45–60.

Hazeleger W, Wang X, Severijns C, Ştefănescu S, Bintanja R, Sterl A, Wyser K, Semmler T, Yang S, van den Hurk B, van Noije T, van der Linden E, van der Wiel K. 2012. EC-Earth V2: description and validation of a new seamless Earth system prediction model. *Climate Dynamics* **39**: 2611–2629.

IPCC. 2013. Summary for policymakers. Climate change 2013: the physical science basis. In *Contribution of Working Group I to the Fifth Assessment Report of the Intergovernmental Panel on Climate Change*, Stocker TF, Qin D, Plattner GK, Tignor M, Allen SK, Boschung J, Nauels A, Xia Y, Bex V, Midgley PM (eds). Cambridge University Press: Cambridge, UK and New York, NY.

Jones PW. 1999. First- and second-order conservative remapping schemes for grids in spherical coordinates. *Monthly Weather Review* **127**: 2204–2210, doi: 10.1175/1520-0493.

Kendon EJ, Roberts NM, Senior CA, Roberts MJ. 2012. Realism of rainfall in a very high-resolution regional climate model. *Journal of Climate* **25**(17): 5791–5806.

Kendon EJ, Roberts NM, Fowler HJ, Roberts MJ, Chan SC, Senior CA. 2014. Heavier summer downpours with climate change revealed by weather forecast resolution model. *Nature Climate Change* **4**: 570–576.

Kotlarski S, Keuler K, Christensen OB, Colette A, Deque M, Gobiet A, Goergen K, Jacob D, Lüthi D, van Meijgaard E, Nikulin G, Schär C, Teichmann C, Vautard R, Warrach-Sagi K, Wulfmeyer V. 2014. Regional climate modeling on European scales: a joint standard evaluation of the EURO-CORDEX RCM ensemble. *Geoscientific Model Development* **7**(4): 1297–1333.

Mearns LO, Gutowski WJ, Jones R, Leung LY, McGinnis S, Nunes AMB, Qian Y. 2009. A regional climate change assessment program for North America. *Eos* **90**(36): 311–312.

Prein AF, Gobiet A, Suklitsch M, Truhetz H, Awan NK, Keuler K, Georgievski G. 2013. Added value of convection permitting seasonal simulations. *Climate Dynamics* **41**(9–10): 2655–2677.

Prein AF, Langhans W, Fosser G, Ferrone A, Ban N, Goergen K, Keller M, Tölle M, Gutjahr O, Feser F, Brisson E, Kollet S, Schmidli J, van Lipzig NPM, Leung R. 2015. A review on regional convection-permitting climate modeling: demonstrations, prospects, and challenges. *Reviews of Geophysics* **53**: 323–361.

Ritter B, Geleyn JF. 1992. A comprehensive radiation scheme for numerical weather prediction models with potential applications in climate simulations. *Monthly Weather Review* **120**(2): 303–325.

Rockel B, Will A, Hense A. 2008. The regional climate model COSMO-CLM (CCLM). *Meteorologische Zeitschrift* **17**(4): 347–348.

Schär C, Ban N, Fischer EM, Rajczak J, Schmidli J, Frei C, Giorgi F, Karl TR, Kendon EJ, Klein Tank AMG, O'Gorman PA, Sillmann J, Zhang X, Zwiers FW. 2016. Percentile indices for assessing changes in heavy precipitation events. *Climatic Change* **137**: 201–216, doi: 10.1007/s10584-016-1669-2.

Steppeler J, Doms G, Schättler U, Bitzer HW, Gassmann A, Damrath U, Gregoric G. 2003. Meso-gamma scale forecasts using the nonhydrostatic model LM. *Meteorology and Atmospheric Physics* **82**(1–4): 75–96.

Tabari H, Agha KA, Willems P. 2014. A perturbation approach for assessing trends in precipitation extremes across Iran. *Journal of Hydrology* **519**: 1420–1427.

Tabari H, Taye MT, Willems P. 2015. Water availability change in central Belgium for the late 21st century. *Global and Planetary Change* **131**: 115–123.

Tiedtke M. 1989. A comprehensive mass flux scheme for cumulus parameterization in large-scale models. *Monthly Weather Review* **117**(8): 1779–1800.

Wagner S, Berg P, Schädler G, Kunstmann H. 2013. High resolution regional climate model simulations for Germany: part II – projected climate changes. *Climate Dynamics* **40**: 415–427.

Walther A, Jeong JH, Nikulin G, Jones C, Chen D. 2013. Evaluation of the warm season diurnal cycle of precipitation over Sweden simulated by the Rossby Centre regional climate model RCA3. *Atmospheric Research* **119**: 131–139.

Weisman ML, Skamarock WC, Klemp JB. 1997. The resolution dependence of explicitly modeled convective systems. *Monthly Weather Review* **125**(4): 527–548.

Wicker LJ, Skamarock WC. 2002. Time-splitting methods for elastic models using forward time schemes. *Monthly Weather Review* **130**(8): 2088–2097.

Willems P, Olsson J, Arnbjerg-Nielsen K, Beecham S, Pathirana A, Gregersen IB, Madsen H, Nguyen VTV. 2012. *Impacts of Climate Change on Rainfall Extremes and Urban Drainage*. IWA Publishing: London.

17

Statistically related coupled modes of South Asian summer monsoon interannual variability in the tropics

F. S. Syed[1],* and F. Kucharski[2,3]

[1]Department of Meteorology, COMSATS Institute of Information Technology, Islamabad, Pakistan
[2]Earth System Physics Section, Abdus Salam International Centre for Theoretical Physics, Trieste, Italy
[3]Center of Excellence for Climate Change Research/Department of Meteorology, King Abdulaziz University, Jeddah, Saudi Arabia

*Correspondence to:
F. S. Syed, Department of
Meteorology, COMSATS Institute
of Information Technology, Park
Road, Chak Shahzad, Islamabad
44000, Pakistan.
E-mail: faisal.met@gmail.com

Abstract

Statistically coupled patterns of South Asian Summer Monsoon (SASM) interannual variability in the tropical oceans have been explored. Maximum covariance analysis (MCA) performed between global tropical sea surface temperature (SST) and SASM precipitation shows that El-Niño southern oscillation (ENSO) is the leading mode in the tropics, whereas the eastern pole of the Indian Ocean Dipole contributes to the second global mode and is the leading mode in the Indian Ocean. South tropical Atlantic SST variability is contributing to the second and third mode in the tropics and is the leading mode in the tropical Atlantic MCA coupled with SASM.

Keywords: interannual variability; monsoon; tropical Atlantic; IOD; ENSO

1. Introduction

Even though the interannual standard deviation of South Asian summer monsoon (SASM) accounts only for 10% of the total rainfall (Turner and Annamalai, 2012), these variations have significant impacts on agricultural production and water availability in this region. A part of the interannual variations is due to internal atmospheric variability and may thus limit the predictability of SASM (Sperber *et al.*, 2001; Saha *et al.*, 2011).

A large part of the predictability is believed to be related to forcing from the tropical oceans (Pacific, Indian, and Atlantic), and the fact that tropical oceans act as slowly varying boundary conditions on the atmosphere provides confidence in the potential predictability of SASM. El-Niño southern oscillation (ENSO) is probably the most important driver of SASM variability because of its global impacts through teleconnections (e.g. Trenberth *et al.*, 1998; Alexander *et al.*, 2002; Deser *et al.*, 2004). However, the effect of ENSO on SASM has become weaker in the recent decades (Krishna *et al.*, 1999; Yadav *et al.*, 2009). Some studies have suggested that the Indian Ocean Dipole (IOD), which has also an influence on SASM (e.g. Pokhrel *et al.*, 2012; Chaudhari *et al.*, 2013; Cherchi and Navarra, 2013), has modified the ENSO teleconnection to SASM (Ashok *et al.*, 2004; Ashok and Saji, 2007). Kucharski *et al.* (2007, 2008) linked the weakening of the ENSO–SASM relationship to the covariabilities of SSTs in the southern equatorial Atlantic with ENSO, which show an anticorrelation in the recent decades. The topical Atlantic influence on SASM is due to a Gill-type quadrupole response in streamfunction to a Atlantic positive (negative) heating that leads

to high (low) pressure over the Arabian Sea and India thus induces low-level divergence (convergence) and reduced (increased) SASM precipitation (Kucharski *et al.*, 2009; Wang *et al.*, 2009).

Maximum covariance analysis (MCA; e.g. Bretherton *et al.*, 1992; Newman and Sardeshmukh, 1995) is a robust tool to estimate coupled modes of variability. MCA has been recently applied to the SASM rainfall and SSTs (Mishra *et al.*, 2012; referred to as M12 in the following). M12 found that the first mode was related to the ENSO influence on SASM rainfall and showed that an overall drying is related to El Nino-type SSTs in the eastern Pacific region. The second MCA rainfall mode identified by M12 showed a more complex pattern with increased rainfall in southern India and reduced rainfall in the northeastern parts. The coupled SST pattern is related to warming mainly in the eastern Indian Ocean and South China Sea, but also in the Arabian Sea and Bay of Bengal. M12 argued that such an SST pattern is related to the previous year ENSO. On the other hand, Rao *et al.* (2010) argued that such a pattern might be a representative of the IOD. Unlike the results of M12, some recent studies (e.g. Chaudhari *et al.*, 2015) have pointed out that the second empirical orthogonal function (EOF) of precipitation is related to the IOD.

The purpose of this study is to interpret the MCA modes between SASM precipitation and tropical SSTs further, also taking into account the resent results of a possible tropical Atlantic influence on the SASM.

2. Data and methods

For precipitation, Asian precipitation – highly resolved observational data integration toward evaluation of water resources (APHRODITE) is used (Yatagai *et al.*, 2012) with horizontal resolution of $0.5° \times 0.5°$. Also, the global ($1° \times 1°$) HadISST sea surface temperature (SST) data set (Rayner *et al.*, 2003) is used. For the large-scale circulation (geopotential heights and winds) characteristics, the monthly mean reanalysis data of NCEP/NCAR (Kalnay *et al.*, 1996) with horizontal resolution of $2.5° \times 2.5°$ are used.

All analyses are performed on the seasonal mean JJAS (June to September) data from 1951 to 2007. The decadal signal is removed by subtracting an 11-year running mean. For the statistical significance of correlations, a Student's *t*-test is used.

MCA is performed to identify coupled patterns between SASM precipitation and tropical SSTs. Some of the issues related to MCA analysis are discussed by Newman and Sardeshmukh (1995). They found that the analysis gives robust results for extracting coupled signals and has small systematic bias. In this study, we investigate the influence of SSTs on SASM precipitation, therefore we focus on the heterogeneous correlation maps. The correlation coefficients between the expansion coefficients (ECs) time series of the two fields for a particular mode are also calculated. The squared covariance fraction (SCF) indicates the percentage of covariance explained by the particular mode and is calculated following the study by Bretherton *et al.* (1992).

3. Results and discussions

Figure 1(a) shows the spatial structure of the mean JJAS precipitation over South Asia. The mean precipitation is above 10 mm day^{-1} over the Western Ghats and the adjoining areas of Bay of Bengal. Other maxima of rainfall can be seen over central India, here mean rainfall is between 6 and 10 mm day^{-1}. The monsoon penetrates into northern Pakistan along the foothills of Himalayas but the mean rainfall remains below 6 mm day^{-1} over northern India and Pakistan.

3.1. SST-based indices in the tropical Oceans and their relationship with SASM precipitation

The NINO3.4 index is used to classify SST variability in the Pacific related to ENSO. It is defined as the average SST anomalies in the region 5°S–5°N and 170°–120°W. Positive (negative) values of NINO3.4 index correspond to El-Niño- (La-Nina) type conditions in the Pacific. The IOD index is defined as the SST anomalies difference between the tropical western Indian Ocean (50°–70°E, 10°S–10°N) and the tropical southeastern Indian Ocean (90°–110°E, 10°–0°S) following the study by Saji *et al.* (1999). A positive IOD index characterizes warmer than normal water in the tropical western Indian Ocean and cooler than normal water in the tropical eastern Indian Ocean. Figure 1(b)

and (e) shows the correlation between NINO3.4 and SASM precipitation and tropical SSTs, respectively. The SSTs in the Pacific show a typical ENSO pattern but the correlations also show significant positive (negative) values in the western (eastern) Indian Ocean which resembles the IOD pattern (Figure 1(f)). The correlation between NINO3.4 and IOD indices is 0.44 for the period considered which is significant at 95% level. The IOD index also shows significant correlation with SSTs in the Pacific (typical ENSO structure) (Figure 1(f)). However, the response of NINO3.4 and IOD is different in the SASM precipitation. NINO3.4 is negatively correlated with SASM precipitation over the whole region, whereas IOD is negatively correlated with the precipitation over Western Ghats and northeastern India and positively correlated with the precipitation over central India (e.g. Chaudhari *et al.*, 2013). We have also assessed the sensitivity of the above results with respect to alternative ENSO and IOD definitions. The results turn out to be very similar if the NINO3 index is used instead of the NINO3.4 index. For IOD, we have tested the average negative SST in the eastern pole (90°–110°E, 10°–0°S), which has been used alternatively as IOD definition in, e.g. Rao *et al.*, 2010, and may be viewed as the region where IOD develops first. We will refer to this index as IOD$_{east}$ in the following. We have verified that the JJAS IOD$_{east}$ index is highly correlated (0.75) with the canonical IOD index in its peak season (September to November) season. This result is consistent with the study of Krishnamurthy and Kirtman (2003). The results for the IOD$_{east}$ correlation with rainfall over the SASM region are similar to that for the canonical IOD definition. However, the corresponding SST pattern (see Supporting Information) is dominated by the eastern Indian Ocean cooling and shows less covariability with eastern Pacific SST anomalies.

The south tropical Atlantic index (STAI) is an indicator of the SST anomalies in the Gulf of Guinea, the eastern tropical South Atlantic Ocean. It is calculated with SST anomalies in the region (30°W–10°E, 20°–0°S) as suggested by Enfield *et al.*, 1999. Positive values of STAI correspond to warmer than normal SSTs in the eastern tropical South Atlantic Ocean. The correlation of STAI with SASM precipitation shows a dipole structure with negative correlations over central India and north western parts of South Asia. STAI does not show any significant correlations with tropical Indian and Pacific SSTs, indicating that the variability in the south tropical Atlantic is independent of ENSO.

All correlations calculated in Figure 1 are contemporaneous for seasonal JJAS means. This assumption has been shown to be valid using lead-lag correlations between an Indian monsoon rainfall index and ENSO as well as STAI in the study by Kucharski *et al.*, 2007 and Kucharski *et al.*, 2008. Also for the IOD influence, this assumption is typically made (e.g. Rao *et al.*, 2010).

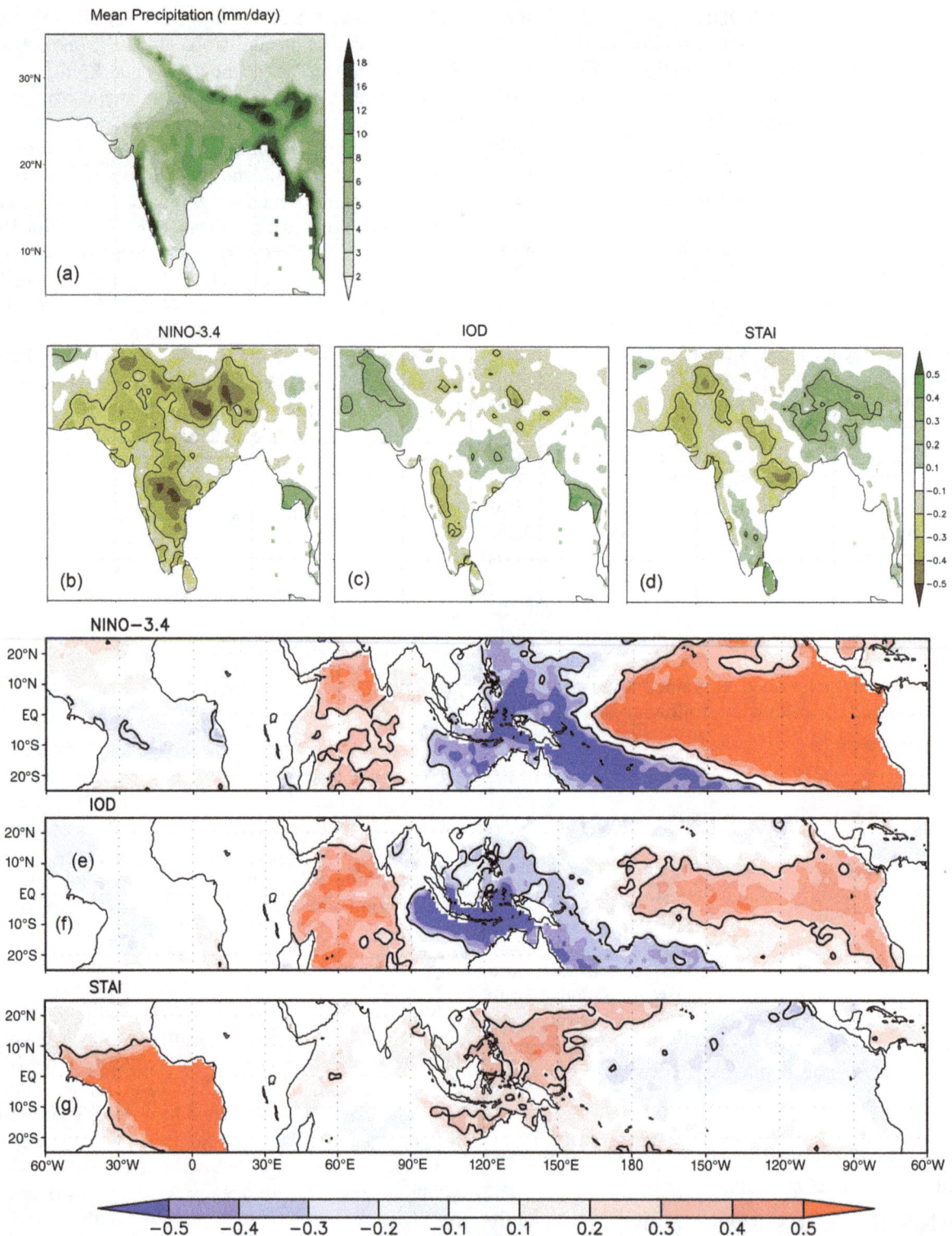

Figure 1. (a) Seasonally averaged SASM JJAS precipitation (mm day^{-1}) climatology (1951–2007) based on APHODITE data set. Correlation coefficients between NINO3.4, IOD and STAI and (b)–(d) SASM precipitation (e)–(g) SST. Black contours indicate the correlation values at 95% significance.

3.2. Coupled patterns of SASM precipitation variability in the global tropics

We first consider the leading patterns of SASM precipitation variability and their relationship with global tropical SSTs. Figure 2 shows the leading three modes of MCA heterogeneous correlation maps, between global tropical SSTs and precipitation over South Asia. The correlations between the EC of global tropical SSTs and SASM precipitation for the first, second and third mode of MCA are 0.65, 0.61 and 0.63, respectively, with SCF of 64.3, 12.1 and 9.8%, respectively, and are well separated from the following modes (which have SCF of about 2%). As expected, ENSO is

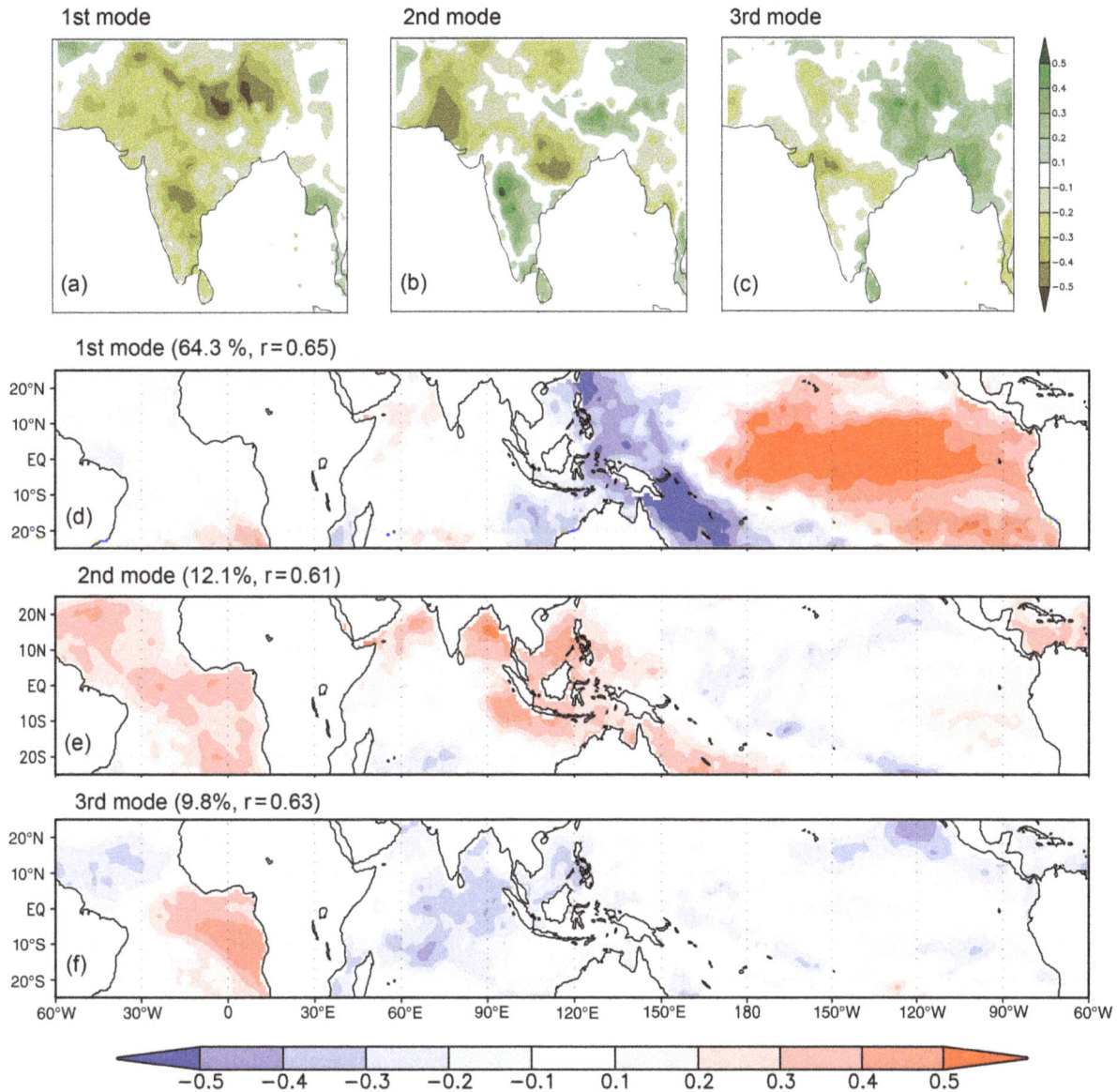

Figure 2. Coupled patterns of SASM precipitation and global tropical (0°–360°E, 20°S–20°N) SST variability based on MCA for 57 years (1951–2007). The patterns indicated by color shading are heterogeneous correlation coefficients for leading three modes between (a)–(c) SST expansion coefficient time series and precipitation (d)–(f) precipitation expansion coefficient time series and SST. The SFC and temporal correlation (*r*) between the SAM and SST expansion coefficients are indicated at the top in (d)–(f).

the leading mode of variability in the tropics coupled with SASM precipitation, which explains the largest covariance. EC of global tropical SSTs first mode has a correlation of 0.97 with NINO3.4 index. The corresponding negative anomalies in SASM precipitation (Figure 2(a)) is in agreement with previous studies, i.e. SASM precipitation tends to be generally suppressed during El-Niño years. This result is also in agreement with the study of Mishra *et al.* (2012).

The second mode of MCA shows that the above normal SSTs over tropical Atlantic, Arabian Sea, the Bay of Bengal and the South China Sea (Figure 2(e)) are coupled with below normal rainfall in central India and above normal rainfall over southern India to the east of the Western Ghats (Figure 2(b)). The EC of global tropical SSTs second mode is correlated with the IOD_east index (−0.45). M12 have found a similar

second mode and have linked this mode to the previous year's ENSO indexes.

The third mode for global tropical SSTs has spatial patterns in SASM precipitation and Atlantic SSTs (Figure 2(c) and (f)) similar to STAI correlation patterns (Figure 1(d) and (g)). The correlation of the SST EC and STAI is 0.46. This clearly shows that the third mode of variability in the SASM precipitation is related to Atlantic SSTs. Interestingly, the second mode is also correlated with the STAI (correlation = 0.5). If one combines SST modes 2 and 3 as $STAI_{rec} = a\ EC2 + b\ EC3$, where a and b are the correlations of EC2 and EC3 with STAI, respectively, then this reconstructed $STAI_{rec}$ has a correlation of 0.77 with STAI (see Supporting Information). The correlation of $STAI_{rec}$ with SASM rainfall and with SSTs indeed confirms the similarity

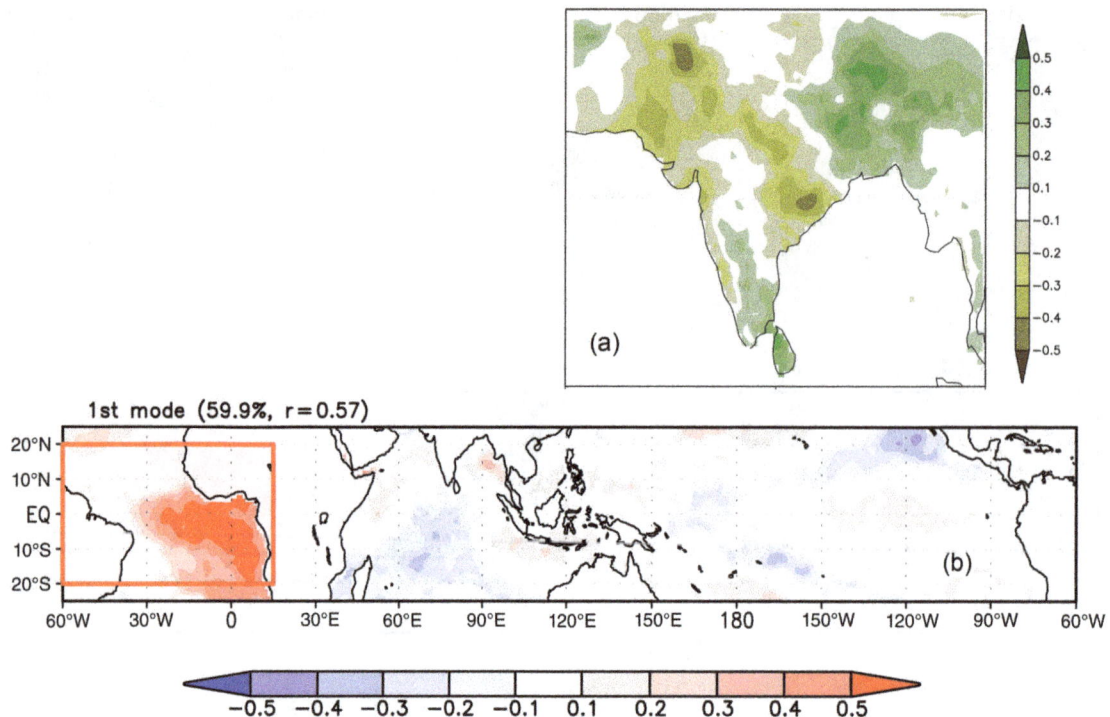

Figure 3. Coupled patterns of SASM precipitation (60°–100°E, 5°–35°N) and tropical Atlantic (60°W–15°E, 20°S–20°N) SST variability based on MCA on 57 years (1951–2007). The patterns indicated by color shading are heterogeneous correlation coefficients for leading mode between (a) SST expansion coefficient time series and SASM precipitation (b) vice versa. The SFC and temporal correlation (*r*) between the South Asian monsoon and SST expansion coefficients are indicated at the top in (b). The rectangle shows the MCA domain in Atlantic.

with the STAI SST pattern and influence on SASM rainfall (see Supporting Information and Figure 1(c) and (f)). This demonstrates that the STAI influence is mainly distributed in modes 2 and 3 of the global MCA. The discovery that the Atlantic is contributing to the statistically coupled modes of variability leads us to the further investigation discussed in the next section.

To further understand the second mode in the above MCA and the role of IOD in the SASM interannual variability, we repeated the analysis by taking SSTs only over tropical Indian Ocean (30°–120°E and 20°S–20°N) and SASM precipitation. The leading mode in this case looks similar to the second mode obtained in the global tropical SST pattern (Figure 2(e)), and it accounts for 43.7% of the SCF in this regional domain with high correlation between ECs of SST and SASM precipitation. Although the study of Mishra *et al.* (2012) linked a similar mode to the previous winter ENSO, the rainfall pattern over the SASM region resembles that of the (negative) IOD (see Figure 1(c)), and the SST pattern resemble that of the (negative) eastern IOD pole (see Supporting Information). The correlation between negative SST EC1 and IOD$_{east}$ is 0.8. The second mode shows a dipole structure in rainfall over the SASM region and a cooling in the southern Indian region. However, there are also anomalies in the Atlantic region that resemble the STAI pattern (see Figure 1(g)). Also, the rainfall pattern of the second mode resembles the STAI influence (Figure 1(d)). It is therefore likely that

this second regional Indian Ocean mode is strongly influenced by forcings from the Atlantic region.

The third mode in this regional MCA with SCF 13.8% has weak anomalies in the Indian Ocean region but shows large negative values in the eastern Pacific region that resemble that of ENSO. Also, the rainfall mode in the SASM region shows overall positive values that are consistent with the (negative) ENSO influence.

3.3. SASM coupled mode of variability in the tropical Atlantic

The regional MCA over tropical Atlantic and SASM precipitation (Figure 3(a) and (b)) shows that the variability in the SST anomalies in the Gulf of Guinea is the leading mode in the tropical Atlantic coupled with SASM precipitation. The SFC is 59.9% and the correlation of SST EC with STAI is 0.98. This mode is well separated from the second mode (SCF = 22.7%). The spatial patterns of precipitation (Figure 3(a)) over South Asia are also very similar to the reference correlation map of STAI (Figure 1(d)). For consistency with the STAI correlation maps (Figure 1(d) and (g)), the EC time series of both SST and precipitation were multiplied with −1 before plotting the heterogeneous correlation maps shown in Figure 3.

Spatial patterns obtained by regressing the seasonally averaged 850 hPa geopotential heights (m) and winds (m s^{-1}) on to the standardized EC (leading mode) of the precipitation shows significant positive geopotential anomalies and the related anticyclonic circulation

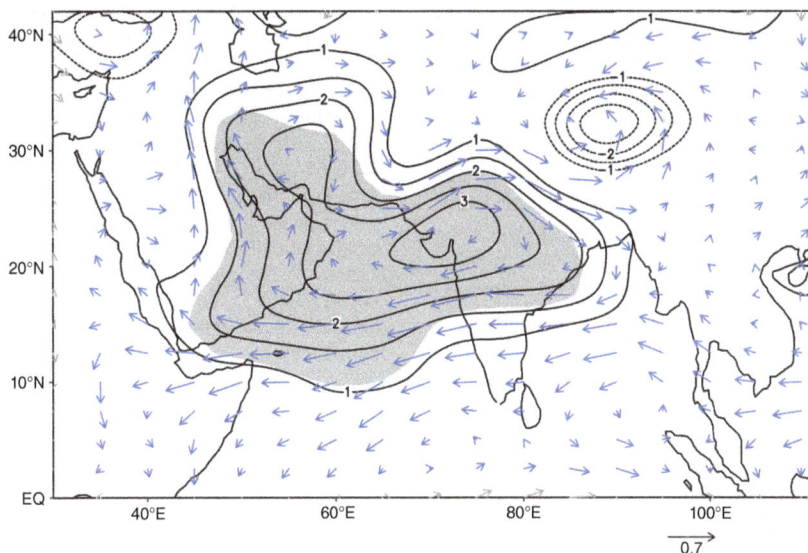

Figure 4. Spatial patterns obtained by regressing the seasonally averaged 850 hPa geopotential heights (m) and winds (m s⁻¹) on to the standardized expansion coefficient (leading mode) of the precipitation for MCA between tropical Atlantic and SASM precipitation. The regression values with 95% significance are shaded.

over the Arabian Sea, which cause decreased precipitation over central India and the north western parts of South Asia related to positive SST anomalies over the tropical Atlantic. The negative geopotential anomalies and related circulation can be linked with increased precipitation over Bangladesh and northeastern India. We have tested the above results using different reanalysis produces and found that the main features are robust. As outlined in the Section 1, the physical mechanism for the tropical Atlantic influence on the SASM region relies on a Gill-type response to the SST-induced heating anomaly. The response to a positive heating anomaly is a large-scale quadrupole in stream function and vorticity (Kucharski *et al.*, 2009). One center of this response is located over the Arabian Sea/Indian region with a low-level negative vorticity anomaly, leading to low-level divergence and reduced rainfall over the large parts of west and central India. Also, the Somali jet is weakened as part of this response. This response is further reinforced as the surface pressure adjusts to the decreased heating local (rainfall), resulting in a high-pressure anomaly over western-central India, as shown in Figure 4. The opposite response over the northeastern parts of the SASM region could be related to a compensating convergence induced by the divergence over central-western India.

4. Conclusions

We have shown that ENSO is the leading mode with explained covariance 64.3% of a MCA performed between tropical SSTs and SASM precipitation, whereas eastern pole of the IOD contributed to the second global mode and is the leading mode when the MCA is performed regionally between Indian Ocean SSTs and SASM. This is consistent with previous findings.

As a new result, we have shown that Atlantic Ocean SST variability is contributing to MCA modes 2 and 3 of global tropical SST with SASM. The south tropical Atlantic appears as the well-separated leading mode of tropical Atlantic SST with SASM precipitation. Increased (decreased) precipitation over the Western Ghats, central India and the northwestern parts and decreased (increased) precipitation over the northeastern parts of South Asia are related to negative (positive) SST anomalies over the tropical South Atlantic in this coupled mode. The influence of negative (positive) tropical South Atlantic SSTs on the SASM region can be explained as Gill-type response that induces cyclonic (anticyclonic) circulation anomalies over the Arabian Sea and central-western Indian region. This mode is also associated with increased (decreased) pressure and thus decreased (increased) precipitation over Bangladesh and adjoining areas.

This studies shows that predicting SST variability in the tropical Atlantic region may provide a crucial contribution to SASM predictability. This is particularly of interest because the tropical Atlantic continues to be a region with severe model biases (Richter *et al.*, 2014).

Acknowledgements

The support of Dr. Jin Ho Yoo in the MCA analysis is highly appreciated. We thank the two anonymous reviewers for their constructive suggestions that helped to improve the manuscript.

Supporting information

The following supporting information is available:

Figure S1. Correlation coefficients between an eastern IOD pole index (negative average SST anomalies in the region 90°–110°E,

10°–0°S; referred to as IOD$_{east}$ the main text) and (a) SASM precipitation (b) SST. Black contours show the correlation values at 95% significance.

Figure S2. Coupled patterns of SASM precipitation and tropical Indian Ocean (30°–120°E and 20°S–20°N) SST variability based on MCA for 57 years (1951–2007). The patterns indicated by color shading are heterogeneous correlation coefficients for leading three modes between (a)–(c) SST expansion coefficient time series and precipitation; (d)–(f) precipitation expansion coefficient time series and SST. The SFC and temporal correlation (r) between the SAM and SST expansion coefficients are indicated at the top in (d)–(f).

Figure S3. Correlation of the reconstructed south tropical Atlantic index, STAI$_{rec}$ (see main text, Section 3.2 for definition) (a) with SAESM rainfall and (b) with SSTs.

Figure S4. Time series (1951–2007), STAI (Black), EC2 (Green), EC3 (Blue) and STAI$_{rec}$ (Red). EC2, EC3 and STAI$_{rec}$ are normalized by 100.

References

Alexander MA, Blade I, Newman M, Lanzante JR, Lau NC, Scott JD. 2002. The atmospheric bridge: the influence of ENSO teleconnections on air-sea interaction over the global oceans. *Journal of Climate* **15**: 2205–2231.

Ashok K, Saji NH. 2007. On impacts of ENSO and Indian Ocean Dipole events on the sub-regional Indian summer monsoon rainfall. *Natural Hazards* **42**(2): 273–285, doi: 10.1007/s11069-006-9091-0.

Ashok K, Guan Z, Saji NH, Yamagata T. 2004. Individual and combined influences of the ENSO and Indian Ocean Dipole on the Indian summer monsoon. *Journal of Climate* **17**: 3141–3155.

Bretherton CS, Smith C, Wallace JM. 1992. An intercomparison of methods for finding coupled patterns in climate data. *Journal of Climate* **5**(6): 541–560.

Chaudhari HS, Pokhrel S, Mohanty S, Saha SK. 2013. Seasonal prediction of Indian summer monsoon in NCEP coupled and uncoupled model. *Theoretical and Applied Climatology* **114**: 459–477.

Chaudhari HS, Pokhrel S, Mohanty S, Saha SK, Dhakate A, Hazra A. 2015. Improved depiction of Indian summer monsoon in latest high resolution NCEP climate forecast system reanalysis. *International Journal of Climatology* **35**: 3102–3119.

Cherchi A, Navarra A. 2013. Influence of ENSO and of the Indian Ocean Dipole on the Indian summer monsoon variability. *Climate Dynamics* **41**(1): 81–103.

Deser C, Phillips AS, Hurrell JW. 2004. Pacific interdecadal climate variability: linkages between the tropics and the North Pacific during boreal winter since 1900. *Journal of Climate* **17**: 3109–3124.

Enfield DB, Mestas AM, Mayer DA, Cid-Serrano L. 1999. How ubiquitous is the dipole relationship in tropical Atlantic sea surface temperatures? *Journal of Geophysical Research-Oceans* **104**: 7841–7848.

Kalnay E, Kanamitsu M, Kistler R, Collins W, Deaven D, Gandin L, Iredell M, Saha S, White G, Woollen J, Zhu Y, Leetmaa A, Reynolds R, Chelliah M, Ebisuzaki W, Higgins W, Janowiak J, Mo KC, Ropelewski C, Wang J, Jenne R, Joseph D. 1996. The NCEP/NCAR 40-Year Reanalysis Project. *Bulletin of the American Meteorological Society* **77**: 437–471.

Krishna KK, Rajagopalan B, Cane MA. 1999. On the weakening relationship between the Indian monsoon and ENSO. *Science* **264**: 2156–2159.

Krishnamurthy V, Kirtman B. 2003. Variability of the Indian Ocean: relation to monsoon and ENSO. *Journal of the Royal Meteorological Society* **129**: 1623–1646.

Kucharski F, Bracco A, Yoo JH, Molteni F. 2007. Low-frequency variability of the Indian monsoon–ENSO relationship and the tropical Atlantic: the 'weakening' of the 1980s and 1990s. *Journal of Climate* **20**: 4255–4266.

Kucharski F, Bracco A, Yoo JH, Molten F. 2008. Atlantic forced component of the Indian monsoon interannual variability. *Geophysical Research Letters* **35**: L04706, doi: 10.1029/2007GL033037.

Kucharski F, Bracco A, Yoo JH, Tompkins A, Feudale L, Ruti P, Dell'Aquila A. 2009. A Gill-Matsuno-type mechanism explains the tropical Atlantic influence on African and Indian monsoon rainfall. *Quarterly Journal of the Royal Meteorological Society* **135**: 569–579.

Mishra V, Smoliak BV, Lettenmaier DP, Wallace JM. 2012. A prominent pattern of year-to-year variability in Indian summer monsoon rainfall. *Proceedings of National Academy of Sciences of United States of America* **109**(19): 7213–7217.

Newman M, Sardeshmukh PD. 1995. A caveat concerning singular value decomposition. *Journal of Climate* **8**: 352–360.

Pokhrel S, Chaudhari HS, Saha SK, Dhakaye A, Yadav RK, Salunke K, Mahapatra S, Rao SA. 2012. ENSO, IOD and the Indian summer monsoon in NCEP climate forecast system. *Climate Dynamics* **39**: 2143–2165.

Rao SA, Chaudhari HS, Pokrel S, Goswami BN. 2010. Unusual central Indian drought of summer monsoon 2008: role of southern tropical Indian Ocean warming. *Journal of Climate* **23**: 5163–5174.

Rayner NA, Parker DE, Horton EB, Folland CK, Alexander LV, Rowell DP, Kent EC, Kaplan A. 2003. Global analyses of sea surface temperature, sea ice, and night marine air temperature since the late nineteenth century. *Journal of Geophysical Research* **108**: D144407, doi: 10.1029/2002JD002670.

Richter I, Xie SP, Behera SK, Doi T, Masumoto Y. 2014. Equatorial Atlantic variability and its relation to mean state biases in CMIP5. *Climate Dynamics* **42**: 171–188.

Saha SK, Halder S, Kumar KK, Goswami BN. 2011. Pre-onset land surface processes and 'internal' interannual variabilities of the Indian summer monsoon. *Climate Dynamics* **36**: 2077–2089.

Saji NH, Goswami BN, Vinayachandran PN, Yamagata T. 1999. A dipole mode in the tropical Indian Ocean. *Nature* **401**: 360–363.

Sperber KR, Brankovic C, Déqué M, Frederiksen CS, Graham R, Kitoh A, Kobayashi C, Palmer T, Puri K, Tennant W, Volodin E. 2001. Dynamical Seasonal Predictability of the Asian Summer Monsoon. *Monthly Weather Review* **129**: 2226–2248.

Trenberth KE, Branstator GW, Karoly D, Kumar A, Lau NC, Ropelewski C. 1998. Progress during TOGA in understanding and modeling global teleconnections associated with tropical sea surface temperatures. *Journal of Geophysical Research* **14**: 291–324.

Turner AG, Annamalai H. 2012. Climate change and the South Asian summer monsoon. *Nature Climate Change* **2**: 587–595, doi: 10.1038/nclimate1495.

Wang C, Kucharski F, Barimalala R, Bracco A. 2009. Teleconnections of the tropical Atlantic to the tropical Indian and Pacific Oceans: a review of recent findings. *Meteorologische Zeitschrift* **18**: 445–454, doi: 10.1127/0941-2948/2009/0394.

Yadav RK. 2009. Changes in the large-scale features associated with the Indian summer monsoon in the recent decades. *International Journal of Climatology* **29**: 117–133.

Yatagai A, Kamiguchi K, Arakawa O, Hamada A, Yasutomi N, Kitoh A. 2012. APHRODITE: Constructing a Long-Term Daily Gridded Precipitation Dataset for Asia Based on a Dense Network of Rain Gauges. *Bulletin of the American Meteorological Society* **93**: 1401–1415.

18

Decadal trends of the annual amplitude of global precipitation

Bin Wang, Xiaofan Li,* Yanyan Huang and Guoqing Zhai

School of Earth Sciences, Zhejiang University, Hangzhou, China

Correspondence to:
X. Li, School of Earth Sciences,
Zhejiang University, 38 Zheda
Road, Hangzhou, Zhejiang
310027, China.
E-mail: xiaofanli@zju.edu.cn

Abstract

In this study, decadal trends of the annual amplitude of global precipitation are compared in Climate Prediction Center (CPC) Merged Analysis of Precipitation (CMAP), Global Precipitation Climatology Project (GPCP), and National Centers for Environmental Prediction (NCEP) reanalysis data sets. The analysis reveals decreasing trends in the CMAP and reanalysis data and a flat trend in the GPCP data. The decreasing trends are mainly associated with the increasing trend of low annual minimum precipitation rate in the CMAP data and high annual minimum precipitation rate in the reanalysis data. The trend in the GPCP data is flat because of the balance between decreasing trends along equatorial oceans and increasing trends over subtropical oceans.

Keywords: decadal trends; annual amplitude; global precipitation rate

1. Introduction

Precipitation is important in regulating global hydrological cycles at multiple temporal and spatial scales. There are no precipitation observations over the vast open ocean until the reliable satellite precipitation retrievals were available in late 1970. The rain gauge observations over lands and global satellite precipitation retrievals are merged to form global precipitation date sets with reasonable spatial resolutions for weather and climate studies. The rain gauge-satellite precipitation data were developed based on a series of intercomparison and validation studies (Spencer, 1993; Kondragunta and Gruber, 1997; Ebert and Manton, 1998; Krajewski et al., 2000; Adler et al., 2001). Among the merged rain gauge-satellite precipitation data, the CMAP [Climate Prediction Center (CPC) Merged Analysis of Precipitation] data (e.g. Xie and Arkin, 1997) and the GPCP (Global Precipitation Climatology Project) data (e.g. Adler et al., 2003; Huffman et al., 2009) are the two of the well-recognized precipitation data sets in meteorological research communities. Gruber et al. (2000) carried out a comparison analysis of the two data sets for the period of July 1987–December 1998 and found a good agreement for the average seasonal cycle over both lands and oceans. Yin et al. (2004) compared the GPCP and CMAP Monthly Precipitation Products for the period 1979–2001 and their EOF analysis showed that the first leading modes of the two data sets are identical.

The information from modeling has been added to form a new data set such as the National Centers for Environmental Prediction (NCEP)/National Center for Atmospheric Research (NCAR) (Kalnay et al., 1996). Janowiak et al. (1998) conducted a comparison study between the NCEP/NCAR reanalysis data and the GPCP data for the period of 1988–1995 and found a good agreement on large-scale but substantial difference on regional scales. The precipitation is a good measure for climate studies of the global system because it is associated with dynamic process, water cycle, and cloud microphysics.

Global precipitation variability corresponds to large-scale background circulation changes. The long-term precipitation changes may be caused by both natural and human-induced processes (e.g. Bates and Jackson, 2001) such as the global warming climate (e.g. Jones and Moberg, 2003). Thus, precipitation change may be associated with the warm climate (e.g. Dai and Trenberth, 2004; Liu et al., 2012). The GPCP data show a flat trend of global precipitation (Gu et al., 2007), whereas the CMAP data reveal a decreasing trend (Yin et al., 2004). The decreasing trend of global precipitation in the CMAP data is disputable because of an artifact of input data change and atoll sampling error over the tropical ocean (Yin et al., 2004). The GPCP data show relative bias error estimates of 10–20% over the tropical Pacific (Alder et al., 2012), but they offer the highest correlation and lowest monthly deviations with reference to the rain-gauge atoll station data provided by the Pacific Rainfall Database (PACRAIN) (Pfeifroth et al., 2013). Because global precipitation data only are available for several decades and the global circulation modeling may contain large uncertainties, in particular, in the modeling of cloud and precipitation properties (e.g. Cess et al., 1997), it may be hard to establish coherent long-term trends of global precipitation from available observational and modeling precipitation data sets.

The objective of this study is to analyse similarities and differences in decadal trends of the annual amplitude of global precipitation and associated precipitation statistics between the data sets of CMAP and GPCP and NCEP reanalysis. The annual amplitude of global precipitation rate is defined as the difference between the annual maximum and minimum monthly mean precipitation rate. We conduct such comparison study in the annual amplitude for the following reasons. First, climate changes such as global warming may have important impacts on precipitation extremes (e.g. Emori and Brown, 2005; Kharin and Zwiers, 2005), which may lead to certain decadal trends of precipitation extremes such as the annual amplitude at annual time scale. Second, Li *et al.* (2015) used this data set to analyse the decadal trends of global precipitation and associated precipitation statistics and found that divergent decadal trends are associated with the differences in precipitation statistics. The decadal trends of the annual amplitude are important aspect of decadal trends of global precipitation. Third, the precipitation statistics associated with the annual amplitude may be different from that related to annual mean global precipitation, which have different sensitivity to data quality. The organization of this study is as follows. The data are briefly discussed in Section 2. The results are presented in Section 3. A summary is given in Section 4.

2. Data

The observational precipitation data used in this study include the CMAP (Xie and Arkin, 1997) and GPCP (Huffman *et al.*, 2009) data. The GPCP data is one of Global Energy and Water Exchanges (GEWEX) global analyses of the water and energy cycle organized by the GEWEX Radiation Panel. Both CMAP and GPCP data were constructed by merging gauge and satellite estimates and have the horizontal resolution of 2.5° latitude by 2.5° longitude. The satellite retrievals in the CMAP data include Geostationary Operational Environmental Satellite (GOES) Precipitation Index (GPI), Outgoing Longwave Radiation (OLR) Precipitation Index (OPI), Special Sensor Microwave/Imager (SSM/I) scattering and SSM/I emission and Microwave Sounding Unit (MSU). The satellite retrievals in the GPCP data include the SSM/I retrievals, merged geosynchronous- and low-Earth-orbit infrared data, the OPI data, and the estimate from Television Infrared Observation Satellite Operational Vertical Sounder (TOVS) and Advanced Infrared Sounders (AIRS).

The modeling data is the reanalysis data (R-1) developed by a joint project between the NCEP and the NCAR, which involves the recovery of land surface, ship, rawinsonde, pibal, aircraft, satellite, and other data (Kalnay *et al.*, 1996; Kistler *et al.*, 2001). The updated version R2 (Kanamitsu *et al.*, 2002) is used in this study. The updates include new physics and observed soil moisture forcing and collection of previous errors. The precipitation in the reanalysis data is from the model integration forward for 6 h after the model is initialized

with other observational variables such as winds, temperature, and moisture. The horizontal resolution is 1.905° latitude by 1.875° longitude.

Adler *et al.* (2012) showed an estimated error bar of ±7% for GPCP long-term global precipitation. Although the reanalysis data were processed with the same assimilation system, the data may include possible artificial trends from increasing input data (e.g. increasing aircraft data) and data quality improvement (e.g. increasing vertical resolution of radiosonde).

3. Results

To study precipitation statistics for the annual amplitude of global precipitation, monthly mean grid-scale precipitation data during 1979–2008 are used to accumulate precipitation amount and to divide by total precipitation amount at the precipitation-rate interval of $0.3 \, \text{mm} \, \text{d}^{-1}$ for the annual maximum and minimum monthly mean precipitation rate, respectively (Figure 1). For both maximum and minimum monthly mean precipitation rates, high precipitation rates (e.g. $10 \, \text{mm} \, \text{h}^{-1}$) have more contributions to total precipitation in the reanalysis data than in the two observational data, which is consistent to that found in Li *et al.* (2015) in their analysis of decadal trends of annual and global

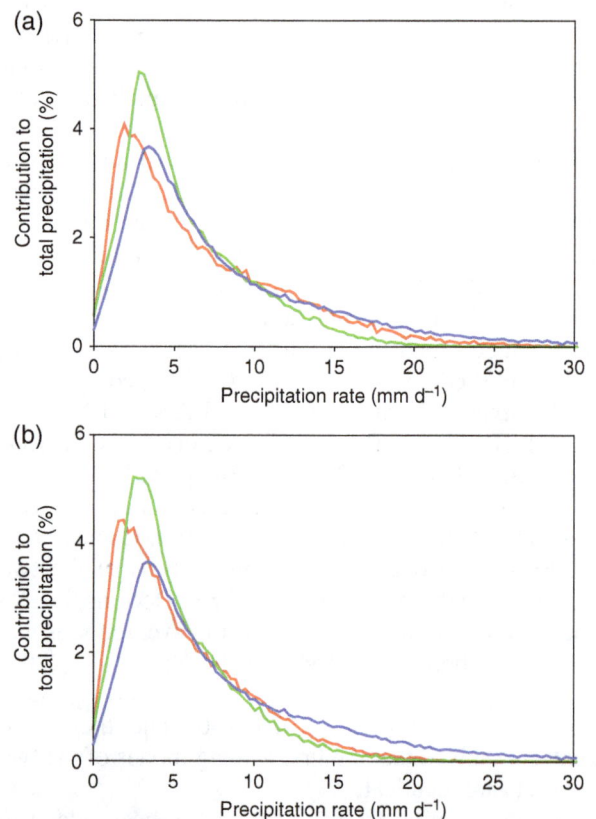

Figure 1. Contribution (%) to total precipitation from grid-scale data of annual (a) maximum and (b) minimum precipitation rate at the precipitation-rate bin of $0.3 \, \text{mm} \, \text{d}^{-1}$ from 1979 to 2008 as function of precipitation rate in the CMAP (red), GPCP (green), and NCEP reanalysis (blue) data.

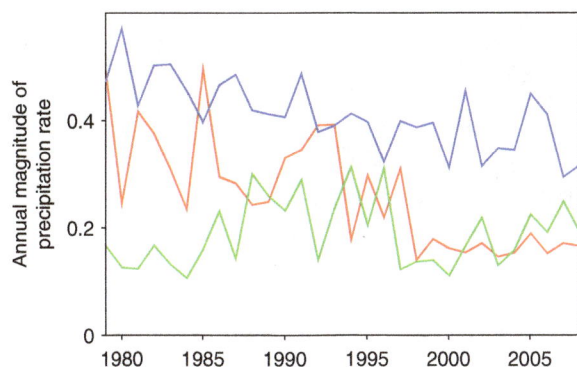

Figure 2. Time series of annual maximum difference of global-mean precipitation rate calculated using the CMAP (red), GPCP (green), and NCEP reanalysis (blue) monthly data from 1979 to 2008. Unit in mm d^{-1}.

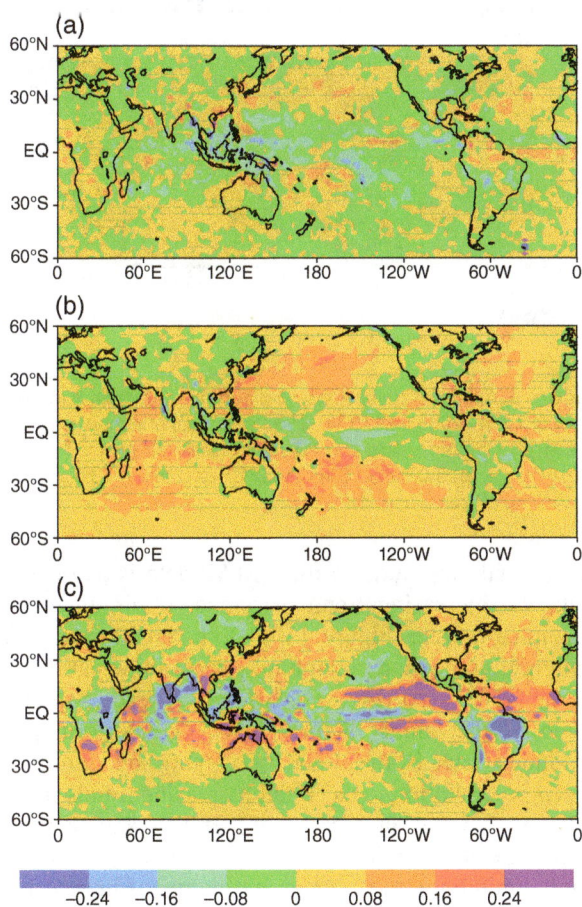

Figure 3. Spatial distributions of linear trends of annual maximum difference of precipitation rate (mm d^{-1} y^{-1}) in the period of 1979–2008 calculated using (a) CMAP, (b) GPCP, and (c) NCEP reanalysis data.

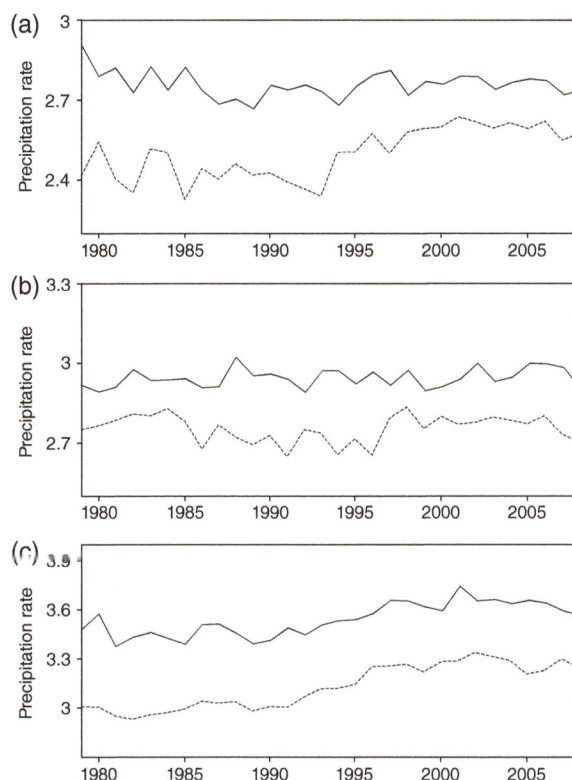

Figure 4. Time series of annual maximum (solid) and minimum (dashed) monthly mean precipitation rates calculated using (a) CMAP, (b) GPCP, and (c) NCEP reanalysis monthly data from 1979 to 2008. Unit in mm d^{-1}.

mean precipitation rate. The reanalysis data produce much stronger decadal signals than the two other observational data do, in particular, over the Intertropical Convergence Zone (ITCZ) and South Pacific Convergence Zone (SPCZ) and surrounding tropical oceanic areas. Strong precipitation could be a common problem in numerical modeling (e.g. Sui *et al.*, 1998) in the reanalysis data where the recycling of moisture is

too large in most models and the lifetime of moisture is too short (Trenberth *et al.*, 2011). On the other hand, the CMAP and GPCP data may underestimate oceanic mean precipitation compared with the merged precipitation estimate from the CloudSat, Tropical Rainfall Measuring Mission (TRMM) and Aqua during 2007–2009 (Behrangi *et al.*, 2014). The CMAP data also reveal larger contributions to total precipitation than the GPCP data do, which may be due to the fact that the two observational data sets use different satellite measurements and different retrieval algorithms.

The analysis of the annual amplitude of globally averaged precipitation rate reveals decreasing trends for the CMAP and reanalysis data and a flat trend for the GPCP data (Figure 2). The linear trends of the annual amplitude of globally averaged precipitation rate and associated linear correlation coefficients are −0.0088 and 0.74 mm d^{-1} year^{-1} for the CMAP data, 0.0010 and 0.06 mm d^{-1} year^{-1} for the GPCP data, and −0.0054 and 0.72 mm d^{-1} year^{-1} for the reanalysis data. Thus, the decreasing trends of the annual amplitude of globally averaged precipitation rate in the CMAP and reanalysis data are statistically significant. The annual amplitude is larger in the CMAP data than in the GPCP data, whereas it is smaller in the CMAP data than in the reanalysis data.

To examine the spatial contribution to linear trend of the annual amplitude of global mean precipitation rate, the spatial distribution of linear trend of the annual

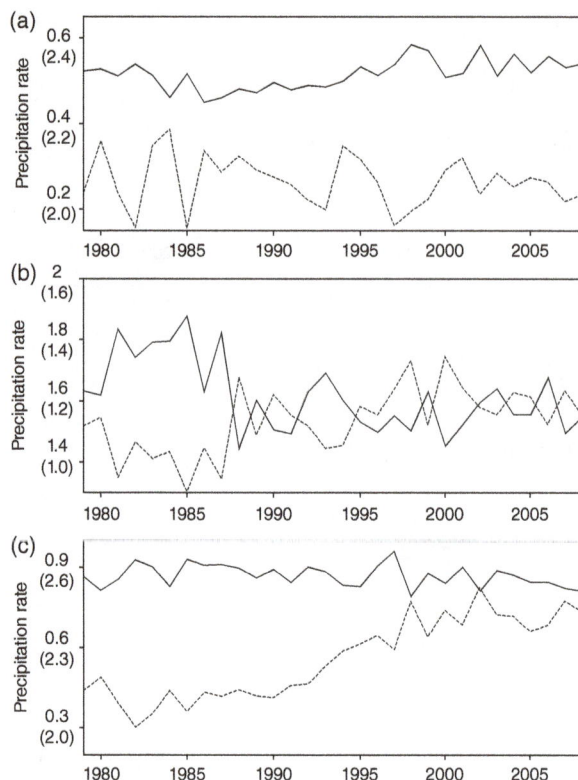

Figure 5. Time series of contributions to the annual minimum global average from grid-scale precipitation data of (a) lower (higher) than 2.1 mm d^{-1} in the CMAP data, (b) 5.0 mm d^{-1} in the GPCP data and (c) 3.3 mm d^{-1} in the NCEP reanalysis data denoted by solid (dashed) line. The plotting scales of precipitation rate for dashed lines are shown by numbers in brackets.

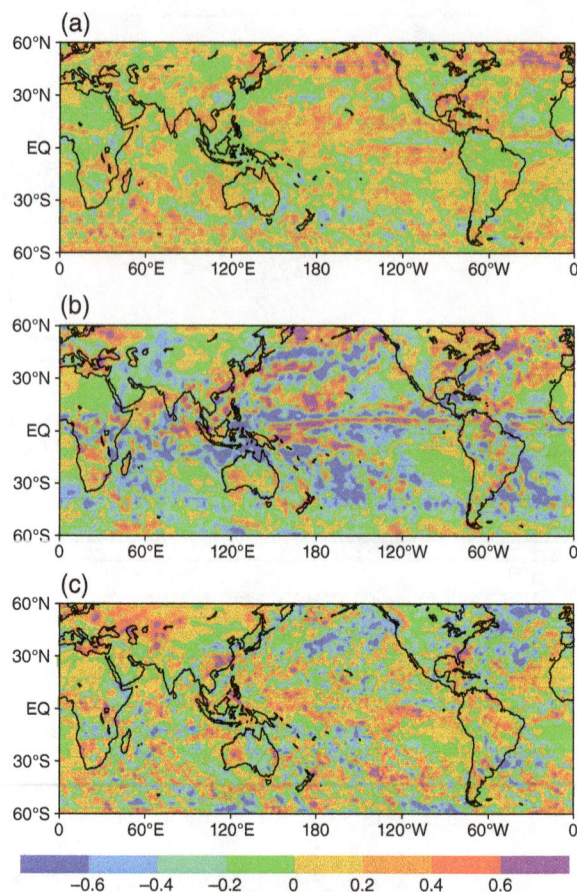

Figure 6. Spatial distributions of difference between the averages of 1994–2008 and 1979–1993 of annual minimum monthly mean precipitation rate for (a) lower than 2.1 mm d^{-1} in the CMAP data, (b) 5.0 mm d^{-1} in the GPCP data and (c) 3.3 mm d^{-1} in the NCEP reanalysis data.

amplitude of precipitation rate is shown in Figure 3. The negative linear trends of the annual amplitude along equatorial Indian and Pacific Oceans mainly contribute to the decreasing trends of global mean annual amplitude in the CMAP data (Figure 3(a)). Compared with those in the CMAP data, the negative linear trends of the annual amplitude along equatorial Indian and Pacific Oceans are weaker in the GPCP data (Figure 3(b)). The positive linear trends occur over subtropical oceanic areas. As a result, the negative linear trends are balanced by the positive linear trends in global mean calculations, which causes the flat trend for global mean annual amplitude in the GPCP data. The positive linear trends of the annual amplitude appear along the central and eastern branches of ITCZ and SPCZ and the Indian Ocean off the equator whereas the negative linear trends occur along the equator (Figure 3(c)). Thus, the decreasing trends of global mean annual amplitude is because of the fact that the negative linear trends are stronger than the positive linear trends.

Because the annual amplitude of global mean precipitation rate is defined as difference between annual maximum and minimum global mean precipitation rate, the decadal trends of annual maximum and minimum global mean precipitation rate are examined separately. Because the annual maximum mean precipitation rate shows a weak decreasing trend, the decreasing trend of

the annual amplitude in the CMAP data is associated with the increasing trend of annual minimum mean precipitation rate (Figure 4(a)). Both annual maximum and minimum mean precipitation rates show flat trends in the GPCP data (Figure 4(b)), which lead to the flat trend of the annual amplitude of global mean precipitation rate. Both annual maximum and minimum mean precipitation rates reveal increasing trends in the reanalysis data, but increasing trend of annual maximum mean precipitation rate is smaller than that of annual minimum mean precipitation rate (Figure 4(c)). Thus, the decreasing trend of the annual amplitude corresponds to the increasing trend of annual minimum mean precipitation rate.

The annual minimum precipitation rate can be further analysed by the low- and high-precipitation rates from grid-scale data. Following Li *et al.* (2015), the grid-scale monthly mean precipitation-rate data for the decadal analysis can be divided by 2.1 mm day^{-1} for the CMAP data, 5.0 mm day^{-1} for the GPCP data, and 3.3 mm h^{-1} for the reanalysis data. In the CMAP data, the annual minimum mean precipitation rate of lower than 2.1 mm day^{-1} reveals the increasing trend, in particular, during the period of 1986–2008, whereas that of higher than 2.1 mm day^{-1} has the flat

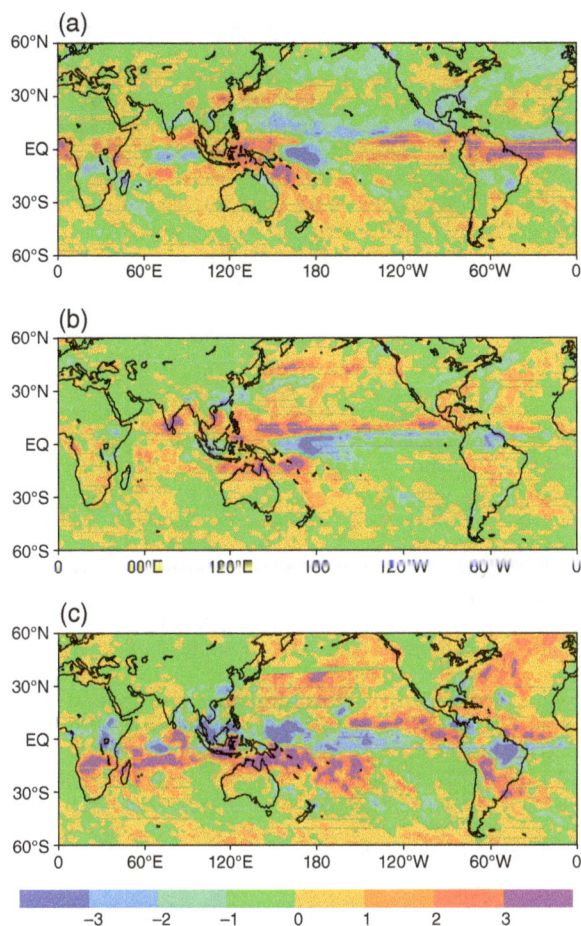

Figure 7. Spatial distributions of difference between the averages of 1994–2008 and 1979–1993 of annual minimum monthly mean precipitation rate for (a) higher than 2.1 mm d^{-1} in the CMAP data, (b) 5.0 mm d^{-1} in the GPCP data and (c) 3.3 mm d^{-1} in the NCEP reanalysis data.

trend (Figure 5(a)). To examine the spatial distribution of the decadal trend, we take spatial distribution of difference in annual minimum monthly mean precipitation rate between temporal averages from the periods 1994–2008 to 1979–1993. The increasing trend of the annual minimum mean precipitation rate of lower than 2.1 mm day^{-1} is primarily associated with the positive difference over mid-latitude oceans (Figure 6(a)), whereas the flat trend of higher than 2.1 mm day^{-1} corresponds mainly to the balance between the positive and negative difference along the equator (Figure 7(a)). Thus, the decreasing trend of the annual amplitude results from the increasing trend of annual minimum global-mean precipitation rate of lower than 2.1 mm day^{-1}. In the GPCP data, the annual minimum mean precipitation rates of lower and higher than 5.0 mm day^{-1} increase and decrease around early 1980s and decrease and increase around 1987–1988, respectively, but their variations are out of phase (Figure 5(b)). The annual minimum mean precipitation rates of lower and higher than 5.0 mm day^{-1} have flat trends after 1989. The decreasing trend from 1979–1993 to 1994–2008 for lower than 5.0 mm day^{-1} is primarily related to the negative difference (Figure 6(b)),

whereas that for higher than 5.0 mm day^{-1} corresponds mainly to the positive difference over the ITCZ and SPCZ (Figure 7(b)). As a result, the annual minimum mean precipitation rate has the flat trend. Because the annual minimum mean precipitation rate of lower than 3.3 mm day^{-1} has a flat trend, the decreasing of the annual amplitude is related to the increasing trend of annual minimum mean precipitation rate of higher than 3.3 mm day^{-1} (Figure 5(c)). The flat trend of annual minimum mean precipitation rate of lower than 3.3 mm day^{-1} is mainly associated with the balance between the negative difference over the oceans and positive difference over the land in the mid-latitudes (Figure 6(c)). The increasing trends of annual minimum mean precipitation rate of higher than 3.3 mm day^{-1} is because of the fact that the positive differences over the off-equatorial areas are stronger than the negative differences along the equator (Figure 7(c)).

4. Summary and discussions

In this study, monthly mean precipitation rates from the CMAP, GPCP, and NCEP reanalysis data from 1979 to 2008 are compared for decadal trends of annual maximum difference (AMD) of global precipitation. The annual amplitude is defined as the difference between annual maximum and minimum monthly mean precipitation rate. The analysis of the annual amplitude of globally averaged precipitation rate shows decreasing trends for the CMAP and reanalysis data but they are different. First, the annual amplitude is larger in the reanalysis data than in the CMAP data. Second, the decreasing trend of the annual amplitude is related to the increasing trend of weak annual minimum mean precipitation rate in the CMAP data, whereas it corresponds to the increasing trend of strong annual minimum mean precipitation rate. The trend of the annual amplitude in the GPCP data is flat because the decreasing linear trends along equatorial Indian and Pacific Oceans are balanced by the increasing linear trends occur over subtropical oceanic areas.

As an artifact of input data change and atoll sampling error over the tropical oceans may lead to the decreasing trend of global precipitation in the CMAP data (Yin *et al.*, 2004), it may have impacts on the decadal trend of AMD. Unlike the CMAP and reanalysis data, the GPCP data has the flat trend of AMD. The difference may be due to the fact that the decadal trend of annual minimum mean precipitation rate in the GPCP data is flat but they show increasing trends in both the CMAP and reanalysis data. Because to the lack of rain gauge measurement over the vast ocean, the oceanic precipitation over the ITCZ and SPCZ, the major contributor to global precipitation, relies on satellite retrievals. The satellite precipitation retrievals cannot be properly validated with the observational data, in particular, at the decadal timescale. The decadal trend of global precipitation may be checked with the analysis of possible physical processes. The process study may provide the theoretical guidance of decadal trend of global precipitation in a united physical framework. Therefore, the coherent decadal trend of global precipitation from the modeling and observational data requires the improvement of both numerical modeling and satellite retrievals.

Acknowledgements

The authors thank Dr P.-P. Xie at Climate Precipitation Center, NOAA for the CMAP precipitation data and the two anonymous reviewers for their constructive comments. B. Wang and X. Li were supported by National Natural Science Foundation of China under Grant No. 41475039 and National Key Basic Research Program of China under Grant No. 2015CB953601. Y. Huang was supported by the National Natural Science Foundation of China under Grant No. 4150050060 and the China Postdoctoral Science Foundation-funded project under Grant No. 2015M570500. G. Zhai was supported by the National Natural Science Foundation of China (41175047) and the National Key Basic Research and Development Project of China (2013CB430100).

References

Adler RF, Kidd C, Pretty G, Morissssey M, Goodman HM. 2001. Intercomparison of global precipitation products: the third Precipitation Intercomparison project (PIP-3). *Bulletin of the American Meteorological Society* **82**: 1377–1396.

Adler RF, Huffman GJ, Chang A, Ferraro R, Xie P-P, Janowiak J, Rudolf B, Schneider U, Curtis S, Bolvin D, Gruber A, Susskind J, Arkin P, Nelkin E. 2003. The version 2 Global Precipitation Climatology Project (GPCP) monthly precipitation analysis (1979–present). *Journal of Hydrometeorology* **4**: 1147–1167.

Adler RF, Gu G, Huffman GJ. 2012. Estimating climatological bias errors for the Global Precipitation Climatology Project (GPCP). *Journal of Applied Meteorology and Climatology* **51**: 84–99.

Bates JJ, Jackson DL. 2001. Trends in upper-tropospheric humidity. *Geophysical Research Letters* **28**: 1695–1698.

Behrangi A, Stephens G, Adler RF, Huffman GJ, Lambrigtsen B, Lebsock M. 2014. An update on the oceanic precipitation rate and its zonal distribution in light of advanced observations from space. *Journal of Climate* **27**: 3957–3965.

Cess RD, Zhang M, Potter G, Alekseev V, Barker H, Bony S, Colman R, Dazlich D, Del Genio A, Déqué M, Dix M, Dymnikov V, Esch M, Fowler L, Fraser J, Galin V, Gates W, Hack J, Ingram W, Kiehl J, Kim Y, Le Treut H, Liang X-Z, McAvaney B, Meleshko V, Morcrette J, Randall D, Roeckner E, Schlesinger M, Sporyshev P, Taylor K, Timbal B, Volodin E, Wang W, Wang W, Wetherald R. 1997. Comparison of the seasonal change in cloud-radiative forcing from atmospheric general circulation models and satellite observations. *Journal of Geophysical Research* **102**(D14): 16,593–16,603.

Dai A, Trenberth KE. 2004. The diurnal cycle and its depiction in the Community Climate System Model. *Journal of Climate* **17**: 930–951.

Ebert EE, Manton MJ. 1998. Performance of satellite rainfall estimation algorithms during TOGA COARE. *Journal of Atmospheric Sciences* **55**: 1537–1557.

Emori S, Brown S. 2005. Dynamic and thermodynamic changes in mean and extreme precipitation under changed climate. *Geophysical Research Letters* **32**, doi: 10.1029/2005GL023272.

Gruber A, Su X, Kanamitsu M, Schemm J. 2000. The comparison of two merged rain gauge-satellite precipitation datasets. *Journal of Atmospheric and Oceanic Technology* **12**: 755–770.

Gu G, Adler RF, Huffman GJ, Curtis S. 2007. Tropical rainfall variability on interannual-to-interdecadal and longer time scales derived from the GPCP monthly product. *Journal of Climate* **20**: 4033–4046.

Huffman GJ, Adler RF, Bolvin DT, Gu G. 2009. Improving the global precipitation record: GPCP version 2.1. *Geophysical Research Letters* **36**: L17808, doi: 10.1029/2009GL040000.

Janowiak JE, Gruber A, Kondragunta CR, Livezey RE, Huffman GJ. 1998. A comparison of the NCEP-NCAR reanalysis precipitation and the GPCP rain gauge-satellite combined dataset with observational error considerations. *Journal of Climate* **11**: 2960–2979.

Jones PD, Moberg A. 2003. Hemispheric and large-scale surface air temperature variations: an extensive revision and an update to 2001. *Journal of Climate* **16**: 206–223.

Kalnay E, Kanamitsu M, Kistler R, Collins W, Deaven D, Gandin L, Iredell M, Saha S, White G, Woollen J, Zhu Y, Leetmaa A, Reynolds R, Chelliah M, Ebisuzaki W, Higgins W, Janowiak J, Mo KC, Ropelewski C, Wang J, Jenne R, Joseph D. 1996. The NCEP/NCAR 40-Year Reanalysis Project. *Bulletin of the American Meteorological Society* **77**: 437–471.

Kanamitsu M, Ebisuzaki W, Woollen J, Yang S-K, Hnilo JJ, Fiorino M, Potter GL. 2002. NCEP-DOE AMIP-II Reanalysis (R-2). *Bulletin of the American Meteorological Society* **83**: 1631–1643.

Kharin V, Zwiers F. 2005. Estimating extremes in transient climate change simulations. *Journal of Climate* **18**: 1156–1173.

Kistler R, Kalnay E, Collins W, Saha S, White G, Woollen J, Chelliah M, Ebisuzaki W, Kanamitsu M, Kousky V, van den Dool H, Jenne R, Fiorino M. 2001. The NCEP-NCAR 50-Year Reanalysis: monthly means CD-ROM and documentation. *Bulletin of the American Meteorological Society* **82**: 247–268.

Kondragunta C, Gruber A. 1997. Intercomparison of spatial and temporal variability of various precipitation estimates. *Advances in Space Science* **19**: 457–460.

Krajewski WF, Ciach GJ, McCollum JR, Bacotiu C. 2000. Initial validation of the Global Precipitation Climatology Project over the United States. *Journal of Applied Meteorology* **39**: 1071–1087.

Li X, Zhai G, Gao S, Shen X. 2015. Decadal trends of global precipitation in recent 30 years. *Atmospheric Science Letters* **16**: 22–26.

Liu C, Allan RP, Huffman GJ. 2012. Co-variation of temperature and precipitation in CMIP5 models and satellite observations. *Geophysical Research Letters* **39**: L13803, doi: 10.1029/2012GL052093.

Pfeifroth U, Mueller R, Ahrens B. 2013. Evaluation of satellite-based and reanalysis precipitation data in the tropical pacific. *Journal of Applied Meteorology and Climatology* **52**: 634–644.

Spencer RW. 1993. Global oceanic precipitation from the MSU during 1979–91 and comparisons to other climatologies. *Journal of Climate* **6**: 1301–1326.

Sui C-H, Li X, Lau K-M. 1998. Radiative-convective processes in simulated diurnal variations of tropical oceanic convection. *Journal of the Atmospheric Sciences* **55**: 2345–2357.

Trenberth KE, Fasullo JT, Mackaro J. 2011. Atmospheric moisture transports from ocean to land and global energy flows in reanalyses. *Journal of Climate* **24**: 4907–4924.

Xie P, Arkin PA. 1997. Global precipitation: a 17-year monthly analysis based on gauge observations, satellite estimates and numerical model outputs. *Bulletin of the American Meteorological Society* **78**: 2539–2558.

Yin X, Gruber A, Arkin P. 2004. Comparison of the GPCP and CMAP merged gauge–satellite monthly precipitation products for the period 1979–2001. *Journal of Hydrometeorology* **6**: 1207–1222.

19

Interdecadal change of the active-phase summer monsoon in East Asia (Meiyu) since 1979

S.-Y. Simon Wang,[1,2,]* Yen-Heng Lin[2] and Chi-Hua Wu[3]

[1] Utah Climate Center, Utah State University, Logan, UT, USA
[2] Department of Plants, Soils, and Climate, Utah State University, Logan, UT, USA
[3] Research Center for Environmental Changes, Academia Sinica, Taipei, Taiwan

*Correspondence to:
S.-Y. Simon Wang, Department of Plants, Soil, and Climate, Utah State University, 4820 Old Main Hill, Logan, UT 84322-4820, USA
E-mail: simon.wang@usu.edu

Abstract

The timing of active-phase East Asian summer monsoon (Meiyu) undergoes a marked shift since 1979. Diagnostic analysis indicates that active convection over Taiwan has occurred later in the season, from late May to early June, with a tendency of increasingly intense rainfall. This timing shift of convection results from a southward migration of Meiyu rainband, driven by an upper-level cyclonic anomaly over eastern China and a lower-level anticyclonic anomaly in the subtropical Western Pacific. Together, these two circulation patterns enhance both the moisture transport and baroclinic forcing. The role of Western Pacific warming and anthropogenic greenhouse gases in these changes is suggested.

Keywords: Meiyu; SST warming; East Asian monsoon; subtropical high

1. Introduction

The East Asian summer monsoon (EASM) undergoes an active–break–revival sequence and the associated migration of the rainbands makes the timing of each phase geographically unique (Chen *et al.*, 2004). This distinct lifecycle of EASM regulates rainfall and water supply in several Asian countries, including Taiwan. Located in the central region of EASM, Taiwan covers $36\,000\,\text{km}^2$ of complex terrain with a population approaching 24 millions (location shown in Figure 1(a), inset). The active phase of EASM (Meiyu) produces the first influx of substantial water for agricultural, industrial and residential uses. Wang and Chen (2008) indicated that the active-phase EASM (interchangeable with Meiyu hereafter) contributes to ~60% of Taiwan's early-summer rainfall. The phases of EASM relative to Taiwan are displayed in Figure 1(a) by the outgoing longwave radiation (OLR) averaged within $119°-122°E$, $21°-25°N$; here, OLR is shown as departure from $235\,\text{W}\,\text{m}^{-2}$ to approximate convective rainfall regime, denoted as ΔOLR ($=235 - \text{OLR}$). This feature is critical because, as of April 2015, Taiwan underwent the most severe drought in its 67 years of recorded history and yet, the arrival of Meiyu mitigated the drought situation. However, predicting the timing and strength of active-phase EASM at longer range (>2 week) remains a challenge, making drought adaptation and planning difficult (M.-M. Lu, Central Weather Bureau, 2015, personal communication).

The Meiyu rainband is driven by the mid-tropospheric warm advection and transient eddies that are steered by the westerly jet, and these circulations induce instability and adiabatic ascent while the tropical warm pool supplies the moisture (Chen *et al.*, 2004; Sampe and Xie, 2010). Previous studies have indicated that interannual variability of the EASM circulations is linked to the Tibetan Plateau thermal conditions and India Ocean sea surface temperature (SST) anomalies (Li and Yanai, 1996; Zhao *et al.*, 2010; Liu and Wang, 2011; Hu and Duan, 2015). These processes are complicated by the varying mid-tropospheric temperature advection within the Meiyu rainband (Kosaka *et al.*, 2011; Okada and Yamazaki, 2012). However, few studies have focused on the interdecadal variability of Meiyu. Among these, Li *et al.* (2010) found that EASM has shifted southward since 1958 probably due to the meridional asymmetric warming between the South China Sea (SCS) and East Asian continent. Luo and Zhang (2015) reported that peak Meiyu rainfall in southern China has tended to arrive later since 1993 due to weakened low-level southwesterly winds. Focusing on Taiwan, Huang and Chen (2014) observed a transition of Meiyu rainfall from the predominately frontal regime to an increase in the diurnal convection regime. Regardless, a mechanistic explanation of the Meiyu's interdecadal variation is lacking; this is analyzed herein.

2. Data

The following data sets are utilized: (1) the 1.0°-resolution daily OLR Version 1.2 produced by the National Oceanic and Atmospheric Administration (NOAA) (Lee and NOAA CDR program, 2011) from 1979 to 2014 with the missing values during May–June

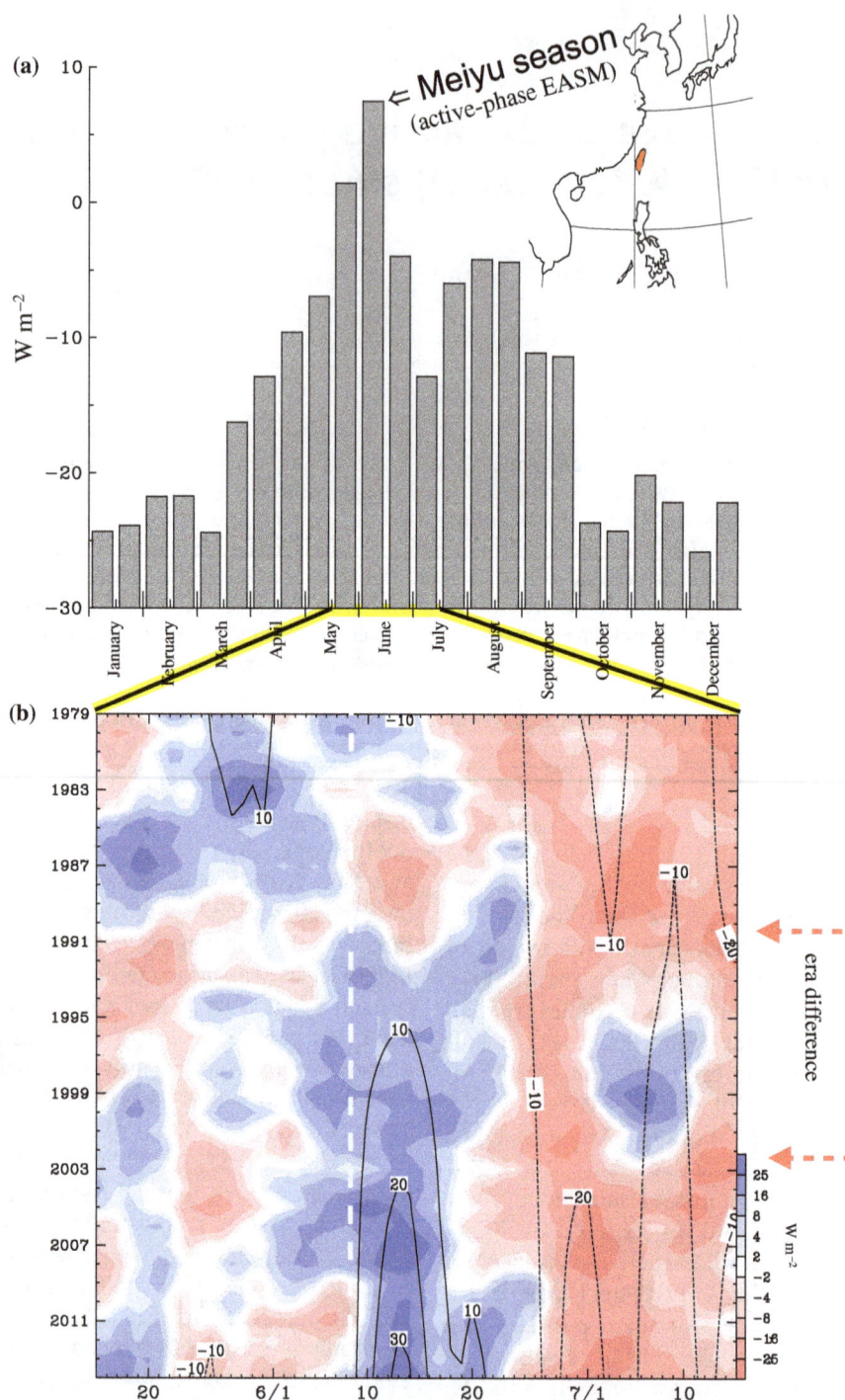

Figure 1. (a) Long-term 15-day evolution of ΔOLR (235 W m^{-2}-OLR) averaged in Taiwan (119°−122°E, 21°−25°N) from 1980 to 2010, following Wang and Chen (2008). The inset map depicts the geographical location of Taiwan (red). (b) Yearly distribution of daily ΔOLR applied with a 5-day moving average (shadings) overlaid with the linear trend contours from 1979 to 2014. A white and two yellow dashed lines indicate the period difference of the Meiyu as referred in the text.

1985 filled by the NOAA interpolated OLR (Liebmann and Smith, 1996), (2) the monthly NOAA Extended Reconstructed SST Version 3b (Smith *et al.*, 2008), (3) the ECMWF post-1979 reanalysis data at a 1.0° resolution (ERA-Interim) (Dee *et al.*, 2011) and (4) the ECMWF 40-year Reanalysis (ERA-40) from 1958 to 1979 at a 2.5° resolution (Uppala *et al.*, 2005) to merge with ERA-Interim for a longer-term analysis (i.e. for Figure 4).

3. Results

The long-term change in the active-phase EASM is examined by analyzing the daily ΔOLR in Taiwan from mid-May to mid-July (x-axis) for each year from 1979 to 2014 (y-axis); this is plotted in Figure 1(b). The use of ΔOLR compensates for the lack of long, stable record of daily precipitation. Here, ΔOLR is subject to a 5-day and 5-year running mean to focus on the

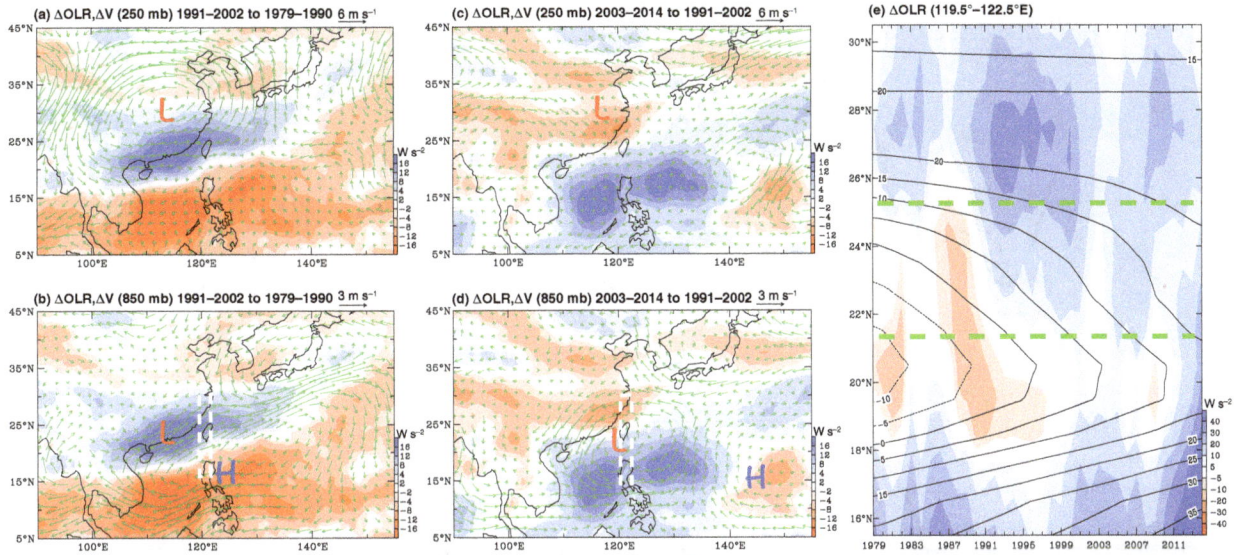

Figure 2. Differences of ΔOLR (shadings) and (a) 250- and (b) 850-mb winds vectors between 1991–2002 and 1979–1990 for the 7 June–20 June period. (c–d) Same as (a–b) except for the differences between 2003–2014 and 1999–2002. (e) ΔOLR latitude (y-axis) and year (x-axis) distribution across the vicinity of Taiwan (white box in (b) and (d)) during the 7 June–20 June periods overlaid with the linear trend contours from 1979 to 2014. The latitudinal extent of Taiwan is shown by the green dashed lines.

predominant intraseasonal variability that drives the EASM lifecycle (Chen *et al.*, 2004). The peak of ΔOLR has undergone a timing shift from mostly late May before the 1990s to predominantly early June. There is also a tendency for ΔOLR to become stronger and more concentrated in mid-June (10th–15th) after 2003. To illustrate this change, we compute the linear trend of ΔOLR for each day from 1979 to 2014 and superimpose it on Figure 1(b) as contours. Apparently, ΔOLR has decreased by 20 W m⁻² in late May accompanied by an increase of 30 W m⁻² in mid-June, estimated from the linear trend. Noteworthy is the change in the convective time span that has reduced from 3 weeks before 2003 to less than 2 weeks afterwards, suggesting more intense rainfall occurring within a shorter period of time. This feature echoes the finding of Huang and Chen (2014) that frontal rainfall regime in Taiwan has gradually been replaced by diurnal convection regime in May and June.

The large-scale circulation and precipitation anomalies associated with the timing shift of Meiyu are examined by two epoch differences of the 250- and 850-hPa winds and ΔOLR: (1) between 1991–2002 and 1979–1990 to depict the timing shift and (2) between 2003–2014 and 1991–2002 to depict the precipitation intensification (these periods are indicated by arrows in Figure 1(b)), in June. The circulation and ΔOLR anomalies during 7th–20th June are plotted in Figure 2. In the earlier period, a robust upper-level cyclonic anomaly forms over eastern China (Figure 2(a), 'L') while a marked low-level anticyclonic anomaly extends from the SCS across the Philippine Sea (Figure 2(b), 'H'). Combined, these circulations induce strong southwesterly flows coupled with upper-level westerlies, promoting baroclinic instability in and around Taiwan. Correspondingly, a substantial increase in ΔOLR is observed in the northern SCS stretched across Taiwan,

signifying an intensification of frontal rainfall regime. These circulation anomalies possibly reflect a stationary wave pattern superimposed on the westerly jet stream that was found to be influenced by the mechanical effect of the Tibetan Plateau (Wu and Chou, 2013).

For the latter period (after 2003), Figure 2(c) shows that the upper-level westerly winds enhance slightly, while a low-level cyclonic circulation appears in the vicinity of Taiwan (Figure 2(d), 'L'). Combined, these circulation changes delineate a meridional migration of Meiyu in the context of interdecadal variation. The change in ΔOLR is also substantial as it is shifted further south adjacent of the Philippines covering only the southern part of Taiwan. To clarify this implication, we plot in Figure 2(e) the latitude-time section of ΔOLR across Taiwan during 7th–20th June. Apparently, positive ΔOLR north of Taiwan has migrated southward from 26° to 20°N. Consequently, what used to be a relatively dry spell in Taiwan (i.e. between 18° and 24°N) has become increasingly convective in recent years. As is shown in Figure S1, Supporting Information, the earlier period of 24 May–6 June undergoes a decrease in convective activity as a result of this ΔOLR migration. These results provide a geographical reference for the timing change of Meiyu.

In order to connect the reported timing shift with the large-scale circulation change, we adopt a method designed to delineate the yearly evolution of a daily variable, following Wang *et al.* (2014). This method uses the empirical orthogonal function (EOF) of the covariance matrix of ΔOLR over Taiwan, by treating ΔOLR's daily interval as eigenvalue and its yearly interval as eigencoefficient. After applying a 5-day moving average (to capture the predominant intraseasonal variability of EASM), we obtain a set of EOFs representing the daily variation of ΔOLR and a set of principal

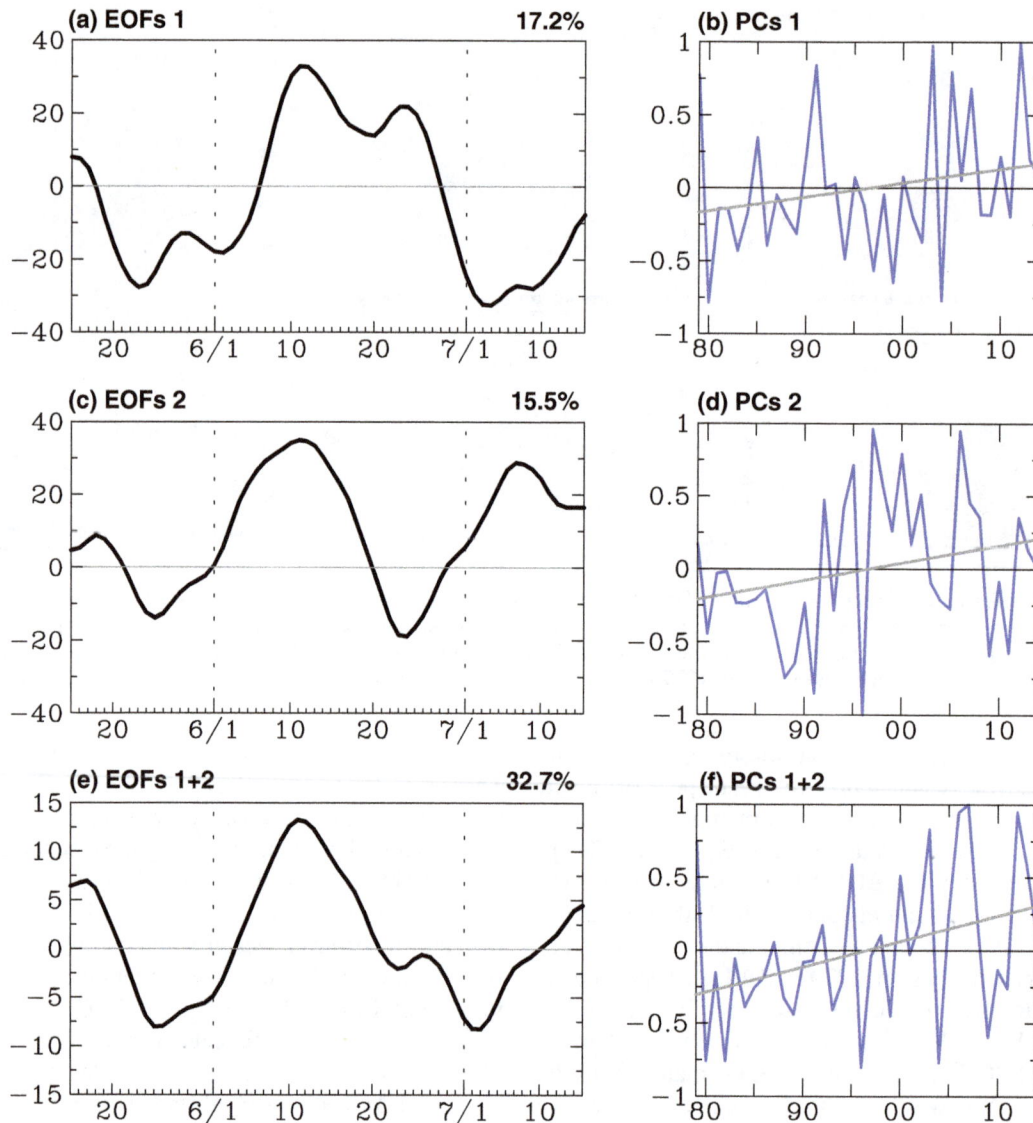

Figure 3. The EOF analysis of daily ΔOLR from 1979 to 2014 for (a) EOFs 1, (b) PCs 1, (c) EOFs 2, (d) PCs 2, (e) reconstructed EOFs 1 + 2 and (f) reconstructed PCs 1 + 2. A 5-day moving average is applied prior to the EOF analysis.

components (PCs) representing the yearly variation. The first two EOFs are shown in Figure 3(a)–(d) representing the amplification of the temporally displaced ΔOLR, constituting collectively 32.7% of the total variance. The EOF 1 (Figure 3(a)) and EOF 2 (Figure 3(c)) show positive values in mid-June with an increasing trend of the PCs (Figure 3(b) and (d)), suggesting a tendency for enhanced convective activity and its timing shift in Taiwan. Next, we combine these two leading modes to reconstruct the ΔOLR changes in Taiwan while filtering out less relevant signals (Van den Dool, 2007). The combinations of EOFs/PCs 1 + 2 are shown in Figure 3(e) and (f). The distribution of EOFs 1 + 2 indicates maximum ΔOLR in mid-June and minimum ΔOLR in late May, and this feature has intensified as shown by the increasing trend in PCs 1 + 2 (significant at $p < 0.05$). Consequently, PCs 1 + 2 form an index enabling us to compare the change in subseasonal variability against interannual variations of any given variable.

By regressing PCs 1 + 2 upon the eddy streamfunction field (i.e. removing the zonal mean) for the month of June, the resultant regression coefficients depict the anomalous circulations accompanying the increased ΔOLR in Taiwan during mid-June. Figure 4(a) and (b) show such circulation patterns at 250 and 850 hPa, depicting westerly (southwesterly) anomalies that prevail over Taiwan at the upper (lower) level. By comparison with Figure 2, these circulation features correspond well with the upper trough in eastern China and the Western Pacific anticyclone. The similarity of circulation patterns between Figure 4(a) and (b) and Figure 2 also suggests that the anomalous circulations leading ΔOLR to become more active in mid-June resulted from two sources: deepening of the upper-level trough northwest of Taiwan and strengthening of the anticyclone in the subtropical western Pacific. In Figure S2, we show the individual regressions of PCs 1 and 2 and their resultant circulation pattern, which reveal similar synoptic processes.

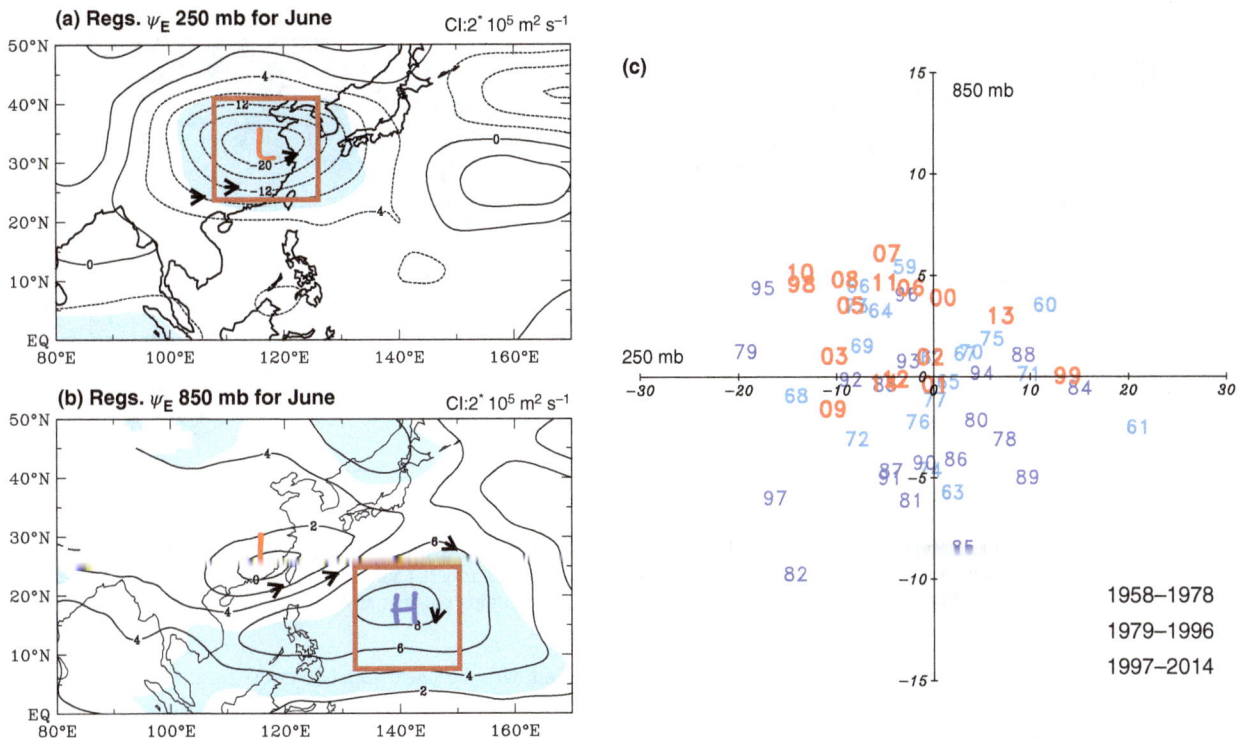

Figure 4. Regression coefficients of the eddy streamfunction (ψ_E) June monthly mean regressed with PCs 1 + 2 for (a) 250 and (b) 850 hPa. Shadings indicate significance at $p < 0.05$; arrows were added to illustrate the anomalous flow direction. H's and L's indicate anticyclonic and cyclonic anomalies, respectively. (c) Scatter diagram of the June eddy streamfunction (ψ_E) values at 250 and 850 hPa averaged from the respective domains outlined in (a) and (b) red boxes, with the last two digits of each year color-coded as indicated in the lower right.

We next compare the interannual variations of the June stream function between the upper-level cyclone and the lower-level anticyclone, using values averaged from their center areas (domain outlined in Figure 4(a) and (b)). The variations of these two circulation features are not correlated ($r < 0.16$), as illustrated by the scatter plot of Figure 4(c). In other words, the circulation patterns in response to PCs 1 + 2 (i.e. increased ΔOLR) could only appear in the second quadrant of the scatter diagram, i.e. when negative 250 hPa values (trough) and positive 850 hPa values (ridge) coexisted. However, by adding the years onto the scatters, there is a discernible change in that the concurrence of the strengthened Western Pacific anticyclone with the deepened eastern China cyclone has increased after 1997 (indicated as red). This result is intriguing in that, although these two levels of circulation do not correlate, in the long run they have become increasingly cohesive in producing precipitation along the SCS-Taiwan corridor in the month of June.

4. Discussions and conclusion

A tendency has been observed in June for the low-level anticyclonic anomaly in subtropical Western Pacific and upper-level cyclonic anomaly in eastern China to occur together more frequently. This feature promotes frontal instability and subsequent convection in early June over Taiwan, delaying its Meiyu season. For the upper level,

previous studies analyzing the change in mid-latitude stationary waves have noted an amplified short-wave regime and associated increases in weather extremes (Screen and Simmonds, 2013; Teng *et al.*, 2013; Wang *et al.*, 2013b; Screen and Simmonds, 2014). Other research (Wang *et al.*, 2013b; Cho *et al.*, 2015) has indicated an intensification of the Eurasia-South Asia short-wave train in the month of June (Yasunari *et al.*, 1991; Ding and Wang, 2005). This reported wave pattern consists of (from west to east) a deepened trough in western Nepal, a strengthened ridge over Bhutan and an enhanced trough over eastern China – these are shown in Figure 5. The cause of this changing wave-train pattern is under debate, and our testing of SST regression with PCs 1 + 2 (not shown) does not reveal any robust linkage with any known climate mode. However, the SST in subtropical Western Pacific has tended to warm by 40% associated with the 30-W m^{-2} increase of ΔOLR in June, based on linear regression. This increase in local SST coincides with the ongoing warming trend in the Western Pacific.

The variation of lower-level circulations in the Western Pacific has been widely documented. Yet, most studies only focused on the typical summer season of June–August, rather than the seasonal transition of May or June. Nevertheless, those studies have uniformly found a link between the strengthened North Pacific subtropical anticyclone and the increased SST under the anthropogenic global warming. The strengthened subtropical anticyclone adds thermal contrast between land

(a) Reg. PC1+2 with June ψ_E 250 mb (contour) and trend of June ψ_E 250 mb (shaded)

(b) Reg. PC1+2 with June ψ_E 250 mb (contour) and trend of June ψ250 mb (shaded)

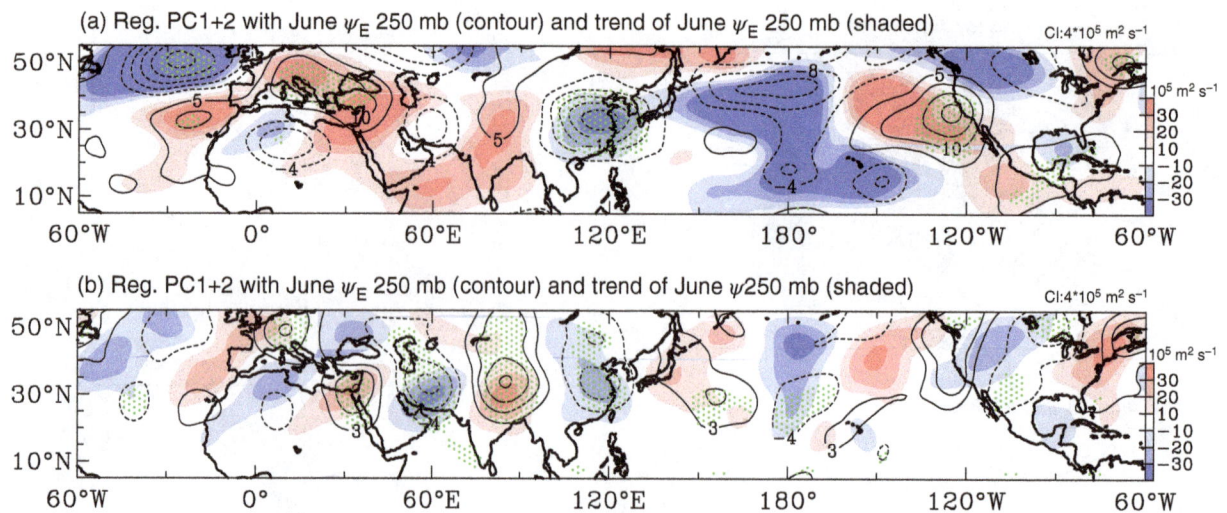

Figure 5. (a) The regression pattern of June ψ_E 250 hPa with PCs 1 + 2 (contours) overlaid with the linear trend slopes of the June ψ_E (shadings). Green-dotted areas indicate significance at $p < 0.05$ for the regression. (b) Same as (a) but for the short-wave regime (i.e. zonal wavenumbers 5 and beyond) applied in each field, following the trending pattern as depicted by Wang et al. (2013b).

and ocean (Li et al., 2012) and further warms the northern Indian Ocean (He and Zhou, 2015) while enhancing thermal contrast in the subtropical Western Pacific (Wang et al., 2013a). These reported changes in oceanic thermal property and land-sea contrast have a detectable anthropogenic footprint and could be linked to the finding of this study.

To reconcile with the previous findings, we did conduct a preliminary analysis using the historical single-forcing experiment of the Community Climate System Model Version 4 (CCSM4) derived from the CMIP5 archive (Taylor et al., 2009). By reproducing Figure 1 using daily precipitation output of CCSM4, which is shown in Figure S3, it is observed that only the anthropogenic greenhouse gases (GHG) forcing simulates the timing shift of the active-phase EASM in a way similar to the observation. Neither the natural forcing nor the aerosol forcing generated any persistent change in the occurrence of peak rainfall. The preliminary result of Figure S3 suggests a possibility that anthropogenic GHG can influence the timing change of Meiyu rainfall in Taiwan. Subsequent analysis using the full archive of CMIP5 outputs will be the focus of future study.

Acknowledgements

Yen-Heng Lin was supported by the Utah State University Presidential Doctoral Research Fellows (PDRF) program. Chi-Hua Wu was supported by the Ministry of Science and Technology of Taiwan under grant NSC 100-2119-M-001-029-MY5 and MOST 104-2111-M-001-001.

Supporting information

The following supporting information is available:

Figure S1. Same as Figure 2(e) but for 24 May−6 June period ΔOLR latitude (y-axis) and year (x-axis) distribution (shadings)

overlaid with the linear trend contours from 1979 to 2014. The green lines show latitude range of Taiwan.

Figure S2. Same as Figure 4(a) and (b) but regressed with PCs 1 for (a) 250 and (b) 850 mb; PCs 2 for (c) 250and (d) 850 mb.

Figure S3. Same as Figure 1, but for CCSM4 historical simulation precipitation in recent 36 years with different forcing: (a,b) anthropogenic greenhouse gases (GHG), (c,d) natural including solar and volcanic forcing (Nat), and (e,f) anthropogenic aerosol (Aero). The yearly distribution of daily precipitation is the departure from seasonal means. Notice the rather weak Meiyu phase of rainfall than the observation, as well as the peak rainfall shift in (b) that is coincident with Figure 1(b).

References

Chen T-C, Wang S-Y, Huang W-R, Yen M-C. 2004. Variation of the East Asian summer monsoon rainfall. Journal of Climate **17**(4): 744–762.

Cho C, Li R, Wang S-Y, Yoon J-H, Gillies RR. 2015. Anthropogenic footprint of climate change in the June 2013 northern India flood. Climate Dynamics , doi: 10.1007/s00382-015-2613-2.

Dee DP, Uppala SM, Simmons AJ, Berrisford P, Poli P, Kobayashi S, Andrae U, Balmaseda MA, Balsamo G, Bauer P, Bechtold P, Beljaars ACM, van de Berg L, Bidlot J, Bormann N, Delsol C, Dragani R, Fuentes M, Geer AJ, Haimberger L, Healy SB, Hersbach H, Hólm EV, Isaksen L, Kållberg P, Köhler M, Matricardi M, McNally AP, Monge-Sanz BM, Morcrette J-J, Park B-K, Peubey C, de Rosnay P, Tavolato C, Thépaut J-N, Vitart F. 2011. The ERA-Interim reanalysis: configuration and performance of the data assimilation system. Quarterly Journal of the Royal Meteorological Society **137**(656): 553–597.

Ding Q, Wang B. 2005. Circumglobal teleconnection in the northern hemisphere summer. Journal of Climate **18**(17): 3483–3505.

He C, Zhou T. 2015. Responses of the western North Pacific subtropical high to global warming under RCP4. 5 and RCP8. 5 scenarios projected by 33 CMIP5 models: the dominance of tropical Indian Ocean–tropical western Pacific SST gradient. Journal of Climate **28**(1): 365–380.

Hu J, Duan A. 2015. Relative contributions of the Tibetan Plateau thermal forcing and the Indian Ocean Sea surface temperature basin mode to the interannual variability of the East Asian summer monsoon. Climate Dynamics, doi: 10.1007/s00382-015-2503-7.

Huang WR, Chen KC. 2014. Trends in pre-summer frontal and diurnal rainfall activities during 1982–2012 over Taiwan and Southeast

China: characteristics and possible causes. *International Journal of Climatology* **35**: 2608–2619.

Kosaka Y, Xie S-P, Nakamura H. 2011. Dynamics of interannual variability in summer precipitation over East Asia. *Journal of Climate* **24**(20): 5435–5453.

Lee HT, NOAA CDR Program. 2011. NOAA Climate Data Record (CDR) of Daily Outgoing Longwave Radiation (OLR), Version 1.2. NOAA National Climatic Data Center: Silver Spring, MD, doi:10.7289/V5SJ1HH2.

Li C, Yanai M. 1996. The onset and interannual variability of the Asian summer monsoon in relation to land-sea thermal contrast. *Journal of Climate* **9**(2): 358–375.

Li J, Wu Z, Jiang Z, He J. 2010. Can global warming strengthen the East Asian Summer Monsoon? *Journal of Climate* **23**: 6696–6705.

Li W, Li L, Ting M, Liu Y. 2012. Intensification of Northern Hemisphere subtropical highs in a warming climate. *Nature Geoscience* **5**(11): 830–834.

Liebmann B, Smith CA. 1996. Description of a complete (interpolated) outgoing longwave radiation dataset. *Bulletin of the American Meteorological Society* **77**: 1275–1277.

Liu X, Wang Y. 2011. Contrasting impacts of spring thermal conditions over Tibetan Plateau on late-spring to early-summer precipitation in southeast China. *Atmospheric Science Letters* **12**(3): 309–315.

Luo X, Zhang Y. 2015. Interdecadal change in the seasonality of rainfall variation in South China. *Theoretical and Applied Climatology* **119**(1–2): 1–11.

Okada Y, Yamazaki K. 2012. Climatological evolution of the Okinawa baiu and differences in large-scale features during May and June. *Journal of Climate* **25**(18): 6287–6303.

Sampe T, Xie S-P. 2010. Large-scale dynamics of the Meiyu-Baiu Rainband: environmental forcing by the westerly jet. *Journal of Climate* **23**(1): 113–134.

Screen JA, Simmonds I. 2013. Exploring links between Arctic amplification and mid-latitude weather. *Geophysical Research Letters* **40**(5): 959–964.

Screen JA, Simmonds I. 2014. Amplified mid-latitude planetary waves favour particular regional weather extremes. *Nature Climate Change* **4**: 704–709.

Smith TM, Reynolds RW, Peterson TC, Lawrimore J. 2008. Improvements to NOAA's historical merged land–ocean surface temperature analysis (1880–2006). *Journal of Climate* **21**(10): 2283–2296.

Taylor KE, Stouffer RJ, Meehl GA. 2009. A summary of the CMIP5 experiment design. http://cmip3pcmdi.llnl.gov/cmip5/docs/Taylor_CMIP5_design.pdf (accessed 1 May 2015).

Teng H, Branstator G, Wang H, Meehl GA, Washington WM. 2013. Probability of US heat waves affected by a subseasonal planetary wave pattern. *Nature Geoscience* **6**(12): 1056–1061.

Uppala SM, Kållberg PW, Simmons AJ, Andrae U, Bechtold VDC, Fiorino M, Gibson JK, Haseler J, Hernandez A, Kelly GA, Li X, Onogi K, Saarinen S, Sokka N, Allan RP, Andersson E, Arpe K, Balmaseda MA, Beljaars ACM, Berg LVD, Bidlot J, Bormann N, Caires S, Chevallier F, Dethof A, Dragosavac M, Fisher M, Fuentes M, Hagemann S, Hólm E, Hoskins BJ, Isaksen L, Janssen PAEM, Jenne R, Mcnally AP, Mahfouf J-F, Morcrette J-J, Rayner NA, Saunders RW, Simon P, Sterl A, Trenberth KE, Untch A, Vasiljevic D, Viterbo P, Woollen J. 2005. The ERA-40 re-analysis. *Quarterly Journal of the Royal Meteorological Society* **131**(612): 2961–3012.

Van den Dool H. 2007. *Chapter 5:* Empirical Methods in Short-Term Climate Prediction. Oxford University Press: Oxford, UK and New York, NY; 215pp.

Wang S-Y, Chen T-C. 2008. Measuring East Asian Summer Monsoon rainfall contributions by different weather systems over Taiwan. *Journal of Applied Meteorology and Climatology* **47**(7): 2068–2080.

Wang B, Xiang B, Lee J-Y. 2013a. Subtropical high predictability establishes a promising way for monsoon and tropical storm predictions. *Proceedings of the National Academy of Sciences* **110**(8): 2718–2722.

Wang S-Y, Davies RE, Gillies RR. 2013b. Identification of extreme precipitation threat across midlatitude regions based on short-wave circulations. *Journal of Geophysical Research: Atmospheres* **118**(19): 2013JD020153.

Wang S-Y, Gillies RR, Dool H. 2014. On the yearly phase delay of winter intraseasonal mode in the western United States. *Climate Dynamics* **42**(5–6): 1649–1664.

Wu CH, Chou MD. 2013. Tibetan Plateau westerly forcing on the cloud amount over Sichuan Basin and the early Asian summer monsoon. *Journal of Geophysical Research, [Atmospheres]* **118**(14): 7558–7568.

Yasunari T, Kitoh A, Tokioka T. 1991. Local and remote responses to excessive snow mass over Eurasia appearing in the northern spring and summer climate – a study with the MRI GCM. *Journal of the Meteorological Society of Japan* **69**(4): 473–487.

Zhao P, Yang S, Yu R. 2010. Long-term changes in rainfall over eastern China and large-scale atmospheric circulation associated with recent global warming. *Journal of Climate* **23**(6): 1544–1562.

20

Tropical cyclogenesis associated with four types of winter monsoon circulation over the South China Sea

Lei Wang,[1,2,*] Ronghui Huang[1] and Renguang Wu[1]

[1] Center for Monsoon System Research, Institute of Atmospheric Physics, Chinese Academy of Sciences, Beijing, China
[2] Key Laboratory of Global Change and Marine-Atmospheric Chemistry, Xiamen, China

*Correspondence to:
L. Wang, Center for Monsoon
System Research, Institute of
Atmospheric Physics, Chinese
Academy of Sciences, P. O. BOX
2718, Zhongguancun Bei'ertiao
No. 6, Beijing, 100190, China.
E-mail: wl@mail.iap.ac.cn

Abstract

In boreal winter during the period 1958–2013, more than two third of tropical cyclone (TC) genesis over the South China Sea (SCS) is found to occur in specific atmospheric environmental fields associated with four types of East Asian winter monsoon circulation which are named the monsoon gyre (MG), the easterly, the reverse-oriented monsoon trough (RMT), and the monsoon confluence (MC), respectively. The first two types account for about 80% of TC geneses. Before TC formation over the SCS, lower-level positive relative vorticity and humidity anomalies are accompanied by mid-troposphere ascent and upper-level divergence anomalies, which are favorable for TC genesis. These anomalies are the most significant in the MG type. Moreover, the eddy kinetic energy (EKE) growth from the barotropic energy conversion contributes beneficially to the evolving of incipient disturbances to a TC over the SCS. In all four types, the meridional wind convergence and the zonal wind shear play an important role in the EKE growth.

Keywords: tropical cyclogenesis; East Asian winter monsoon; circulation type

I. Introduction

Tropical cyclone (TC) genesis over the western North Pacific (WNP) is a hot topic. TC formation is closely related to environmental fields, such as low-level relative vorticity, mid-level relative humidity, and so on (Emanuel, 2003; Richard and Zhou, 2012). Besides, TC genesis often occurs in the east part of the monsoon gyre (Lander, 1994). A subsequent study (Ritchie and Holland, 1999) indicates that there are four low-level circulation types favorable for TC genesis and 70% of tropical cyclogenesis occurs along the monsoon shear line and the monsoon convergence zone. A recent study (Feng et al., 2014) investigated large-scale circulation patterns favorable for TC genesis over the WNP. Moreover, the dynamic effect of the large-scale circulation on TC genesis from the perspective of the energy conversion is increasingly concerned. The barotropic energy transformation in the basic flow of the lower troposphere can provide synoptic-scale disturbances for TC genesis (Maloney and Hartmann, 2001). TC frequency is modulated by synoptic-scale disturbances related to the barotropic energy conversion in the monsoon trough over the WNP (Wu et al., 2012). The variation in the location and intensity of the monsoon trough lead to the conversion from mean kinetic energy to eddy kinetic energy, which can influence TC genesis (Wu et al., 2014a).

Although TCs generated over the South China Sea (SCS) are not much, they have an important influence on countries near the SCS (Wang et al., 2013). In contrast with the WNP, the large-scale circulation feature over the SCS is more complicated. The SCS is located in the WNP warm pool where the atmosphere–ocean interaction is active (Huang et al., 2003; He and Wu, 2010; Wu et al., 2014b) and the large-scale atmospheric circulation not only provides appropriate low-level relative vorticity and vertical shear of the horizontal wind for tropical cyclogenesis (Lander, 1994; Briegel and Frank, 1997; Ritchie and Holland, 1999) but also favors the rapid intensification of a TC (Chen et al., 2015). Besides, the SCS is a conjunction region of atmospheric circulations from East Asia, the Indian Ocean, and the Pacific Ocean and is directly influenced by the East Asian monsoon in the lower troposphere. The East Asian monsoon plays a key role in TC genesis over the SCS (Wang et al., 2007).

It is worth noting that the large-scale circulation of the East Asian winter monsoon (EAWM) favorable for TC genesis over the SCS during boreal winter is seldom concerned. The EAWM is established in East Asia after the abrupt change in East Asian atmospheric circulation in October (Yeh et al., 1959). Along the coast of East Asia, the EAWM flow is channeled down into the SCS (Krishnamurti et al., 1973; Lau and Chang, 1987; Chen et al., 2005). The convective activity over the SCS is closely related to the EAWM (Chang et al., 1979, 2003; Compo et al., 1999; Wang and Chen, 2010). In the winter of the El Niño year, there is an anomalous anticyclone near Philippines (Wang et al., 2000) with the southerly anomaly over the offshore region of the East Asia (Zhang et al., 1996). Tropical cyclogenesis over the SCS often occurs during October through December when the EAWM flow occupies gradually the low-level

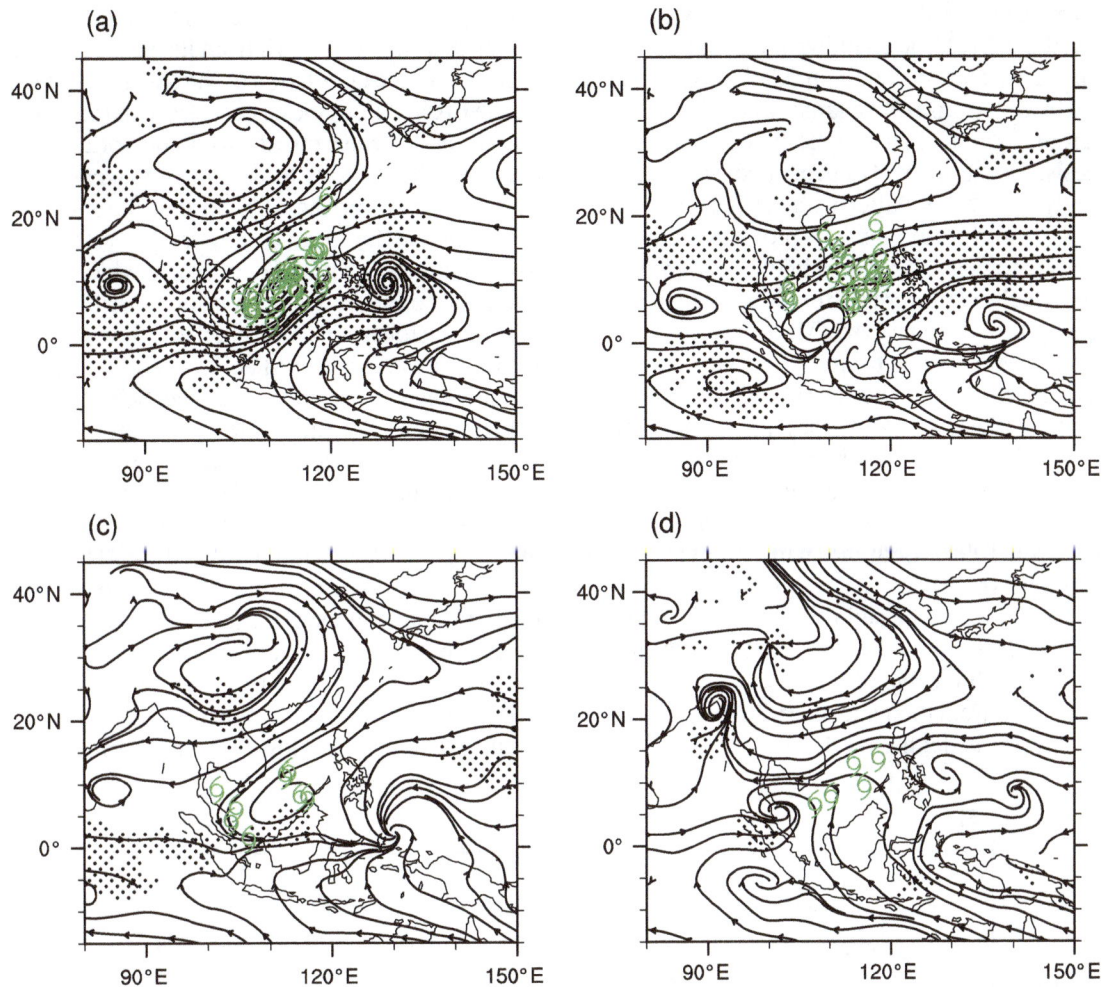

Figure 1. Composites of 850 hPa streamline at −72 h from tropical cyclogenesis in (a) MG, (b) easterly, (c) RMT, and (d) MC type from October to December during 1958–2013. Typhoon symbols show TC genesis locations. Stippling denotes regions where the difference between four circulation types and the climatic mean from October to December during 1958–2013 is significant above the 90% confidence level according to the Student's *t* test.

over the SCS (Wang *et al.*, 2007). The present study classifies the EAWM circulation into four types and investigates their dynamic effect on TC genesis over the SCS, respectively.

2. Data and methodology

The Japanese 55-year Reanalysis (JRA-55) data (6 hourly and 1.25° latitude/longitude grid) from the Japan Meteorological Agency is used to analyze dynamic characteristics of EAWM circulation types favorable for tropical cyclogenesis over the SCS (100–120°E, 0–30°N). Besides, the TC best track data (including 6 hourly TC position and intensity) from the International Best Tracks Archive for Climate Stewardship (IBTrACS) project is used to identify the location and time of TC geneses over the SCS.

The TC genesis location and time are derived from the first record of its track in the IBTrACS data. A composite analysis is exerted to reveal EAWM circulation types favorable for TC formation and explore the associated environmental fields. In addition, the band-pass filter is applied to extract synoptic-scale disturbances in order to investigate the role of the barotropic energy transformation in TC genesis.

3. Winter monsoon circulation types associated with tropical cyclogenesis

TC genesis over the SCS is considered to be influenced by the EAWM if the 72-h mean meridional wind before TC genesis in both area 1 (105–120°E, 10–25°N) and area 2 (115–130°E, 25–40°N) shows northerly. About 95 TCs were generated over the SCS from October to

Table 1. The number and percentage of tropical cyclogenesis in four East Asian winter monsoon circulation types over the SCS (0°–30°N, 100–120°E) from October to December during 1958–2013.

	MG	Easterly	RMT	MC
Number	32	22	8	5
Percentage	47.8%	32.8%	11.9%	7.5%

December during 1958–2013, of which 67 were influenced by the EAWM, accounting for about 70.5% of the total. The winter monsoon circulation type may be identified by the feature of the large-scale (the $10° \times 10°$ region around the TC) streamline field at -72 h from TC genesis. Because the TC generated over the SCS can quickly land ashore after its genesis (Wang *et al.*, 2013), this identification method provides support for the in situ TC forecast. As shown in Figure 1, the winter monsoon circulation favorable for TC genesis includes four types: the monsoon gyre (MG), the easterly, the reverse-oriented monsoon trough (RMT), and the monsoon confluence (MC). The number of TCs from the MG, easterly, RMT, and MC types accounts for about 47.8, 32.8, 11.9, and 7.5%, respectively, of the total from all winter monsoon circulation types (Table 1). The monsoon gyre is a kind of special monsoon circulation (Lander, 1994). Both the monsoon gyre and the disturbance at its east part can develop into TC. Previous studies provide different definitions for the monsoon gyre (Lander, 1994; Wu *et al.*, 2013). These differences mainly lie in the life cycle of a monsoon gyre. In view of the complex circulation and short TC life cycle over the SCS, the TC is defined as the one generated in the MG type if the streamline field around it shows an obvious closed vortex at -72 h from its genesis. As shown in Figure 1(a), the circulation of the MG type is modulated by the EAWM flow, the easterly from the central-eastern Pacific and the eastward cross-equatorial flow influenced by the Coriolis force, which makes strong shear flow existing over the SCS so that the monsoon gyre forms in the lower troposphere. Near this monsoon gyre, there are a lot of TC geneses, which account for about half of all in winter monsoon circulation types.

From October to December, the low-level easterly from the central-eastern Pacific is an important part of the low-level circulation over the SCS. This easterly contains multiple time scale perturbations which provide initial perturbation kinetic energy for TC genesis. If the streamline field around a TC shows an obvious easterly at -72 h from its genesis, this one is defined as the TC generated in the easterly type. The circulation of the easterly type near the equator is often accompanied by small and medium scale disturbances, which is favorable for tropical cyclogenesis (Figure 1(b)).

In boreal winter, the trough line of the monsoon trough over the SCS generally shows northwest-southeast direction. However, it is sometimes along

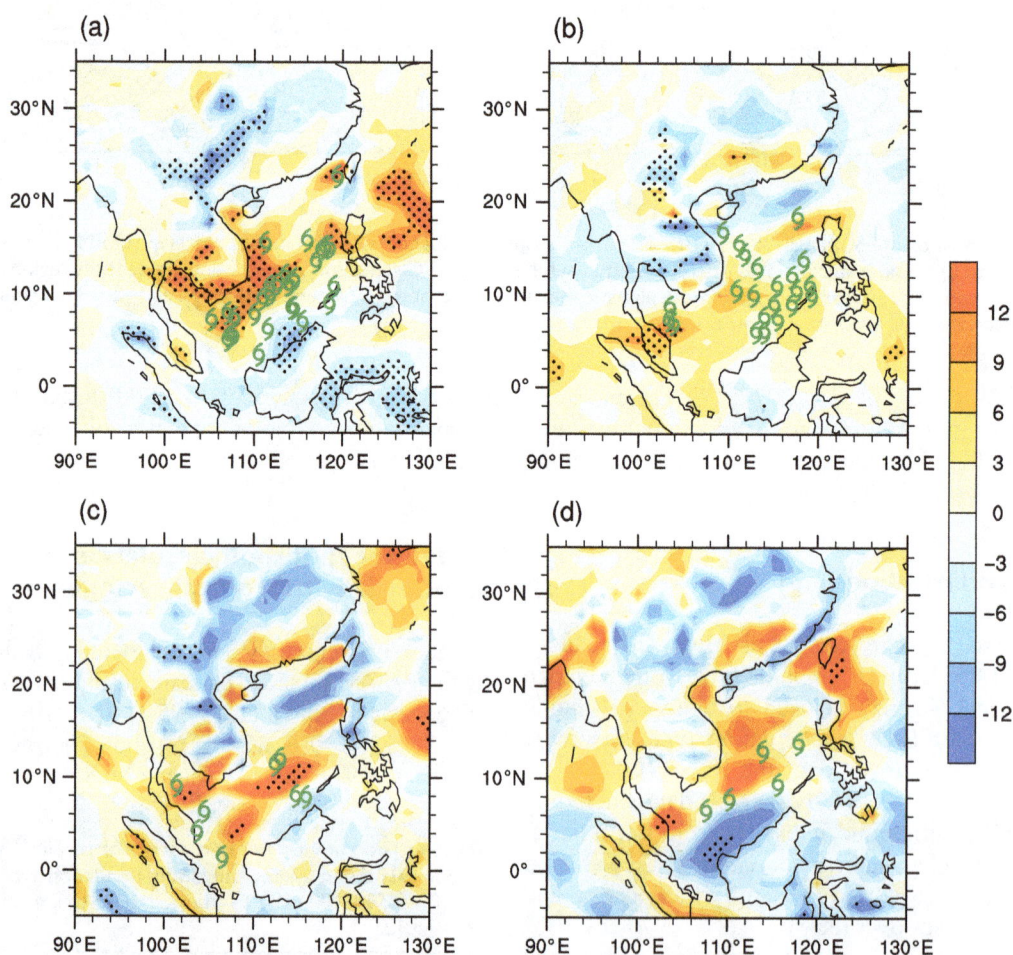

Figure 2. Composites of 850 hPa vorticity anomalies (the 72-h mean before tropical cyclogenesis minus the climatic mean from October to December during 1958–2013; 10^{-6} s^{-1}; shaded) in (a) MG, (b) easterly, (c) RMT, and (d) MC type. Typhoon symbols show TC genesis locations. Stippling denotes regions where the anomalies are significant above the 90% confidence level according to the Student's *t* test.

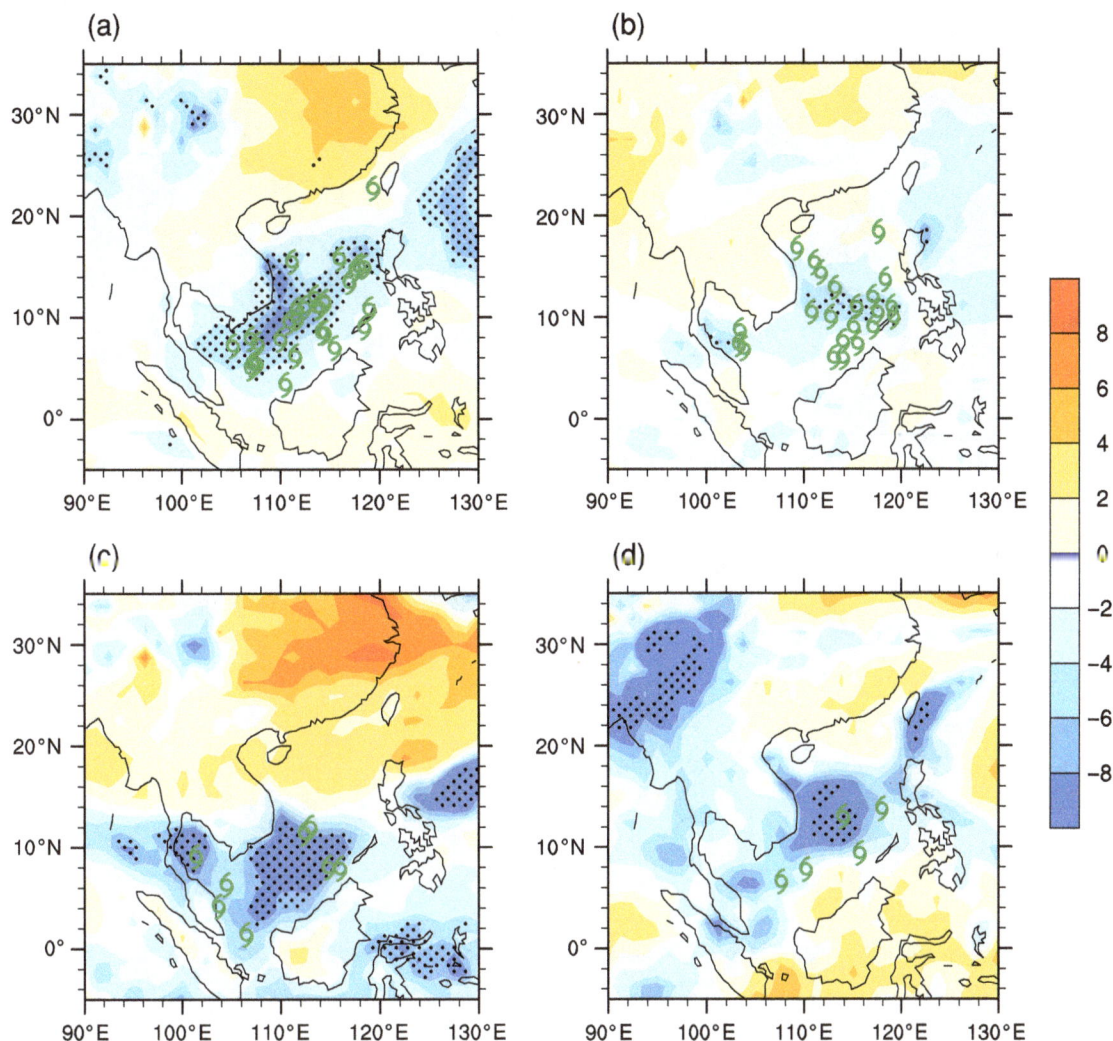

Figure 3. Composites of 500 hPa omega anomalies (the 72-h mean before tropical cyclogenesis minus the climatic mean from October to December during 1958–2013; 10^{-2} Pa s^{-1}; shaded) in (a) MG, (b) easterly, (c) RMT, and (d) MC type. Typhoon symbols show TC genesis locations. Stippling denotes regions where the anomalies are significant above the 90% confidence level according to the Student's t test.

northeast-southwest direction and is favorable for TC genesis. Thus, the TC generated in the monsoon trough with northeast-southwest direction is defined as the one of RMT type. As shown in Figure 1(c), the EAWM flow goes southwestward via Vietnam, then eastward via the Kalimantan Island, which leads to a monsoon trough with the northeast-southwest direction forming over the SCS. This RMT type provides the horizontal wind shear and vertical vorticity for tropical cyclogenesis.

When the EAWM flow prevails over the SCS, the low-level circulation over the SCS is dominated by northeasterly, easterly and cross-equatorial flow. These airflows converge over the SCS. Thus, the TC generated in this kind of monsoon circulation is defined as the one of MC type. As shown in Figure 1(d), compared with the RMT type, the cross-equatorial flow is stronger and extends to the southern-central SCS. It converges with the EAWM flow and the easterly over the southern-central SCS, which provides plenty of perturbation kinetic energy for tropical cyclogenesis.

4. Environmental fields in winter monsoon circulation types

The low-level flow provides dynamical and thermo-dynamic conditions for tropical cyclongenesis (Gray, 1979). From October to December, the EAWM has an important influence on environmental fields, such as low-level vorticity, mid-troposphere relative humidity, and so on (Wang *et al.*, 2007). These key environmental fields play a crucial role in tropical cyclogenesis over the SCS.

In the MG type, tropical cyclogenesis mainly occurs over the southern-central SCS where there are suitable environmental fields. There is an apparently positive vorticity anomaly in the lower tropo-sphere over the SCS, especially in the central SCS (Figure 2(a)). Corresponding to the location of the vorticity anomaly, a mid-tropospheric ascent anomaly and an upper-level positive divergence anomaly are also obvious (Figures 3(a) and 4(a)). These dynamic conditions are accompanied by a positive relative

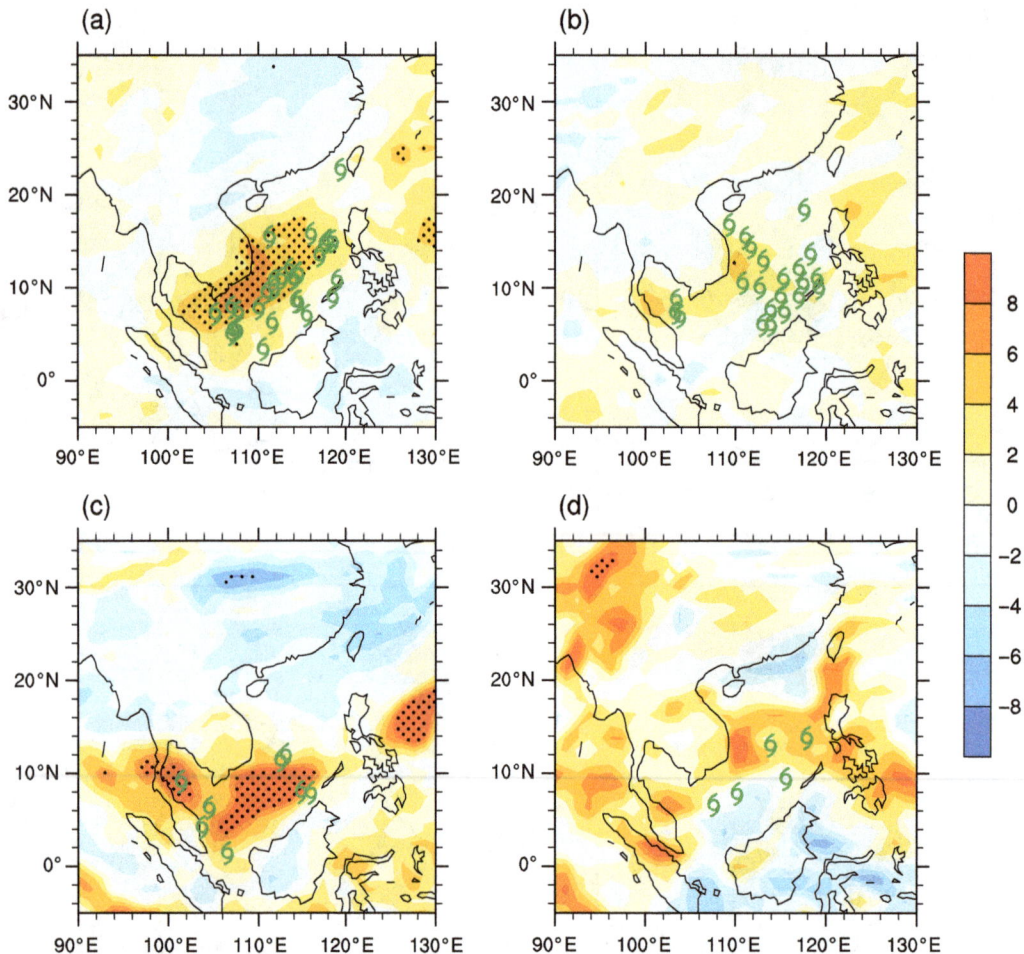

Figure 4. Composites of 200 hPa divergence anomalies (the 72-h mean before tropical cyclogenesis minus the climatic mean from October to December during 1958–2013; 10^{-6} s^{-1}; shaded) in (a) MG, (b) easterly, (c) RMT, and (d) MC type. Typhoon symbols show TC genesis locations. Stippling denotes regions where the anomalies are significant above the 90% confidence level according to the Student's t test.

humidity anomaly in the lower troposphere over the SCS (Figure 5(a)), which is favorable for TC formation.

In the easterly type, the lower-level positive vorticity anomaly, mid-tropospheric ascent anomaly, upper-level positive divergence anomaly, and positive relative humidity anomaly in the lower troposphere are weaker compared with the MG type, especially in the central SCS (Figures 2(b), 3(b), 4(b) and 5(b)). However, these environmental fields are still favorable for tropical cyclogenesis. TCs in the easterly type are mainly generated in the eastern-central SCS (Figure 2(b)). In the RMT type, the strong positive vorticity anomaly, mid-tropospheric ascent anomaly, positive divergence anomaly, and positive relative humidity anomaly occur over the southern SCS where all of TC geneses in this type are observed (Figures 2(c), 3(c), 4(c) and 5(c)). In contrast with the RMT type, four environmental field anomalies favorable for tropical cyclogenesis in the MC type mainly exist in the northern-central SCS where TC geneses in this type mainly occur (Figures 2(d), 3(d), 4(d) and 5(d)).

In addition, low-level convergence also shows apparent positive anomalies in these types over the SCS (figure not shown). Although vertical wind shear in these types seems to be a little strong (figure not shown), other environmental fields mentioned above basically determine dynamic and thermodynamic conditions favorable for TC genesis over the SCS. This feature shows that vertical wind shear is not a key factor influencing TC genesis in EAWM types over the SCS.

5. The synoptic-scale disturbance in winter monsoon circulation types

The barotropic energy transformation in the basic flow of the lower troposphere can increase the eddy kinetic energy (EKE), which provides the synoptic-scale disturbance for TC genesis (Maloney and Hartmann, 2001). The EKE can be computed from the eddy winds (u', v'):

$$\text{EKE} = \frac{\left(\overline{u'^2} + \overline{v'^2}\right)}{2}$$

where the overbar denotes a 72-h mean of before tropical cyclogenesis and prime indicates a time-mean perturbation which is obtained by exerting a 3- to 8-day bandpass filter.

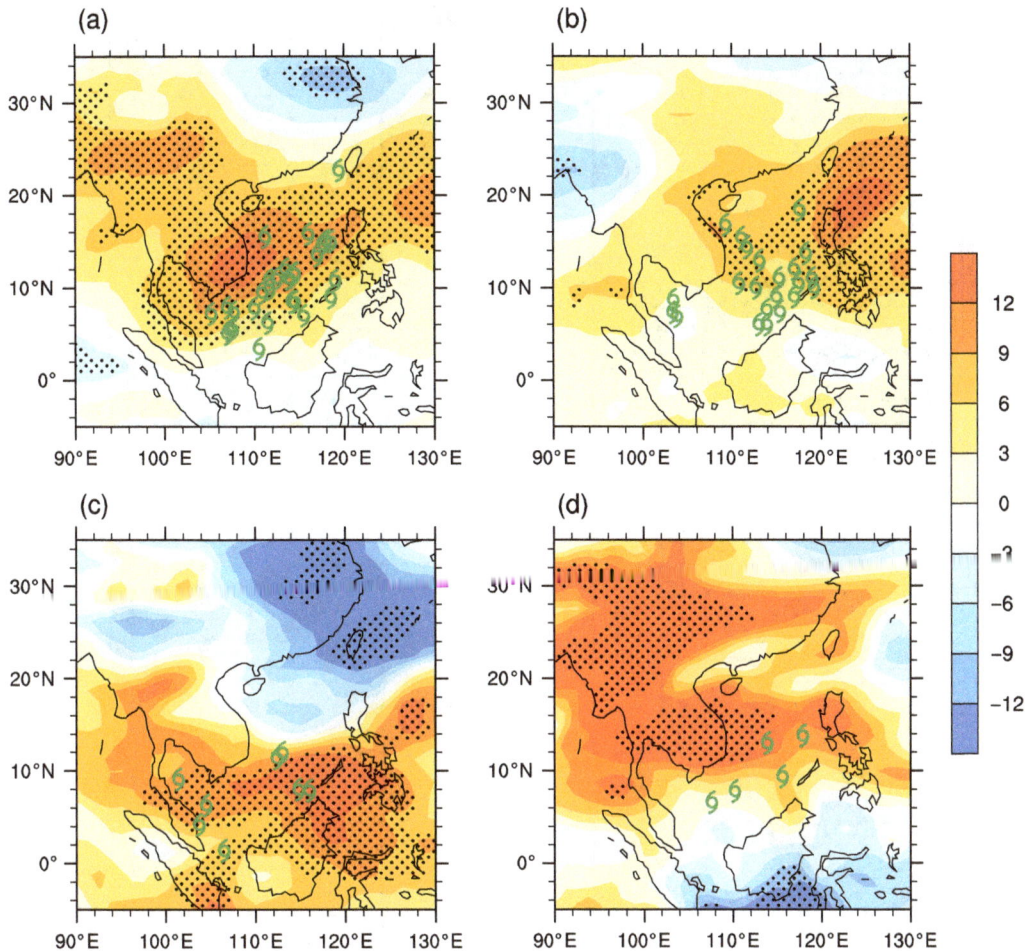

Figure 5. Composites of 500–700 hPa relative humidity anomalies (the 72-h mean before tropical cyclogenesis minus the climatic mean from October to December during 1958–2013; %; shaded) in (a) MG, (b) easterly, (c) RMT, and (d) MC type. Typhoon symbols show TC genesis locations. Stippling denotes regions where the anomalies are significant above the 90% confidence level according to the Student's t test.

Recent studies show that the growth of the EKE comes mainly from the transformation from the mean kinetic energy to the EKE (Chen and Sui, 2010; Hsu et al., 2011). As shown in the following equation (Maloney and Hartmann, 2001):

$$\frac{\partial K'_{\text{baro}}}{\partial t} = -\overline{u'v'}\frac{\partial \overline{u}}{\partial y} - \overline{u'v'}\frac{\partial \overline{v}}{\partial x} - \overline{u'^2}\frac{\partial \overline{u}}{\partial x} - \overline{v'^2}\frac{\partial \overline{v}}{\partial y}$$

where u, v and K indicate the zonal wind, the meridional wind and the EKE, respectively, overbar denotes a 72-h mean before tropical cyclogenesis, and prime a perturbation which is obtained by a 3- to 8-day band-pass filter, the tendency of barotropic energy conversion term ($\partial K'_{\text{baro}}/\partial t$) is calculated with 850-hPa wind data and contributed from the zonal wind shear term, the meridional wind shear term, the zonal wind convergence term, and the meridional wind convergence term.

As shown in Figure 6, the EKE growth over the SCS occurs in all of winter monsoon circulation types to different extent. Meanwhile, TCs are generated basically in the region where the EKE growth is obvious. Figure 6(a) shows that anomalous northerly converges with anomalous southwesterly over the central SCS

in the MG type. The EKE from the barotropic energy conversion of the basic flow is obvious over the SCS. TC geneses are focused in the region where the EKE growth is fast. In the easterly type, anomalous easterly dominates most parts of the SCS and the region of the EKE growth in the easterly type is smaller than that of the MG type and is mainly located in the southern-central SCS where TC formation often occurs (Figure 6(b)). In the RMT type, the EKE increases obviously over the southern SCS where there is a reverse-oriented monsoon trough and TC geneses are observed (Figure 6(c)). As shown in Figure 6(d), the central SCS is a convergent region of anomalous northerly, westerly, and cross-equatorial flow. In this convergent region, there are TC geneses near the area of the EKE growth. In addition, most of TC geneses occur in region 1 (100–110°E, 5–10°N) and region 2 (110–120°E, 5–20°N), where the meridional wind convergence term and the zonal wind shear term make a key contribution to the EKE growth. Therefore, the convergence and shear related to the EAWM play an important role in the EKE growth favorable for TC genesis.

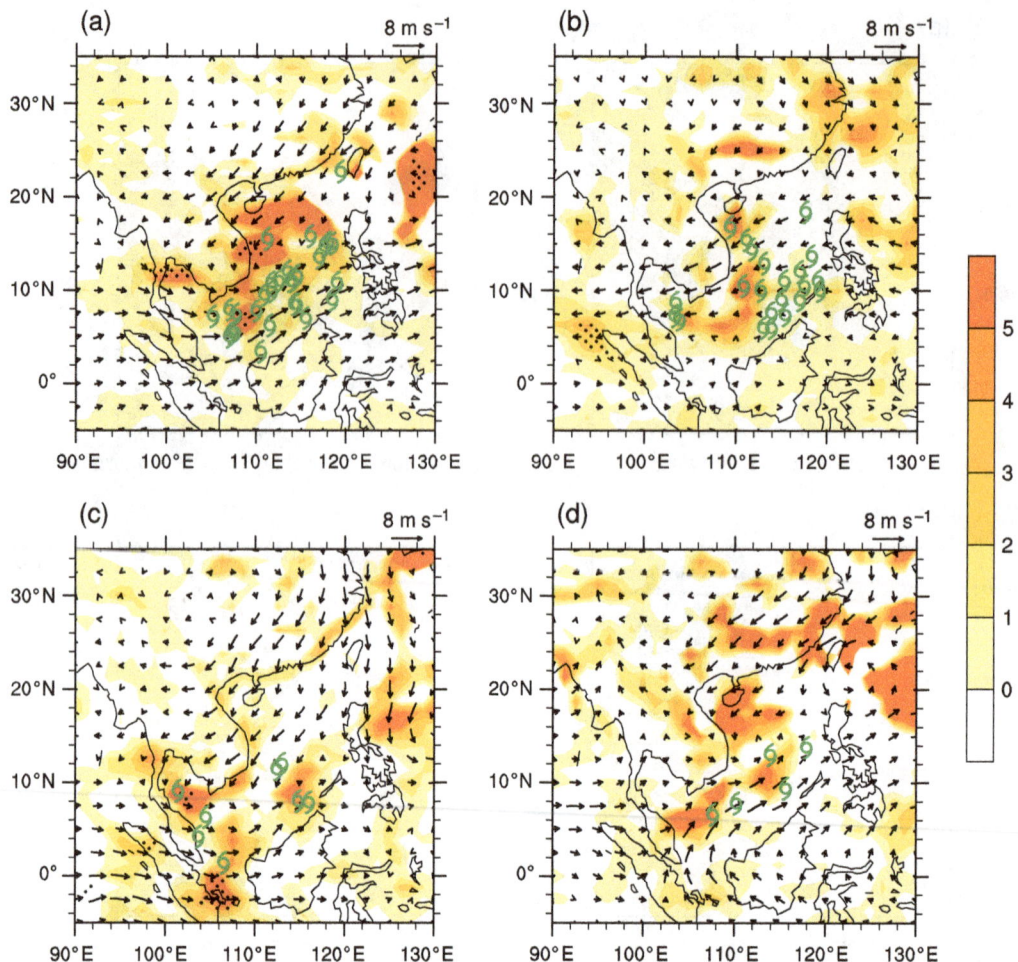

Figure 6. Composites of anomalies (the 72-h mean before tropical cyclogenesis minus the climatic mean from October to December during 1958–2013) of 850 hPa EKE time change rate (10^{-5} m^2 s^{-3}; shaded) due to the barotropic energy conversion and 850 hPa winds (m s^{-1}) in (a) MG, (b) easterly, (c) RMT, and (d) MC type. Typhoon symbols show TC genesis locations. Stippling denotes regions where the anomalies of the EKE are significant above the 90% confidence level according to the Student's t test.

6. Summary

Four types of EAWM circulation favorable for tropical cyclogenesis over the SCS are defined from October to December during 1958–2013, namely the MG, easterly, RMT and MC. The number of TC geneses in these circulation types accounts for about 70.5% of all in the same period over the SCS. Compared with the climatic mean from October to December during 1958–2013, four types of EAWM circulation provide more beneficial environmental fields for TC formation. An apparent positive vorticity anomaly is located in the lower troposphere over the SCS. This vorticity anomaly is accompanied by a mid-tropospheric ascent anomaly and an upper-level positive divergence anomaly. Meanwhile, a positive relative humidity anomaly is obvious in the lower troposphere over the SCS. Moreover, these circulation types over the SCS play an important role in the EKE growth which is beneficial to evolving of incipient disturbances to TCs. From the perspective of the barotropic energy conversion, the meridional wind convergence term and the zonal wind shear term make a key contribution to the EKE growth in the tropical

cyclogenesis region. These types of EAWM circulation provide a support for the important influence of the EAWM on TC genesis over the SCS in boreal winter.

It is a remarkable fact that the vertical wind shear is not a key factor for tropical cyclogenesis in four types of EAWM circulation over the SCS in boreal winter. In view of few TC geneses occurring in the MC type while its distinct circulation feature from other circulation types, this type is classified but not emphatically analyzed. In addition, anomalies of the mid-tropospheric omega and the lower troposphere relative humidity documented above are strongly significant, while anomalies of the EKE time change rate due to the barotropic energy conversion in basic flow are less significant. This feature may be due to the limited quantities of tropical cyclogenesis over the SCS in boreal winter or the influence of other factors which need to be further investigated.

Acknowledgements

This study is supported by the National Natural Science Foundation of China (Grant No. 41375065), the Open Fund of the Key Laboratory of Global Change and Marine-Atmospheric

Chemistry (Grant No. GCMAC1304), and the National Natural Science Foundation of China (Grant No. 41461164005). The authors are grateful for the two anonymous reviewers for their constructive comments.

References

Briegel LM, Frank WM. 1997. Large-scale influences on tropical cyclogenesis in the western North Pacific. *Monthly Weather Review* **125**: 1397–1413.

Chang CP, Erickson JE, Lau KM. 1979. Northeasterly cold surges and near-equatorial disturbances over the winter MONEX area during December 1974. Part I: Synoptic aspects. *Monthly Weather Review* **107**: 812–829.

Chang CP, Liu CH, Kuo HC. 2003. Typhoon Vamei: an equatorial tropical cyclone formation. *Geophysical Research Letters* **30**: 1150, doi: 10.1029/2002GL016365.

Chen GH, Sui CH. 2010. Characteristics and origin of quasi-biweekly oscillation over the western North pacific during boreal summer. *Journal of Geophysical Research* **115**: D14113, doi: 10.1029/2009JD013389.

Chen W, Yang S, Huang RH. 2005. Relationship between stationary planetary wave activity and the East Asian winter monsoon. *Journal of Geophysical Research* **110**: D14110.

Chen XM, Wang YQ, Zhao K. 2015. Synoptic flow patterns and large-scale characteristics associated with rapidly intensifying tropical cyclones in the South China Sea. *Monthly Weather Review* **143**: 64–87, doi: 10.1175/MWR-D-13-00338.1.

Compo GP, Kiladis GN, Webster PJ. 1999. The horizontal and vertical structure of east Asian winter monsoon pressure surges. *Quarterly Journal of the Royal Meteorological Society* **125**: 29–54.

Emanuel K. 2003. Tropical cyclones. *Annual Review of Earth and Planetary Sciences* **31**: 75–104.

Feng T, Chen GH, Huang RH, Shen XY. 2014. Large-scale circulation patterns favorable to tropical cyclogenesis over the western North Pacific and associated barotropic energy conversions. *International Journal of Climatology* **34**: 216–227.

Gray WM. 1979. Hurricanes: Their formation, structure and likely role in the tropical circulation. In *Meteorology over the Tropical Oceans*, Shaw DB (ed). Royal Meteorological Society: Reading, UK; 155–218.

He ZQ, Wu RG. 2010. Seasonality of interannual atmosphere–ocean interaction in the South China Sea. *Journal of Oceanography* **69**: 699–712, doi: 10.1007/s10872-013-0201-9.

Hsu PC, Li T, Tsou CH. 2011. Interactions between boreal summer intraseasonal oscillations and synoptic-scale disturbances over the western North Pacific. Part I: energetics diagnosis. *Journal of Climate* **24**: 927–941.

Huang RH, Zhou LT, Chen W. 2003. The progresses of recent studies on the variabilities of the East Asian monsoon and their causes. *Advances in Atmospheric Sciences* **20**: 55–69.

Krishnamurti TN, Kanamitsu M, Koss WJ, Lee JD. 1973. Tropical east–west circulations during the northern winter. *Journal of the Atmospheric Sciences* **30**: 780–787.

Lander MA. 1994. An exploratory analysis of the relationship between tropical storm-formation in the western North Pacific and ENSO. *Monthly Weather Review* **122**: 636–651.

Lau KM, Chang CP. 1987. Planetary scale aspects of the winter monsoon and atmospheric teleconnections. In *Monsoon Meteorology*, Chang CP, Krishnamurti TN (eds). Oxford University Press; 161–202.

Maloney ED, Hartmann DL. 2001. The Madden–Julian Oscillation, barotropic dynamics, and North Pacific tropical cyclone formation. Part I: observations. *Journal of the Atmospheric Sciences* **58**: 2545–2558.

Richard CYL, Zhou W. 2012. Changes in western Pacific tropical cyclones associated with the El Niño-Southern Oscillation cycle. *Journal of Climate* **25**: 5864–5878, doi: 10.1175/JCLI-D-11-00430.1.

Ritchie EA, Holland GJ. 1999. Large-scale patterns associated with tropical cyclongenesis in the western Pacific. *Monthly Weather Review* **127**: 2027–2043.

Wang L, Chen W. 2010. How well do existing indices measure the strength of the East Asian winter monsoon? *Advances in Atmospheric Sciences* **27**: 855–870, doi: 10.1007/s00376-009-9094-3.

Wang B, Wu RG, Fu XH. 2000. Pacific-East Asian teleconnection: how does ENSO affect East Asian climate? *Journal of Climate* **13**: 1517–1536.

Wang GH, Su JL, Ding YH, Chen DK. 2007. Tropical cyclone genesis over the South China Sea. *Journal of Marine Systems* **68**: 318–326.

Wang L, Huang RH, Wu RG. 2013. Interdecadal variability in tropical cyclone frequency over the South China Sea and its association with the Indian Ocean sea surface temperature. *Geophysical Research Letters* **40**: 768–771, doi: 10.1002/GRL.50171.

Wu L, Wen ZP, Huang RH, Wu RG. 2012. Possible linkage between the monsoon trough variability and the tropical cyclone activity over the western North Pacific. *Monthly Weather Review* **140**: 140–150, doi: 10.1175/MWR-D-11-00078.1.

Wu LG, Zong HJ, Jia L. 2013. Observational analysis of tropical cyclone formation associated with Monsoon Gyres. *Journal of the Atmospheric Sciences* **70**: 1023–1034.

Wu L, Wen ZP, Li T, Huang RH. 2014a. ENSO-phase dependent TD and MRG wave activity in the western North Pacific. *Climate Dynamics* **42**: 1217–1227.

Wu RG, Chen W, Wang GH, Hu KM. 2014b. Relative contribution of ENSO and East Asian winter monsoon to the South China Sea SST anomalies during ENSO decaying years. *Journal of Geophysical Research* **119**: 5046–5064, doi: 10.1002/2013JD02109.

Yeh TC, Tao SY, Li MT. 1959. The abrupt change of circulation over the Northern Hemisphere during June and October. In *The Atmosphere and the Sea in Motion*, Bolin B (ed). Rockefeller Institute Press; 249–267.

Zhang RH, Sumi A, Kimoto M. 1996. Impact of El Niño on the East Asian monsoon: A diagnostic study of the '86/87 and '91/92 events. *Journal of the Meteorological Society of Japan* **74**: 49–62.

21

Two typical modes in the variabilities of wintertime North Pacific basin-scale oceanic fronts and associated atmospheric eddy-driven jet

Liying Wang,[1] Xiu-Qun Yang,[1]* Dejian Yang,[1] Qian Xie,[2] Jiabei Fang[1] and Xuguang Sun[1]

[1]CMA-NJU Joint Laboratory for Climate Prediction Studies, and Jiangsu Collaborative Innovation Center of Climate Change, School of Atmospheric Sciences, Nanjing University, Nanjing, China
[2]College of Meteorology and Oceanography, National University of Defense Technology, Naning, China

*Correspondence to:
X.-Q. Yang, School of Atmospheric Sciences, Nanjing University, Nanjing 210023, China.
E-mail: xqyang@nju.edu.cn

Abstract

The role of oceanic fronts in the midlatitude air–sea interaction remains unclear. This study defines new indexes to quantify the intensity and location of two basin-scale oceanic frontal zones in the wintertime North Pacific, i.e. the subtropical and subarctic frontal zones (STFZ, SAFZ). With these indexes, two typical modes, which are closely related to two large-scale sea surface temperature (SST) anomaly patterns resembling Pacific Decadal Oscillation (PDO) and North Pacific Gyre Oscillation (NPGO), respectively, are found in the oceanic front variabilities as well as in their associations with the midlatitude atmospheric eddy-driven jet. Corresponding to an PDO-like SST anomaly pattern, an enhanced STFZ occurs with a southward shifted SAFZ, which is associated with enhanced overlying atmospheric front, baroclinicity and transient eddy vorticity forcing, thus with an intensification of the westerly jet; and vice versa. On the other hand, corresponding to an NPGO-like SST pattern, an enhanced SAFZ occurs with a northward shifted STFZ, which is associated with a northward shift of the atmospheric front, baroclinicity, transient eddy vorticity forcing, and westerly jet; and vice versa. These results suggest that the basin-scale oceanic frontal zone is a key region for the midlatitude air–sea interaction in which the atmospheric transient eddy's dynamical forcing is a key player in such an interaction.

Keywords: Basin scale oceanic front; sea surface temperature; eddy-driven jet; transient eddy forcing; Pacific Decadal Oscillation; North Pacific Gyre Oscillation

1. Introduction

Many previous studies have identified typical modes of the midlatitude North Pacific sea surface temperature (SST) variabilities, such as Pacific Decadal Oscillation (PDO) (Mantua *et al.*, 1997) and North Pacific Gyre Oscillation (NPGO) (Di Lorenzo *et al.*, 2008), and their associations with the atmosphere (Miller *et al.*, 1994; Trenberth and Hurrell, 1994). However, how the midlatitude large-scale SST anomaly can affect the atmosphere remains a challenging issue (Peng and Whitaker, 1999; Fang and Yang, 2011; Frankignoul *et al.*, 2011; Liu, 2012). Recent studies have suggested that such an impact might be related to the midlatitude oceanic front (Qiu and Chen, 2005, 2010; Taguchi *et al.*, 2007; Ceballos *et al.*, 2009; Fang and Yang, 2016). There exist two oceanic frontal zones in the wintertime North Pacific, i.e. the subtropical frontal zone (STFZ) and subarctic frontal zone (SAFZ), respectively (Nakamura *et al.*, 2008; Nakamura and Yamane, 2010). The oceanic front could maintain midlatitude storm track activities by restoring sharp cross-frontal gradient of surface air temperature, via an oceanic baroclinic adjustment mechanism (Nakamura *et al.*, 2008; Taguchi *et al.*, 2009; Brayshaw *et al.*, 2011). Dynamical diagnoses and numerical experiments indicate that the oceanic front and associated atmospheric transient eddy activities could also crucial in the unstable midlatitude air–sea interaction (Fang and Yang, 2016; Wang *et al.*, 2016; Yao *et al.*, 2016).

It is necessary to further characterize the midlatitude oceanic front and its variability with observational analyses. Previous studies have successfully indentified the regional features of SAFZ with the SST gradient maximum and its latitude in the western Pacific (e.g. Taguchi *et al.*, 2012). Considering the large-scale features of SST anomalies in the North Pacific (such as PDO or NPGO) and the zonally-elongated variabilities of the intense meridional SST gradients associated with STFZ and SAFZ in the entire North Pacific, it is of great interest to characterize the feature in the variabilities of the North Pacific oceanic fronts on the zonal-basin scale and their associations with the midlatitude atmosphere.

In this paper, based on two SST datasets with different resolutions, we define four new indexes to quantify the intensity and location of the basin-scale STFZ and SAFZ during winter in North Pacific, respectively. With these indexes, we examine the typical modes of the two oceanic fronts and their associations with overlying atmospheric eddy-driven jet. Section 2 describes

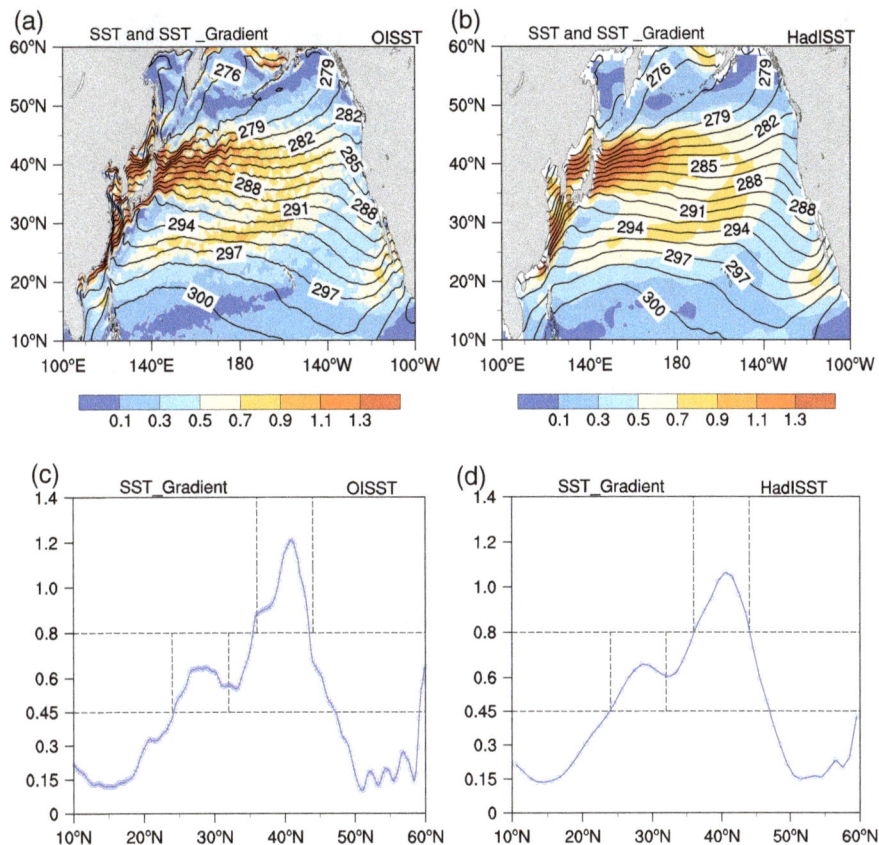

Figure 1. Climatological distributions of wintertime SST (contours, Unit: K) and its meridional gradient (absolute value, shading, Unit: 10^{-5} Km^{-1}) in North Pacific for (a) OISST during 1982–2010 and (b) HadISST during 1960–2010, and the climatological SST meridional gradient (Unit: 10^{-5} Km^{-1}) zonally-averaged over 145°E ~ 145°W as plotted a function of latitude for (c) OISST and (d) HadISST. The horizontal blue lines signify the empirically-given critical value for front index definition, which is 0.45 for STFZ and 0.8 for SAFZ, respectively. The vertical blue lines indicate the latitudinal range of each frontal zone for front index definition, which is 24° ~ 32°N for STFZ, and 36° ~ 44°N for SAFZ, respectively.

the datasets used. In Section 3, we firstly present a definition of basin-scale front indexes and validate it with two different resolution SST datasets. In terms of these indexes, we then examine the relationship between the oceanic front and large scale SST variabilities, and the associated atmospheric anomalies. The final section is devoted to conclusions.

2. Data

Two SST datasets with different resolutions are used to analyze oceanic front variabilities. One is the Optimum Interpolation SST Version 2 data (OISST) for 1982–2011 with relatively fine spatial resolution of 0.25° and temporal resolution of 1 day, which is developed by the National Oceanic and Atmospheric Administration (NOAA)'s National Climatic Data Center. The other is the Hadley Centre Global Sea Ice and SST (HadISST) monthly mean SST data for 1960–2011, with relatively coarse global spatial resolution of 1.0°. The monthly mean time series of the PDO and NPGO indexes for 1960–2011 are also used to investigate relationship between oceanic front variabilities and large scale SST variability modes. The PDO index is taken

from NOAA Climate Prediction Center and the NPGO index is provided by E. Di Lorenzo. Atmospheric anomalies associated with oceanic front variabilities are examined with the National Centers for Environmental Prediction/National Center for Atmospheric Research (NCEP/NCAR) daily reanalysis dataset for 1960–2011 with a global resolution of 2.5°, as well as with the NCEP Climate Forecast System Reanalysis (CFSR) dataset for 1982–2010 with a global resolution of 0.5°. In this study, all the datasets are averaged over December, January and February (DJF) to represent wintertime mean for 1982–2010 for OISST and CFSR datasets and 1960–2010 for other datasets.

3. Results

3.1. Definition of basin-scale oceanic front indexes

The climatological distributions of wintertime SST and its meridional gradient in North Pacific demonstrate that large SST meridional gradients are located in the midlatitude North Pacific, especially in two separated frontal zones, i.e. the STFZ near 30°N and the SAFZ near 40°N (Figures 1(a) and (b)). Roughly, the two datasets give similar distributions in SST and its

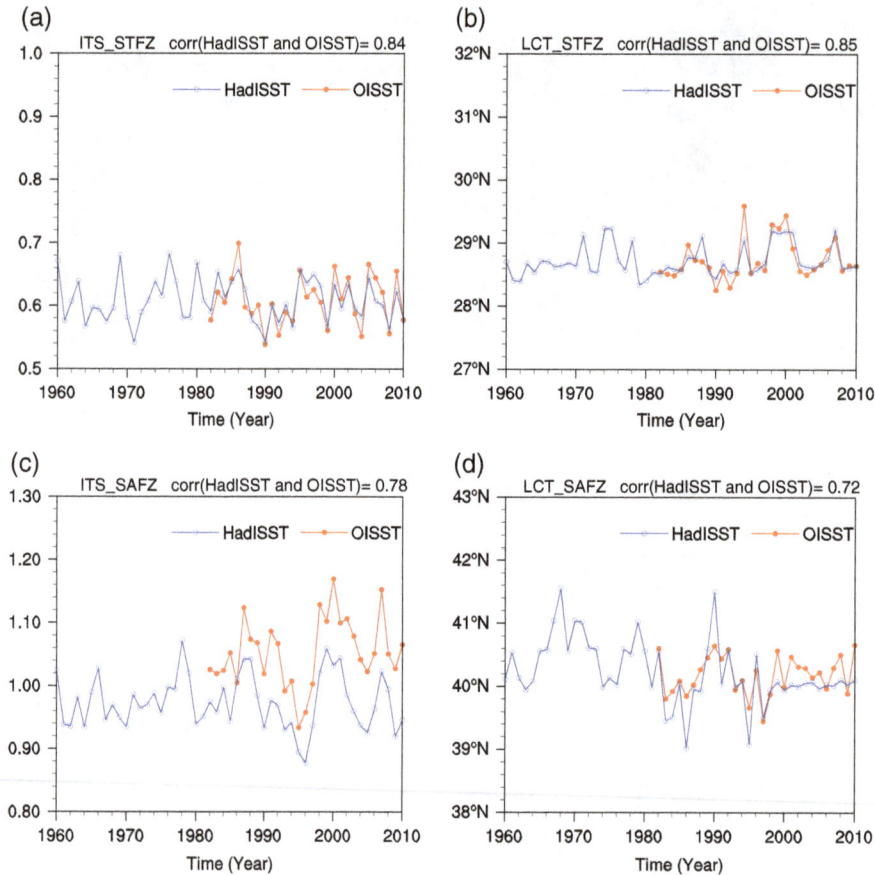

Figure 2. (a) Time series of the intensity index of STFZ (Units: 10^{-5} Km^{-1}) for HadISST during 1960–2010 (blue line) and for OISST during 1982–2010 (red line). (b) The latitudinal location index of STFZ (Units: °N) for HadISST during 1960–2010 (blue line) and for OISST during 1982–2010 (red line). (c), (d) are the same as (a), (b), but for SAFZ. The correlation coefficients between the front indexes for OISST and HadISST during 1982–2010 are annotated in each figure. Coefficient for statistical significance at 95% level is 0.37.

gradient. The gradient in the western SAFZ based on OISST dataset with higher resolution (Figure 1(a)) is larger than that based on HadISST dataset (Figure 1(b)), probably due to the resolution difference. Since we focus on the basin-scale characteristics of the fronts, the SST gradients are zonally-averaged over the whole North Pacific basin, as illustrated in Figures 1(c) and (d). From 20°N to higher latitudes, the SST gradient becomes stronger and reaches its first maximum at 28.5°N, indicating the STFZ. After slightly weakening, the SST gradient increases again and reaches its second maximum (also the strongest) at about 40.5°N, indicating the SAFZ.

We define two new indexes to quantify the intensity and location, respectively, of each basin-scale oceanic frontal zone, using the SST meridional gradient zonally-averaged over 145°E ~ 145°W, as follows. Given an oceanic frontal zone (say, 24° ~ 32°N for STFZ or 36° ~ 44°N for SAFZ, as seen in Figure 1), its intensity for each winter is defined as.

$$ITS = \sum_{i=1}^{N} G_i / N \qquad (1)$$

where G_i is the value of zonally-averaged SST meridional gradient that is no less than an empirically-given

critical value (here, 0.45×10^{-5} Km^{-1} for STFZ, and 0.8×10^{-5} Km^{-1} for SAFZ) at the i-th latitudinal grid point within the zone, and N is the number of total grid points that satisfy the criteria above. The intensity index defined reflects an average of the SST meridional gradient within a frontal zone. Furthermore, the location of a front, as a function of latitude, is defined as.

$$LCT = \sum_{i=1}^{N} \left(G_i \times LAT_i \right) / \sum_{i=1}^{N} G_i \qquad (2)$$

where LAT_i is the latitude at the i-th grid point within the front zone. Obviously, this definition reflects a weighted-average of LAT_i with respect to G_i, indicating that the location of a front is mainly determined by larger SST meridional gradients within the frontal zone.

In terms of the definitions above, the time series of zonally-averaged intensity and location of two fronts are computed with two SST datasets, as shown in Figure 2. Overall, the two datasets exhibit consistent variabilities in both the zonally-averaged intensity and location during the common period 1982–2010 for two fronts, except for the intensity of SAFZ which is slightly weaker in HadISST than in OISST, probably due to the resolution difference. Despite the shortcoming,

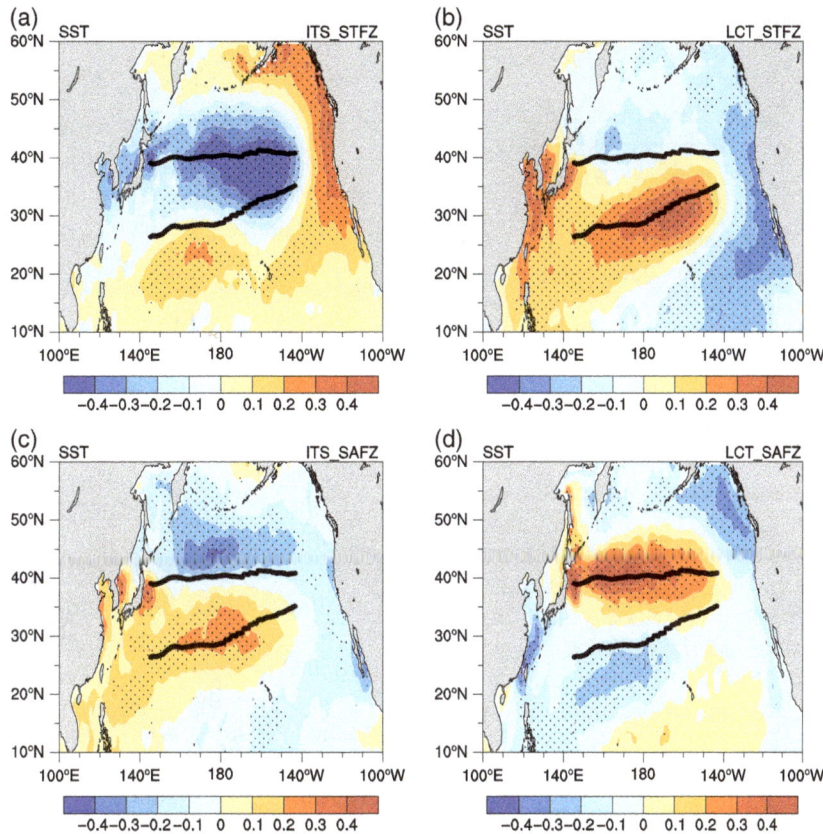

Figure 3. Spatial distributions of the regressed SST anomalies (shading, Units: K) upon (a) the intensity index of STFZ, (b) the location index of STFZ, (c) the intensity index of SAFZ, and (d) the location index of SAFZ, based on HadISST data for winter of 1960–2010. Areas with statistical significance no less than 95% level are dotted. The climatological locations of STFZ and SAFZ are denoted by black lines.

the indexes based on HadISST data can characterize the North Pacific basin-scale oceanic front long-term variabilities for its longer time spanning. Note that the analyses based on the HadISST and NCEP/NCAR reanalysis datasets for 1960–2010 and the OISST and CFSR datasets for 1982–2010 are respectively made in this study, and consistent results are obtained. In the following subsections, the results based on the HadISST and NCEP/NCAR reanalysis datasets for 1960–2010 are only presented.

3.2. Relationship between the oceanic front and SST variabilities

Correlation coefficients between any two front characteristic indexes are computed to investigate the internal relation of oceanic front variabilities. As shown in Table S1 (Supporting information), there is no significant relationship between the intensity and location variability of either STFZ or SAFZ. Nevertheless, there is a significant negative correlation between the location of SAFZ and the intensity of STFZ, indicating that STFZ gets stronger when SAFZ moves southward; and vice versa. Meanwhile, there is a significant positive correlation between the intensity of SAFZ and the location of STFZ, suggesting that the enhancement of SAFZ is accompanied with the northward movement of STFZ; and vice versa. The two types of correlations between

the STFZ and SAFZ variabilities can be further found to be closely associated with different large-scale SST anomaly patterns in North Pacific.

The negative correlation between the intensity of STFZ and the location of SAFZ can be identified by the regressed SST anomalies in Figures 3(a) and (d). The SST anomaly pattern regressed upon the intensity of STFZ (Figure 3(a)) is quite similar to that regressed upon the location of SAFZ (Figure 3(d)), but for the opposite sign. The negative SST anomaly centered around 40°N (the climatological position of SAFZ) tends to induce a southward shift of the strongest SST meridional gradient (i.e. SAFZ); meanwhile, associated with the southward shift of SAFZ, the negative SST anomaly north of 30°N and the positive SST anomaly south of it tend to increase the SST meridional gradient around 30°N (i.e. the intensity of STFZ); and vice versa. This type of SST anomaly pattern exhibits much resemblance to PDO, the leading EOF mode of the SST anomaly north of 20°N in North Pacific. This relationship is further testified by a significant, positive correlation of 0.7 between the STFZ's intensity and PDO indexes, and a significant, negative correlation of −0.53 between the SAFZ's location and PDO indexes, as shown in Table S1. This suggests that when PDO is in its positive phase, SAFZ shifts southward, but STFZ gets stronger; and vice versa. Thus, the intensity of STFZ together with the location of SAFZ is in well correspondence with the interannual and decadal variability of PDO.

On the other hand, the positive correlation between the location of STFZ and the intensity of SAFZ can be identified by the regressed SST anomalies in Figures 3(b) and (c). The SST anomaly pattern regressed upon the location of STFZ shown (Figure 3(c)) is rather similar to that regressed upon the intensity of SAFZ (Figure 3(b)). The positive SST anomaly around 30°N tends to induce a poleward movement of the large meridional SST gradient in STFZ; meanwhile, the positive SST anomaly south of 40°N together with the negative SST anomaly north of 40°N tends to strengthen SAFZ; and vice versa. This type of SST anomaly pattern reflects NPGO, the second leading EOF mode of the sea surface height anomaly or the SST anomaly in North Pacific (Di Lorenzo et al., 2008). As shown in Table S1, the coefficients between the SAFZ's intensity (or STFZ's location) and NPGO indexes are significantly positive, suggesting that the positive phase of NPGO is accompanied with the intensification of SAFZ and northward movement of STFZ; and vice versa.

3.3. Associated atmospheric anomalies

Owing to the significant cross correlation between the intensity of one front and the location of another, as found above, the atmospheric anomalies are regressed only upon the intensity and the location of STFZ, respectively. Figure 4 shows the altitude-latitude distributions of the regressed atmospheric anomalies zonally-averaged over 145°E ~ 145°W for the meridional air temperature gradient, baroclinicity represented by the maximum Eady growth rate (Hoskins and Valdes, 1990), zonal wind speed, and tendency of zonal wind speed forced by the diabatic heating and transient eddy forcing. As the meridional gradient is taken of the quasi-geostrophic potential vorticity (QGPV) equation used in Lau and Holopainen (1984) and Fang and Yang (2016), then the tendency of seasonal mean zonal wind speed ($\overline{u}$) induced by the diabatic heating (F_1), the transient eddy thermal forcing (F_2) and the transient eddy vorticity forcing (F_3) can be written as,

$$
\left[\frac{1}{f}\nabla^2 + f\frac{\partial}{\partial p}\left(\frac{1}{\sigma_1}\frac{\partial}{\partial p}\right)\right]\frac{\partial \overline{u}}{\partial t}
$$

$$
= \underbrace{\frac{1}{\partial y}\left[\frac{\partial}{\partial p}\left(\frac{R}{\sigma_1 p}\overline{Q}_d\right)\right]}_{F1}
$$

$$
+ \underbrace{\frac{1}{\partial y}\left[\frac{\partial}{\partial p}\left(\frac{R}{\sigma_1 p}\overline{Q}_{eddy}\right)\right]}_{F2} - \underbrace{\frac{1}{f}\frac{\partial \overline{F}_{eddy}}{\partial y}}_{F3} \qquad (3)
$$

where the over bar denotes the seasonal average and the prime denotes the transient, synoptic (2–7 days) eddy,

$\overline{Q}_d$ is the diabatic heating, $\overline{Q}_{eddy}$ is the transient eddy thermal forcing, $\overline{Q}_{eddy} = -\nabla \cdot \left(\overline{V'T'}\right)$, and $\overline{F}_{eddy}$ is the transient eddy vorticity forcing, $\overline{F}_{eddy} = -\nabla_h \cdot \left(\overline{V'\varsigma'}\right)$. For comparison, the climatological distributions for those atmospheric variables are also shown with contours in Figure 4. Climatologically, the meridional air temperature gradient (Figures 4(a) and (b)) and baroclinicity (Figures 4(c) and (d)) are the strongest in the lower- and mid-troposphere at midlatitudes, while the westerly jet prevails with center at about 200 hPa over 30° ~ 40°N with equivalent barotropic structure (Figures 4(e) and (f)). Compared with the diabatic heating (Figures 4(g) and (h)) and the transient eddy thermal forcing (Figures 4(i) and (j)), only the transient eddy vorticity forcing (Figures 4(k) and (l)) is substantially positive slightly north of the jet center, acting to reinforce the midlatitude jet. Note that the wintertime jet over the midlatitude North Pacific is basically a merging of the eddy-driven jet and the subtropical jet.

We find that the atmospheric anomalies regressed upon the intensity of STFZ are thoroughly in phase with their corresponding climatologies over 30° ~ 40°N, indicating their variabilities only in intensity, as shown with shading in the left panels of Figure 4. When the intensity of STFZ is stronger than usual, which occurs with a southward shifted SAFZ, both the meridional air temperature gradient (atmospheric front) and the baroclinicity below 300 hPa are considerably increased (Figures 4(a) and (c)), which favors more transient baroclinic eddy activities in the vicinity of 35°N. As a result, the transient eddy vorticity forcing induced zonal wind tendency gets stronger with equivalent barotropic structure over 30° ~ 40°N (Figure 4(k)), which tends to enforce a positive anomaly of zonal wind speed appearing at the center of jet and thus intensify the westerly jet (Figure 4(e)). However, the diabatic heating (Figure 4(g)) and the transient eddy thermal forcing (Figure 4(i)) tend to decrease the upper level jet and increase lower level westerly, with a baroclinic structure.

On the other hand, the atmospheric anomalies regressed upon the location of STFZ are characterized by a meridional dipole pattern with zero contour around 35°N, indicating their variabilities only in location, as shown with shading in the right panels of Figure 4. When STFZ goes northward, which occurs with an enhanced SAFZ, both the meridional air temperature gradient and the atmospheric baroclinicity below 300 hPa exhibit positive anomalies over 40° ~ 50°N and negative anomalies over 20° ~ 30°N (Figures 4(b) and (d)). In comparison with their climatologies, such an anomaly pattern implies a northward shift of the atmospheric front and baroclinic zone. Consequently, the transient eddy vorticity forcing is strikingly enhanced along 40° ~ 50°N (Figure 4(l)), especially in the upper troposphere, giving rise to a northward movement of the westerly jet, as shown in Figure 4(f). In this case,

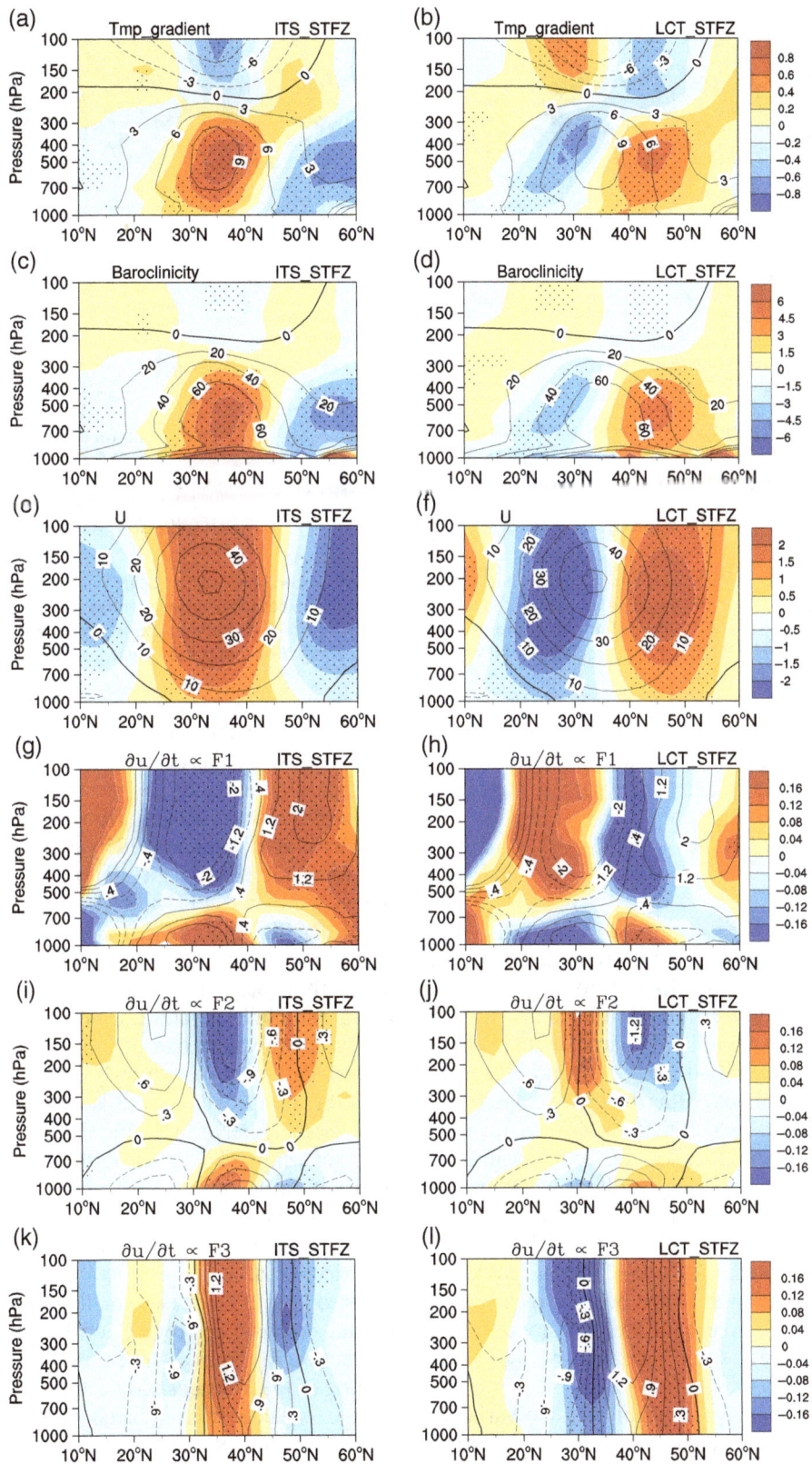

Figure 4. Altitude-latitude distributions of the climatologies (contours) and the atmospheric anomalies (shading) zonally-averaged over 145°E ~ 145°W regressed upon the intensity index of STFZ (left panels) and upon the location index of STFZ (right panels) for (a, b) the air temperature meridional gradient (Units: 10^{-5} Km^{-1}), (c, d) the baroclinicity or maximum Eady growth rate (Units: 10^{-5} ms^{-2}), (e, f) the zonal wind speed (Units: ms^{-1}), (g, h) the tendency of zonal wind speed forced by diabatic heating (F1, Units: 10^{-5} ms^{-2}), (i, j) the tendency of zonal wind speed forced by transient eddy thermal forcing (F2, Units: 10^{-5} ms^{-2}), and (k, l) the tendency of zonal wind speed forced by transient eddy vorticity forcing (F3, Units: 10^{-5} ms^{-2}). The front indexes are based on the HadISST data for winter of 1960–2010. Areas with statistical significance no less than 95% level are dotted.

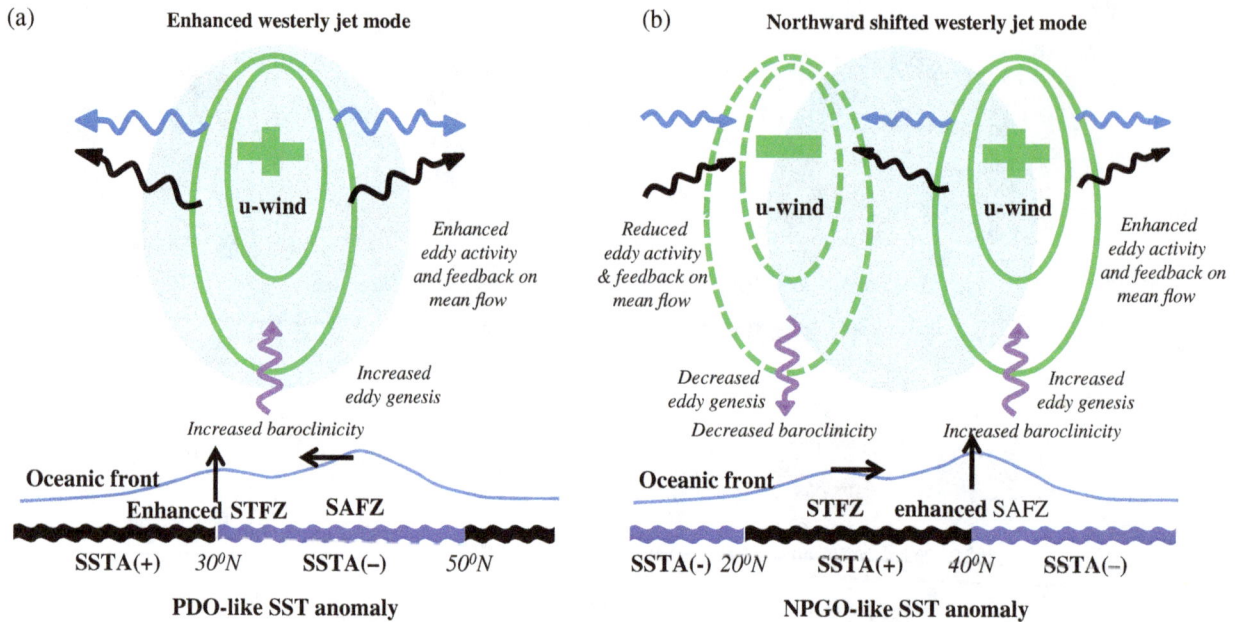

Figure 5. Schematic diagram of two typical modes in the variabilities of wintertime North Pacific basin-scale oceanic fronts and associated atmospheric eddy-driven jet: (a) the enhanced westerly jet mode corresponding to an enhanced STFZ and an southward shifted SAFZ, caused by the PDO-like SST anomaly, and (b) the northward shifted westerly jet mode corresponding to an enhanced SAFZ and an northward shifted STFZ, caused by the NPGO-like SST anomaly. Note that the climatological westerly jet is shaded.

the diabatic heating (Figure 4(h)) and the transient eddy thermal forcing (Figure 4(j)) are also characterized by a meridional dipole pattern, but with a baroclinic structure.

4. Summary

This study characterizes the variabilities of two zonally-basin-scale oceanic frontal zones (i.e. STFZ and SAFZ) in the wintertime North Pacific as well as their associations with the midlatitude atmosphere, especially with the eddy-driven jet. New characteristic indexes are defined with zonally-averaged SST meridional gradient over 145°E–145°W to quantify the intensity and the location of each basin-scale frontal zone, respectively, and further validated with the fine resolution OISST data for 1982–2010 and the coarse resolution HadISST data for 1960–2010. These indexes measured by two different resolution SST datasets are almost the same over the overlapping period (1982–2010), except for the intensity of SAFZ which is slightly weak when measured by the coarse resolution data.

Cross-correlation analyses exhibit that the variabilities of two basin-scale oceanic fronts in the wintertime North Pacific are not independent of each other. There is a significant negative (positive) cross correlation between the intensity of STFZ (SAFZ) and the location of SAFZ (STFZ), indicating that STFZ (SAFZ) gets stronger when SAFZ (STFZ) moves southward (northward), and vice versa. The two types of cross correlations between STFZ and SAFZ are closely associated with two different modes of large scale SST

variabilities in the wintertime North Pacific, i.e. the PDO and NPGO, respectively.

Regression analyses demonstrate that the two types of the basin-scale oceanic front variabilities identified above are also characterized by two types of the overlying atmospheric anomalies, as summarized in Figure 5. When there is a PFO-like positive SST anomaly in North Pacific (Figure 5(a)), STFZ gets stronger while SAFZ goes southward, which is associated with significantly enhanced lower- and mid-level atmospheric front, baroclinicity, and transient eddy vorticity forcing with equivalent barotropic structure, thus with an intensification of the westerly jet; and vice versa. On the other hand, when there is an NPGO-like positive SST anomaly in North Pacific (Figure 5(b)), SAFZ gets stronger while STFZ goes northward, which is associated with a northward shift of the atmospheric front, the baroclinic zone, the transient eddy vorticity forcing strikingly in the upper troposphere, and the westerly jet; and vice versa.

These results suggest that the basin-scale oceanic frontal zone is a key region for the large-scale air–sea interaction in the midlatudes, in which the dynamical forcing of the transient eddy activity that is determined by the oceanic front-induced low-level atmospheric meridional temperature gradient and baroclinicity is a key player in such an interaction. Note that, although this study is based on observational analyses, the major findings presented here agree well with those results from higher resolution numerical experiments (Wang *et al.*, 2016; Yao *et al.*, 2016), and support the hypothesis for the unstable midlatitude air–sea interaction proposed by Fang and Yang (2016).

Acknowledgement

This work is jointly supported by the National Natural Science Foundation of China under grants 41330420, 41621005 and 41375074.

Supporting information

The following supporting information is available:

Table S1. Correlation coefficients between two oceanic front indexes, and between the front indexes and the PDO and NPGO indexes

References

Brayshaw DJ, Hoskins B, Blackburn M. 2011. The basic ingredients of the North Atlantic storm track. Part II: sea surface temperatures. *Journal of the Atmospheric Sciences* **68**: 1784–1805 https://doi.org/10.1175/2011JAS3674.1.

Ceballos LI, Di Lorenzo E, Hoyos CD, Taguchi B. 2009. North Pacific gyre oscillation synchronizes climate fluctuations in the eastern and western boundary systems. *Journal of Climate* **22**: 5163–5174. https://doi.org/10.1175/2009JCLI2848.1.

Di Lorenzo E, Schneider N, Cobb KM, Franks PJS, Chhak K, Miller AJ, McWilliams JC, Bograd SJ, Arango H, Curchitser E, Powell TM, Rivière P. 2008. North Pacific gyre oscillation links ocean climate and ecosystem change. *Geophysical Research Letters* **35**: L08607. https://doi.org/10.1029/2007GL032838.

Fang J, Yang XQ. 2011. The relative roles of different physical processes in the unstable midlatitude ocean-atmosphere interactions. *Journal of Climate* **24**: 1542–1558. https://doi.org/10.1175/2010JCLI3618.1.

Fang J, Yang XQ. 2016. Structure and dynamics of decadal anomalies in the wintertime midlatitude North Pacific ocean-atmosphere system. *Climate Dynamics* **47**: 1989–2007. https://doi.org/10.1007/s00382-015-2946-x.

Frankignoul C, Sennéchael N, Kwon YO, Alexander MA. 2011. Influence of the meridional shifts of the Kuroshio and the Oyashio extensions on the atmospheric circulation. *Journal of Climate* **24**: 762–777. https://doi.org/10.1175/2010JCLI3731.1.

Hoskins BJ, Valdes PJ. 1990. On the existence of storm tracks. *Journal of the Atmospheric Sciences* **47**: 1854–1864.

Lau NC, Holopainen E. 1984. Transient eddy forcing of the time-mean flow as identified by geopotential tendencies. *Journal of the Atmospheric Sciences* **41**: 313–328.

Liu Z. 2012. Dynamics of interdecadal climate variability: a historical perspective. *Journal of Climate* **25**: 1963–1995.

Mantua NJ, Hare SR, Zhang Y, Wallace JM, Francis RC. 1997. A Pacific interdecadal climate oscillation with impacts on salmon production. *Bulletin of the American Meteorological Society* **78**: 1069–1079.

Miller AJ, Cayan DR, Barnett TP, Graham NE, Oberhuber JM. 1994. The 1976–77 climate shift of the Pacific Ocean. *Oceanography* **7**: 21–26.

Nakamura M, Yamane S. 2010. Dominant anomaly patterns in the near-surface baroclinicity and accompanying anomalies in the atmosphere and oceans. Part II: North Pacific basin. *Journal of Climate* **23**: 6445–6467. https://doi.org/10.1175/2010JCLI3017.1.

Nakamura H, Sampe T, Goto A, Ohfuchi W, Xie SP. 2008. On the importance of midlatitude oceanic frontal zones for the mean state and dominant variability in the tropospheric circulation. *Geophysical Research Letters* **35**: L15709. https://doi.org/10.1029/2008GL034010.

Peng S, Whitaker JS. 1999. Mechanism determining the atmospheric response to midlatitude SST anomalies. *Journal of Climate* **12**: 1393–1408.

Qiu B, Chen S. 2005. Variability of the Kuroshio extension jet, recirculation gyre, and mesoscale eddies on decadal time scales. *Journal of Physical Oceanography* **35**: 2090–2103.

Qiu B, Chen S. 2010. Eddy-mean flow interaction in the decadally modulating Kuroshio extension system. *Deep Sea Research Part II Topical Studies in Oceanography* **57**: 1098–1110. https://doi.org/10.1016/j.dsr2.2008.11.036.

Taguchi B, Xie SP, Schneider N, Nonaka M, Sasaki H, Sasai Y. 2007. Decadal variability of the Kuroshio extension: observations and an eddy-resolving model hindcast. *Journal of Climate* **20**: 2357–2377.

Taguchi B, Nakamura H, Nonaka M, Xie SP. 2009. Influences of the Kuroshio/Oyashio extensions on air-sea heat exchanges and storm-track activity as revealed in regional atmospheric model simulations for the 2003/04 cold season. *Journal of Climate* **22**: 6536–6560. https://doi.org/10.1175/2009JCLI2910.1.

Taguchi B, Nakamura H, Nonaka M, Komori N, Kuwano-Yoshida A, Takaya K, Goto A. 2012. Seasonal evolutions of atmospheric response to decadal SST anomalies in the North Pacific subarctic frontal zone: observations and a coupled model simulation. *Journal of Climate* **25**: 111–139. https://doi.org/10.1175/JCLI-D-11-00046.1.

Trenberth KE, Hurrell JW. 1994. Decadal atmosphere-ocean variations in the Pacific. *Climate Dynamics* **9**: 303–319. https://doi.org/10.1007/BF00204745.

Wang LY, HB H, Yang XQ, Ren XJ. 2016. Atmospheric eddy anomalies associated with the wintertime North Pacific subtropical front strength and their influences on the seasonal-mean atmosphere. *Science China Earth Sciences* **59**: 2022–2036. https://doi.org/10.1007/s11430-016-5331-7.

Yao Y, Zhong Z, Yang XQ. 2016. Numerical experiments of the storm track sensitivity to oceanic frontal strength within the Kuroshio/Oyashio extensions. *Journal of Geophysical Research - Atmospheres* **121**: 2888–2900. https://doi.org/10.1002/2015JD024381.

22

Entrainment and droplet spectral characteristics in convective clouds during transition to monsoon

Sudarsan Bera,* G. Pandithurai and Thara V. Prabha

Indian Institute of Tropical Meteorology, Pune, India

Correspondence to:
S. Bera, Indian Institute of Tropical Meteorology, Dr. Homi Bhabha Road, Pashan, Pune 411008, India.
E-mail: sbera.cat@tropmet.res.in

Abstract

In situ observations in growing deep cumulus in polluted and clean environments during the transition to monsoon over Indian peninsula are used to investigate entrainment effects on droplet size distribution and the width of possible cloud cores. Pre-monsoon clouds indicate reduction in spectral width in the diluted cloud volumes due to lateral entrainment whereas monsoon cloud has higher spectral width. Relative dispersion is observed to increase with the distance from cloud core of monsoon cloud while for pre-monsoon clouds it remained almost constant. Enhanced entrainment in polluted pre-monsoon clouds is responsible for narrow cloud core compared to monsoon cloud.

Keywords: entrainment; spectral characteristics; CAIPEEX

1. Introduction

Cumulus clouds are widely prevalent over tropical region and contribute significantly to the global radiative forcing and hydrological cycle (Morrison and Grabowski, 2007). Studying their characteristics is important to understand the cloud development and warm rain process (Devenish *et al.*, 2012). Entrainment of dry air in cumulus clouds has been a subject of numerous studies due to its importance in several dynamical and microphysical processes (Paluch and Baumgardner, 1989; Grabowski, 2007). Entrainment in cumulus clouds impacts microphysical parameters significantly and aerosol pollution can modify the entrainment response on cloud properties (Jiang *et al.*, 2006). One of the fundamental problems in cloud physics is to understand the rapid onset of warm rain process (Tölle and Krueger, 2014), which can be either linked to droplet size distribution (DSD) broadening by entrainment-mixing mechanisms (Lasher-Trapp *et al.*, 2005) or to the microphysics of less diluted cloud core (Blyth *et al.*, 2005). Broadening of DSD produced by entrainment-mixing is responsible for accelerating coalescence growth (Berry and Reinhardt, 1974) and is also important for radiative properties of the clouds (Lasher-Trapp *et al.*, 2005). Representation of entrainment and impact on microphysics accurately in climate models is needed (Tölle and Krueger, 2014) and *in situ* observations has great value in this respect, especially over south Asia where these processes in monsoon clouds are less understood.

The width of cloud core decreases in polluted environment due to enhanced evaporation rate (Seigel, 2014). Possibility of evaporation-entrainment feedback in polluted cumulus clouds is also suggested (Jiang *et al.*, 2006; Small *et al.*, 2009) and lateral entrainment may impact width of cloud core where adiabatic cloud properties are preserved. It could be hypothesized that lateral entrainment dilutes the cloud edges but an inner core remained with maximum liquid water content (LWC) and droplet number concentration, which is quite different for a pre-monsoon and a monsoon cloud. So, the cloud volume can be separated in two parts: an inner core (less diluted) and a shell (highly diluted) surrounding the core. The strength of entrainment dictates the possible width of the inner core.

Entrainment broadens the DSD and produces dispersion in droplet size spectra (Lu *et al.*, 2008). Spectral width (σ) is the standard deviation of DSD and relative dispersion (ε) is the ratio of spectral width to mean radius ($<r>$) (Lu *et al.*, 2008). Dispersion effect can offset the Twomey cooling (Twomey, 1974) in polluted clouds (Liu and Daum, 2002) and sometimes up to 39% (Pandithurai *et al.*, 2012) using *in situ* observations. Relative dispersion (varies within a narrow range of 0.25−0.35 for warm convective clouds observed over Istanbul region) is not correlated with the droplet number concentration (N_d) or LWC of the clouds but the variance of relative dispersion changes with N_d and LWC (Tas *et al.*, 2015). Higher variance was associated with low N_d and LWC typically at cloud edges, due to lateral entrainment. Most of the studies discussed lateral entrainment in shallow cumulus or stratocumulus and very less attention is given to the deep cumulus with a few (3−4) kilometer depth. Shallow cumulus is dominated by cloud top entrainment (Paluch, 1979) whereas lateral entrainment is more significant in deep cumulus (Böing *et al.*, 2014). Deep cumulus that forms a significant component of the monsoon cloud systems might exhibit dilution much inside of the clouds due to larger vertical extend unlike the shallow cumulus where dilution is limited to outer cloud shell driven by buoyancy gradient (Heus and Jonker, 2008; Small *et al.*, 2009). The dynamics of deep cumulus is also greatly

Table 1. Summary of the flights (15, 16, 21 and 22 June) which are been used in the article. Number of clouds and number of samples correspond to warm cloud region only and the same have been presented in the article. Sub-cloud aerosol number concentration (N_a) is measured by PCASP. CDP observed mean (warm cloud) droplet number concentration (N_d) and droplet effective diameter (D_e) are presented here. Environmental RH corresponds to out cloud samples. Sub-cloud updraft is observed at 100 m below cloud base by AIMMS probe.

Flight date	No. of clouds	No. of samples	Cloud base (m)	N_a (cm^{-3})	N_d (cm^{-3})	D_e (µm)	RH (%) in boundary layer	RH (%) in cloudy layer	Sub-cloud updraft (m s^{-1})
15 June	14	186	2813	936.9 ± 117.3	332.7 ± 118.2	12.14 ± 1.64	59.8 ± 12.6	74.0 ± 4.1	1.77 ± 1.12
16 June	18	192	2709	1178.4 ± 226.8	501.1 ± 242.6	12.26 ± 1.48	57.2 ± 15.4	73.4 ± 3.3	1.68 ± 1.32
21 June	14	156	2452	967.5 ± 195.4	417.3 ± 169.5	13.06 ± 1.67	63.5 ± 12.3	78.4 ± 7.8	1.57 ± 1.21
22 June	15	180	2062	330.5 ± 69.3	186.5 ± 35.2	16.88 ± 2.83	71.2 ± 11.9	77.8 ± 6.4	1.82 ± 1.35

influenced by environmental thermodynamics (Prabha et al., 2011). It was also noted that rain drop formation process in these clouds is more favorable in the cloud core region (Khain et al., 2013). This study is aimed to understand the cloud edge entrainment and core dispersion characteristics while transition to monsoon happened over the Indian peninsula. The effect of entrainment on DSD and spectral dispersion in deep cumulus clouds is discussed as a function of distance from the cloud core.

2. Experiment and methodology

The study used the *in situ* cloud observations of growing cumulus taken during the Cloud Aerosol Interaction and Precipitation Enhancement Experiment phase-I (CAIPEEX-I), conducted over peninsular India during June 2009. Each flight corresponds to about 2 h of observation at 1 Hz (100 m) but only cloud penetration data are mainly used for detailed microphysical analysis. Number of clouds and number of samples used in the study are indicated in Table 1. A detailed description of the instruments and overview of the experiment is provided by Kulkarni et al. (2012). The case studies (15, 16, 21 and 22 June) considered here are deep cumulus-congestus clouds observed over Hyderabad (17.45°N, 78.46°E) during transition to summer monsoon period. The same observations are been used as representative of pre-monsoon (15, 16 June) and monsoon (22 June) clouds (Prabha et al., 2011, 2012; Khain et al., 2013; Bera et al., 2016). Cloud droplet probe (CDP) and a hot-wire sensor are used to measure droplet size and LWC, respectively. Passive cavity aerosol spectrometer probe (PCASP) measured aerosol number concentration (range 0.1–3.0 µm). Temperature and humidity are measured by AIMMS-20 probe. Several horizontal penetrations lasting few seconds are conducted at different levels at and above cloud base. The observations presented here are from growing cumulus which is conditionally probed to fulfill the objective of the experiment. The growth stage of cloud is also validated by observing incloud vertical velocity from cloud base to 1 km above and found more than 80% of the cloud penetration data having updraft, as described by Gerber et al. (2008) to select growing cumulus. Aircraft observations taken at a frequency of

1 Hz corresponds to about 100 m spatial resolution. The cloud samples were selected from the warm region of cloud, without any precipitation size drops in DSD as also verified with the help of a cloud imaging probe (as described by Prabha et al., 2011). A threshold droplet number concentration (N_d) of 10 cm^{-3} as followed by Rangno and Hobbs (2005) is used to identify the incloud samples. Adiabatic liquid water is calculated by using the linear relationship $LWC_{ad} = C_w h$, where C_w is the condensation rate and h is the height above cloud base (Brenguier, 1991). Adiabatic fraction (AF) is the ratio of LWC and LWC_{ad}. Cloud core width is determined by the cloudy region with highest LWC and AF (>0.4). Distance from highest LWC point to cloud edge ($N_d < 10$ cm^{-3}) is used to normalize the distance (between 0 and 1).

Mixing fraction of cloudy mass (χ), i.e the ratio of adiabatic cloudy mass at cloud base and total mass after entrainment (adiabatic cloudy air + dry entrained air) at height level z above cloud base, is used to estimate fractional entrainment rate (λ) by following the approach of Lu et al. (2012a) and detailed description of this methodology is provided in Appendix S1. Entrainment rate (λ) is estimated as

$$\lambda = -\frac{\ln \chi}{h}, \tag{1}$$

where χ is the mixing fraction of the cloudy air at height level h above the cloud base. We used integrated mixing fraction (χ) between cloud base and the observation level and averaged fractional entrainment rate (cf. Appendix S1 for details). Mixing fraction approach for estimating entrainment rate has advantage as measurement of incloud temperature and water vapor is not required that may have errors due to wetting of sensors. This method can be applied in remote-sensing technique to estimate entrainment rate. This approach also connects the estimation of entrainment rate directly to the microphysical effect of entrainment (Lu et al., 2014), and could be intercompared with similar estimates from remote sensing.

3. Results and discussion

The study discusses the effect of entrainment on microphysics and core width of cumulus clouds developing under varying aerosol and thermodynamic conditions (transition to monsoon). The large-scale meteorological conditions are depicted in Figure S1, Supporting Information. The wind vectors at 850-mb pressure level show that south-westerly wind dominated over Hyderabad on 22 June (as an indication of monsoon onset). Vertical variation of potential temperature (θ) and water vapor mixing ratio (q_v) from radiosonde measurements are shown in Figure S2. Boundary layer temperature decreases and moisture content increases after the onset of monsoon. It may be noted that depth of the atmospheric boundary layer (ABL) and cloud base height (Table 1) decreases as monsoon sets, compared to pre-monsoon days due to high moisture transport and low incoming solar radiation reducing surface heating. Thermodynamic and microphysics of pre-monsoon and monsoon clouds are discussed by Prabha et al. (2011) and Bera et al. (2016). The summary of the flights which is presented in Table 1 shows highest sub-cloud aerosol number concentration (N_a) on 16 June (1178 cm^{-3}) and after onset N_a reduced to 330 cm^{-3} on 22 June. Other two cases (15 and 21 June) are moderately polluted. RH in the boundary layer increases progressively with the monsoon onset although mid-layer RH remains closely same. Monsoon cloud (22 June) has the largest effective diameter and least droplet number concentration (Table 1). Sub-cloud updraft is similar ($1.5-1.8 \pm 1.2 \, \mathrm{ms^{-1}}$) in all clouds. The information of cloud samples and number of clouds during each flight used in the study is presented in Table 1.

The DSD as a function of droplet diameter (D) during the four flights is shown in Figure 1. DSD is shown for entire flight (cloud base to cloud top) which is not used for the rest of the analysis as we are focusing on warm and non-precipitating part of the clouds. The altitude of flight observation, corresponding effective diameter (D_e), distance from cloud core and spectral width of DSD are depicted on the top of each DSD at 1 Hz. Relatively small droplets ($D < 30 \, \mu m$) are observed in pre-monsoon clouds (15, 16 June) with narrow spectral width compared to the clouds on 21 and 22 June, which produced larger droplets ($D > 30 \, \mu m$) and broadened DSDs. Precipitation size droplets ($D_e > 24 \, \mu m$) are formed above 4 km altitude on 22 June (which is not considered for later part) and characterized by broader DSDs in contrast to other three clouds. The threshold for rain drops to form is $D_e > 24 \, \mu m$ is reported by Pinsky and Khain (2002) and Rosenfeld et al. (2002). The dash line (black) on the top panel of Figure 1(a)–(d) indicates the height level, below which incloud data are used for the main analysis of the study. The figure provides valuable information of spectral width as a function of distance from cloud core at different altitudes. This shows that spectral width and the distance from cloud core are in negative correlation for pre-monsoon clouds (higher spectral width at cloud core) but positively correlated for monsoon cloud (indicated vertical black line in the sub plot). Information of droplet number concentration (N_d) averaged within the warm

region of cloud is provided in Table 1. N_d is observed to be higher in all clouds except for 22 June (monsoon cloud). Entrainment of dry air and subsequent mixing dilutes the cloudy mass. Due to dilution both LWC and N_d reduces. Figure 2(a) shows observed N_d in highly diluted (AF < 0.4) and slightly diluted (AF > 0.4) cloud samples at different altitude levels. We see that N_d significantly reduces in the highly diluted region compared to slightly diluted region in case of pre-monsoon clouds but the changes are very less in monsoon cloud. Another aspect is that the variance of N_d (represented by the error bars) at different altitude levels is much higher in the highly diluted part of the pre-monsoon clouds. These variances are associated with the evaporation of droplets in the highly diluted cloud samples. Most of these cases show a continued increase in N_d above the cloud base (mainly in the slightly diluted part of the warm cloud, cf. Figure 2(a)). This is attributed to the incloud activation as already discussed in the previous study (Prabha et al., 2011). In spite of significant reduction in droplet number concentration, effective radius of droplets does not show any significant changes between highly diluted and slightly diluted cloud volumes in both pre-monsoon or monsoon clouds (figure not presented).

3.1. Impact of lateral entrainment on cloud core width

For studying entrainment effect, we have selected only long cloud passes with a minimum pass time of 6 s (600 m length) as followed by Small et al. (2009). Probability distribution of AF (Figure 2(b)) shows that polluted clouds are more adiabatic (AF > 0.7 exists in pre-monsoon clouds only) in cloud core region but the probability of higher AF (>0.4) decreases sharply which suggests a narrow convective core (samples number is less). In case of the monsoon cloud, the probability of higher AF (>0.4) remained almost constant at higher value which is in contrast to the pre-monsoon clouds where rapid decrease is seen. This result suggests that lateral entrainment impacts the core width of pre-monsoon and monsoon clouds differently. A new analysis is presented to investigate the entrainment effect on cloud core width based on maximum LWC. AF is varied as a function of distance (normalized) from highest LWC region (Figure 3). Convective core is retained with maximum LWC (or AF) and as we move from convective core to cloud edges, LWC (or AF) decreases gradually. The width of cloud core is the distance over which AF remained high (>0.4). Cloud penetrations at different altitude levels (color code represents height above cloud base) are considered with a threshold penetration time of 10 s which is about 1 km penetration length so that convective core exists in between the cloud edges. Number of clouds used in the analysis (Figure 3(a)–(d)) indicated by marker legend correspond to each cloud pass. One of the important results is a sharp reduction of AF within first 30–40% distance is observed in case of polluted pre-monsoon

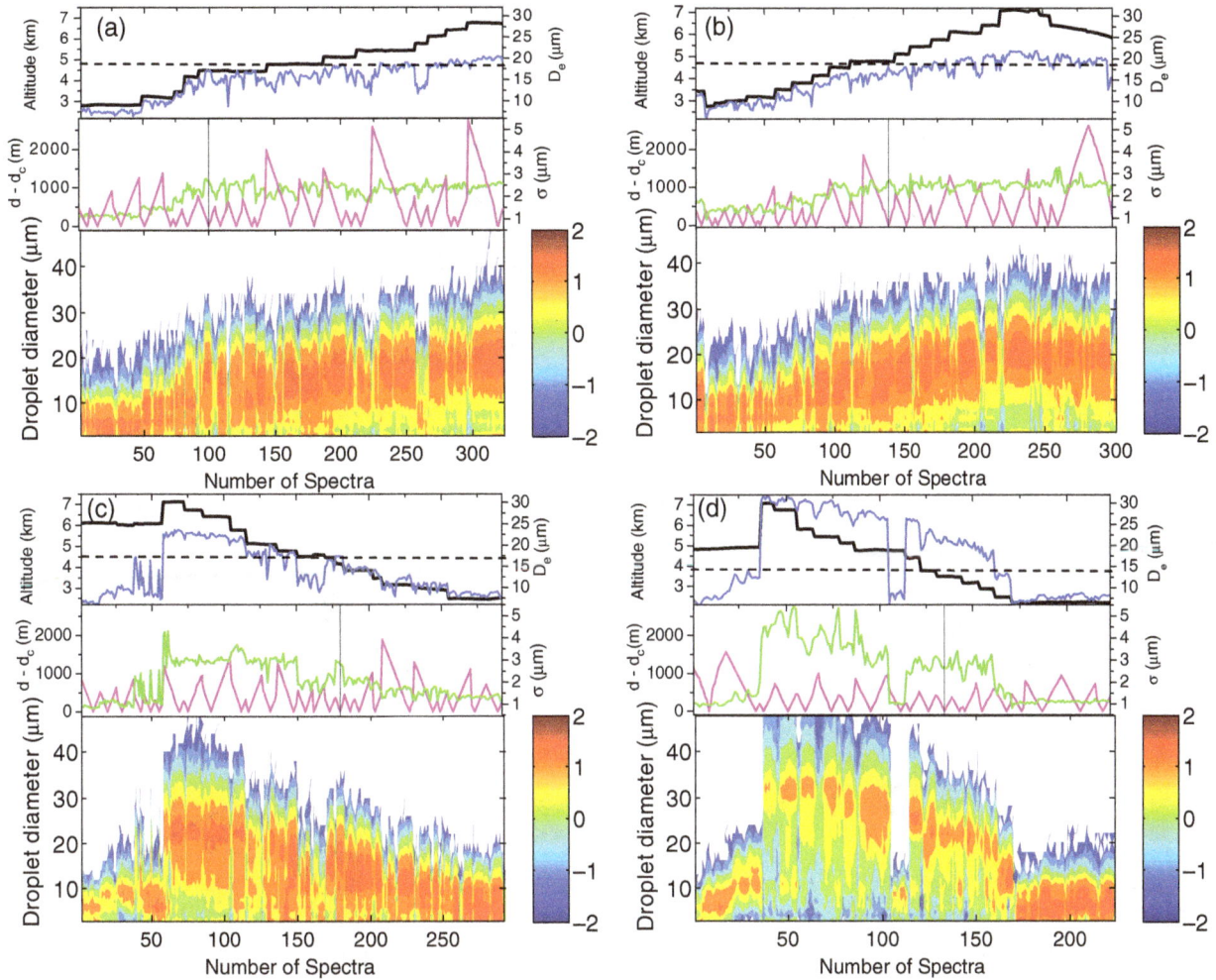

Figure 1. Droplet size distribution (DSD) of four research flight observations. (a)–(d) represents 15, 16, 21 and 22 June, respectively. Droplet number concentration per unit size (dN/dD) is presented in the log scale by the color code shown in the right panel. Altitude of observation (red line), effective diameter (D_e; blue line), distance from cloud core ($d - d_c$, magenta line) and spectral width of DSD (green line) is plotted on the top sub panel of each plot.

clouds but monsoon cloud shows a slower decrease of AF and indicates a wider core possibly due to weaker entrainment. The linear fit (black line) also indicate similar results where AF decreases more rapidly (higher negative slope) with distance from core in pre-monsoon clouds but in case of monsoon cloud, AF decreases slowly (lower negative slope) with distance from core. As the distance is normalized by core to edge distance, the actual width of cloud core may vary for different cloud penetration. For a 2 km cloud penetration length the width of cloud core will be around 600–800 m in pre-monsoon cases. It may also be noted that the AF in the core region is higher in the pre-monsoon clouds compared to the monsoon cloud. The monsoon cloud is also less adiabatic at lower elevation and more adiabatic in the core region at 2 km height above cloud base (this is observed in selected cloud penetration samples but is not an overall feature), however, in a pre-monsoon cloud, throughout the warm region, the high adiabatic region (though narrow) may be noted.

Figure 3(e)–(h) shows the PDF for AF > 0.4 and AF < 0.4 as a function of the distance from highest LWC point. The probability of higher AF (>0.4) is reduced drastically as the distance increases in case of polluted pre-monsoon clouds. This essentially implies that less-diluted region (or cloud core) exists in a very narrow region close to the highest LWC point. So, these cumuli are mostly diluted even several hundred meters inside of the cloud due to lateral entrainment. But monsoon cloud (Figure 3(h)) has relatively wider core so that higher AF (>0.4) can exists even far away from the highest LWC point.

3.2. Droplet spectral characteristics

Now, we will discuss the effect of entrainment on droplet spectral width (σ) and relative dispersion (ε), which are important parameters used in cloud models. Figure 4(a)–(d) shows joint probability distribution of spectral width and distance from highest LWC point. Spectral width in monsoon cloud is observed higher than polluted pre-monsoon clouds which is in agreement with previous studies (Prabha *et al.*, 2011, 2012; Bera *et al.*, 2016). Higher values of spectral width are also observed in a region of highest LWC (also reported by Prabha *et al.*, 2012) and as we

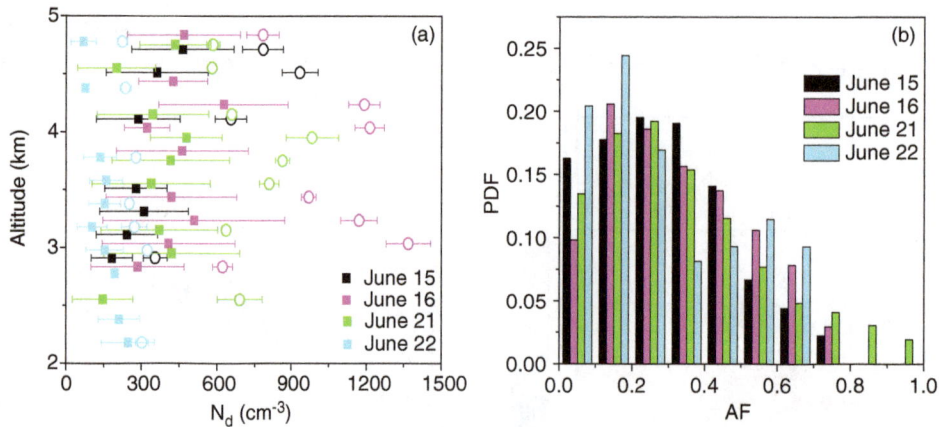

Figure 2. (a) Droplet number concentration (N_d) of four research flights (represented by different colors) for highly diluted region (solid marker) and slightly diluted region (empty marker). Error bars represents standard deviations at different height levels. (b) Probability distribution function (PDF) of AF.

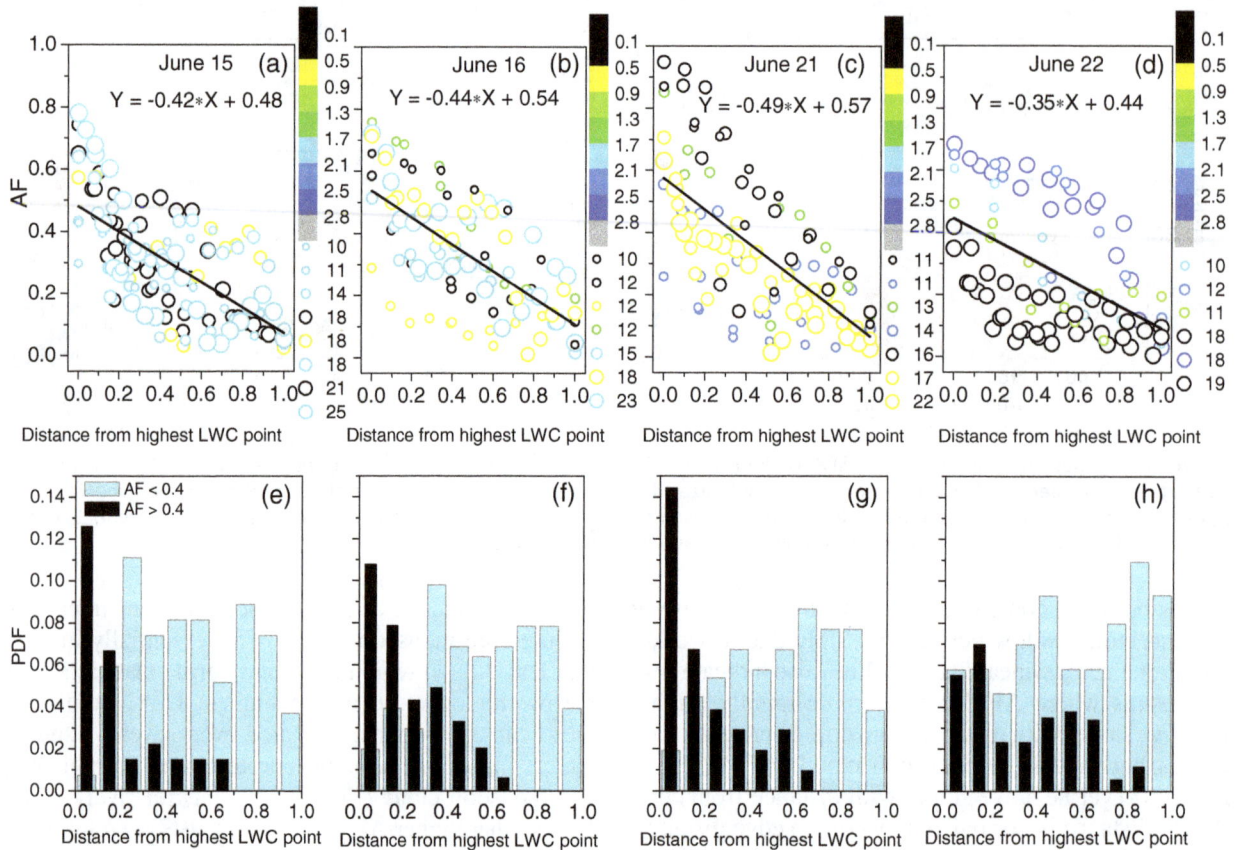

Figure 3. (a)–(d) Variation of AF with distance from highest LWC point. Here the distance is normalized by cloud core to edge distance. Color code represents height above cloud base in kilometer and the size of markers represents cloud penetration length (in seconds) as shown in the right panel. Linear fit of scatter points is represented by black straight line and corresponding linear fit equation is shown on the top. (e)–(h) PDF of AF > 0.4 (black bars) and AF < 0.4 (cyan bars) (respectively for 15, 16, 21 and 22 June) as a function of distance from highest LWC point.

move towards cloud edges spectral width decreases in all clouds except the monsoon cloud (22 June). The changes of spectral width at cloud edges (highly diluted region) of pre-monsoon clouds are mainly associated with lateral entrainment but collision-coalescence may contribute to DSD change in monsoon cloud where larger droplets are present. The result suggests opposite response (narrowing of DSD) of deep cumulus clouds

in dry pre-monsoon condition compared to the shallow cumulus where broadened DSDs are observed due to entrainment mixing (Tölle and Krueger, 2014 and reference therein). The monsoon cloud, however, shows contrasting behavior where higher spectral width is seen for samples far away from highest LWC point. The appearance of higher σ in the highly diluted cloud edge region of monsoon cloud could be caused by

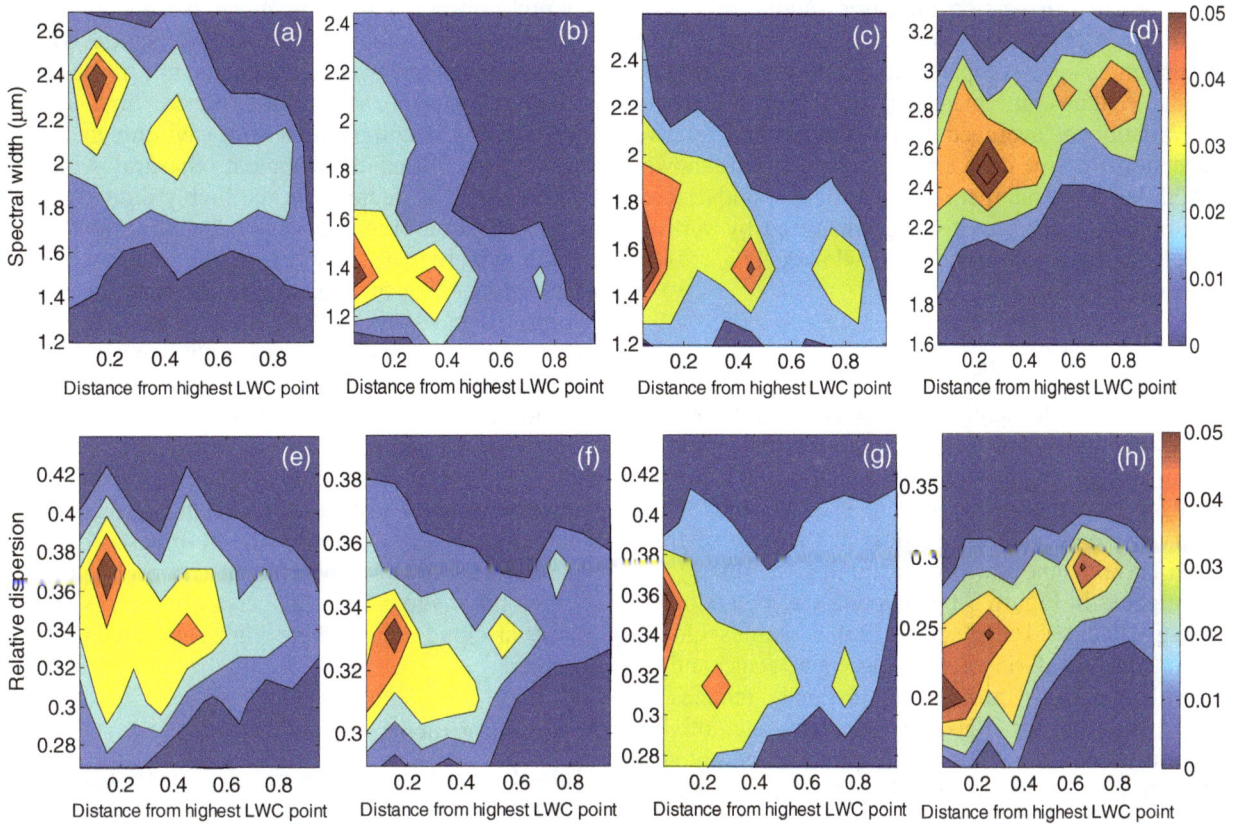

Figure 4. (a)–(d) Contour plot of joint probability distribution of spectral width (σ) and distance from highest LWC point. (e)–(h) Joint probability distribution of relative dispersion (ε) and distance from highest LWC point. Here the distance is normalized by cloud core to edge distance.

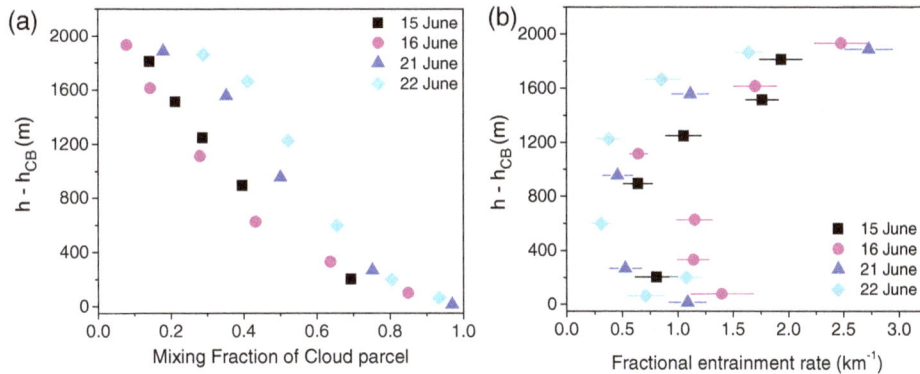

Figure 5. Vertical variation of mixing fraction of cloud parcel (a) and fractional entrainment rate (b). Error bars represent measurement uncertainties.

collision-coalescence of the larger droplets (Devenish *et al.*, 2012) which are not present in pre-monsoon clouds (Prabha *et al.*, 2011). As the environment is moist and aerosols near clouds are pre-moistened as illustrated by (Konwar *et al.*, 2015), there is partial evaporation of droplets, which broaden the spectra in downdrafts and may activate in cloud updrafts. The presence of collision-coalescence in monsoon deep convective cloud is reported by Prabha *et al.* (2011) and Khain *et al.* (2013). It is also evident from Figure 4 that probability is higher only close to highest LWC region in pre-monsoon clouds but there exists higher probability even close to cloud edge region in monsoon

cloud. Relative dispersion (Figure 4(e)–(h)) in polluted pre-monsoon clouds is higher than observed in monsoon cloud which is in agreement with the previous study by Pandithurai *et al.* (2012). Inside the cloud core relative dispersion has a range of 0.3–0.4 in pre-monsoon clouds and 0.2–0.3 in monsoon clouds. In all pre-monsoon cases, relative dispersion remained almost constant with distance from highest LWC point which is in agreement with Tas *et al.* (2015), who reported that relative dispersion does not vary with LWC or N_d. It should be pointed out that the higher variances noted are due to the cloud edge entrainment. However, monsoon cloud has distinct feature where relative dispersion increases significantly with the distance from cloud core although mean relative dispersion is less compared to pre-monsoon clouds. Similar features are presented in Figure S3 where the variation of averaged spectral width and relative dispersion with the distance from cloud core is shown. In addition to entrainment-mixing effect, cloud droplet spectral dispersion can also be influenced by vertical velocity and aerosol variations (Pawlowska *et al.*, 2006; Lu *et al.*, 2012b).

3.3. Entrainment rate

The above results suggest that entrainment impacts the cloud core width and microphysical parameters distinctly in pre-monsoon and monsoon clouds. Changes in entrainment rate are used to explain the observed cloud features. Vertical distribution of fractional entrainment rate (λ) and mixing fraction (χ) is investigated in Figure 5 by adapting the methodology as suggested by Lu *et al.* (2012a). Cumulus clouds entrain dry air during its development stage and as a result mixing fraction of cloudy mass reduces as it deepens. So, for higher entrainment rate, mixing fraction becomes smaller. Mixing fraction at any specific level above cloud base indicates the degree of mixing of cloudy volume with the environmental air in comparison with the cloud base. It is almost same at cloud base height (ideally equal to 1.0) but as depth increases clouds are getting mixed with their close environment and fraction of cloudy mass decreases. At any height level, pre-monsoon polluted clouds have lower fraction of cloudy mass. Thus the pre-monsoon clouds are relatively more mixed (χ is smaller) with the environment than the monsoon cloud. Vertical distribution of fractional entrainment rate (λ) also indicates similar results (Figure 5). Higher entrainment rate is observed in polluted pre-monsoon clouds and lowest entrainment rate is seen in less polluted monsoon cloud (22 June) at any height levels except the cloud base. The results discussed above support the argument that polluted pre-monsoon clouds exhibit stronger entrainment which impacts cloud microphysical parameters significantly. The entrainment rate observed here is comparable with the previous studies (Gerber *et al.*, 2008; Lu *et al.*, 2012a).

4. Conclusion

The entrainment and spectral characteristics of clouds during the transition to monsoon period over southern peninsular region were investigated using *in situ* airborne observations. The entrainment of unsaturated environmental air decreases N_d significantly in the highly diluted regions compared to less diluted volume in pre-monsoon cases. Lateral entrainment produces higher variance of N_d due to droplet evaporation in most diluted part. Analysis based on the highest LWC point indicates that less-diluted region of

the pre-monsoon clouds is much narrow compared to the monsoon cloud where a wider cloud core is found. Enhancement of entrainment rate is observed in polluted pre-monsoon clouds where high aerosol concentration in combination with environmental conditions favor cloud development. Spectral width of DSD is found to decrease in the highly diluted region of pre-monsoon clouds associated with droplet evaporation which is in contrast to the shallow clouds where broadening of DSD due to entrainment-mixing is reported in many previous studies. But after transition to monsoon, it is observed that spectral width increases in the highly diluted part which can be due to collision-coalescence process active in monsoon cloud as well as the partial evaporation and activation (as shown by Prabha *et al.*, 2011). Relative dispersion (ε) remained nearly constant with varying distance from cloud core in case of polluted pre-monsoon clouds but opposite effect is observed in less polluted monsoon cloud where relative dispersion increases significantly at cloud edges where higher spectral width is observed.

Acknowledgements

IITM and the CAIPEEX projects are fully funded by the Ministry of Earth Sciences (MoES), Government of India, New Delhi. Authors acknowledge the dedication of many IITM scientists in carrying out CAIPEEX experiment and data collection.

Supporting information

The following supporting information is available:

Appendix S1. Calculation of entrainment rate.

Figure S1. Contour plot of sea level pressure (SLP) and wind vector at 850 mb pressure level. The location of experiment (Hyderabad) is marked by black square.

Figure S2. Vertical profile of environmental potential temperature (θ) and water vapor mixing ratio (q_v) from radiosonde observation.

Figure S3. Variation of (a) averaged spectral width, (b) averaged relative dispersion, (c) standard deviation of spectral width and (d) standard deviation of relative dispersion as a function of distance from highest LWC point. Four flights are represented by different colors and markers as indicated in (a).

References

Bera S, Prabha TV, Grabowski WW. 2016. Observations of dynamics-microphysics interactions in convective clouds over the Indian sub-continent (under review).

Berry EX, Reinhardt RL. 1974. An analysis of cloud drop growth by collection. Part II: single initial distributions. *Journal of Atmospheric Science* **31**: 1825–1831.

Böing SJ, Jonker HJJ, Nawara WA, Siebesma AP. 2014. On the deceiving aspects of mixing diagrams of deep cumulus convection. *Journal of Atmospheric Science* **71**: 56–68.

Blyth AM, Lasher-Trapp SG, Cooper WA. 2005. A study of thermals in cumulus clouds. *Quarterly Journal of the Royal Meteorological Society* **131**: 1171–1190.

Brenguier J-L. 1991. Parameterization of condensation process: a theoretical approach. *Journal of Atmospheric Science* **48**: 264–282.

Devenish BJ, Bartello P, Brenguier J-L, Collins LR, Grabowski WW, IJzermans RHA, Malinowski SP, Reeks MW, Vassilicos JC, Wang L-P, Warhaft Z. 2012. Droplet growth in warm turbulent clouds. *Quarterly Journal of the Royal Meteorological Society* **138**: 1401–1429.

Gerber H, Frick G, Jensen J, Hudson J. 2008. Entrainment, mixing, and microphysics in trade-wind cumulus. *Journal of the Meteorological Society of Japan* **86A**: 87–106.

Grabowski WW. 2007. Representation of turbulent mixing and buoyancy reversal in bulk cloud models. *Journal of Atmospheric Science* **64**: 3666–3680.

Heus T, Jonker HJJ. 2008. Subsiding shells around shallow cumulus clouds. *Journal of Atmospheric Science* **65**: 1003–1018, doi: 10.1175/2007JAS2322.1.

Jiang H, Xue H, Teller A, Feingold G, Levin Z. 2006. Aerosol effects on the lifetime of shallow cumulus. *Geophysical Research Letters* **33**: L14806.

Khain A, Prabha TV, Benmoshe N, Pandithurai G, Ovchinnikov M. 2013. The mechanism of first raindrops formation in deep convective clouds. *Journal of Geophysical Research* **118**: 9123–9140.

Konwar M, Panicker AS, Axisa D, Prabha TV. 2015. Near-cloud aerosols in monsoon environment and its impact on radiative forcing. *Journal of Geophysical Research: Atmospheres* **120**: 1445–1457, doi:10.1002/2014JD022420.

Kulkarni JR, Maheshkumar RS, Morwal SB, Padma Kumari B, Konwar M, Deshpande CG, Joshi RR, Bhalwankar RV, Pandithurai G, Safai PD, Narkhedkar SG, Dani KK, Nath A, Nair S, Sapre VV, Puranik PV, Kandalgaonkar S, Mujumdar VR, Khaladkar RM, Vijayakumar R, Thara VP, Goswami BN. 2012. The Cloud Aerosol Interactions and Precipitation Enhancement Experiment (CAIPEEX): overview and preliminary results. *Current Science* **102**: 413–425.

Lasher-Trapp SG, Cooper WA, Blyth AM. 2005. Broadening of droplet size distributions from entrainment and mixing in a cumulus cloud. *Quarterly Journal of the Royal Meteorological Society* **131**: 195–220.

Liu Y, Daum PH. 2002. Indirect warming effect from dispersion forcing. *Nature* **419**: 580–581.

Lu M-L, Feingold G, Jonsson HH, Chuang PY, Gates H, Flagan RC, Seinfeld JH. 2008. Aerosol-cloud relationships in continental shallow clouds. *Journal of Geophysical Research* **113**: D15201.

Lu C, Liu Y, Yum SS, Niu S, Endo S. 2012a. A new approach for estimating entrainment rate in cumulus clouds. *Geophysical Research Letters* **39**: L04802.

Lu C, Liu Y, Niu S, Vogelmann AM. 2012b. Observed impacts of vertical velocity on cloud microphysics and implications for aerosol indirect effects. *Geophysical Research Letters* **39**: L21808.

Lu C, Liu Y, Niu S, Endo S. 2014. Scale dependence of entrainment-mixing mechanisms in cumulus clouds. *Journal of Geophysical Research: Atmospheres* **119**: 13,877–13,890.

Morrison H, Grabowski WW. 2007. Comparison of bulk and bin warm-rain microphysics models using a kinematic framework. *Journal of Atmospheric Science* **64**: 2839–2861.

Paluch IR. 1979. The entrainment mechanism in Colorado cumuli. *Journal of Atmospheric Science* **36**: 2467–2478.

Paluch IR, Baumgardner DG. 1989. Entrainment and fine scale mixing in a continental convective cloud. *Journal of Atmospheric Science* **46**: 261–278.

Pawlowska H, Grabowski WW, Brenguier J-L. 2006. Observations of the width of cloud droplet spectra in stratocumulus. *Geophysical Research Letters* **33**: L19810.

Pandithurai G, Dipu S, Prabha TV, Maheskumar RS, Kulkarni JR, Goswami BN. 2012. Aerosol effect on droplet spectral dispersion in warm continental cumuli. *Journal of Geophysical Research* **117**: D16202.

Pinsky M, Khain A. 2002. Effects of in-cloud nucleation and turbulence on droplet spectrum formation in cumulus clouds. *Quarterly Journal of the Royal Meteorological Society* **128**(580): 501–533.

Prabha TV, Khain A, Maheskumar RS, Pandithurai G, Kulkarni JR, Konwar M, Goswami BN. 2011. Microphysics of pre-monsoon and monsoon clouds. *Journal of Atmospheric Science* **68**: 1882–1901.

Prabha TV, Patade S, Pandithurai G, Khain A, Axisa D, Pradeep Kumar P, Maheshkumar RS, Kulkarni JR, Goswami BN. 2012. Spectral width of pre-monsoon and monsoon clouds over Indo-Gangetic valley. *Journal of Geophysical Research* **117**: D20205.

Rangno AL, Hobbs PV. 2005. Microstructures and precipitation development in cumulus and small cumulonimbus clouds over the warm pool of the tropical Pacific Ocean. *Quarterly Journal of the Royal Meteorological Society* **131**: 639–673.

Rosenfeld D, Lahav R, Khain A, Pinsky M. 2002. The role of sea spray in cleansing air pollution over ocean via cloud processes. *Science* **297**(5587): 1667.

Seigel RB. 2014. Shallow cumulus mixing and subcloud-layer responses to variations in aerosol loading. *Journal of Atmospheric Science* **71**: 2581–2603.

Small JD, Chuang PY, Feingold G, Jiang H. 2009. Can aerosol decrease cloud lifetime? *Geophysical Research Letters* **36**: L16806.

Tas E, Teller A, Altaratz O, Axisa D, Bruintjes R, Levin Z, Koren I. 2015. The relative dispersion of cloud droplets: its robustness with respect to key cloud properties. *Atmospheric Chemistry and Physics* **15**: 2009–2017.

Tölle MH, Krueger SK. 2014. Effects of entrainment and mixing on droplet size distributions in warm cumulus clouds. *Journal of Advances in Modeling Earth Systems* **6**: 281–299, doi: 10.1002/2012MS000209.

Twomey S. 1974. Pollution and planetary albedo. *Atmospheric Environment* **8**: 1251–1256.

23

Trends in extreme precipitation indices across China detected using quantile regression

Lijun Fan[1,*] and Deliang Chen[2]

[1] Key Laboratory of Regional Climate-Environment Research for Temperate East Asia (RCE-TEA), Institute of Atmospheric Physics, Chinese Academy of Sciences, Beijing, China
[2] Regional Climate Group, Department of Earth Sciences, University of Gothenburg, Sweden

*Correspondence to:
L. Fan, Room 212, 40# Hua Yan Li, Qi Jia Huo Zi, Institute of Atmospheric Physics, Chinese Academy of Sciences, Chao Yang District, P. O. Box 9804, Beijing 100029, China. E-mail: fanlj@tea.ac.cn

Abstract

For China, long-term changes are detected not only in the means of eight extreme precipitation indices, but also in their distribution shapes by quantile regression. This resulted in different trends for the means and other aspects of the index distributions. The differences between changes in the means and upper/lower extremes vary with region and index. A noteworthy feature is that changes in upper tails of the index distributions across a broad area, especially in the south, are at a much higher rate than mean trends estimated by the traditional linear regression model. This has practical implications for disaster risk management.

Keywords: extreme precipitation indices; quantile trend; quantile regression; China

1. Introduction

In recent decades, changes in climatic extremes have attracted widespread attention, owing to their huge impact on human life, environment, economy, and society (IPCC, 2012; Chen et al., 2015). Numerous studies of trends in extreme precipitation events have been made for various regions of the world (Jones et al., 1999; Klein Tank et al., 2006; Vincent and Mekis, 2006; Donat et al., 2013).

In China, a number of studies on extreme precipitation indices have been conducted (Wang and Zhou, 2005; Zhai et al., 2005; Fu et al., 2008; You et al., 2011). These studies have reported trends in several precipitation indices over the last few decades. Annual total precipitation, average wet-day precipitation, maximum 1- and 5-day precipitation, and number of heavy precipitation days show weak increasing trends, whereas a decreasing trend has been observed for number of consecutive dry days (CDD). For all precipitation indices except number of CDD, stations in the Yangtze River Basin and southeastern and northwestern China have the largest positive trends, whereas stations in the Yellow River Basin, central China, and the Sichuan Basin have the largest negative trends.

Most of the studies above have paid much attention to the mean trend of extreme precipitation indices, e.g. that estimated by linear regression models based on ordinary least squares regression (LSR). Although LSR is a commonly used method in studying linear trends, they have limitations. For example, they only provide information on linear trend of the mean condition of the indices, but trends of other aspects (e.g. upper tails) of the index distributions, which are generally more valuable than the mean trend in climate risk studies (Tareghian and Rasmussen, 2012), cannot be properly addressed.

Quantile regression (QR) was developed as an extension to LSR (Koenke and Basset, 1978). This method has the ability to estimate slopes of changes not only in the mean but in all parts of the distribution of a time series. Therefore, this method can provide a more complete picture of long-term temporal trend in those series (Barbosa, 2008).

In contrast to past research, this study focuses on investigating slopes of trends in different quantiles of the conditional distributions of eight extreme precipitation indices in China, particularly those in the upper tails of those distributions, using QR. Although this method has been recently introduced in climate change studies (Barbosa, 2008; Tareghian and Rasmussen, 2012; Lee et al., 2013; Wasko and Sharma, 2014), to the author's knowledge, its application to extreme precipitation indices in China is new. Comparison between quantile and mean trends of the indices can aid understanding of the merits of QR. More importantly, application of this method is helpful for comprehensive understanding of long-term trends of precipitation extremes in China, in terms of extent and magnitude. Furthermore, investigation of changes in upper tails of the distributions of precipitation extremes, which often cause great climate disasters, is more interesting and useful in climate risk assessment, in comparison with the central tendency of precipitation extremes.

The structure of this article is as follows. The data and method used are described in Section 2. Results of quantile trends are presented in Section 3. Key findings are summarized in Section 4.

2. Data and method

2.1. Data

An observed daily precipitation data set from 825 stations in China for the period 1961–2013 was collected from the China Meteorological Data Sharing Service System. This data set is quality controlled, including check and correction of internal and spatial consistency and potential outliers (http://data.cma.cn for details). A homogeneous long-term climate series is the basis in climate change research. It is defined as one time series where variations are caused by variations in weather and climate only. Unfortunately, most climatic time series are affected by a number of non-climatic factors (e.g. instrument change, different observing practices, and station relocations) that make these data unrepresentative of the actual climate variation with time (Peterson et al., 1998). Zhang et al. (2012) found that homogeneity of the monthly precipitation data is satisfactory at most stations in China using the RHtest method established by Wang et al. (2007) to be used to test the homogeneity of yearly, monthly, and daily time series and to adjust the series with lag-1 autocorrelation error and multiple change points. They also found that relocation of stations is one of the main causes of inhomogeneity in the precipitation data. Wijngaard et al. (2003) concluded that inhomogeneity could be caused by true local climate variations and may impact the analysis of trends and variability in climatic extremes. Considering that our analysis concentrates on upper-tail trends of precipitation extremes, we only removed stations with obvious relocation and incomplete time series for the period 1961–2013. Finally, we retained a total of 578 stations for use in the analysis.

2.2. Definition of extreme precipitation indices

Eight indices of precipitation extremes, which are recommended by the Expert Team on Climate Change Detection Monitoring and Indices (Klein Tank et al., 2006), were selected. These are numbers of days with heavy [Precipitation (Pr) > 10 mm] and very heavy (Pr > 20 mm) precipitation (R10MM and R20MM), maximum numbers of consecutive wet days (CWD) (Pr ≥ 1 mm) and CDD, total precipitation on very wet (Pr > 95th percentile) and extreme wet (Pr > 99th percentile) days (R95PTOT and R99PTOT, mm), and finally, maximum 1- and 5-day precipitation (RX1DAY and RX5DAY, mm). A detailed description of the indices is in Donat et al. (2013). The indices were calculated using the RclimDex software package, developed at the Meteorological Service of Canada (available from http://etccdi.pacificclimate.org/software.shtml). The 95th and 99th percentiles of precipitation are calculated during the period of 1961–1990, which is defined as the latest global normal period currently used for climate reference by World Meteorological Organization (http://public.wmo.int for details). The indices were calculated on an annual basis at each station.

2.3. Quantile regression

QR is generally regarded as an extension of LSR. Within the LSR framework, the response variable (Y) is linearly related to time (t): $Y = f(\alpha, \beta, t) = \alpha t + \beta$, with α denoting the linear slope and β the constant intercept. That is, the variable Y can be described as a function of time t and parameters α and β. The two parameters are assessed from the ordinary least squares estimate of the expected value of response variable Y conditional on t [$E(Y|t)$], and thereby calculated by minimizing the sum of squared residuals

$$\sum_i \left[y_i - f\left(\alpha, \beta, t_i\right) \right]^2 \qquad (1)$$

Let Y be a random variable with cumulative distribution function $F_y(y)$. The τth quantile of Y is defined $Q_Y(\tau)$ such that $P[Y \leq Q_Y(\tau)] = \tau$, where, $\tau \in [0, 1]$. The quantile function $Q_Y(\tau)$ is defined as $Q_Y(\tau) = F_Y^{-1}(\tau)$. Considering the conditional distribution Y given $T=t$, the conditional quantile function is defined as $Q_Y(\tau|t)$. Therefore, QR can be interpreted in a similar way, by replacing $E(Y|t)$ with the quantile of the response variable $Q_Y(\tau|t)$. For quantile τ, the linear QR model can be written as

$$Y(\tau|t) = f'\left(\alpha_\tau, \beta_\tau, t\right) = \alpha_\tau t + \beta_\tau + \xi \qquad (2)$$

where α_τ is the quantile slope and β_τ is the intercept for each τ. ξ is the error with the expectation of zero. Instead of LSR, the two parameters are estimated from the conditional quantile function by minimizing the sum of asymmetrically weighted absolute residuals.

$$\sum_i \rho_\tau \left[y_i - f'\left(\alpha_\tau, \beta_\tau, t_i\right) \right] \qquad (3)$$

where ρ_τ is the tilted absolute value function (Barbosa, 2008).

Detailed description of the QR method can be found in the literature (Barbosa, 2008; Tareghian and Rasmussen, 2012; Lee et al., 2013).

2.4. Trend analysis

The QR method was used to estimate slopes of trends in quantiles 0.01–0.99 (in steps of 0.01) of the annual extreme precipitation indices for all stations. The 'quantreg' package within R software was run for QR analysis. The mean trend was calculated by LSR, based on ordinary least squares. Confidence intervals (CIs) for the slopes of mean and quantiles were assessed using ordinary bootstrap method. The adjusted bootstrap percentile intervals were generated using the 'boot' package within R software (Davison and Kuonen, 2002). Statistical significances were defined using whether zero lies in the CI at the 5% level. Regional time series of the indices were calculated as their arithmetic average over all the stations. Trends in quantiles 0.05, 0.50 and 0.95 for each index were compared with the mean trend to investigate their similarities and differences. Finally, we chose slopes

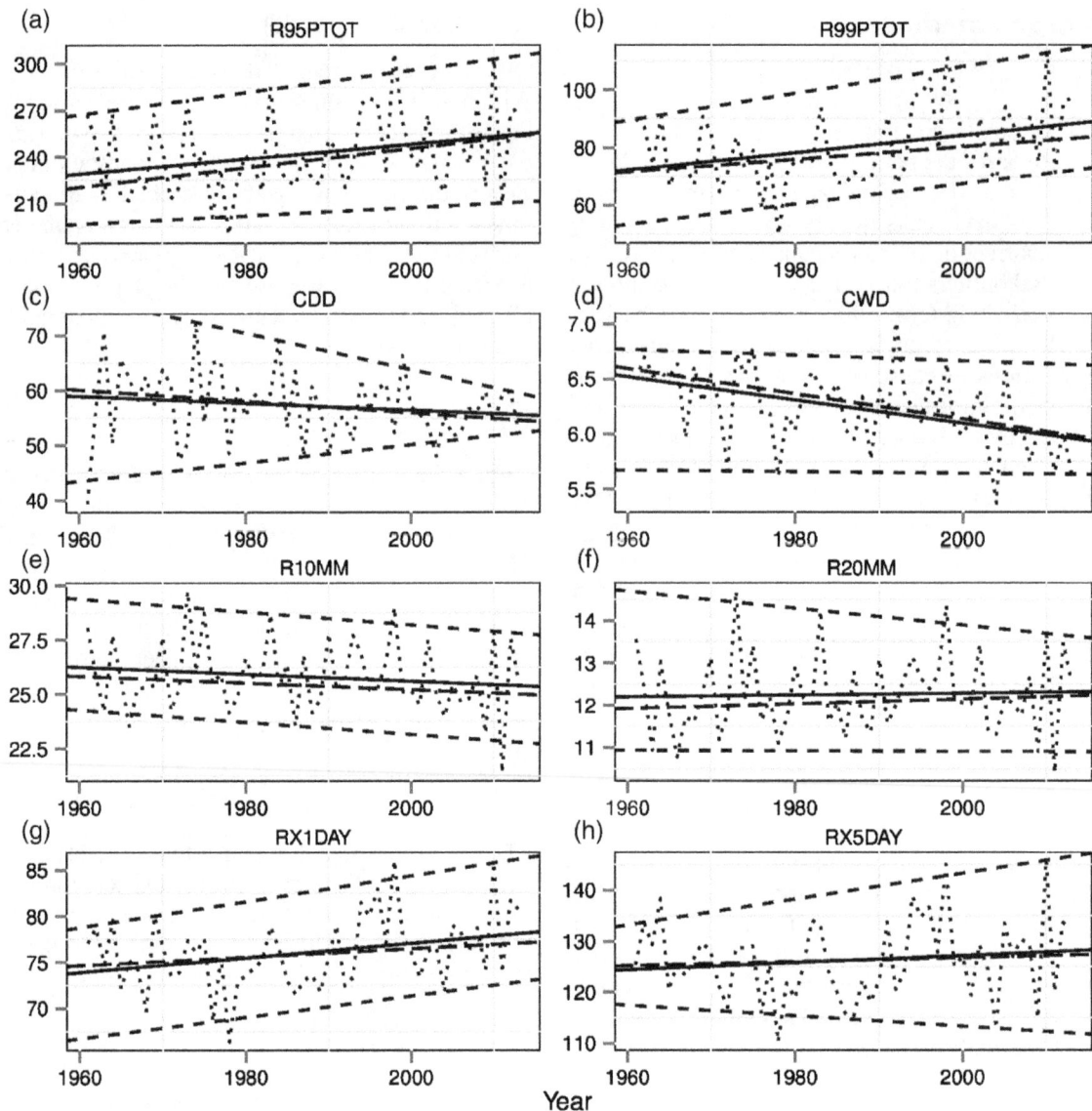

Figure 1. Regional mean time series (dotted curves) of eight extreme precipitation indices for all China, and their trends of the mean (solid lines) and 0.05 (lower dashed lines), 0.50 (long dash lines), and 0.95 (upper dashed lines) quantiles. Units of vertical axis in panels (a), (b), (g), and (h) are mm; panels (c)–(f) are days.

in quantile 0.95 to illustrate spatial pattern of slope of trend in the upper tail of the distribution of each index. Here, all calculations are done at the annual time step.

3. Results

3.1. Quantile trends of regional precipitation indices

Figure 1 displays regional mean time series of the eight extreme precipitation indices in China, along with mean trends estimated by LSR and trends in 0.05 (lower), 0.50 (median), and 0.95 (upper) quantiles estimated by QR. These correspond respectively to 5% (minima), 50% (median), and 95% (maxima) of the ordered observations. Table 1 lists trends in Figure 1 and their corresponding significance levels. The mean and median

trends can be understood as a measurement of central tendency for the indices, whereas those of the lower and upper quantiles reflect long-term trend in the lower and upper tails of the conditional distributions of the indices. As can be seen in the figure, trends of the median for all indices are very close to those of the mean, whereas the upper and lower quantiles show significantly different trends from the mean slopes. Specifically, for very wet days (R95PTOT) and extremely wet days (R99PTOT), there are increasing trends in the upper quantile, with larger slopes than those of the lower, median, and mean quantiles (Figure 1(a) and (b)). For consistency of dry days, CDD shows no substantial change in central tendency with time, but there is evidence that its distribution narrows (Figure 1(c)). For CWD, there is a significantly decreasing trend in the median and mean. However, trend lines in the lower and upper quantiles are almost parallel, suggesting that although the mean

Table I. Quantile and mean trends of precipitation indices during 1961–2013 for China as a whole.

Indices	Units	Mean	Trends Quantiles 0.05	0.50	0.95
R95PTOT	mm decade^{-1}	4.83	2.66	6.27	**7.16**
R99PTOT	mm decade^{-1}	**3.03**	**3.54**	2.20	**4.68**
CDD	days decade^{-1}	−0.63	1.66	−1.04	**−3.30**
CWD	days decade^{-1}	**−0.11**	−0.01	**−0.11**	−0.03
R10MM	days decade^{-1}	−0.16	−0.28	−0.15	−0.30
R20MM	days decade^{-1}	0.02	−0.01	0.06	−0.20
RX1DAY	mm decade^{-1}	**0.80**	**1.16**	0.46	**1.40**
RX5DAY	mm decade^{-1}	0.69	−1.03	0.40	**2.54**

Significant trends (at the 5% significance level) are indicated in bold.

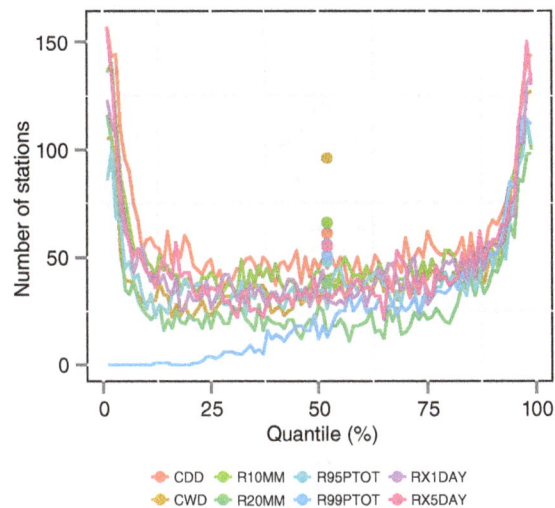

Figure 2. Number of stations at which trends in the mean (point) and quantiles 0.01–0.99 in increments of 0.01 (line) for an individual index are statistically significant at the 5% level.

trend decreased, trends in the extreme quantiles did not change dramatically (Figure 1(d)). For both RX1DAY and RX5DAY, the upper-quantile slopes are significant and greater than those of the median and mean (Figure 1(g) and (h)). For R10MM and R20MM, there are no significant changes in the mean and quantiles (Figure 1(e) and (f)).

3.2. Number of stations with significant trends on quantile

If trend of an index in a given quantile for a given station is significant at the 5% level, we consider that the index in this quantile has a significant impact on this station. The more the number of stations with significant trends, the wider the impact of this index in a given quantile on the country. Thus, we investigated the dependence of number of stations with significant trends on quantile for each index. The QR method was implemented for a range of quantiles 0.01–0.99 (in steps of 0.01) for all stations and indices. As can be seen in Figure 2, number of stations at which trends are statistically significant (at the 5% significance level) varies by quantile. In particular, for quantiles greater than 0.90 or less than 0.10, number of stations rapidly increases with quantile. Compared with ~50 stations having significant mean trends (dots), upper quantiles show significant trends at ~150 stations. This finding implies that change in the upper and lower quantiles of the distributions of the extreme precipitation indices has a more widespread influence than that of their means across the country.

3.3. Spatial patterns of upper-quantile trends of precipitation extremes

In this section, we examine spatial patterns of upper-quantile trends of the conditional distributions of the indices. Although long-term changes in quantile 0.99 of the index distributions has a much wider influence over the entire region (Figure 2), we used quantile 0.95 to reflect changes in upper tails of the index distribution shapes to obtain statistically stable and reliable results. Upper-quantile trend slopes

were compared with mean trends to examine their similarities and differences in terms of spatial pattern. To facilitate description of our result, we separated China into seven regions: northwestern (NW), southwestern (SW), southern (S), central (C), eastern (E), northern (N), and northeastern (NE) (Figure 3(a)).

For R95PTOT and R99PTOT, upper-quantile and mean trends display consistently spatial patterns across the country. Increasing trends were detected in northwestern, northeastern, eastern and southern China. There are declining trends in northern and central China and the western part of southern China, in line with You *et al.* (2011). However, upper-quantile trends have greater magnitudes and are statistically significant at more stations than the mean trends. The strongest increasing trend in upper quantile was seen in eastern and southern China, with magnitudes up to 200 mm decade^{-1} for R95PTOT (Figure 3(a) and (b)), and 150 mm decade^{-1} for R99PTOT. Comparing these with the mean trend for R95PTOT (<50 mm decade^{-1}), we found that rates of changes in upper quantile greatly strengthened. Since most Chinese people are living in the east, this significant increasing trend of the most sever extreme precipitation has put a great pressure on the society. On the other hand, it also raises an important scientific question about the causes of this change and how this will evolve in the future. A recent study (Ou *et al.*, 2013) compared the observed trends for the means in a number of extreme precipitations in China, which reveals that while the temporal trend in the extreme precipitation for western China is well captured by most models, the trends of the extreme precipitation in eastern China are poorly captured by most models. This is especially true for the so-called southern flood and northern drought pattern. Eastern China is strongly influenced by Asian summer monsoon and human activities. Further studies are needed to identify processes that are responsible for this change and to reliably project the future changes.

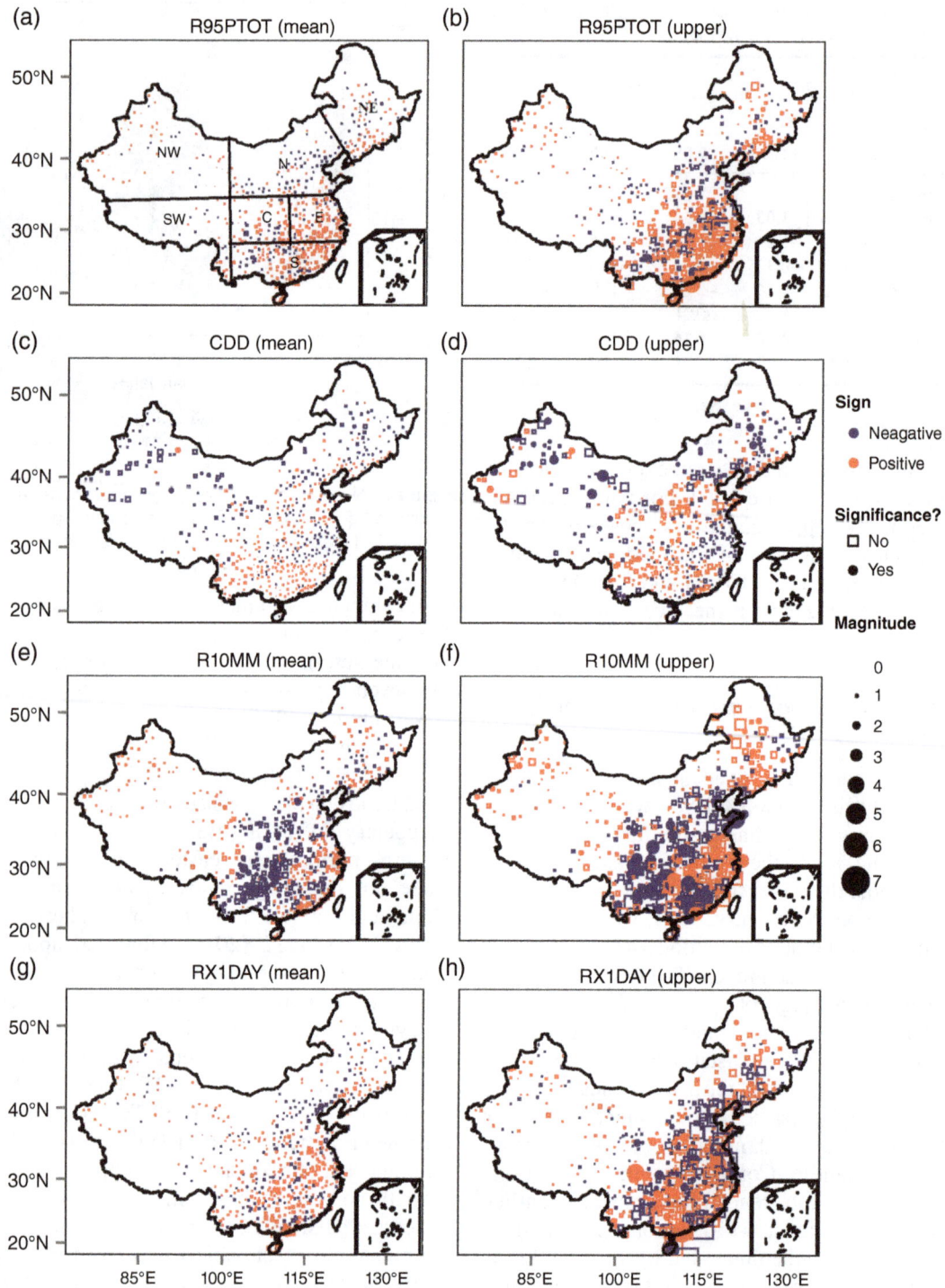

Figure 3. Comparison of mean and upper (0.95) quantile trends for four precipitation indices. Units of panels (a) and (b) are 50 mm decade^{-1}; (c) and (d) 10 days decade^{-1}; (e) and (f) days decade^{-1}; (g) and (h) 10 mm decade^{-1}. To facilitate discussion, the region is separated into seven subregions in panel (a). Positive and negative values denote increasing and decreasing trends, respectively.

For CDD, there are decreasing trends in the upper quantile in northwestern, northeastern, and eastern regions while there are increasing trends in northern, central, and southern regions, in accordance with the spatial distribution of mean trends estimated by LSR. However, significant changes were detected by LSR only at stations in the northwest and northeast. There are larger magnitudes in upper-quantile trends at most stations in China, notably in the northwest, with trend ~20 days decade^{-1}, compared to the mean trend with <10 days decade^{-1} (Figure 3(c) and (d)). For CWD, the differences are minor between the upper-quantile and mean trends in the north, where weak or no trends were observed. Interestingly, great differences are visible between these in the central and south areas. There are decreasing trends in central and southern China and rising trends in eastern China for the two types of trends. However, the largest trends in the upper quantile are up to 4 days decade^{-1}, relative to those of the mean (<1 day decade^{-1}).

For R10MM, pronounced decreasing trends were detected by LSR, mainly in central China and the western part of southern China, at a rate of ~2 days decade^{-1}. For the upper quantile, there are larger negative trends (~5 days decade^{-1}), mainly in central China and the western part of southern China, and positive ones (~5 days decade^{-1}) in eastern China (Figure 3(e) and (f)). The R20MM results are similar to those of R10MM, but with weaker trends (~3 days decade^{-1}) in the upper quantile.

For RX1day, there are no evident trends in the mean state. In contrast, the upper quantile has much stronger positive (in central and southern China) and negative (in northern China and eastern coast of eastern China) trends, with magnitudes up to 70 mm decade^{-1}, along with more stations with significant change (Figure 3(g) and (h)). The findings for RX5day agree with those of RX1day, but trends in the upper quantile reach 100 mm decade^{-1} for the former.

4. Summary and conclusion

The QR method was used to study trends in various quantiles of the conditional distributions of eight indices of precipitation extremes in China. Major findings are as follows.

1. The distributions of precipitation extremes in China have experienced variable changes in recent decades. For China as a whole, the entire distribution of CDD narrows with time, while that of RX5DAY widens. CWD shows a remarkable decline in the mean, but no obvious trends in the upper and lower quantiles. R95PTOT, R99PTOT, and RX1DAY show increases in the mean and three quantiles, with the upper quantile having a much higher trend. R10MM and R20MM do not have significant trend in the mean and three quantiles.

2. The number of stations with significant trends varies depending on quantile for each index. More stations with obvious changes were detected in the lower and upper tails of the distributions of the indices. This suggests that change in the extreme parts of the distributions of precipitation extremes occurs over a much broader region of China than those of the mean.

3. Change in maxima of the indices exhibits spatial patterns coherent with those of the means estimated by LSR. However, rates of such change are dramatically higher than those of the mean. The strengthened magnitude in rate varies by index and region. There are decreasing trends of CDD in northwestern, northeastern, and eastern China, and increasing ones in northern, central, and the western region of southern China. Region-wide, there are larger magnitudes of upper-quantile trends of CDD than those

of the mean. There are increases for the other seven indices in northwestern, northeastern, and eastern China, but declines in northern, central, and southern China. However, the most remarkable changes in the rate of the upper-quantile trend concentrate in the south of China, relative to that of the mean trend.

Compared with LSR, the QR method provides a more complete picture of long-term trend of the conditional distributions of precipitation extremes. Change of precipitation extremes appears not only in the mean but also in various parts of their distributions. This indicates that analysis of the mean variability of extreme climate events is far from adequate in the study of extreme climate change. Given its advantage over LSR, QR should be considered for research into long-term climatic trend and risk assessment.

Our findings reveal that changes in upper extremes of the distributions of the extreme precipitation indices, which very often cause serious weather and climate risks, have occurred in a broader area of China and at a much higher rate than previously believed. No matter what processes are behind this change, we need to take the identified differences between upper-quantile and mean long-term trend of precipitation extremes seriously.

Acknowledgements

This work was jointly sponsored by the National Basic Research Programme of China '973' Programme (2012CB956203) and National Natural Science Foundation of China (41275110).

References

Barbosa SM. 2008. Quantile trends in Baltic Sea level. *Geophysical Research Letters* 35: L22704, doi: 10.1029/2008GL035182.

Chen D, Walther A, Moberg A, Jones PD, Jacobeit J, Lister D. 2015. *European Trend Atlas of Extreme Temperature and Precipitation*. Springer: Dordrecht; Heidelberg; New York, NY, and London; 178, doi: 10.1007/978-94-017-9312-4.

Davison AC, Kuonen D. 2002. An introduction to the bootstrap with applications in R. *Statistical Computing & Statistical Graphics Newsletter* 13(1): 6–11.

Donat MG, Alexander LV, Yang H, Durre I, Vose R, Dunn RJH, Willett KM, Aguilar E, Brunet M, Caesar J, Hewitson B, Jack C, Klein Tank AMG, Kruger AC, Marengo J, Peterson TC, Renom M, Oria Rojas C, Rusticucci M, Salinger J, Elrayah AS, Sekele SS, Srivastava AK, Trewin B, Villarroel C, Vincent LA, Zhai P, Zhang X, Kitching S. 2013. Updated analyses of temperature and precipitation extreme indices since the beginning of the twentieth century: the HadEX2 dataset. *Journal of Geophysical Research: Atmospheres* 118: 1–16, doi: 10.1002/jgrd.50150.

Fu J, Qian W, Lin X, Chen D. 2008. Trends in graded precipitation in China from 1961–2000. *Advances in Atmospheric Sciences* 25: 267–278.

IPCC. 2012. Managing the risks of extreme events and disasters to advance climate change adaptation. In A Special Report of Working Groups I and II of the Intergovernmental Panel on Climate Change, Cambridge University Press, Cambridge and New York, NY.

Jones PD, Horton EB, Folland CK, Hulme M, Parker DE, Basnett TA. 1999. The use of indices to identify changes in climatic extremes. *Climatic Change* 42: 131–149.

Klein Tank AMG, Peterson TC, Quadir DA, Dorji S, Zou X, Tang H, Santhosh K, Joshi UR, Jaswal AK, Kolli RK, Sikder AB, Deshpande NR, Revadekar JV, Yeleuova K, Vandasheva S, Faleyeva M,

Gomboluudev P, Budhathoki KP, Hussain A, Afzaal M, Chandrapala L, Anvar H, Amanmurad D, Asanova VS, Jones PD, New MG, Spektorman T. 2006. Changes in daily temperature and precipitation extremes in central and south Asia. *Journal of Geophysical Research* **111**: D16105, doi: 10.1029/2005JD006316.

Koenke R, Basset GW. 1978. Regression quantiles. *Econometrica* **46**: 33–50.

Lee K, Baek H, Cho C. 2013. Analysis of changes in extreme temperature using quantile regression. *Asia-Pacific Journal of Atmospheric Sciences* **49**(3): 313–323, doi: 10.1007/s13143-013-0030-1.

Ou T, Chen D, Linderholm HW, Jeong JH. 2013. Evaluation of global climate models in simulating extreme precipitation in China. *Tellus A* **65**: 19799, doi: 10.3402/tellusa.v65i0.19799.

Peterson TC, Easterling DR, Karl TR, Groisman P, Nicholls N, Plummer N, Torok S, Auer I, Boehm R, Gullett D, Vincent L, Heino R, Tuomenvirta H, Mestre O, Szentimrey T, Salinger J, Forland EJ, Hanssen-Bauer I, Alexandersson H, Jones P, Parker D. 1998. Homogeneity adjustments of *in situ* atmospheric climate data: a review. *International Journal of Climatology* **18**: 1493–1517.

Tareghian R, Rasmussen P. 2012. Analysis of Arctic and Antarctic sea ice extent using quantile regression. *International Journal of Climatology* **33**: 1079–1086, doi: 10.1002/joc.3491.

Vincent LA, Mekis E. 2006. Changes in daily and extreme temperature and precipitation indices for Canada over the twentieth century. *Atmosphere-Ocean* **44**(2): 177–193.

Wang Y, Zhou L. 2005. Observed trends in extreme precipitation events in China during 1961–2001 and the associated changes in large-scale circulation. *Geophysical Research Letters* **32**: L07070, doi: 10.1029/2005GL022574.

Wang X, Wen Q, Wu Y. 2007. Penalized maximal *t* test for detecting undocumented mean change in climate data series. *Journal of Applied Meteorology and Climatology* **46**(6): 916–931, doi: 10.1175/JAM2504.1.

Wasko C, Sharma A. 2014. Quantile regression for investigating scaling of extreme precipitation with temperature. *Water Resources Research* **50**: 3608–3614, doi: 10.1002/2013WR015194.

Wijngaard JB, Klein Tank AMG, Können GP. 2003. Homogeneity of 20th century European daily temperature and precipitation series. *International Journal of Climatology* **23**: 679–692, doi: 10.1002/joc.906.

You Q, Kang S, Aguilar E, Pepin N, Flugel W, Yan Y, Xu Y, Zhang Y, Huang J. 2011. Changes in daily climate extremes in China and their connection to the large scale atmospheric circulation during 1961–2003. *Climate Dynamics* **36**: 2399–2417, doi: 10.1007/s00382-009-0735-0.

Zhai P, Zhang X, Wan H, Pan X. 2005. Trends in total precipitation and frequency of daily precipitation extremes over China. *Journal of Climate* **18**: 1096–1108.

Zhang G, He J, Zhou Z, Cao L. 2012. Homogeneity study of precipitation data over China using RHtest method (in Chinese). *Meteorological Science and Technology* **40**: 914–921.

24

Multi-model analysis of the Atlantic influence on Southern Amazon rainfall

Jin-Ho Yoon*

Atmospheric Sciences and Global Change Division, Pacific Northwest National Laboratory, Richland, WA, USA

*Correspondence to:
J.-H. Yoon, Atmospheric Science and Global Change Division, Pacific Northwest National Laboratory, P.O. Box 999, MSIN:K9-30, 902 Battelle Blvd., Richland, WA 99352, USA.
E-mail: jin-Ho.Yoon@pnnl.gov

Abstract

Amazon rainfall is subject to year-to-year fluctuation resulting in drought and flood in various intensities. A major climatic driver of the interannual variation of the Amazon rainfall is El Niño/Southern Oscillation. Also, the sea surface temperature over the Atlantic Ocean is identified as an important climatic driver on the Amazon water cycle. Previously, observational data sets were used to support the Atlantic influence on Amazon rainfall. Here, it is found that multiple global climate models do reproduce the Atlantic–Amazon link robustly. However, there exist differences in rainfall response, which primarily depends on the climatological rainfall amount.

Keywords: Amazon rainfall; Atlantic SST variability; CMIP5

1. Introduction

Amazon rainfall variability has been extensively studied from many different angles. Generally, the sea surface temperature (SST) variability over the tropical Pacific Ocean, i.e. El Niño and Southern Oscillation (ENSO) is a major climatic driver of Amazon rainfall fluctuation on interannual timescales. ENSO modulates the atmospheric branch of the global hydrological cycle – convergence of atmospheric water vapor in the Tropics, and hence the regional water cycle over Amazon (Walker, 1928; Kousky *et al.*, 1984). For example, anomalous warm (cold) water over the tropical Pacific Ocean, i.e. El Niño (La Niña) condition induces downward (upward) motion and reduced (more) rainfall over the Amazon through atmospheric teleconnections.

In addition to the ENSO, an additional climatic driver, Atlantic SST variability, has recently been identified as an important contributor in determining the fate of the Amazon water cycle. Both the southern and northern Atlantic SST affect the regional water budget and the severity of the burning season of the Amazon basin (Uvo *et al.*, 2000; Marshall *et al.*, 2001; Ronchail *et al.*, 2002; Yoon and Zeng, 2010; Chen *et al.*, 2011; Fernandes *et al.*, 2011). For example, less (more) rainfall over the Amazon basin is linked to warm (cold) SST in the tropical North Atlantic, which is closely related to an anomalously northward (southward) migration of the Inter-Tropical Convergence Zone (ITCZ) (e.g. Moron *et al.*, 1995; Enfield, 1996).

The possibility of a tropical Atlantic influence on the Amazon's rainfall was highlighted after two severe droughts occurred in 2005 and 2010 (Marengo *et al.*, 2008; Zeng *et al.*, 2008; Marengo *et al.*, 2011). The severity of these drought events is once in a century

event or even worse (Marengo *et al.*, 2008; Marengo *et al.*, 2011). In 2005, significant changes of rainfall and river levels were not observed in the northeastern Amazon, where severe droughts occurred historically due to extreme El Niño events (1926, 1983 and 1998). On the other hand, the Atlantic SST had very severe impact on the Southern Amazon basin (SAB), particularly during its dry season (May–September) in 2005 (Zeng *et al.*, 2008). In 2010, although the drought was initiated by the El Niño, it became more intense during dry season due to the warming of the tropical North Atlantic Ocean (Marengo *et al.*, 2011).

Another interesting implication of the Atlantic–Amazon link can be found where the change in the tropical Atlantic Ocean SST and its variability could play a key role in determining the future of Amazon rainfall (e.g. Cox *et al.*, 2008; Good *et al.*, 2008; Cook *et al.*, 2012). For example, the warmer tropical North Atlantic and the colder tropical South Atlantic Oceans would produce a favorable condition for the future drying in the SAB. Similarly, the paleoclimatic evidence indicates that the North Atlantic SST can modulate the Amazon rainfall in the longer timescales (Mosblech *et al.*, 2012).

Considering the importance of the Atlantic–Amazon link, it is an interesting research question how this link is simulated by different global climate models. Also, by analyzing and comparing simulations by multiple climate models, potential deficiencies in the global climate models can be identified and suggestions for improvement can be made to the model development community.

Here, two different model experiments are utilized: (1) the idealized experimental sets with the Atmospheric Global Climate Models (AGCMs) by the US Climate Variability and Predictability (CLIVAR)

Table 1. Description of the experimental designs. More details can be found in Table 1 of Schubert *et al.* (2009).

	Warm Atlantic	Normal Atlantic	Cold Atlantic
Warm Pacific	PwAw	PwAn	PwAc
Normal Pacific	PnAw	PnAn	PnAc
Cold Pacific	PcAw	PcAn	PcAc

Table 2. A brief description of the AGCMs used by the US CLIVAR working group for drought.

Model	Resolution	Reference
AM2.1	2 × 2.5, L25	Delworth et al. (2006)
GFS	T62 (~2 × 2), L64	Campana and Kaplan (2005)
NSIPP-1	3 × 3.75, L34	Bacmeister et al. (2000)
CCM3.0	T42 (~2.8 × 2.8), 18 hybrid levels	Kiehl et al. (1998)
CAM3.5	T85, 27 hybrid levels	Collins et al. (2006), Neale et al. (2008)

drought working group, summarized in Table 1, and (2) the historical experiments with the Coupled Global Climate Models (CGCMs) in the Coupled Model Inter-comparison Project phase (CMIP5; Taylor *et al.*, 2012). The purposes of this study are to evaluate how well the global climate model simulates the Atlantic–Amazon link and to understand a potential mechanism to cause any difference in the simulated Atlantic–Amazon link by each global climate model. A brief discussion on methods is given in Section 2. Results and concluding remarks are in Sections 3 and 4, respectively.

2. Methods

In this study, we focus the SAB covering Southern Amazonia and central Brazil (65°–50°W, 20°–5°S) with clear-cut rainy and dry seasons (Cook *et al.*, 2012). The basin receives rainfall more than 6 mm day^{-1} during its rainy season (December–March) and around or less than 1 mm day^{-1} during its dry season (May–September) in accordance with ITCZ movement. Climatologically, evaporation dominates precipitation over the Southern Amazon during dry season (Figure S1, Supporting Information), indicating diverging water vapor flux. More interestingly, the SAB is the region most sensitive to tropical Atlantic SST variability (Yoon and Zeng, 2010). Also, this region is vulnerable to future changes in the hydrological cycle (Cook *et al.*, 2012).

We utilize five sets of the idealized experiments from different AGCMs that were originally designed for the North American Drought study by the US CLIVAR working group (Schubert *et al.*, 2009). The same idealized SST conditions (Table 1) are used to force the five different AGCMs (Table 2). Run with warm (cold) Atlantic SST is compared with the normal Atlantic condition with respect to the different SST conditions in the Pacific Ocean. Using these sensitivity experiments, we can assess how Atlantic SST can affect the Southern Amazon rainfall regardless of the Pacific SST conditions.

Also, the CGCMs in the CMIP5 (Taylor *et al.*, 2012) are used to evaluate the Atlantic–Amazon link. Historical experiments by the 39 CGCMs are analyzed (Table 3). In the CGCM experiments, it is difficult to separate the Atlantic influence on Amazon rainfall due to complex interplay with the tropical Pacific SST, similar to the observational data. Thus, the ENSO influence was removed by using linear regression in Yoon and

Zeng (2010). Here, the partial correlation is used to isolate the Atlantic–Amazon link without the ENSO. The mathematical expression to obtain a partial correlation (ρ) can be written as follows:

$$\rho_{\text{P(SAB)SST(NATL)}\times\text{SST(NINO3.4)}}$$

$$= \frac{\rho_{\text{P(SAB)SST(NATL)}} - \rho_{\text{P(SAB)SST(NINO3.4)}} \times \rho_{\text{SST(NATL)SST(NINO3.4)}}}{\sqrt{\begin{array}{c}1 - \rho_{\text{P(SAB)SST(NINO3.4)}}^2 \\ \times \rho_{\text{SST(NATL)SST(NINO3.4)}}^2\end{array}}} \quad (1)$$

where NATL (SATL) represent the tropical North (South) Atlantic Ocean following (Enfield and Mayer, 1997; Yoon and Zeng, 2010). The North (South) Atlantic Ocean covers 60°–20°W, 5°–20°N (35°W–10°E, 20°S–EQ). The observed gridded precipitation data set is from the Climate Research Unit (Harris *et al.*, 2014). Observed SST is from the Hadley Center (Rayner *et al.*, 2006). Derived evaporation is obtained from the offline land surface model.

3. Results

Figure 1 and Figure S2 confirm that all five AGCMs simulate less precipitation over the SAB in response to the warm anomalous SST over the North Atlantic Ocean during dry season. This result is robust regardless of the SST conditions over the tropical Pacific Ocean. The only exception is found with the Global Forecast System (GFS) model for neutral Pacific Ocean conditions. The response in the Geophysical Fluid Dynamics Laboratory (GFDL) is weaker in general. Overall, the response is most obvious in NASA's Seasonal-to-interannual Prediction Program (NSIPP) and Community Atmosphere Model (CAM3.5), and less in GFS and GFDL models. To figure out what causes the different response between models, we further analyze the mean characteristics of rainfall.

Figure 2 clearly demonstrates that a wetter model, such as NSIPP or CAM3.5, produces much stronger rainfall response to the SST anomalies over the Atlantic Ocean compared to a drier one, such as CAM3 or GFDL. These results support a hypothesis that the differences in precipitation response to warm or cold Atlantic SST are likely driven by those in the dry season rainfall amount over the region. Although the

Table 3. List of the CGCMs from the CMIP5 used in this study.

Model name	Institute
ACCESS1-0, ACCESS1-3	Commonwealth Scientific and Industrial Research Organization, Australia (CSIRO) and Bureau of Meteorology, Australia (BOM)
BCC-CSM-1.1, BCC-CSM-1.1m	Beijing Climate Center, China Meteorological Administration
BNU-ESM	College of Global Change and Earth System Science, Beijing Normal University
CMCC-CESM, CMCC-CM, CMCC-CMS	Centro Euro-Mediterraneo per I Cambiamenti Climatici
CanESM2	Canadian Centre for Climate Modeling and Analysis
CCSM4	National Center for Atmospheric Research
CNRM-CM5	Centre National de Recherches Meteorologiques/Centre Europeen de Recherche et Formation Avancees en Calcul Scientifique, France
GFDL-CM3, GFDL-ESM2G, GFDL-ESM2M	NOAA, Geophysical Fluid Dynamics Laboratory
CSIRO-Mk3.6.0	Commonwealth Scientific and Industrial Research Organization/Queensland Climate Change Centre of Excellence (CSIRO-QCCCE)
FGOALS-g2	Institute of Atmospheric Physics, Chinese Academy of Sciences
FIO-ESM	The First Institute of Oceanography, SOA, China
Had-GEM2, Had-GEM2-CC, Had-GEM2-ES	Met Office Hadley Centre (additional HadGEM2-ES realizations contributed by Instituto Nacional de Pesquisas Espaciais)
INMCM4	Institute for Numerical Mathematics
IPSL-CM5A-LR, IPSL-CM5A-MR, IPSL-CM5B-LR	Institute Pierre-Simon Laplace
MIROC5	Atmosphere and Ocean Research Institute (The University of Tokyo), National Institute for Environmental Studies and Japan Agency for Marine-Earth Science and Technology
MIROC-ESM, MIROC-ESM-CHEM	Japan Agency for Marine-Earth Science and Technology, Atmosphere and Ocean Research Institute (The University of Tokyo) and National Institute for Environmental Studies
MPI-ESM-LR, MPI-ESM-MR	Max Planck Institute for Meteorology (MPI-M)
MRI-CGCM3	Meteorological Research Institute
NorESM1-M, NorESM1-ME	Norwegian Climate Centre (NCC)
CESM1, CESM1-BGC, CESM1-CAM5, CESM1-CAM5-FV2	National Science Foundation, Department of Energy, National Center for Atmospheric Research

climatological rainfall amount, to some extent, determines the intensity of rainfall response to the Atlantic SST, it is not the only condition. As seen in GFS and CAM3.5 of which the dry season rainfall amount are close to each other, however, the anomalous rainfall amounts in response to the warm and cold Atlantic SST are different in its intensity. In this case, other factors such as the atmospheric physical parameterizations are likely responsible for the differences. Evaporation and the convergence of water vapor fluxes are also analyzed (Figure S3). Interestingly, evaporation does show similar behavior as precipitation, while the convergence of water vapor flux presents more complex response. It is noted here that the number of models in this idealized experiment is quite limited so the result is not conclusive.

The above result is based on the AGCMs where the same SST condition is prescribed for different models. Next, can the coupled climate models (CGCMs) reproduce the Atlantic–Amazon link? To answer this question, 39 CGCMs in the CMIP5 are analyzed with a partial correlation coefficient with removing an influence of the ENSO. Although it assumes linear relationship, it provides a practical measure how the Atlantic SST can affect the Amazon rainfall without extra experiments, such as those in the Figure 1 with only Atlantic or Pacific SST conditions.

Figure 3 shows the partial correlation between rainfall over the SAB and the tropical North Atlantic SST with the NINO3.4 SST influence removed. It is found that the majority of the models, 35 of 39 models, do produce a clear Atlantic–Amazon link: warm (cold) tropical North Atlantic and less (more) rainfall over the SAB during dry season. However, its intensity varies greatly among different models.

The part (b) of the Figure 3 shows the dry season rainfall over the SAB. Unlike Figure 2, no clear relationship between the dry season mean (Figure 3(b)) and anomalous rainfall, measured in terms of partial correlation (Figure 3(a)), is found. Evaporation and the convergence of water vapor flux have been analyzed for all 39 CGCMs (Figure S4). Similarly as AGCMs case, evaporation tends to follow precipitation, while the convergence of water vapor flux does show more complex behavior. This is somewhat expected because the CGCMs produce their own SST and its variability, which are different in various ways. Also, the atmospheric teleconnection pattern simulated by the CGCMs is different. For example, each CGCM exhibits very different performance in simulating the ENSO (Guilyardi *et al.*, 2012; Bellenger *et al.*, 2014).

4. Concluding remarks

The Atlantic influence on the Amazon rainfall has a couple of important implications. First, it is only recently identified as a major climatic driver of the regional water cycle of the Amazon (e.g. Marengo *et al.*, 2008; Yoon and Zeng, 2010; Chen *et al.*, 2011;

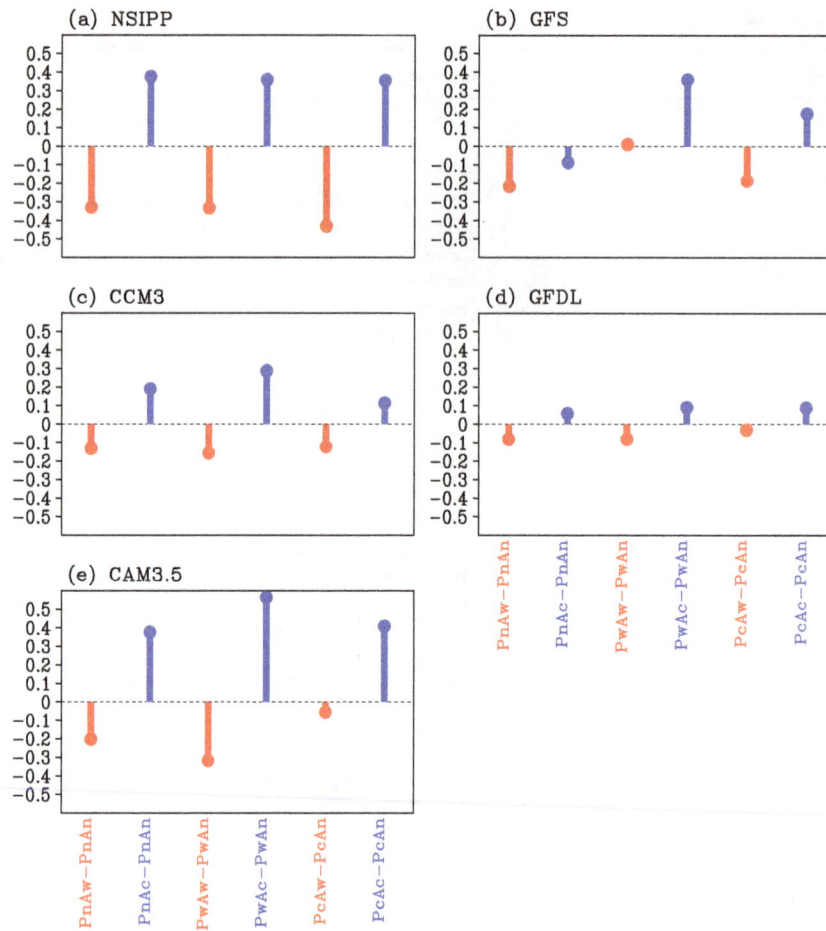

Figure 1. Precipitation difference (unit: mm day^{-1}) averaged over the Southern Amazon basin (65°−50°W, 20°−5°S) during its dry season (May−September) simulated by the five AGCMs forced by the idealized sea surface temperature conditions. More details are in the Table 1 and Schubert *et al.* (2009)). All cases are compared with the normal Atlantic condition. For example, Warm Pacific and Warm Atlantic is compared with Warm Pacific and Normal Atlantic (PwAw−PwAn) and so on.

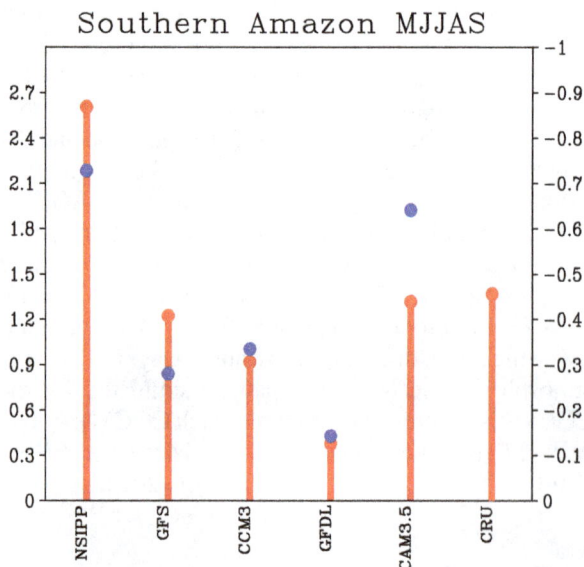

Figure 2. Dry season mean rainfall (unit: mm day^{-1}) of the SAB (red dots) and rainfall difference between Warm Atlantic and Cold Atlantic (blue dots), which is averaged of PwAw−PwAc, PnAw−PnAc and PcAw−PcAc. The right most histogram indicates an observational value from the Climate Research Unit (Harris *et al.*, 2014).

Fernandes *et al.*, 2011; Marengo *et al.*, 2011). It is partly because the influence of ENSO on Amazon rainfall is much stronger than that of Atlantic SST. Therefore, separating the signal of the Atlantic–Amazon from that of ENSO is rather challenging (Yoon and Zeng, 2010). Second, it has been used to explain the future of the Amazon water cycle (Cox *et al.*, 2008; Good *et al.*, 2008; Cook *et al.*, 2012). Therefore, it is important for the global climate models to reproduce the Atlantic–Amazon link realistically, not just in terms of rainfall amount but also associated atmospheric condition, such as evaporation and water vapor flux.

All the AGCMs with idealized SST conditions simulate drier (wetter) Southern Amazon during May−September in response to the warm (cold) North Atlantic Ocean. Despite the fact that it is a robust signal, there is difference in the amount of rainfall change due to the Atlantic SST by the different AGCMs, which depends on the individual model's capability to simulate the dry season rainfall amount properly, which can be contributed by multiple factors, such as spatial resolution and physical parameterizations. In other words, some models with very little dry season rainfall climatologically do simulate no significant response to the anomalous Atlantic SST conditions.

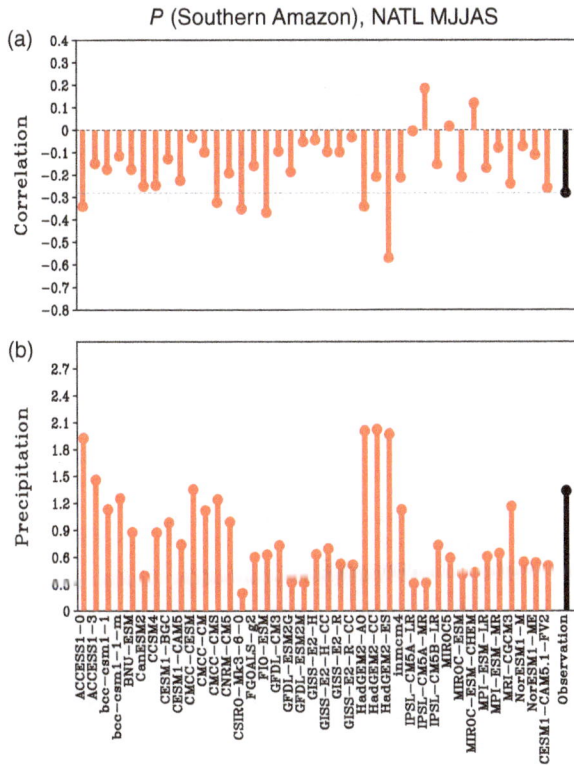

Figure 3. Partial correlation of rainfall over the SAB and SST over the tropical North Atlantic Ocean removing influence of ENSO from historical simulation of 39 coupled climate models in CMIP5 (a) in the period of 1901–2005. Black dot is for observation with precipitation from CRU and SST from Hadley Center in the period of 1951–2010. Dry season mean rainfall amount is shown in (b).

This relationship between the dry season rainfall amount and the Atlantic–Amazon link becomes more complicated in case of the fully coupled climate models. It is found that the majority of the CGCMs, 35 of 39 models, do produce right signal: warm (cold) tropical North Atlantic and less (more) rainfall over the SAB. However, its intensity changes greatly among different models. Further research is required to understand what make climate models produce different response. Even though the climatological rainfall amount is potentially one of the conditions that likely determine the sensitivity of the Atlantic–Amazon link in the AGCMs, it is not the only mechanism in the CGCMs. In case of the CGCMs, the differences in simulated tropical SST variability could be an important factor.

Evaporation and the convergence of water vapor flux, other terms in the atmospheric water vapor budget (Equation S1), are also analyzed for both AGCMs and CGCMs (Figures S1–S4). It is interesting to observe that evaporation tends to behave in a similar manner as precipitation. In other words, more evaporation is linked to stronger anomalous rainfall response to warm or cold Atlantic Ocean SST. However, the convergence of water vapor flux and rainfall response does not show a systematic behavior (Figures S3 and S4).

In applying the climate projection to a specific region such as the SAB, it is desired to use caution whether the individual model meet certain conditions, such as the climatological precipitation amount. If a CGCM produces less precipitation than what is observed, it is likely to have suppressed climate variability and change signals, and vice versa. Therefore, the climate model development community needs to focus on the climatological features of the regional water cycle first.

Acknowledgements

J.-H. Yoon is supported by the Office of Science of the US Department of Energy. Historical simulations in the CMIP5 were originally processed by Dr Jung Choi at Seoul National University and kindly shared. Editorial suggestions and internal review by Dr Kyo-Sun Sunny Lim at PNNL is valuable to improve the manuscript. PNNL is operated for the Department of Energy by Battelle Memorial Institute under Contract DEAC05-76RLO1830.

Supporting information

The following supporting information is available:

Appendix S1. Additional information about evaporation and convergence of water vapor flux is provided in the Supporting information.

References

Bacmeister J, Schubert SD, Suarez M, Pegion PJ. 2000. Atlas of Seasonal Means Simulated by the NSIPP 1 Atmospheric GCM. Technical Report Series on Global Modeling and Data Assimilation 104606, Vol. 17, 194 pp.

Bellenger H, Guilyardi E, Leloup J, Lengaigne M, Vialard J. 2014. ENSO representation in climate models: from CMIP3 to CMIP5. *Climate Dynamics* **42**(7–8): 1999–2018.

Campana K, Caplan P. 2005. Technical Procedure Bulletin for T382 Global Forecast System. http://www.emc.ncep.noaa.gov/gc_wmb/Documentation/TPBoct05/T382.TPB.FINAL.htm (accessed 15 October 2014).

Chen Y, Randerson JT, Morton DC, DeFries RS, Collatz GJ, Kasibhatla PS, Giglio L, Jin YF, Marlier ME. 2011. Forecasting fire season severity in South America using sea surface temperature anomalies. *Science* **334**(6057): 787–791.

Collins WD, Bitz CM, Blackmon ML, Bonan GB, Bretherton CS, Carton JA, Chang P, Doney SC, Hack JJ, Henderson TB, Kiehl JT, Large WG, McKenna DS, Santer BD, Smith RD. 2006. The community climate system model version 3 (CCSM3). *Journal of Climate* **19**(11): 2122–2143.

Cook B, Zeng N, Yoon JH. 2012. Will Amazonia dry out? Magnitude and causes of change from IPCC climate model projections. *Earth Interactions* **16**: 27.

Cox PM, Harris PP, Huntingford C, Betts RA, Collins M, Jones CD, Jupp TE, Marengo JA, Nobre CA. 2008. Increasing risk of Amazonian drought due to decreasing aerosol pollution. *Nature* **453**(7192): 212–217.

Delworth TL, Broccoli AJ, Rosati A, Stouffer RJ, Balaji V, Beesley JA, Cooke WF, Dixon KW, Dunne J, Dunne KA, Durachta JW, Findell KL, Ginoux P, Gnanadesikan A, Gordon CT, Griffies SM, Gudgel R, Harrison MJ, Held IM, Hemler RS, Horowitz LW, Klein SA, Knutson TR, Kushner PJ, Langenhorst AR, Lee H-C, Lin S-J, Lu J, Malyshev SL, Milly PCD, Ramaswamy V, Russell J, Schwarzkopf MD, Shevliakova E, Sirutis JJ, Spelman MJ, Stern WF, Winton M, Wittenberg AT, Wyman B, Zeng F, Zhang R. 2006. GFDL's CM2 global coupled climate models. Part I: formulation and simulation characteristics. *Journal of Climate* **19**(5): 643–674.

Enfield DB. 1996. Relationships of inter-American rainfall to tropical Atlantic and Pacific SST variability. *Geophysical Research Letters* **23**(23): 3305–3308.

Enfield DB, Mayer DA. 1997. Tropical Atlantic sea surface temperature variability and its relation to El Nino Southern Oscillation. *Journal of Geophysical Research, Oceans* **102**(C1): 929–945.

Fernandes K, Baethgen W, Bernardes S, Defries R, Dewitt DG, Goddard L, Lavado W, Lee DE, Padoch C, Pinedo-Vasquez M, Uriarte M. 2011. North tropical Atlantic influence on western Amazon fire season variability. *Geophysical Research Letters* **38**: 5.

Good P, Lowe JA, Collins M, Moufouma-Okia W. 2008. An objective tropical Atlantic sea surface temperature gradient index for studies of south Amazon dry-season climate variability and change. *Philosophical Transactions of the Royal Society B: Biological Sciences* **363**(1498): 1761–1766.

Guilyardi E, Bellenger H, Collins M, Ferrett S, Cai W, Wittenberg A. 2012. A first look at ENSO in CMIP5. *CLIVAR Exchanges* **17**: 4.

Harris I, Jones PD, Osborn TJ, Lister DH. 2014. Updated high-resolution grids of monthly climatic observations – the CRU TS3.10 dataset. *International Journal of Climatology* **34**(3): 623–642.

Kiehl JT, Hack JJ, Bonan GB, Boville BA, Williamson DL, Rasch PJ. 1998. The national center for atmospheric research community climate model: Community Climate Model (CCM3). *Journal of Climate* **11**(6): 1131–1149.

Kousky VE, Kagano MT, Cavalcanti IFA. 1984. A review of the Southern Oscillation – oceanic-atmospheric circulation changes and related rainfall anomalis. *Tellus A: Dynamic Meteorology and Oceanography* **36**(5): 490–504.

Marengo JA, Nobre CA, Tomasella J, Oyama MD, De Oliveira GS, De Oliveira R, Camargo H, Alves LM, Brown IF. 2008. The drought of Amazonia in 2005. *Journal of Climate* **21**(3): 495–516.

Marengo JA, Tomasella J, Alves LM, Soares WR, Rodriguez DA. 2011. The drought of 2010 in the context of historical droughts in the Amazon region. *Geophysical Research Letters* **38**: 5.

Marshall J, Kushner Y, Battisti D, Chang P, Czaja A, Dickson R, Hurrell J, McCartney M, Saravanan R, Visbeck M. 2001. North Atlantic climate variability: phenomena, impacts and mechanisms. *International Journal of Climatology* **21**(15): 1863–1898.

Moron V, Bigot S, Roucou P. 1995. Rainfall variability in subequatorial America and Africa and relationships with the main sea-surface temperature modes (1951–1990). *International Journal of Climatology* **15**(12): 1297–1322.

Mosblech NAS, Bush MB, Gosling WD, Hodell D, Thomas L, van Calsteren P, Correa-Metrio A, Valencia BG, Curtis J, van Woesik R. 2012. North Atlantic forcing of Amazonian precipitation during the last ice age. *Nature Geoscience* **5**(11): 817–820.

Neale RB, Richter JH, Jochum M. 2008. The impact of convection on ENSO: from a delayed oscillator to a series of events. *Journal of Climate* **21**(22): 5904–5924.

Rayner NA, Brohan P, Parker DE, Folland CK, Kennedy JJ, Vanicek M, Ansell TJ, Tett SFB. 2006. Improved analyses of changes and uncertainties in sea surface temperature measured in situ since the mid-nineteenth century: the HadSST2 dataset. *Journal of Climate* **19**(3): 446–469.

Ronchail J, Cochonneau G, Molinier M, Guyot JL, Chaves AGD, Guimaraes V, de Oliveira E. 2002. Interannual rainfall variability in the Amazon basin and sea-surface temperatures in the equatorial Pacific and the tropical Atlantic Oceans. *International Journal of Climatology* **22**(13): 1663–1686.

Schubert S, Gutzler D, Wang HL, Dai A, Delworth T, Deser C, Findell K, Fu R, Higgins W, Hoerling M, Kirtman B, Koster R, Kumar A, Legler D, Lettenmaier D, Lyon B, Magana V, Mo K, Nigam S, Pegion P, Phillips A, Pulwarty R, Rind D, Ruiz-Barradas A, Schemm J, Seager R, Stewart R, Suarez M, Syktus J, Ting MF, Wang CZ, Weaver S, Zeng N. 2009. A US CLIVAR project to assess and compare the responses of global climate models to drought-related SST forcing patterns: overview and results. *Journal of Climate* **22**(19): 5251–5272.

Taylor KE, Stouffer RJ, Meehl GA. 2012. An overview of CMIP5 and the experiment design. *Bulletin of the American Meteorological Society* **93**(4): 485–498.

Uvo CB, Tolle U, Berndtsson R. 2000. Forecasting discharge in Amazonia using artificial neural networks. *International Journal of Climatology* **20**(12): 1495–1507.

Walker GT. 1928. Ceara (Brazil) famines and the general air movement. *Beitrage Zur Physik der Frein Atmosphere* **14**: 88–93.

Yoon JH, Zeng N. 2010. An Atlantic influence on Amazon rainfall. *Climate Dynamics* **34**(2–3): 249–264.

Zeng N, Yoon JH, Marengo JA, Subramaniam A, Nobre CA, Mariotti A, Neelin JD. 2008. Causes and impacts of the 2005 Amazon drought. *Environmental Research Letters* **3**(1): 014002.

25

A possible linkage of the Western North Pacific summer monsoon with the North Pacific Gyre Oscillation

Wei Zhang[1,2,3]* and Ming Luo[4]

[1]Key Laboratory of Meteorological Disaster of Ministry of Education, Nanjing University of Information Science & Technology, People's Republic of China
[2]Collaborative Innovation Center on Forecast and Evaluation of Meteorological Disasters, Nanjing University of Information Science & Technology, People's Republic of China
[3]Climate Change and Sustainability Laboratory, Shenzhen Research Institute, The Chinese University of Hong Kong, Shatin, Hong Kong, China
[4]Institute of Environment, Energy and Sustainability, The Chinese University of Hong Kong, Shatin, Hong Kong, China

*Correspondence to:
W. Zhang, Key Laboratory of
Meteorological Disaster of
Ministry of Education, and
Collaborative Innovation Center
on Forecast and Evaluation of
Meteorological Disasters,
Nanjing University of Information
Science & Technology, 219
Ningliu Road, Nanjing 210044,
Jiangsu Province, China. E-mail:
wzhang@nuist.edu.cn

Abstract

This study examines possible links between the western North Pacific summer monsoon (WNPSM) and the North Pacific Gyre Oscillation (NPGO). Our research findings are summarized as follows: (1) In addition to the Central Pacific El Niño, the WNPSM also exerts significant modulation on the NPGO. (2) A time lag is detected in the association between the WNPSM and the NPGO. The WNPSM in JJAS (from June to September) has significant impacts on the NPGO in MAM (from March to May) of the following year. (3) The wave train patterns forced by the WNPSM that are being integrated by the ocean mixed layer may be responsible for the lagged response of the NPGO.

Keywords: western North Pacific summer monsoon; North Pacific Gyre Oscillation; North Pacific Oscillation

1. Introduction

The Asian summer monsoon system includes two subsystems: the Indian summer monsoon (ISM) and the East Asian summer monsoon (EASM) (e.g. Chang and Krishnamurti, 1987; Tao and Chen, 1987; Ding and Chan, 2005; Ding, 2007). Wang and Lin (2002) further extended the Asian monsoon system to incorporate the western North Pacific summer monsoon (WNPSM). The ISM and the WNPSM are tropical monsoons whereas the EASM is a subtropical monsoon system (Wang and Lin, 2002; Ding and Chan, 2005). Moreover, the ISM and the EASM are caused by land-sea thermal contrasts whereas the WNPSM originates from meridional SST gradients (Tao and Chen, 1987; Murakami and Matsumoto, 1994; Li and Wang, 2005). The WNPSM is characterized by a wide spectrum of temporal scales, ranging from biweekly, seasonal, interannual, to decadal (Li and Wang, 2005). The WNPSM influences weather and climate in East Asia and even North America by teleconnections such as wave train patterns (e.g. Wang et al., 2001; Li and Wang, 2005; Jiang and Lau, 2008).

The North Pacific Gyre Oscillation (NPGO) is the second empirical orthogonal function (EOF) mode of the sea surface height anomalies in the North Pacific (Di Lorenzo et al., 2008) and it perfectly tracks the second EOF mode of the sea surface temperature (SST) anomalies in the North Pacific (Bond and Harrison, 2000; Bond et al., 2003). Similar to the Pacific Decadal Oscillation (PDO) that is the first EOF mode of the SST anomalies in the North Pacific (Mantua et al., 1997; Zhang et al., 1997), the NPGO also features a pronounced decadal variation. Previous studies have documented that the NPGO is forced by surface wind stress induced by the sea level pressure (SLP) anomalies in the North Pacific (Di Lorenzo et al., 2008, 2010), which are known as the North Pacific Oscillation (NPO) (Walker and Bliss, 1932; Rogers, 1981). Some studies have shown that the NPGO has a strong linkage with the ENSO, especially the CP El Niño (e.g. Lienert and Doblas-Reyes, 2013). The NPGO/NPO is associated with the Pacific meridional mode (PMM) (Chiang and Vimont, 2004). The NPGO and PMM are closely associated with the occurrence of WNP tropical cyclones (Zhang et al., 2013, 2016).

Wang et al. (2001) have reported that the WNPSM induces several marked patterns in 500- and 850-hPa geopotential height and 200-hPa wind fields in regions elongating from the WNP to the eastern North Pacific by wave train patterns. It is thus of great interest to examine whether the WNPSM exerts impacts on the NPGO and its underlying mechanisms. The accomplishment of this study will therefore advance our understanding of the monsoon–NPGO relationship and provide useful references for the prediction of the

Table 1. The correlation between the WNPSM in JJAS and the NPGO index from JJA in the current year to JJA in the following year.

Correlation (partial)	NPGO JJA	NPGO SON	NPGO DJF	NPGO MAM	NPGO JJA
WNPSM (JJAS)	−0.05 (−0.11)	−0.06 (−0.22)	−0.12 (−0.12)	**−0.31* (−0.34*)**	**−0.28* (−0.31*)**
EASM (JJAS)	0.18 (0.08)	0.19 (0.08)	−0.17 (−0.14)	−0.12 (−0.24)	−0.10 (−0.22)
SCSSM (JJAS)	−0.09 (0.03)	−0.15 (−0.01)	−0.02 (−0.02)	−0.19 (−0.10)	−0.18 (−0.08)

Bold-face numbers indicate those at 0.05 level of significance.
*0.05 level of significance.
The numbers in the parentheses indicate those by the partial correlation analysis.

NPGO. Owing to the rising role of the NPGO in shaping the climate system under global warming (Di Lorenzo et al., 2008; Lienert and Doblas-Reyes, 2013; Ding et al., 2015), it is of great significance to predict the NPGO.

The remainder of this article is organized as follows. Section 2 presents the data sets and methodology used in this study. Section 3 discusses analysis results, followed by Section 4 that provides possible underpinning mechanisms. Section 5 presents the discussion and concluding remarks.

2. Data and methodology

The NPGO index is defined as the temporal coefficient of the second EOF of the Sea Surface Height (SSH) anomalies in the North Pacific and is directly obtained from the website http://www.o3d.org/npgo/ for the period 1960–2009 (Di Lorenzo et al., 2008). The Western North Pacific Summer Monsoon (WNPSM) index (Wang and Fan, 1999; Wang et al., 2001) is defined as: WNPSM index = U850(100°–130°E, 5°–15°N) − U850(110°–140°E, 20°–30°N).

The WNPSM index exhibits both the Rossby-wave induced low-level vorticity and the magnitude of the tropical westerly (Wang et al., 2001). The South China Sea Summer Monsoon (SCSSM) index is defined as a seasonal (JJAS) dynamical normalized seasonality (DNS) averaged over the South China Sea monsoon domain (0°–25°N, 100°–125°E) at 925 hPa (e.g. Li and Zeng, 2002, 2003). The EASM index is defined as the differences in area-averaged 200 hPa wind (zonal or meridional) between the region (110–150°E, 40–50°N) and the region (110–150°E, 25–35°N) (Lau et al., 2000). The SST data are derived from the Met Office Hadley Center with a spatial resolution of 1° × 1° (http://www.metoffice.gov.uk/hadobs/hadsst2/). The meteorological variables (e.g. zonal and meridional surface wind stress and SLP) are obtained from the National Centers for Environmental Prediction/National Center for Atmospheric Research (NCEP/NCAR) reanalysis data (Kalnay et al., 1996). The El Niño Modoki index (EMI) is used to measure the SST anomalies in the Central Pacific. It is defined as EMI = $[SSTA]_A + 0.5[SSTA]_B + 0.5[SSTA]_C$, where $[SSTA]_A$, $[SSTA]_B$, and $[SSTA]_C$ indicate the SST anomalies averaged over the regions of (10°S–10°N, 165°E–140°W), (15°S–5°N, 110°–70°W), and

(10°S–20°N, 125°–145°E), respectively. The monthly mixed layer depth (MLD) data are obtained from the Geophysical Fluid Dynamics Laboratory (GFDL) ocean data assimilation project with ~1° spatial resolution (http://www.gfdl.noaa.gov/ocean-data-assimilation-model-output).

The Butterworth filter is applied to extract low-pass signals from the time series of climatic indices (Seager et al., 2007). The Pearson correlation analysis is employed to derive linear associations between two or more variables (Von Storch and Zwiers, 2001). Partial correlation analysis is also utilized to control the influences of other variables when calculating associations (Von Storch and Zwiers, 2001).

3. Analysis results

This study uses time-lagged correlation analysis to examine whether the WNPSM and the NPGO exhibit lead/lag relationships. Results show that the WNPSM leads the NPGO for several months (Table 1). Specifically, the WNPSM in JJAS has an insignificant correlation with the NPGO is significant. However, the correlation becomes significant from spring to summer of the following year. Therefore, there exists a marked time-lagged association between the WNPSM and the NPGO. To consider the autocorrelation, we calculate the 'effective sample size' (Metz, 1991) for the time series. The correlation is still statistically significant at 0.05 significance level under an effective sample size of 40. This time lag appears to be more pronounced after a low-pass filter (>10 years retained) is applied to the WNPSM index and the NPGO index (Figure 1(b)). At multidecadal time scales, a marked time-lagged relationship is built between two indices because the correlation reaches 0.55 at 0.01 level of significance (Figure 1(b)). However, no significant association is detected between the EASM (Lau et al., 2000) and the NPGO and between the SCSSM and the NPGO (Table 1). However, the EASM index defined as an area-averaged seasonally (JJA) DNS at 850 hPa within the East Asian monsoon domain (10°–40°N, 110°–140°E) (Li and Zeng, 2002, 2003) has a correlation coefficient of 0.81 with the western North Pacific monsoon index (WNPMI). If we use the EASM index defined by Li and Zeng (2002), the correlation coefficient between EASM index and the NPGO is 0.177 that still fails to pass the significance test. This suggests that

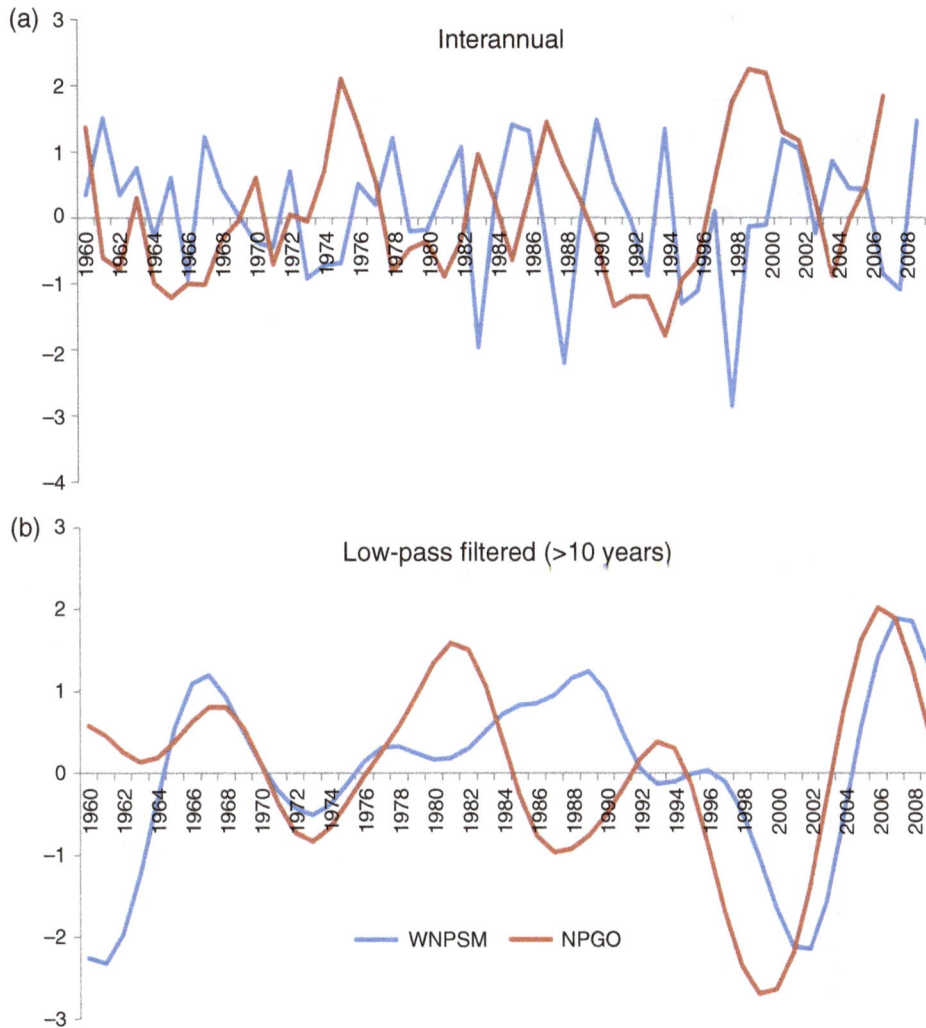

Figure 1. The original interannual (a) and low-pass filtered (b) time series of the WNPSM index (blue solid) and the NPGO index (red solid).

EASM is not significantly associated with the NPGO. Previous research has shown that the WNPSM is caused by meridional SST gradient in the WNP whereas the SCSSM and EASM are caused by land-sea thermal contrasts (Tao and Chen, 1987; Murakami and Matsumoto, 1994; Li and Wang, 2005). Therefore, the SCSSM and EASM have lower chance to affect the weather and climate in the North Pacific. The WNPSM, which is resultant from the SST gradient, may induce a wave train pattern to modulate the NPGO. It is known that the WNPSM is a strong air–sea interaction and is confined in the WNP (Tao and Chen, 1987; Murakami and Matsumoto, 1994; Li and Wang, 2005) whereas NPGO is characterized by strong signals in the North Pacific (e.g. Bond et al., 2003).

The Central Pacific (CP) El Niño is intimately associated with the NPGO (e.g. Di Lorenzo et al., 2008, 2010). Therefore, the partial correlation is obtained between the WNPSM and the NPGO by controlling the CP El Niño index. Partial correlation analysis results are consistent with those derived by the Pearson correlation analysis (Table 1). Therefore, the WNPSM leads the NPGO for around 6 months and such association is

generally independent of that caused by the CP El Niño. To isolate the impacts of CP El Niño, we also remove the linear fit of NPGO with the CP El Niño index when doing regression analysis. The results after removing the linear fit of CP El Niño are largely consistent with those without removing the linear fit. To make the association robust, we analyze the association between the WNPSM in JJA and NPGO. Nonetheless, no significant difference is identified between the WNPSM in JJAS and NPGO and between the WNPSM in JJA and NPGO. Although NPGO has strong decadal variability, this study focuses on the year-to-year variation of the WNPSM-NPGO association.

4. Possible mechanisms

Foregoing analyses have demonstrated that the WNPSM leads the NPGO for around 6 months. Previous studies have shown that the NPO, a striking SLP pattern in the North Pacific, plays a central role in forcing the NPGO (Di Lorenzo et al., 2008, 2010). Wang et al. (2001) has proposed that the anomalies

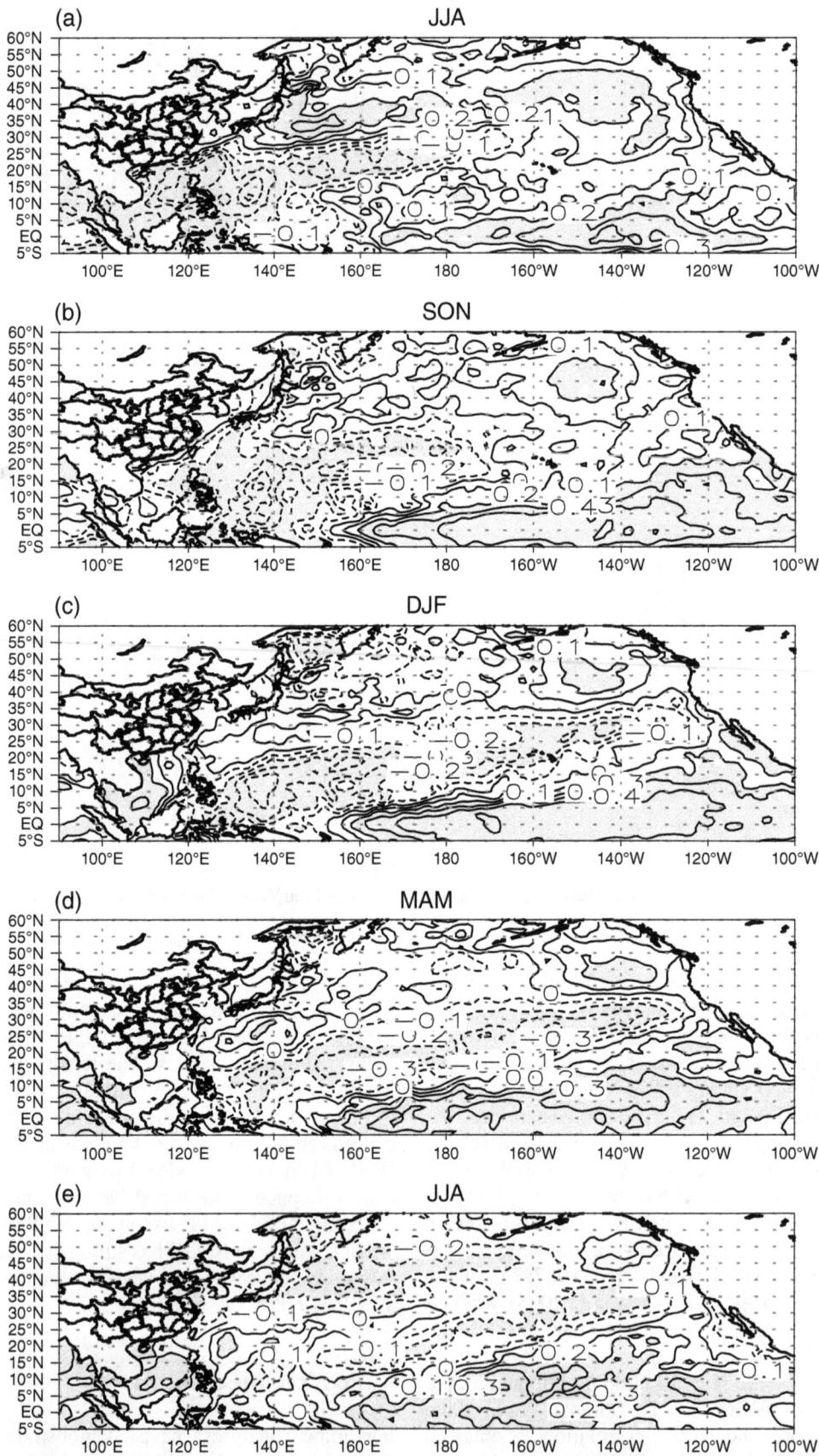

Figure 2. Correlation between the WNPSM in JJA and SST in (a) JJA, (b) SON, (c) DJF, (d) MAM in the following year, and (e) JJA in the following year. The shading represents those at 0.05 level of significance. The rectangle represents the region with strong NPGO signals.

in large-scale circulation are intimately associated with the WNPSM. Such anomalies are featured by two patterns in the North Pacific: one is a low-level cyclonic circulation at 20°N that extends from 100° to 170°E and the other is a prominent anti-cyclonic anomaly along 35°N laying from 110°E to the date line (Wang *et al.*, 2001). Such low-level patterns bear apparent similarity to the SLP patterns of the NPO. This provides insights into the interpretation of the links between the WNPSM and the NPGO. Our conjecture is that the WNPSM forces the 850-/500-hPa geopotential height anomalies in the WNP, which travel to the North Pacific by a wave train. The wave train pattern may affect the NPO pattern, which then drives the NPGO. The heat pattern induced by the wave train pattern may be submerged to deeper layers of the ocean in DJF while it is integrated over time in the North Pacific and then rise up in MAM when the mixed layer becomes shallow. We dissect from these aspects underlying physical mechanisms to interpret the WNPSM-NPGO association.

4.1. WNPSM and SST anomalies in the North Pacific

Figure 2 displays the spatial patterns of correlation between the WNPSM index in JJA and SST from JJA to JJA of the following year. In JJA and SON, the northern part of the NPGO is built (Figure 2(a) and (b)). After 3 months (in DJF), the southern part starts to develop in the key NPGO region (Figure 2(c)). It is shown that the NPGO pattern has been well organized in MAM of the following year (Figure 2(d)). In JJA of the following year, the northern part of the NPGO virtually retreats (Figure 2(e)). Therefore, we argue that the WNPSM tends to force the NPGO in a step-wise manner and the association between the WNPSM and the NPGO peaks in MAM of the following year. Of note is that the CP El Niño yields essential atmospheric forcing on the NPO and NPGO through the atmospheric bridge, as manifested in Di Lorenzo *et al.* (2010). This study indicates that the WNPSM may also play a role in the atmospheric forcing on the NPGO and NPO via the wave train pattern discussed in Wang *et al.* (2001) and Liu and Wang (2013). Figure 2 also shows that the SST patterns have certain similarity with the PDO in the North Pacific. We then examine whether WNPSM has strong associations with the PDO/NPO simultaneously or with certain time lags (Trenberth and Hurrell, 1994; Schneider and Cornuelle, 2005). The correlation between the WMPSM index in JJA has very weak correlation with the PDO index in JJA or the following several months (Table 2).

4.2. WNPSM and mixed layer depth

The lag of the impact of the signal on SSTs in the North Pacific can be analyzed by the annual cycle of the depth of the mixed layer in the North Pacific (Lienert *et al.*, 2011). The WNPSM signal (heat anomaly) may

Table 2. The correlation between WNPSM and PDO from summer to the summer in the following year.

Correlation	PDO JJA	PDO SON	PDO DJF	PDO MAM	PDO JJA
WNPSM (JJAS)	−0.09	−0.1	−0.08	−0.07	−0.08

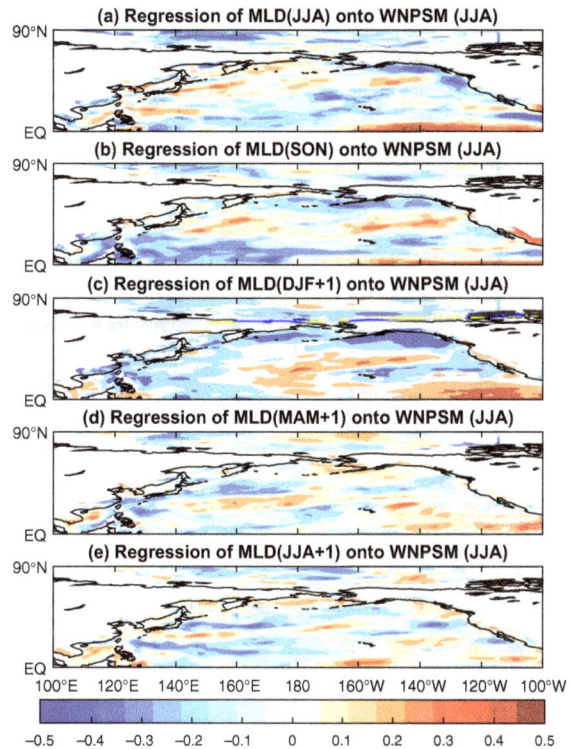

Figure 3. Correlation between the WNPSM in JJA and mixed layer depth averaged in (a) JJA, (b) SON, (c) DJF, (d) MAM in the following year, and (e) JJA in the following year. The shading represents those at 0.05 level of significance.

be submerged to deeper layers of the ocean in DJF while it is integrated over time in the North Pacific and then rise up in MAM when the mixed layer becomes shallow. Figure 3 supports the point that the mixed layer depth changes in the North Pacific are consistent with the forcing of WNPSM in JJA and following several months (Figure 3). In JJA, the regression of MLD onto WNPSM index is characterized by positive anomalies in the central North Pacific and negative anomalies in the northeastern Pacific coast. This spatial pattern lasts for 6–8 months. In MAM of the following year, the pattern reverses in which that there are negative anomalies in the central North Pacific and negative anomalies in the northeastern Pacific (Figure 3). This suggests that the heat restored in the ocean is released when the mixed layer depth becomes shallow in spring.

4.3. WNPSM and NPO

The NPO, a strong extra-tropical circulation in the atmosphere, features an apparent SLP dipole pattern (Rogers, 1981). This study detects a significant correlation between the WNPSM and the NPO at 0.01 level

Table 3. The correlation between the WNPSM index in JJAS and the NPO index.

Correlation	NPO JJA	NPO SON	NPO DJF
WNPSM (JJAS)	−0.74**	−0.29	0.03

**0.01 level of significance.

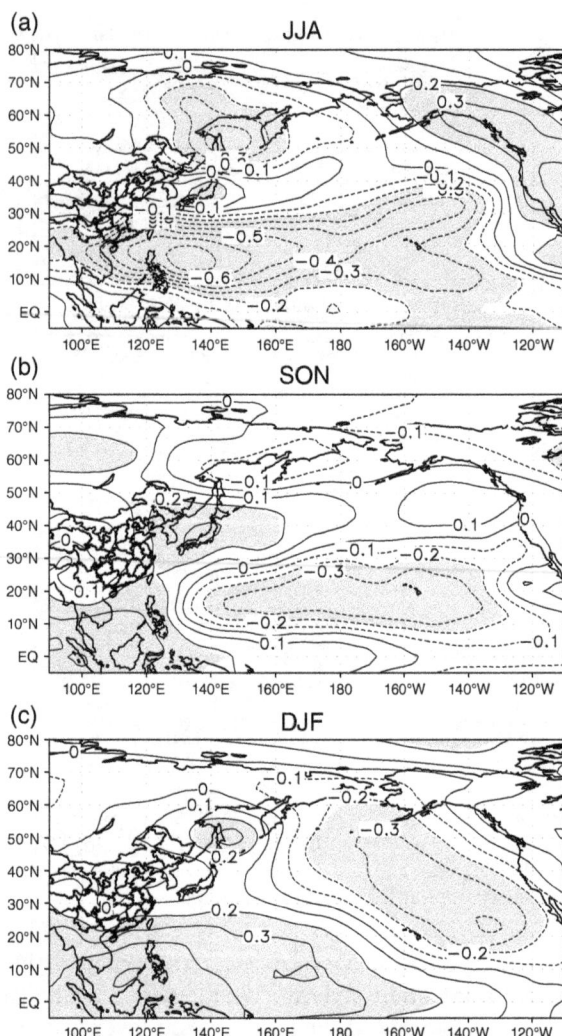

Figure 4. The spatial correlation between WNPSM and 850 hPa geopotential height in (a) JJA, (b) SON, and (c) DJF. The shading represents those at 0.05 level of significance.

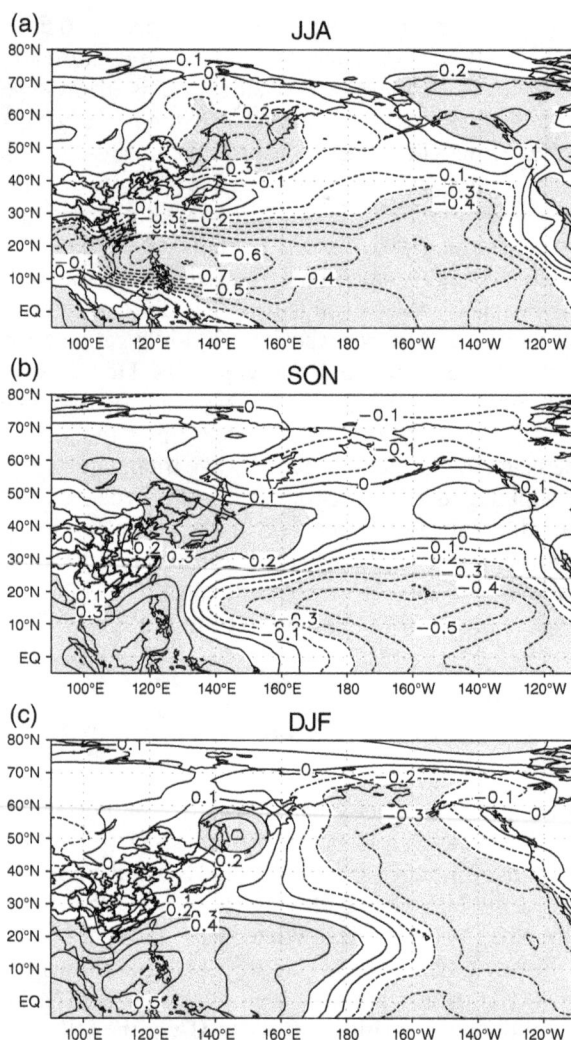

Figure 5. The spatial correlation between WNPSM and SLP in (a) JJA, (b) SON, and (c) DJF. The shading represents those at 0.05 level of significance.

of significance (Table 3). Nonetheless, this significant association becomes quite weak in the following autumn and winter, indicating that the forcing of the WNPSM on the NPO tends to be fast and transient.

Spatial correlation between the WNPSM in JJAS and the 850-hPa geopotential height (GPH850) anomalies displays three distinct patterns in JJA, SON, and DJF, respectively (Figure 4). In JJA, the SLP anomalies covers the region elongating from 100°E to 140°W and centered at around 20°N, which includes the southern pole of the NPO pattern as shown in Furtado *et al.* (2012). It is noteworthy that the 850-hPa geopotential (GPH850) patterns are featured by a meridional wave train pattern from 10° to 70°N anchoring at 140°E (Figure 4(a)). This wave train pattern was first documented in Wang *et al.*

(2001) and has been simulated (Liu and Wang, 2013) using the two-level model of Wang and Xie (1997). Moreover, the remarkable positive GPH850 anomalies are adjacent to the North Pacific (20°–65°N; Bond *et al.*, 2003), bearing a strong resemblance to the northern pole of the NPO. As reported in Di Lorenzo *et al.* (2008), the NPGO has its strongest signal in the Northeast Pacific near the California current. The GPH850 patterns (Figure 4(a)) are adversely consistent with the findings of Di Lorenzo *et al.* (2008). In SON, the elongated GPH850 patterns are confined to a smaller region and its center moves to the Central Pacific. In SON and DJF, the GPH850 patterns close to the California current can no longer be observed (Figure 4(b)).

The NPO is originally defined by SLP anomalies and features a north–south seesaw in SLP over the North Pacific (Walker and Bliss, 1932; Rogers, 1981) and is the atmospheric component of the NPGO (Di Lorenzo *et al.*, 2008). To substantiate our analysis, the correlation between the WNPSM and SLP sufficiently corroborate those between the WNPSM and GPH850 in JJA, SON, and DJF (Figures 4 and 5). The WNPSM

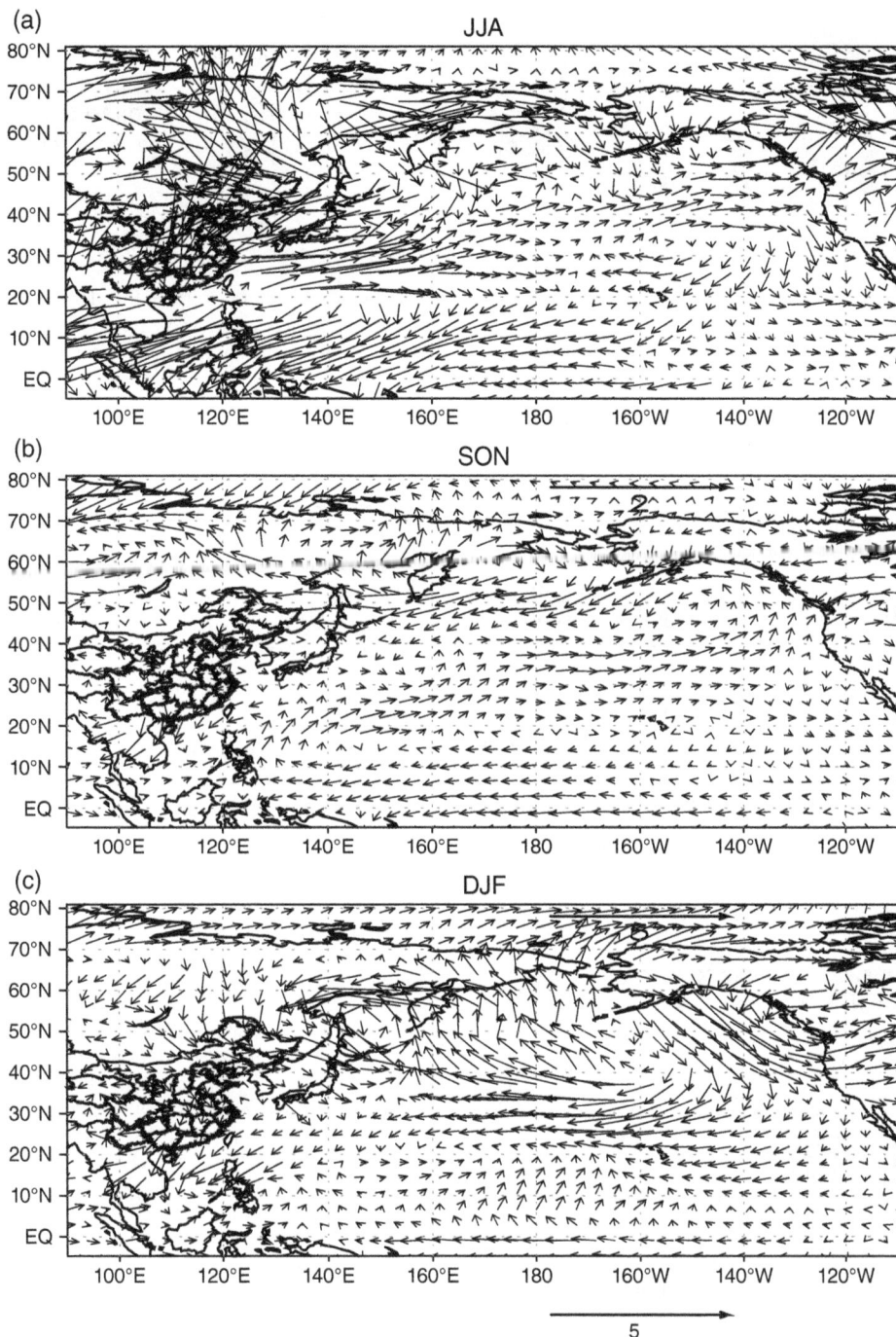

Figure 6. The regressions of zonal and meridional surface wind stress in (a) JJA, (b) SON, and (c) DJF on the WNPSM index in JJAS. The regression is multiplied by a factor of 100 (unit: N m^{-2}). The rectangle represents the key NPGO region in the North Pacific.

has a profound association with SLP anomalies in both the southern and northern poles of NPO in the JJA (Figure 5(a)). Such patterns, however, dissipate in SON and DJF, suggesting that the links between the WNPSM and NPO are built in summer and the forcing of NPO on NPGO is responsible for the SSH and SST patterns in next spring.

4.4. WNPSM and surface wind stress

Foregoing discussions have reflected that the WNPSM exerts pronounced impacts on the SLP and GPH850

anomalies. We further examine possible physical mechanisms underlying the linkage between the WNPSM and surface wind stress anomalies.

Figure 6 exhibits the regressions of zonal and meridional surface wind stress in JJA, SON, and DJF on the WNPSM index in JJAS. Interestingly, a marked dipole surface wind stress pattern is also observed in the key NPGO region (Figure 6). However, this pattern disappears in SON and DJF, echoing the fact that the linkage between the WNPSM and the NPO is built through the wave train pattern in summer, which

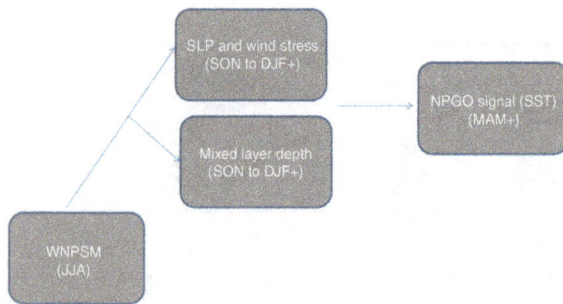

Figure 7. The schematic diagram of the linkage between WNPSM and NPGO.

is responsible for the anomalous NPO patterns. It is widely accepted that the NPO is conducive to surface wind stress anomalies that force the SST and SSH variability and is responsible for the NPGO (Di Lorenzo *et al.*, 2008, 2010).

Figure 7 summarizes the mechanisms underlying the linkage between WNPSM and NPGO. The WNPSM modulates SLP and wind stress in the North Pacific. Such changes in SLP and wind stress can restore heat in the ocean by modulating mixed layer depth in the following months. Moreover, when the MLD gets shallow in the MAM of the following years, the heat in the ocean is released from the ocean surface, leading to changes in the NPGO (Figure 7).

5. Discussion and conclusion

This study has made an attempt to investigate whether and how the WNPSM influences the NPGO. The authors' research findings are summarized as follows:

1. In addition to the CP El Niño, the WNPSM also exerts significant influences on the NPGO. Time-lagged correlation analysis indicates that the WNPSM has a significant association with NPGO.
2. A remarkable time lag is identified between the WNPSM and the NPGO. In other words, the WNPSM in JJAS has significant modulation on the NPGO in MAM of the following year.
3. The influences of the WNPSM on the NPGO arise from the wave train pattern forced by the WNPSM. Such wave train pattern modulates the SLP pattern of the NPO simultaneously, which forces anomalous surface wind stress on the sea surface, causing the NPGO-pattern SSTA in the North Pacific in the following spring.

The WNPSM exhibits a new manner to modulate the NPGO different from that by the CP El Niño. This study confirms the striking role of the NPO in forcing the NPGO and proposes a new way to modulate the NPO by tropical forcing. It is noted that this study proposes a hypothesis on the modulation of WNPSM on NPGO. Further study is required to further verify the physical mechanisms underlying the linkage. Decadal prediction of interannual tropical and North Pacific SST is examined with a dynamical climate model (Lienert

and Doblas-Reyes, 2013). Due to the rising role of the NPGO in modulating weather and climate systems (e.g. Zhang *et al.*, 2013), it is of great significance to predict the NPGO. In addition to the CP El Niño, the WNPSM should be taken into account in the prediction of the NPGO.

Acknowledgements

The authors are grateful to the editor Dr Christopher Holloway and two anonymous reviewers for their insightful comments. This research was supported by the National Natural Science Foundation of China (grant numbers: 41201045, 41401052, 41505035, and 41575078).

References

Bond NA, Harrison DE. 2000. The Pacific Decadal Oscillation, air-sea interaction and central north Pacific winter atmospheric regimes. *Geophysical Research Letters* **27**(5): 731–734.

Bond NA, Overland JE, Spillane M, Stabeno P. 2003. Recent shifts in the state of the North Pacific. *Geophysical Research Letters* **30**(23): 2183.

Chang C-P, Krishnamurti TN. 1987. *Monsoon Meteorology*. Oxford University Press: New York, NY.

Chiang JCH, Vimont DJ. 2004. Analogous Pacific and Atlantic meridional modes of tropical atmosphere–ocean variability. *Journal of Climate* **17**(21): 4143–4158, doi: 10.1175/JCLI4953.1.

Di Lorenzo E, Schneider N, Cobb K, Franks P, Chhak K, Miller A, McWilliams J, Bograd S, Arango H, Curchitser E. 2008. North Pacific Gyre Oscillation links ocean climate and ecosystem change. *Geophysical Research Letters* **35**(8): L08607.

Di Lorenzo E, Cobb K, Furtado J, Schneider N, Anderson B, Bracco A, Alexander M, Vimont D. 2010. Central Pacific El Niño and decadal climate change in the North Pacific Ocean. *Nature Geoscience* **3**(11): 762–765.

Ding YH. 2007. The variability of the Asian summer monsoon. *Journal of the Meteorological Society of Japan* **85B**: 21–54.

Ding Y, Chan JCL. 2005. The East Asian summer monsoon: an overview. *Meteorology and Atmospheric Physics* **89**(1–4): 117–142.

Ding R, Li J, Tseng Y-h, Sun C, Guo Y. 2015. The Victoria mode in the North Pacific linking extratropical sea level pressure variations to ENSO. *Journal of Geophysical Research, [Atmospheres]* **120**: 27–45, doi: 10.1002/2014JD022221.

Furtado J, Di Lorenzo E, Anderson B, Schneider N. 2012. Linkages between the North Pacific Oscillation and central tropical Pacific SSTs at low frequencies. *Climate Dynamics* **39**(12): 2833–2846.

Jiang X, Lau N-C. 2008. Intraseasonal teleconnection between North American and western North Pacific monsoons with 20-day time scale. *Journal of Climate* **21**(11): 2664–2679.

Kalnay E, Kanamitsu M, Kistler R, Collins W, Deaven D, Gandin L, Iredell M, Saha S, White G, Woollen J. 1996. The NCEP/NCAR 40-year reanalysis project. *Bulletin of the American Meteorological Society* **77**(3): 437–471.

Lau KM, Kim KM, Yang S. 2000. Dynamical and boundary forcing characteristics of regional components of the Asian summer monsoon. *Journal of Climate* **13**(14): 2461–2482.

Li T, Wang B. 2005. A review on the western North Pacific monsoon: synoptic-to-interannual variabilities. *Terrestrial, Atmospheric and Oceanic Sciences* **16**(2): 285–314.

Li J, Zeng Q. 2002. A unified monsoon index. *Geophysical Research Letters* **29**(8): 1274.

Li J, Zeng Q. 2003. A new monsoon index and the geographical distribution of the global monsoons. *Advances in Atmospheric Sciences* **20**(2): 299–302.

Lienert F, Doblas-Reyes FJ. 2013. Decadal prediction of interannual tropical and North Pacific sea surface temperature. *Journal of Geophysical Research, [Atmospheres]* **118**: 5913–5922.

Lienert F, Fyfe JC, Merryfield WJ. 2011. Do climate models capture the tropical influences on North Pacific sea surface temperature variability? *Journal of Climate* **24**(23): 6203–6209.

Liu F, Wang B. 2013. Mechanisms of global teleconnections associated with the Asian summer monsoon: an intermediate model analysis. *Journal of Climate* **26**(5): 1791–1806.

Mantua NJ, Hare SR, Zhang Y, Wallace JM, Francis RC. 1997. A Pacific interdecadal climate oscillation with impacts on salmon production. *Bulletin of the American Meteorological Society* **78**(6): 1069–1079.

Metz W. 1991. Optimal relationship of large-scale flow patterns and the barotropic feedback due to high-frequency eddies. *Journal of the Atmospheric Sciences* **48**: 1141–1159.

Murakami T, Matsumoto J. 1994. Summer monsoon over the Asian continent and western North Pacific. *Journal of the Meteorological Society of Japan* **72**(5): 719–745.

Rogers JC. 1981. The North Pacific Oscillation. *Journal of Climatology* **1**(1): 39–57.

Schneider N, Cornuelle BD. 2005. The forcing of the Pacific Decadal Oscillation. *Journal of Climate* **18**: 4355–4373.

Seager R, Ting M, Held I, Kushnir Y, Lu J, Vecchi G, Huang H-P, Harnik N, Leetmaa A, Lau N-C. 2007. Model projections of an imminent transition to a more arid climate in southwestern North America. *Science* **316**(5828): 1181–1184.

Tao S, Chen L. 1987. A review of recent research on the East Asian summer monsoon in China. In *Monsoon Meteorology*, Chang CP, Krishnamurti TN (eds). Oxford University Press: New York, NY; 60–92.

Trenberth KE, Hurrell JW. 1994. Decadal atmosphere–ocean variations in the Pacific. *Climate Dynamics* **9**: 303–319.

Von Storch H, Zwiers FW. 2001. *Statistical Analysis in Climate Research*. Cambridge University Press: Cambridge.

Walker GT, Bliss E. 1932. World weather V. *Memoirs of the Royal Meteorological Society* **4**: 53–84.

Wang B, Fan Z. 1999. Choice of South Asian summer monsoon indices. *Bulletin of the American Meteorological Society* **80**(4): 629–638.

Wang B, Lin Ho. 2002. Rainy season of the Asian-Pacific summer monsoon. *Journal of Climate* **15**(4): 386–398.

Wang B, Xie X. 1997. A model for the Boreal Summer Intraseasonal Oscillation. *Journal of the Atmospheric Sciences* **54**(1): 72–86.

Wang B, Wu R, Lukas R. 1999. Roles of the western North Pacific wind variation in thermocline adjustment and ENSO phase transition. *Journal of the Meteorological Society of Japan* Series 2 **77**: 1–16.

Wang B, Wu RG, Lau KM. 2001. Interannual variability of the Asian summer monsoon: contrasts between the Indian and the western North Pacific-east Asian monsoons. *Journal of Climate* **14**(20): 4073–4090.

Zhang Y, Wallace JM, Battisti DS. 1997. ENSO-like interdecadal variability: 1900–93. *Journal of Climate* **10**(5): 1004–1020.

Zhang W, Leung Y, Min J. 2013. North Pacific Gyre Oscillation and the occurrence of western North Pacific tropical cyclones. *Geophysical Research Letters* **40**: 5205–5211, doi: 10.1002/grl.50955.

Zhang W, Vecchi GA, Murakami H, Villarini G, Jia L. 2016. The Pacific meridional mode and the occurrence of tropical cyclones in the western North Pacific. *Journal of Climate* **29**: 381–398.

26

Delayed effect of Arctic stratospheric ozone on tropical rainfall

Fei Xie,[1] Jiankai Zhang,[2]* Wenjun Sang,[2] Yang Li,[2] Yulei Qi,[3] Cheng Sun,[1] Yang Li[3] and Jianchuan Shu[4]

[1] College of Global Change and Earth System Science, Beijing Normal University, Beijing, China
[2] Key Laboratory for Semi-Arid Climate Change of the Ministry of Education, College of Atmospheric Sciences, Lanzhou University, Lanzhou, China
[3] School of Atmospheric Sciences, Chengdu University of Information Technology, Chengdu, China
[4] Institute of Plateau Meteorology, China Meteorological Administration, Chengdu, China

*Correspondence to:
Dr J. Zhang, College of Atmospheric Sciences, NO. 222 Tianshui South Road, Lanzhou, 730000, Gansu Province, China.
E-mail: jkzhang@lzu.edu.cn

Abstract

The tropical precipitation has a wide effect on the tropical economics and social life. Many studies made efforts to improve the tropical precipitation forecast using tropical climate factors. This study, based on observations, found that Arctic stratospheric ozone (ASO) could exert a significant effect on the tropical precipitation, i.e. there is more (less) rainfall over the eastern Pacific and less (more) precipitation over the western Pacific when the ASO anomalies are lower (larger) than normal. It is because a decrease (increase) in ASO could affect El Niño (La Niña) events and lead to a weakened (enhanced) Walker circulation. Time-slice experiments confirmed that the ASO anomalies can force El Niño–Southern Oscillation-like anomalies of tropical sea surface temperature and subsequent tropical precipitation anomalies. In addition, the ASO variations could also change the occurrence probability of extreme precipitation in the tropics. During the anomalously low (high) ASO events, there are more occurrences of heavier precipitation over the eastern Pacific (western Pacific) and of lighter precipitation over the western Pacific (eastern Pacific). Furthermore, the ASO variations lead tropical rainfall by approximately 21 months, suggesting that the ASO can serve as a potentially effective predictor of tropical rainfall.

Keywords: Arctic stratospheric ozone (ASO); tropical rainfall; ENSO; Walker circulation

I. Introduction

Rainfall variability has an enormous impact on human society through modulating the availability of water resources and its effects on, e.g. agricultural yields, forage, and hydroelectric power generation. Which factors affect rainfall variability have been broadly studied, and this topic has been one of the hottest research topics. There are a large number of studies focusing on the effects of important climate change forcings, e.g. El Niño–Southern Oscillation (ENSO), North Atlantic Oscillation, carbon dioxide doubling, etc. on the global rainfall variability. Recently, there are more and more studies pointed out that the stratospheric ozone also can influence precipitation.

Stratospheric ozone is vital in protecting Earth's life by absorbing harmful solar ultraviolet radiation (Lubin and Jensen, 2002; Chipperfield et al., 2015). It is also a key in controlling of the stratospheric temperature through atmospheric radiative heating that influences stratospheric circulation and chemical composition (Tung, 1986; Haigh, 1994; Ramaswamy et al., 1996; Forster and Shine, 1997), which can even affect tropospheric climate (Baldwin and Dunkerton, 2001; Graf and Walter, 2005; Cagnazzo and Manzini, 2009; Ineson and Scaife, 2009; Reichler et al., 2012; Karpechko et al., 2014; Kidston et al., 2015; Zhang et al., 2016).

There has been a strong decline in Antarctic stratospheric ozone in the past six decades (Solomon, 1990, 1999; Ravishankara et al., 1994, 2009; Dhomse et al., 2006, 2013), which can significantly influence the Southern Hemisphere high and middle latitudes circulation (Son et al., 2008, 2010; Thompson et al., 2011; Gerber and Son, 2014; Waugh et al., 2015), and even affect the Southern Hemisphere extratropical and tropical precipitation (Feldstein, 2011; Kang et al., 2011).

Although Arctic stratospheric ozone (ASO) has not shown such a strong decrease as Antarctic ozone, the amplitude of the inter-annual variability of ASO is comparable with, or even much larger than, that of Antarctic stratospheric ozone (Manney et al., 2011). Thus, the effects of ASO variations on Northern Hemisphere rainfall also deserve investigation.

Smith and Polvani (2014) and Calvo et al. (2015) performed numerical experiments and revealed that the precipitation variability over some regions in the high–middle latitudes of the Northern Hemisphere significantly responses to extreme ASO anomaly events. More recently, Ivy et al. (2017) presented observational evidence for linkages between extreme ASO anomalies in March and Northern Hemisphere precipitation in spring (March–April), suggesting that March ASO is a useful indicator of spring-averaged (March–April) precipitation in specific regions of the

Northern Hemisphere high–middle latitudes. However, it is still unknown whether the effect of ASO on precipitation, also like that of Antarctic stratospheric ozone, can extend from high–middle latitudes to tropics.

The tropical rainfall variability is significantly modulated by ENSO and subsequent variations in the Walker circulation (Lau and Sheu, 1988; Rasmusson and Arkin, 1993; Dai *et al.*, 1997; Camberlin *et al.*, 2004; Power *et al.*, 2013; Chung and Power, 2014; Huang and Xie, 2015; Huang, 2016; Yim *et al.*, 2016). The Walker circulation is a large overturning cell that spans the tropical Pacific Ocean, characterized by rising motion (lower sea level pressure; SLP) over Indonesia and sinking motion (higher SLP) over the eastern Pacific (Bjerknes, 1969; Gill, 1980). Fluctuations in the Walker circulation reflect changes in the location and strength of tropical upwelling and downwelling, which have important impacts on tropical climate (Horel and Wallace, 1981; Kousky *et al.*, 1984; Wang *et al.*, 2013, 2015; Hu *et al.*, 2016). Based on observational analyses and transient simulations, Xie *et al.* (2016) recently found that the ASO anomalies possibly affect ENSO events. Based on this study, there is a possibility that the ASO anomalies can affect tropical rainfall variations.

This work is to found the observed evidences that the tropical rainfall variations is related to the ASO changes by linking the changes in ASO, ENSO, the Walker circulation, and tropical rainfall. Furthermore, the second aim of this paper is to perform a series of sensitive experiments using state-of-the-art climate model to demonstrate that the ASO variability can force the ENSO-like sea surface temperature (SST) anomalies and relevant tropical rainfall variations. It not only further extend the implication of Xie *et al.* (2016) that ASO anomalies affect tropical climate, but also improve the understanding of the characteristic of tropical rainfall variations related to ASO changes.

2. Data, methods, and simulations

In the Northern Hemisphere high latitudes, the variability and depletion of ozone concentrations are most pronounced in the region 60–90°N, at an altitude of 150–50 hPa (Manney *et al.*, 2011). The monthly anomaly of ozone concentration (after removing the climatological mean seasonal cycle) averaged over this region is used as the ASO index. This definition of the ASO index follows Xie *et al.* (2016). Note that hereafter −ASO index means the inverted ASO index. Ozone values are derived from the Stratospheric Water and OzOne Satellite Homogenized (SWOOSH, 1984–2013) dataset (Davis *et al.*, 2016), which is in good agreement with ozone ($r = 0.89$) from the Global Ozone Chemistry and Related trace gas Data Records for the Stratosphere (GOZCARDS, 1979–2012) project (Froidevaux *et al.*, 2015).

The NASA Global Precipitation Climatology Project (GPCP) monthly precipitation dataset from 1979 to the present, which combines observations and satellite precipitation data into a 2.5° × 2.5° global grid (Huffman *et al.*, 2009), is used in this study. SST and SLP are derived from the UK Met Office Hadley Centre for climate prediction and research SST (HadSST) and SLP (HadSLP) field datasets, respectively. Vertical velocity is from the National Centers for Environmental Prediction-Department of Energy (NCEP-DOE) dataset (version 2; NCEP2).

The calculation of the two-tailed Student's t-test and the effective number (N^{eff}) of degrees of freedom (DOF) follows Xie *et al.* (2016). The N^{eff} of DOF is determined by the following approximation:

$$\frac{1}{N^{eff}} \approx \frac{1}{N} + \frac{2}{N}\sum_{j=1}^{N}\frac{N-j}{N}\rho_{XX}(j)\,\rho_{YY}(j),$$

where, N is the sample size, and ρ_{XX} and ρ_{YY} are the autocorrelations of two sampled time series, X and Y, respectively, at time lag j.

The National Center for Atmospheric Research's (NCAR) Community Earth System Model (CESM, version 1.0.6) is used in this study. CESM includes ocean (POP2), land (CLM4), sea ice (CICE), and interactive atmosphere (CAM/WACCM) components, and is a fully coupled global climate model. The Whole Atmosphere Community Climate Model (WACCM, version 4; Marsh *et al.*, 2013) is utilized for the atmospheric component. WACCM4 is a climate–chemistry model with 66 vertical levels extending from the surface to approximately 140 km, and with approximately 1 km vertical resolution in the tropical tropopause and lower stratosphere layers. For our study, we disabled the interactive chemistry of WACCM4. All simulations employed a horizontal resolution of 1.9° × 2.5° (latitude × longitude) for the atmosphere and approximately the same for the ocean. All the forcing data employed in this study are available from the CESM model input data repository. Nine experiments were performed. An overview of all coupled experiments is given in Table 1.

3. Results

Figure 1 shows the correlation coefficients between the global field of rainfall and the −ASO index with the lead of the ASO index ranging from 0 to 21 months, at 3-month intervals. There are almost no significant correlation coefficients between ASO and rainfall variations over the globe while the time lag is less than 15 months (Figures 1(a)–(f)). When the ASO variations lead rainfall by approximately 18 months, however, significant negative correlation coefficients appear over Indonesia and northern South America, and positive correlation coefficients over the tropical eastern Pacific (Figure 1(g)). The patterns of correlation coefficients are most robust for ASO leading rainfall by approximately 21 months (Figure 1(h)). To further probe the relationship between the changes in ASO and tropical rainfall over the region 5°N–5°S and 120–150°E and the region 0–10°S and 150–120°W

Table 1. Fully coupled CESM-WACCM4 experiments with various specified ozone forcing.

Exp[a]	Specified ozone forcing	Other forcing
E_{11} E_{12} E_{13}	Time-slice run as the control experiment. The specified ozone forcing is a 12-month cycle of monthly ozone averaged from 1980 to 2015. Three ensemble simulations using slightly different initial conditions.[b]	Fixed solar constant, fixed greenhouse gas (GHG) values [averages of emissions scenario A2 of the Intergovernmental Panel on Climate Change (WMO, 2003) over the period 1980–2015], volcanic aerosols [from the stratospheric processes and their role in climate (SPARC) chemistry–climate model validation (CCMVal) REF-B2 scenario recommendations], and QBO phase signals with a 28-month zonal wind fixed cycle.
E_{21} E_{22} E_{23}	Same as E_{1x}, except that the ozone in the region 30–90°N at 300–30 hPa[c] is decreased by 15% compared with E_{1x}. Three ensemble simulations using slightly different initial conditions.	Same as E_{1x}.
E_{31} E_{32} E_{33}	Same as E_{1x}, except that ozone in the region 30–90°N at 300–30 hPa is increased by 15% compared with E_{1x}. Three ensemble simulations using slightly different initial conditions.	Same as E_{1x}.

[a]Integration time for E_{1x-3x} is 30 years.
[b]To produce different initial conditions, the parameter <pertlim> is used in the CESM model, which produces an initial temperature perturbation. The magnitude is about e^{-14}.
[c]To avoid the effect of the boundary of ozone change on the Arctic stratospheric circulation simulation, the replaced region (30–90°N, 300–30 hPa) was larger than the region used to define the ASO index (60–90°N, 150–50 hPa).

during the past three decades, their lead–lag correlation is shown in Figure 2. A significant negative (positive) lagged correlation (lag ca. 20 months) is also observed as shown in Figures 2(a) and (b). Figures 1 and 2 imply that depleted ASO may be related to decreased rainfall in Indonesia and northern South America, and increased rainfall in the tropical eastern Pacific, and vice versa for high ASO.

It is well known that ENSO events can influence the Walker circulation across the equator, thereby affecting tropical precipitation (Horel and Wallace, 1981; Kousky et al., 1984; Lau and Sheu, 1988; Rasmusson and Arkin, 1993; Dai et al., 1997). The patterns of correlation between global rainfall and the −ASO index (Figure 1(h)) resemble the ENSO-like anomalies in tropical rainfall that has been noted by previous studies (Lau and Sheu, 1988; Rasmusson and Arkin, 1993; Dai et al., 1997; Camberlin et al., 2004; Power et al., 2013; Chung and Power, 2014; Huang and Xie, 2015; Huang, 2016; Yim et al., 2016). An ASO decrease in prior may cause a warm phase of ENSO activity (Xie et al., 2016); i.e. positive SST anomalies over the eastern Pacific and negative SST anomalies over the western Pacific. Consequently, there is anomalous upward motion over the tropical eastern Pacific and downwelling over Indonesia and northern South America, as suggested by the negative (positive) correlation coefficients between the −ASO index and pressure vertical velocity approximately 21 months later over the eastern Pacific (Indonesia and northern South America) (Figure 3(a)). The SLP change over the tropical Pacific is another indicator of change in the Walker circulation. There are negative correlation coefficients between the −ASO index and SLP over the eastern Pacific and positive correlations over Indonesia and northern South America (Figure 3(b)), indicating falling SLP over the eastern Pacific and rising SLP

over Indonesia and northern South America when ASO decreases. All above analysis supports the hypothesis that, as described in Section 1, ASO depletion results in a warm ENSO phase (Xie et al., 2016) which is associated with a weakened Walker circulation (Heureux et al., 2013; Kociuba and Power, 2015).

Note that a 35-month low-pass filter is applied in Figure 3. On one hand, the interannual variations of both ASO and tropical signals are affected by quasi-biennial-oscillation (QBO). The 35-month low-pass filter would favor to remove the effect of QBO on the correlation between ASO and tropical vertical velocity and SLP. On the other hand, according to the results in Xie et al. (2016), this connection is gradually achieved from extratropics to tropics via ocean–atmosphere dynamic interaction; i.e. North Pacific Oscillation (NPO) anomalies caused by ASO change force the Victoria Mode (VM) of the North Pacific, which further modulates the development of ENSO through the seasonal footprinting mechanism. Finally, the tropical vertical velocity and SLP is changed with regard to the ENSO development. Previous studies have demonstrated that only the low-frequency variations in VM are effective in modulating the development of ENSO through the seasonal footprinting mechanism (Vimont et al., 2001; Ding et al., 2015). This is why a 35-month low-pass filter is applied to tropical vertical velocity and SLP variations in Figure 3. It would highlight the low-frequency variations, which are related to ozone changes, in the vertical velocity and SLP but remove noise-like high-frequency variations.

An ensemble of nine time-slice experiments ($E_{1x}–E_{3x}$) is further performed to simulate the influence of ASO changes on the spatial patterns of tropical rainfall. The results show that decreased (increased)

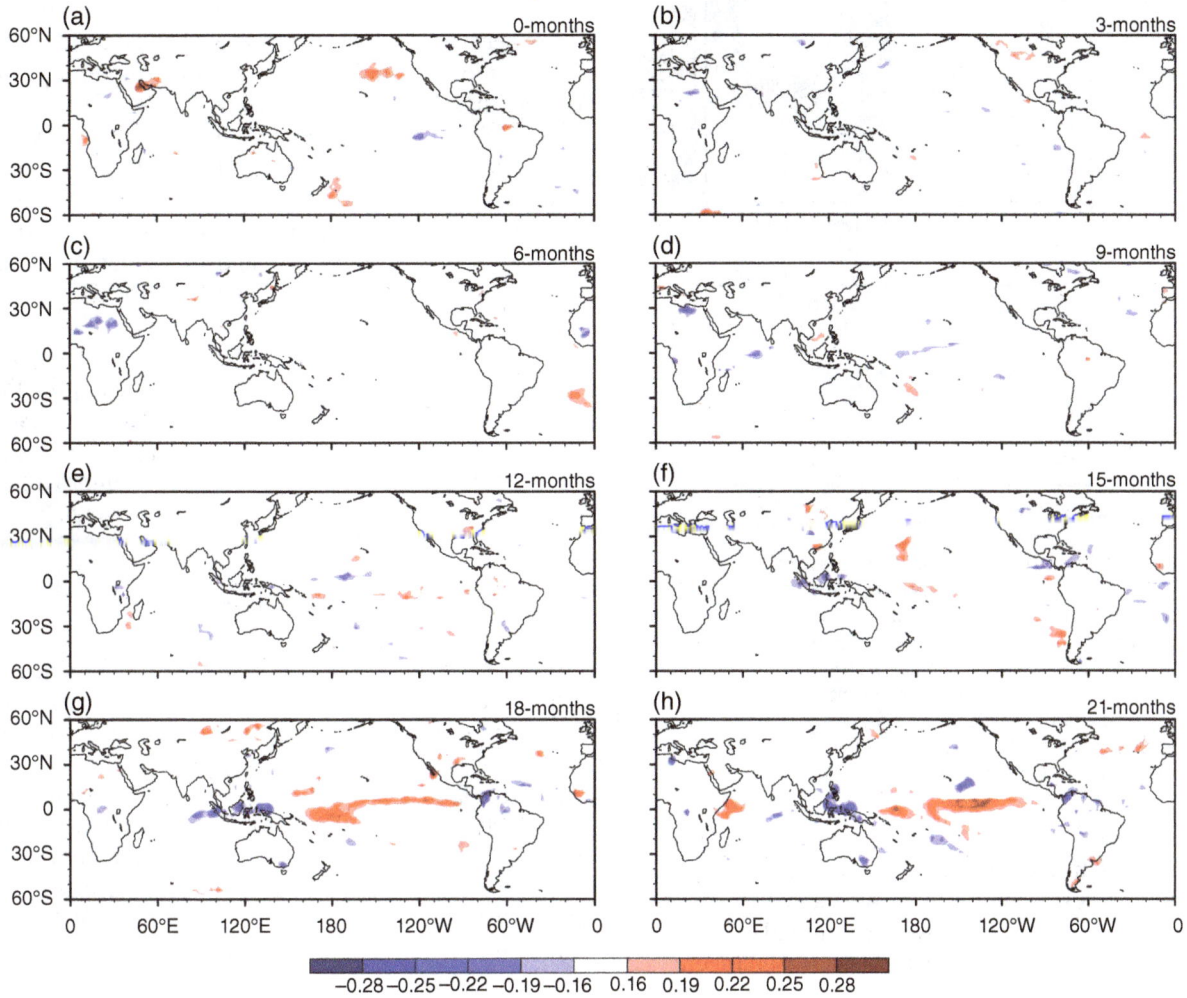

Figure 1. Correlation coefficients between the −ASO index and rainfall, for the ASO index leading rainfall by (a) 0 months, (b) 3 months, (c) 6 months, (d) 9 months, (e) 12 months, (f) 15 months, (g) 18 months, and (h) 21 months. Ozone is based on SWOOSH data. Rainfall is derived from the GPCP data. Only regions with correlations significant at the 95% confidence level are shaded (see Section 2 for details of the statistical significance test).

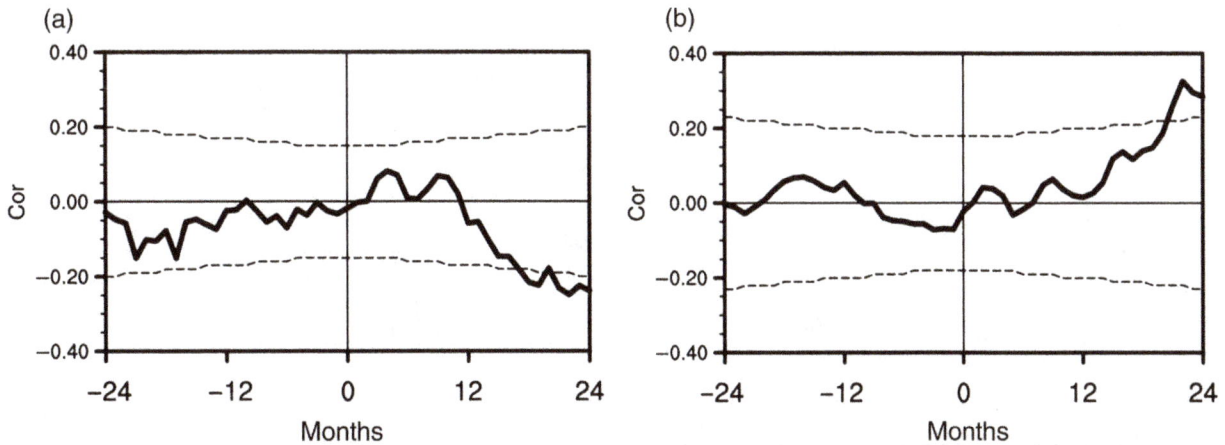

Figure 2. Lead−lag correlation between the monthly −ASO and tropical rainfall variations; the ASO is a time series of ozone averaged over the region 60−90°N at 150−50 hPa based on SWOOSH data; tropical rainfall is averaged in the region 5°N−5°S and 120−150°E for (a) and in the region 0−10°S and 150−120°W for (b) from the GPCP data. The positive months on the x-axis refer to the ASO leading tropical rainfall and negative months refer to tropical rainfall leading the ASO; the dashed lines denote the 90% confidence level.

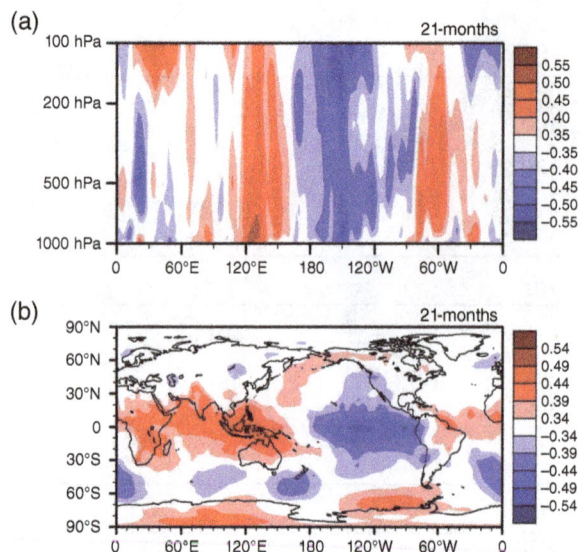

Figure 3. Correlation coefficients between the −ASO index and tropical mean (20°S–20°N) pressure vertical velocity (a) and SLP (b), with the ASO index variations leading by 21 months. Ozone is based on SWOOSH data. Pressure vertical velocity is from the NCEP2 reanalysis data. SLP is from Hadley Center. Only regions with correlations significant at the 95% confidence level are shaded (see Section 2 for details of the statistical significance test). All values are detrended, and a 35-month low-pass filter is applied (see Xie *et al.*, 2016).

ASO events can force positive (negative) SST anomalies in the tropical eastern Pacific representing warm (cold) ENSO events (Figures 4(a) and (b)) and force decreased (increased) convective rainfall in Indonesia and northern South America but increased (decreased) convective rainfall in the tropical eastern Pacific

(Figures 4(c) and (d)). This result is corresponding to above statistical analysis. The time-slice numerical experiments fully support the statistical conclusions based on observations. Figure 5 shows the probability density distribution *versus* precipitation over the region 5°N–5°S and 120–150°E and the region 0–10°S and 150–120°W from the experiments E_{1x}–E_{3x}. It is found that there is a larger occurrence probability of heavier precipitation over the region 0–10°S and 150–120°W (5°N–5°S and 120–150°E) and of lighter precipitation over the region 5°N–5°S and 120–150°E (0–10°S and 150–120°W) during the ASO decrease (increase) events. This result further confirms the ASO impacts on tropical rainfall. Furthermore, this figure implies that the ASO decrease (increase) can also improve (suppress) the occurrence of extreme precipitation events over the West Pacific region and suppress (improve) the occurrence over Middle and East Pacific Ocean.

4. Conclusions and discussions

Observations were used to analyze the influence of ASO changes on tropical rainfall. The atmospheric bridge connecting the stratospheric ozone variations over the polar region, the Walker circulation, and tropical precipitation should be linked to the mechanism by which ASO variations modulate ENSO variability revealed by Xie *et al.* (2016). Twenty-one months after ASO depletion, the warm ENSO anomaly caused by ASO depletion leads to a weakened tropical Walker circulation, and vice versa after an increase in ASO. The anomalous Walker circulation changes the vertical motion in the tropics, thereby affecting tropical

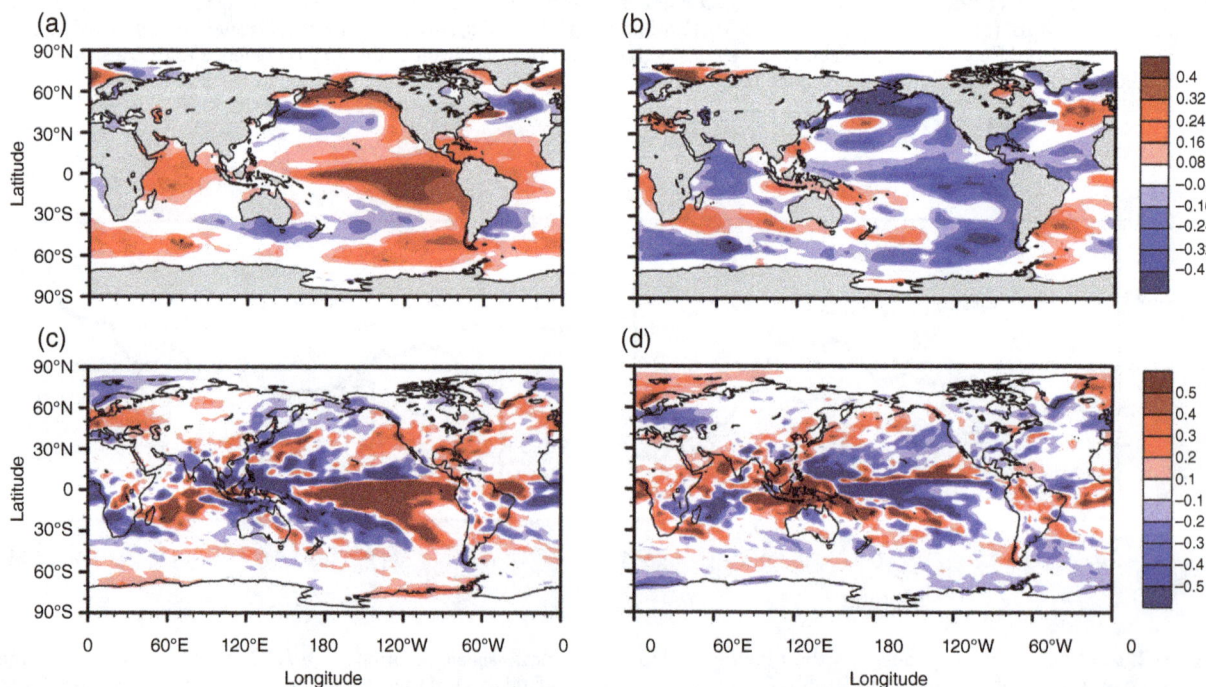

Figure 4. (a) Simulated SST anomalies (°C) caused by a 15% decrease in ASO [($E_{21} + E_{22} + E_{23}$) − ($E_{11} + E_{12} + E_{13}$)]. (b) Same as (a), but caused by a 15% increase in the ASO [($E_{31} + E_{32} + E_{33}$) − ($E_{11} + E_{12} + E_{13}$)]. (c) and (d) Same as (a) and (b), but for simulated convective precipitation rate anomalies (m s^{-1}) × 1.0e8.

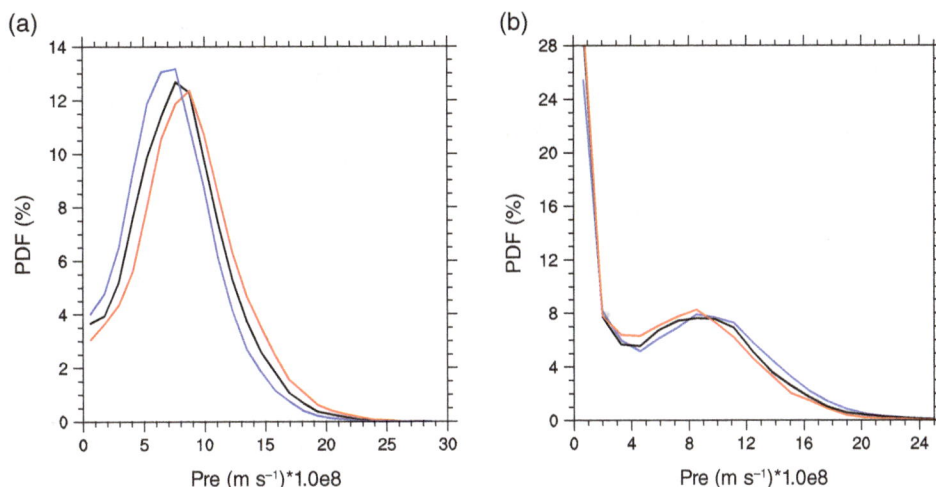

Figure 5. The probability density distribution *versus* precipitation over the region 5°N–5°S and 120–150°E for (a) and the region 0–10°S and 150–120°W for (b). The black, red and blue lines are calculated from E_{1x}, E_{2x}, and E_{3x}, respectively.

rainfall. The transient and ensemble time-slice experiments fully support the results from the statistical analysis of the observations. In particular, the ensemble time-slice experiments further confirm that ASO depletion (increase) indeed forces a warm (cold) ENSO-like SST anomalies as suggested by Xie *et al.* (2016). The simulation also show that more precipitation than normal over the eastern Pacific and less precipitation over Indonesia and northern South America are significantly associated with depleted ASO. The situation is reversed for increased ASO.

An interesting feature in Figure 4 is that the intensities of ENSO-like signs of SST and rainfall anomalies forced by the ASO decrease (Figures 4(a) and (c)) are asymmetry with that forced by the ASO increase (Figures 4(b) and (d)). Figure 6 depicts the sum of SST (rainfall) anomalies caused by the 15% decrease and increase in ASO. It illustrates that the positive SST and rainfall anomalies over center and eastern Pacific (corresponding to El Niño activity) forced by the ASO decrease are larger than the negative anomalies (corresponding to La Niña activity) forced by the ASO increase. The results suggest that the influences of ASO decrease on tropical ENSO-like SST and precipitation anomalies appears to be stronger than those of ASO increase. This may be because the effects of the equal decrease and increase in ASO on Arctic stratospheric circulation are asymmetry (Xie *et al.*, 2008; Hu *et al.*, 2015). There exists an observed phenomenon that the magnitude of El Niño on average tends to be larger than the magnitude of La Niña, and the strongest El Niño is stronger than the strongest La Niña (Burgers and Stephenson, 1999; Kessler, 2002). Factors determining the asymmetry of ENSO have been investigated by many studies (Kang and Kug, 2002; Jin *et al.*, 2003; An and Jin, 2004; Liang *et al.*, 2017). The time-slice experiments in this study (Figures 4 and 6) show a hint of the asymmetry of ENSO related to the ASO changes, which deserves further investigation.

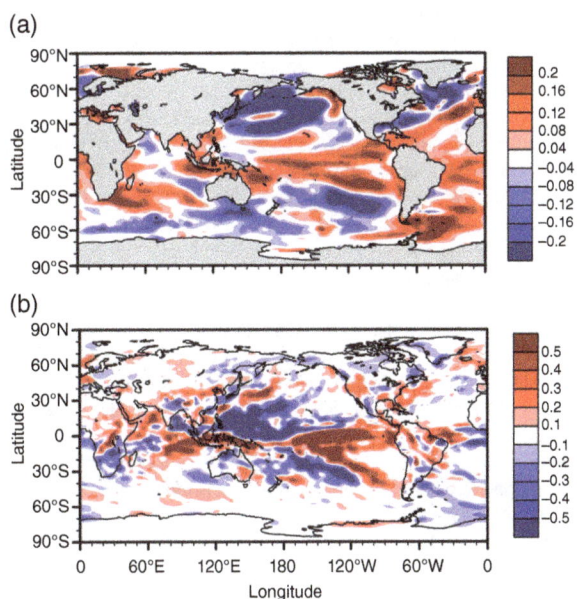

Figure 6. (a) The sum of SST anomalies of Figures 4(a) and (b). (b) The sum of rainfall anomalies of Figures 4(c) and (d).

Acknowledgements

Funding for this project was provided by the National Key Basic Research Program of China (973 Program) (2014CB441202), the National Natural Science Foundation of China (41575039, 41705022), and the Youth Scholars Program of Beijing Normal University. F. Xie and J. Zhang are supported by the Fundamental Research Funds for the Central Universities fron BNU and from LZU (lzujbky-2017-4). We acknowledge the datasets from the SWOOSH, Climate Prediction Center/NOAA, Hadley Centre, and NCEP2. We thank NCAR for providing the CESM model.

References

An SI, Jin FF. 2004. Nonlinearity and asymmetry of ENSO. *Journal of Climate* 17: 2399–2412.

Baldwin MP, Dunkerton TJ. 2001. Stratospheric harbingers of anomalous weather regimes. *Science* 294: 581–584.

Bjerknes J. 1969. Atmospheric teleconnections from the equatorial Pacific. *Monthly Weather Review* **9**: 7163–7172.

Burgers G, Stephenson DB. 1999. The "normality" of El Niño. *Geophysical Research Letters* **26**: 1027–1030.

Cagnazzo C, Manzini E. 2009. Impact of the stratosphere on the winter tropospheric teleconnections between ENSO and the North Atlantic and European Region. *Journal of Climate* **22**: 1223–1238.

Calvo N, Polvani LM, Solomon S. 2015. On the surface impact of Arctic stratospheric ozone extremes. *Environmental Research Letters* **10**: 094003.

Camberlin P, Chauvin F, Douville H, Zhao Y. 2004. Simulated ENSO-tropical rainfall teleconnections in present-day and under enhanced greenhouse gases conditions. *Climate Dynamics* **23**: 641–657.

Chipperfield MP, Dhomse SS, Feng W, McKenzie RL, Velders G, Pyle JA. 2015. Quantifying the ozone and ultraviolet benefits already achieved by the Montreal Protocol. *Nature Communications* **6**: 7233.

Chung C, Power S. 2014. Precipitation response to La Niña and global warming in the Indo-Pacific. *Climate Dynamics* **43**: 3293–3307.

Dai A, Fung IY, Del Genio AD. 1997. Surface observed global land precipitation variations during 1900–1988. *Journal of Climate* **10**: 2943–2962.

Davis SM, Rosenlof KH, Hassler B, Hurst DF, Read WG, Vömel H, Selkirk H, Fujiwara M, Damadeo R. 2016. Stratospheric Water and Ozone Satellite Homogenized (SWOOSH) database: a long-term database for climate studies. *Earth System Science Data* **8**: 461–490.

Dhomse SS, Weber M, Wohltmann I, Rex M, Burrows JP. 2006. On the possible causes of recent increases in northern hemispheric total ozone from a statistical analysis of satellite data from 1979 to 2003. *Atmospheric Chemistry and Physics* **6**: 1165–1180.

Dhomse SS, Chipperfield MP, Feng W, Ball WT, Unruh YC, Haigh JD, Krivova NA, Solanki SK, Smith AK. 2013. Stratospheric O3 changes during 2001–2010: the small role of solar flux variations in a chemical transport model. *Atmospheric Chemistry and Physics* **13**: 10113–10123.

Ding R, Li J, Tseng YH, Sun C, Guo Y. 2015. The victoria mode in the North Pacific linking extratropical sea level pressure variations to ENSO. *Journal of Geophysical Research* **120**: 27–45.

Feldstein SB. 2011. Subtropical rainfall and the Antarctic ozone hole. *Science* **332**: 925–926.

Forster P, Shine K. 1997. Radiative forcing and temperature trends from stratospheric ozone changes. *Journal of Geophysical Research* **102**: 10841–10855.

Froidevaux L, Anderson J, Wang H-J, Fuller RA, Schwartz MJ, Santee ML, Livesey NJ, Pumphrey HC, Bernath PF, Russell JM, McCormick MP. 2015. Global OZone Chemistry And Related trace gas Data records for the Stratosphere (GOZCARDS): methodology and sample results with a focus on HCl, H2O, and O3. *Atmospheric Chemistry and Physics* **15**: 10471–10507.

Gerber EP, Son S. 2014. Quantifying the summertime response of the Austral Jet Stream and Hadley Cell to stratospheric ozone and greenhouse gases. *Journal of Climate* **27**: 5538–5559.

Gill AE. 1980. Some simple solutions for heat-induced tropical circulation. *Quarterly Journal of the Royal Meteorological Society* **106**: 447–462.

Graf HF, Walter K. 2005. Polar vortex controls coupling of North Atlantic Ocean and atmosphere. *Geophysical Research Letters* **32**: L01704.

Haigh JD. 1994. The role of stratospheric ozone in modulating the solar radiative forcing of climate. *Nature* **370**: 544–546.

Heureux ML, Lee S, Lyon B. 2013. Recent multidecadal strengthening of the Walker circulation across the tropical Pacific. *Nature Climate Change* **3**: 571–576.

Horel J, Wallace J. 1981. Planetary-scale atmospheric phenomena associated with the Southern Oscillation. *Monthly Weather Review* **109**: 813–829.

Hu D, Tian W, Xie F, Wang C, Zhang J. 2015. Impacts of stratospheric ozone depletion and recovery on wave propagation in the boreal winter stratosphere. *Journal of Geophysical Research* **120**: 8299–8317.

Hu D, Tian W, Guan Z, Guo Y, Dhomse S. 2016. Longitudinal asymmetric trends of tropical cold-point tropopause temperature and

their link to strengthened walker circulation. *Journal of Climate* **29**: 7755–7771.

Huang P. 2016. Time-varying response of ENSO-induced tropical pacific rainfall to global warming in CMIP5 models. Part I: multimodel ensemble results. *Journal of Climate* **29**: 5763–5778.

Huang P, Xie S-P. 2015. Mechanisms of change in ENSO-induced tropical Pacific rainfall variability in a warming climate. *Nature Geoscience* **8**: 922–926.

Huffman GJ, Adler RF, Bolvin DT, Gu G. 2009. Improving the global precipitation record: GPCP version 2.1. *Geophysical Research Letters* **36**: L17808.

Ineson S, Scaife AA. 2009. The role of the stratosphere in the European climate response to El Niño. *Nature Geoscience* **2**: 32–36.

Ivy DJ, Solomon S, Calvo N, Thompson DWJ. 2017. Observed connections of Arctic stratospheric ozone extremes to Northern Hemisphere surface climate. *Environmental Research Letters* **12**: 024004.

Jin FF, An SI, Timmermann A, Zhao J. 2003. Strong El Niño events and nonlinear dynamical heating. *Geophysical Research Letters* **30**: 1120.

Kang IS, Kug JS. 2002. El Niño and La Niña sea surface temperature anomalies: asymmetry characteristics associated with their wind stress anomalies. *Journal of Geophysical Research* **107**: 4372.

Kang SM, Polvani LM, Fyfe JC, Sigmond M. 2011. Impact of polar ozone depletion on subtropical precipitation. *Science* **332**: 951–954.

Karpechko A, Yu A, Perlwitz J, Manzini E. 2014. A model study of tropospheric impacts of the Arctic ozone depletion 2011. *Journal of Geophysical Research* **119**: 7999–8014.

Kessler WS. 2002. Is ENSO a cycle or a series of events? *Geophysical Research Letters* **29**: 2125.

Kidston J, Scaife A, Hardiman S, Mitchell D, Butchart N, Baldwin M, Gray L. 2015. Stratospheric influence on tropospheric jet streams, storm tracks and surface weather. *Nature Geoscience* **8**: 433–440.

Kociuba G, Power SB. 2015. Inability of CMIP5 models to simulate recent strengthening of the walker circulation: implications for projections. *Journal of Climate* **28**: 20–35.

Kousky VE, Kagano MT, Cavalcanti IF. 1984. A review of the southern oscillation: oceanic–atmospheric circulation changes and related rainfall anomalies. *Tellus A* **36**: 490–504.

Lau KM, Sheu PJ. 1988. Annual cycle, quasi-biennal oscillation, and southern oscillation in global precipitation. *Journal of Geophysical Research* **93**: 10975–10988.

Liang J, Yang XQ, Sun DZ. 2017. Factors determining the asymmetry of ENSO. *Journal of Climate* **30**: 6097–6106.

Lubin D, Jensen EH. 2002. Effects of clouds and stratospheric ozone depletion on ultraviolet radiation trends. *Nature* **3**: 77710–77713.

Manney GL, Santee ML, Rex M, Livesey NJ, Pitts MC, Veefkind P, Nash ER, Wohltmann I, Lehmann R, Froidevaux L, Poole LR, Schoeberl MR, Haffner DP, Davies J, Dorokhov V, Gernandt H, Johnson B, Kivi R, Kyrö E, Larsen N, Levelt PF, Makshtas A, McElroy CT, Nakajima H, Parrondo MC, Tarasick DW, Gathen P, Walker KA, Zinoviev NS. 2011. Unprecedented Arctic ozone loss in 2011. *Nature* **478**: 469–475.

Marsh DR, Mills MJ, Kinnison DE, Lamarque J-F, Calvo N, Polvani LM. 2013. Climate change from 1850 to 2005 simulated in CESM1 (WACCM). *Journal of Climate* **26**: 7372–7391.

Power S, Delage F, Chung C, Kociuba G, Keay K. 2013. Robust twenty-first-century projections of El Niño and related precipitation variability. *Nature* **502**: 541–545.

Ramaswamy V, Schwarzkopf MD, Randel WJ. 1996. Fingerprint of ozone depletion in the spatial and temporal pattern of recent lower-stratospheric cooling. *Nature* **382**: 616–618.

Rasmusson EM, Arkin PA. 1993. A global view of large-scale precipitation variability. *Journal of Climate* **6**: 1495–1522.

Ravishankara AR, Turnipseed A, Jensen NR, Barone S, Mills MJ, Howard CJ, Solomon S. 1994. Do hydrofluorocarbons destroy stratospheric ozone. *Science* **263**: 71–75.

Ravishankara AR, Daniel JS, Portmann RW. 2009. Nitrous oxide (N2O): the dominant ozone-depleting substance emitted in the 21st century. *Science* **326**: 123–125.

Reichler T, Kim J, Manzini E, Kroger J. 2012. A stratospheric connection to Atlantic climate variability. *Nature Geoscience* **5**: 783–787.

Smith KL, Polvani LM. 2014. The surface impacts of Arctic stratospheric ozone anomalies. *Environmental Research Letters* **9**: 074015.

Solomon S. 1990. Antarctic ozone: progress towards a quantitative
understanding. *Nature* **347**: 347–354.

Solomon S. 1999. Stratospheric ozone depletion: a review of concepts
and history. *Reviews of Geophysics* **37**: 275–316.

Son S-W, Polvani LM, Waugh DW, Akiyoshi H, Garcia R, Kinnison D,
Pawson S, Rozanov E, Shepherd TG, Shibata K. 2008. The impact
of stratospheric ozone recovery on the Southern Hemisphere westerly
jet. *Science* **320**: 486–489.

Son S-W, Gerber EP, Perlwitz J, Polvani LM, Gillett NP, Seo K-H,
Eyring V, Shepherd TG, Waugh D, Akiyoshi H, Austin J, Baumgaert-
ner A, Bekki S, Braesicke P, Brühl C, Butchart N, Chipperfield MP,
Cugnet D, Dameris M, Dhomse S, Frith S, Garny H, Garcia R, Hardi-
man SC, Jöckel P, Lamarque JF, Mancini E, Marchand M, Michou M,
Nakamura T, Morgenstern O, Pitari G, Plummer DA, Pyle J, Rozanov
E, Scinocca JF, Shibata K, Smale D, Teyssèdre H, Tian W, Yamashita
Y. 2010. Impact of stratospheric ozone on Southern Hemisphere cir-
culation change: a multimodel assessment. *Journal of Geophysical
Research* **115**: D00M07.

Thompson DWJ, Solomon S, Kushner PJ, England MH, Grise KM,
Karoly DJ. 2011. Signatures of the Antarctic ozone hole in Southern
Hemisphere surface climate change. *Nature Geoscience* **4**: 741–749.

Tung KK. 1986. On the relationship between the thermal structure of
the stratosphere and the seasonal distribution of ozone. *Geophysical
Research Letters* **13**: 1308–1311.

Vimont DJ, Battisti DS, Hirst AC. 2001. Footprinting: a seasonal connection between the tropics and mid-latitudes. *Geophysical Research
Letters* **28**: 3923–3926.

Wang W, Matthes K, Schmidt T, Neef L. 2013. Recent variability of the
tropical tropopause inversion layer. *Geophysical Research Letters* **40**:
6308–6313.

Wang W, Matthes K, Schmidt T. 2015. Quantifying contributions to
the recent temperature variability in the tropical tropopause layer.
Atmospheric Chemistry and Physics **15**: 5815–5826.

Waugh DW, Garfinkel CI, Polvani LM. 2015. Drivers of the recent
tropical expansion in the Southern Hemisphere: changing SSTs or
ozone depletion. *Journal of Climate* **28**: 6581–6586.

Xie F, Tian W, Chipperfield MP. 2008. Radiative effect of ozone
change on stratosphere-troposphere exchange. *Journal of Geophysical
Research* **113**: D00B09.

Xie F, Li J, Tian W, Fu Q, Jin F-F, Hu Y, Zhang J, Wang W, Sun C,
Feng J, Yang Y, Ding R. 2016. A connection from Arctic stratospheric
ozone to El Niño-Southern oscillation. *Environmental Research Letters* **11**: 124026.

Yim BY, Yeh S-W, Sohn B-J. 2016. ENSO-related precipitation and
its statistical relationship with the walker circulation trend in CMIP5
AMIP models. *Atmosphere* **7**: 19.

Zhang J, Tian WS, Chipperfield MP, Xie F, Huang J. 2016. Persistent
shift of the Arctic polar vortex towards the Eurasian continent in recent
decades. *Nature Climate Change* **6**: 1094–1099. https://doi.org/10
.1038/NCLIMATE3136.

27

Revisiting the relationship between the South Asian summer monsoon drought and El Niño warming pattern

Fangxing Fan,[1]* Xiao Dong,[1] Xianghui Fang,[2] Feng Xue,[1] Fei Zheng[1] (ID) and Jiang Zhu[1]

[1] International Center for Climate and Environment Sciences, Institute of Atmospheric Physics, Chinese Academy of Sciences, Beijing, China
[2] Institute of Atmospheric Sciences, Fudan University, Shanghai, China

*Correspondence to:
F. Fan, International Center for
Climate and Environment
Sciences, Institute of Atmospheric
Physics, Chinese Academy of
Sciences, Beijing 100029, China.
E-mail:
fanfangxing@mail.iap.ac.cn

Abstract

Based on the observational and reanalysis data, El Niño warming patterns associated with the South Asian summer monsoon droughts are investigated. While the inverse relationship between the eastern Pacific (EP) type of El Niño-Southern Oscillation (ENSO) and the Indian monsoon rainfall weakened significantly, the correlation between the central Pacific (CP) type of ENSO and the monsoon rainfall strengthened after the late 1970s. Moreover, the drought-producing El Niño warming pattern also exhibits a notable decadal modulation associated with the climate shift. The analysis results indicate that both the EP type of El Niño with positive sea surface temperature (SST) anomalies extended to the date line and the CP type of El Niño with the maximum warming located in the central equatorial Pacific may produce severe droughts over the Indian subcontinent. Although the CP warming is more effective in driving anomalous rising motion in the central equatorial Pacific and consequently producing anomalous subsidence over South Asia, the position and strength of the anomalous ascending and descending branches of the Walker circulation are sensitive to the detailed distributions of tropical SST anomalies and determined by the competing effects of the CP and EP warming.

Keywords: South Asian summer monsoon; eastern Pacific El Niño; central Pacific El Niño; El Niño warming pattern; decadal transition

1. Introduction

El Niño-Southern Oscillation (ENSO) is the dominant mode of climate variability on interannual timescale, which can exert great impacts on the global climate, including the South Asian summer monsoon (SASM). The relationship between ENSO and SASM on both interannual and interdecadal timescales has been extensively addressed by many previous studies (Rasmusson and Carpenter, 1983; Ropelewski and Halpert, 1987; Webster and Yang, 1992; Grove, 1998; Krishnamurthy and Goswami, 2000; Fan *et al.*, 2010). The warm (cold) episodes of ENSO tend to be associated with below-normal (above-normal) Indian monsoon rainfall, and the El Niño index and the All-India monsoon rainfall (AIMR) index (Mooley and Parthasarathy, 1984; Parthasarathy *et al.*, 1994) are significantly and negatively correlated. The recognition of a different type of El Niño, referred to as the central Pacific (CP) El Niño (Ashok *et al.*, 2007; Yu and Kao, 2007; Ashok and Yamagata, 2009; Kao and Yu, 2009; Kug *et al.*, 2009), provides a new way to explore ENSO influences on global and regional climate. Different from the canonical eastern Pacific (EP) El Niño, the CP type of El Niño is characterized by positive sea surface temperature (SST) anomalies concentrated in the central tropical Pacific and is less sensitive to the thermocline variations (Kao and Yu, 2009; Zheng *et al.*, 2014). As the ENSO affects

the SASM through the east–west displacement of the Walker circulation, the change in spatial configurations of SST anomalies leads to the change in the position and strength of the rising and sinking branches of the Walker circulation anomalies, and consequently alters the spatial distributions of precipitation anomalies over South and Southeast Asia (Ropelewski and Halpert, 1987; Palmer *et al.*, 1992).

The climate system experienced notable decadal variation in the late 1970s, which is related to the phase transition of the Pacific Decadal Oscillation (Trenberth, 1990; Mantua and Hare, 2002) and human activity induced global warming. During this climate transition, the conventional inverse relationship between ENSO and SASM has collapsed. Krishna Kumar *et al.* (1999) analyzed the 140-year historical record and found a drop in correlations between Indian summer monsoon rainfall and summer (June-July-August (JJA) mean) Niño-3 SST anomalies after the late 1970s. Although this weakening of the ENSO-Indian monsoon rainfall relationship may be due to natural variability (Mehta and Lau, 1997; Gershunov *et al.*, 2001), some evidence (Krishna Kumar *et al.*, 1999) suggests that anthropogenic global warming might be the root cause. Further investigations (Krishna Kumar *et al.*, 2006) of the detailed spatial distributions of eastern-to-central tropical Pacific SST anomalies related to the Indian monsoon droughts revealed that

El Niño events characterized by CP warming are more effective in producing Indian monsoon failure than those accompanied by EP warming.

Better understanding of the ENSO-monsoon association is of great importance for monsoon predictions. The precipitation data used in Krishna Kumar *et al.* (2006) to illustrate composite spatial patterns only covers the period 1979–2004. Whether the Indian monsoon drought-related El Niño warming pattern underwent a decadal shift is still unclear. The objectives of this study are to identify the spatial characteristics of El Niño warming patterns associated with the Indian monsoon droughts and to examine whether the monsoon drought-inducing El Niños exhibit different warming patterns between pre- and post-late 1970s. The rest of this article is organized as follows. The data and method are described in Section 2. Section 3 shows the main results, followed by conclusions in Section 4,

2. Data and method description

The observational data used in this study include monthly mean SST data provided by the Hadley Centre's sea ice and sea surface temperature dataset (HadISST) (Rayner *et al.*, 2003) and the precipitation dataset with high-resolution grids for Asia developed by the Asian Precipitation – Highly Resolved Observational Data Integration Towards Evaluation of Water Resources (APHRODITE) project (Yatagai *et al.*, 2012). The APHRODITE's precipitation dataset has been used to evaluate the models' performance in reproducing observed precipitation in East Asia (Lin and Zhou, 2015). This study also used precipitation and velocity potential from the Twentieth Century Reanalysis Version 2 (Compo *et al.*, 2011), which was provided by the National Oceanic and Atmospheric Administration's (NOAA's) Office of Oceanic and Atmospheric Research (OAR) Earth System Research Laboratory (ESRL) Physical Science Division (PSD), Boulder, Colorado, from its website (http://www.esrl .noaa.gov/psd/). This long-term reanalysis dataset was generated using an Ensemble Kalman Filter data assimilation method and surface pressure observations were assimilated into the model.

The AIMR, representing the interannual variability of the regionally averaged summer monsoon [June–September (JJAS)] rainfall over India, was calculated based on rain gauge observations from hundreds of stations (Parthasarathy *et al.*, 1995). While the influence of ENSO on the East Asian summer monsoon is significant during the decaying stage of ENSO through an anomalous western North Pacific anticyclone (Wang *et al.*, 2000), the ENSO has an instantaneous impact on the SASM in the developing phase of ENSO by altering the ascending and descending branches of the Walker circulation (Krishna Kumar *et al.*, 1999; Wang *et al.*, 2001). The criterion for selecting strong El Niño events is that the standardized boreal summer mean (JJAS) Niño-3 (5°S–5°N, 90–150°W)

Table 1. Strong El Niño events (standardized Niño-3 index > 1.0) and the associated All-India monsoon rainfall (AIMR) during the shorter (1950–2012) and longer (1871–2012) periods.

Years	Niño-3 SST anomaly (°C)	Standardized Niño-3 SST anomaly index	Type	AIMR (mm)	Standardized AIMR index
Strong El Niño events (1950–2012)					
1951	1.11	1.46	EP	738.7	−1.20
1957	1.18	1.55	EP	788.5	−0.61
1963	0.86	1.13	CP	857.7	0.21
1965	1.22	1.60	EP	709.2	−1.54
1972	1.62	2.12	EP	652.8	−2.21
1976	0.96	1.26	EP	856.6	0.20
1982	1.19	1.56	EP	735.1	−1.24
1983	1.03	1.35	EP	955.6	1.37
1987	1.43	1.88	CP	697.0	−1.69
1997	2.40	3.15	EP	871.4	0.37
Strong El Niño Events (1871–2012)					
1877	1.63	2.42	CP	603.9	−2.93
1888	0.98	1.46	EP	811.4	−0.44
1896	0.82	1.22	EP	828.4	−0.23
1900	0.84	1.25	CP	889.2	0.50
1902	1.45	2.16	CP	791.7	−0.67
1904	0.72	1.08	EP	750.1	−1.17
1905	1.27	1.88	CP	716.3	−1.58
1914	0.80	1.19	EP	897.9	0.60
1918	0.85	1.27	EP	650.6	−2.37
1919	0.74	1.10	CP	884.4	0.44
1925	0.87	1.30	EP	803.5	−0.53
1930	0.97	1.44	EP	804.4	−0.52
1941	0.76	1.13	CP	728.1	−1.44
1951	1.01	1.50	EP	738.7	−1.31
1957	1.09	1.62	EP	788.5	−0.71
1963	0.79	1.17	EP	857.7	0.12
1965	1.16	1.71	EP	709.2	−1.66
1972	1.58	2.34	EP	652.8	−2.34
1976	0.94	1.39	EP	856.6	0.10
1982	1.18	1.75	EP	735.1	−1.35
1983	1.03	1.52	EP	955.6	1.29
1987	1.44	2.13	CP	697.0	−1.81
1997	2.44	3.62	EP	871.4	0.28
2009	0.79	1.18	CP	667.6	−2.16

The EP and CP types of El Niño are identified based on the EP/CP ENSO index method developed by Kao and Yu (2009).

SST anomaly index is greater than +1.0. It is worth noting that we have used the summer rather than winter Niño-3 index in this study to identify El Niño events because we focus on examining the simultaneous relationship between ENSO and SASM. Furthermore, the EP and CP types of El Niño have been differentiated using a previously derived procedure [i.e. the combined regression-Empirical Orthogonal Function (EOF) method] (Kao and Yu, 2009; Yu *et al.*, 2012). To define the EP ENSO index, the SST anomalies regressed on the Niño-4 (5°S–5°N, 160°E–150°W) index were firstly removed from the original SST anomalies, and then the leading principal component of the residual SST anomalies was extracted with the EOF analysis to represent the EP ENSO variability. Similarly, the CP ENSO index was defined as the leading principal

(a)

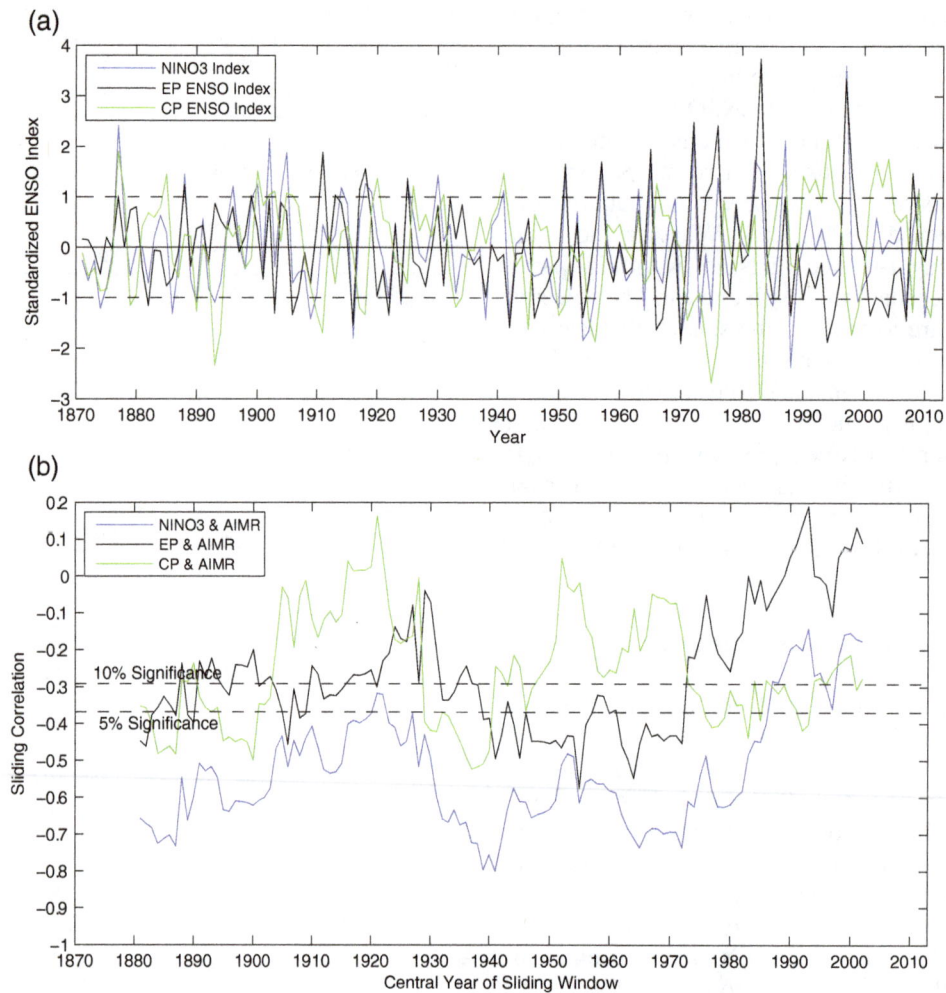

(b)

Figure 1. (a) Interannual time series of the Niño-3 SST anomaly index (blue), the EP ENSO index (red), and the CP ENSO index (green) for boreal summer season (June–September) during the 1871–2012 period. These series have been standardized to have zero mean and unit standard deviation. (b) Sliding correlations on a 21-year moving window between the ENSO indices and the AIMR index during the 1871–2012 period. The horizontal dashed lines indicate the one-tailed $p = 0.05$ and $p = 0.1$ significance levels.

component of the SST anomalies in which the influence from the Niño 1 + 2 (0–10°S, 80–90°W) region had been removed. Strong El Niño events were classified into EP and CP types based on the relative values of the summer mean EP and CP ENSO indices. The selected strong El Niño events and two types of El Niño during two time periods are listed in Table 1.

3. Results

The interannual variations of the standardized summer mean (JJAS) Niño-3 SST anomaly index and the two types of ENSO indices are shown in Figure 1(a). The Niño-3 index is highly correlated with the EP ENSO index ($r = 0.66$) and weakly but significantly correlated with the CP ENSO index ($r = 0.21$) during the 1871–2012 period. Figure 1(b) illustrates the sliding correlations on a 21-year moving window between the ENSO indices and the AIMR index. The sliding correlations between the Niño-3 index and the AIMR index were substantially significant before the 1980s

but weakened considerably thereafter. The sliding correlations between the EP ENSO index and the AIMR index, though relatively weak, are almost in phase with those between the Niño-3 index and the AIMR index. The in-phase oscillations of the two sliding correlation curves imply that the weakening of the ENSO-SASM relationship after the late 1970s may be partly explained by the breakdown in the connection between the EP ENSO and the Indian monsoon rainfall. By contrast, the sliding correlations between the CP ENSO index and the AIMR index strengthened ($r = -0.3 \sim -0.4$) and became marginally significant ($p < 0.1$) in the last three decades of the twentieth century, indicating that the CP type of SST variation could explain 9 ~ 16% of the Indian monsoon rainfall variability during this time interval. It should be noted that the SASM response to CP El Niño warming is sensitive to the specific definition of the CP ENSO index (Garfinkel *et al.,* 2013).

Figure 2 shows the simultaneous relationship between Niño-3 SST anomalies and the Indian monsoon rainfall anomalies in boreal summer season. It is evident that deficient (abundant) Indian monsoon rainfall conditions are more likely to occur in the developing phase of El Niño (La Niña) events. The negative correlations between the Niño-3 index and the AIMR index are statistically significant at the one-tailed $p = 0.001$ level for both the shorter 1950–2012 interval ($r = -0.48$) and longer 1871–2012 interval ($r = -0.53$). However, the occurrence of moderate to strong El Niño events does not necessarily guarantee the Indian monsoon failure. For 10 (24) strong El Niño events (standardized Niño-3 index > 1.0) during the 1950–2012 period (the 1871–2012 period), only half of them produced Indian monsoon droughts (standardized AIMR index < −1.0), whereas the other half were accompanied by normal (−1.0 ≤ standardized AIMR index ≤ 1.0) or even abundant rainfall (standardized AIMR index > 1.0) conditions. Given the important role of the CP warming in the Indian monsoon drought raised by Krishna Kumar *et al.* (2006), it is worth asking whether the drought-inducing El Niños are of the CP type. Somewhat surprisingly, for the shorter 1950–2012 period (Figure 2(a)), there is not a clear split between the EP and CP types of strong El Niños with respect to this consideration. When the time interval is extended back to 1871 (Figure 2(b)), we found a greater tendency for the CP El Niños (62.5%; 5 out of 8) to produce Indian monsoon droughts than the EP El Niños (37.5%; 6 out of 16). However, because of the difficulty in discerning El Niño types from the early record when data are sparse (Giese and Ray, 2011), these results are not sufficient to demonstrate that the CP type of El Niño is the determinative factor in the Indian monsoon failure.

We next examined the differences in composite spatial patterns of SST and precipitation between drought (standardized AIMR index < −1.0) and drought-free (standardized AIMR index ≥ −1.0) El Niño years (Figure 3). Undoubtedly, the composite difference patterns of precipitation are featured by substantially decreased rainfall over the Indian subcontinent (Figure 3(a)). During the 1950–2012 period, the primary differences in composite SST pattern are the greater warming in the central Pacific around 170°W and less warming in the eastern Pacific along the South American coast, consistent with the results in Krishna Kumar *et al.* (2006). Splitting the whole period (1950–2012) into two subperiods (i.e. pre-1979 and post-1979), we found that the composite SST difference patterns corresponding to the drought *versus* drought-free rainfall conditions exhibit contrasting features (Figure 3(b)). The composite SST difference pattern in the later subperiod is quite similar to that in the whole period, though the positive and negative SST anomalies in the central and eastern Pacific are more significant. In contrast, the drought-inducing El Niño pattern in the earlier subperiod is characterized by overall enhanced warming in both central and eastern equatorial Pacific, and the most prominent enhancement of warming appears in the Niño1 + 2 region. Similar composite difference patterns can also be produced by a larger group of El Niño events selected

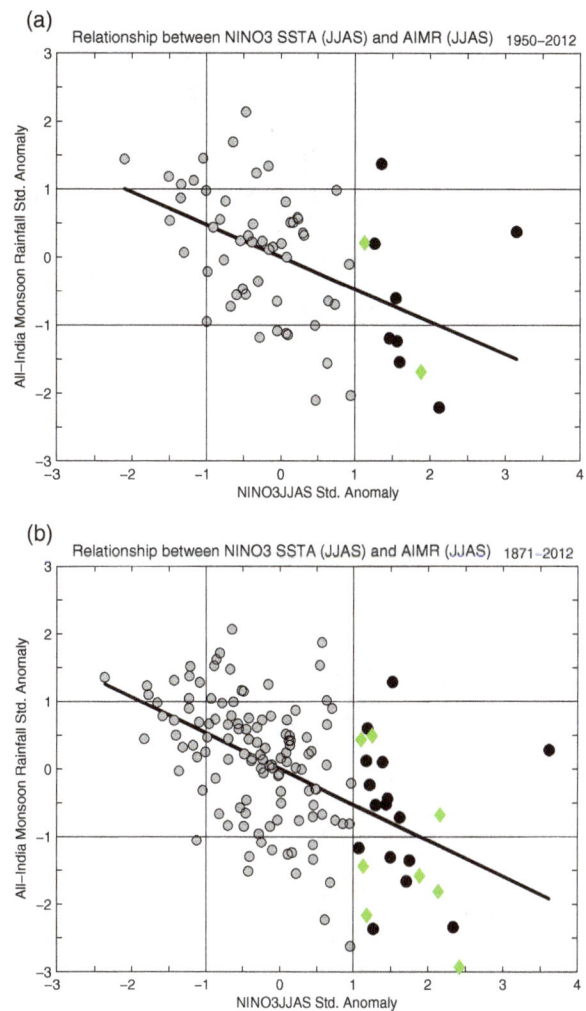

Figure 2. The standardized summer mean (June–September) Niño-3 SST anomaly index and the standardized All-India monsoon rainfall (AIMR) index during (a) the 1950–2012 period and (b) 1871–2012 period. Linear regression of the AIMR index against the Niño-3 index is shown in thick black line. EP and CP El Niño events with standardized Niño-3 SST anomaly index > 1.0 are denoted by red dots and green diamonds, respectively.

with a lower threshold (standardized Niño-3 index exceeding +0.65) (Appendices S1 and S2, Supporting information). These results indicate that the El Niño warming pattern associated with Indian monsoon drought experienced a notable decadal modulation in the late 1970s.

For the purpose of better understanding the underlying physical mechanism, velocity potential anomalies at 200 hPa in response to tropical SST anomalies in strong El Niño years are depicted in Figures 4 and S3. Note that the 200 hPa velocity potential is used as a surrogate for the Walker circulation with negative (positive) anomalies corresponding to ascending (descending) motion anomalies. Corresponding to the aforementioned composite SST difference patterns, composite difference patterns of precipitation and velocity potential at 200 hPa between drought and drought-free El Niño years are featured by enhanced ascending motion

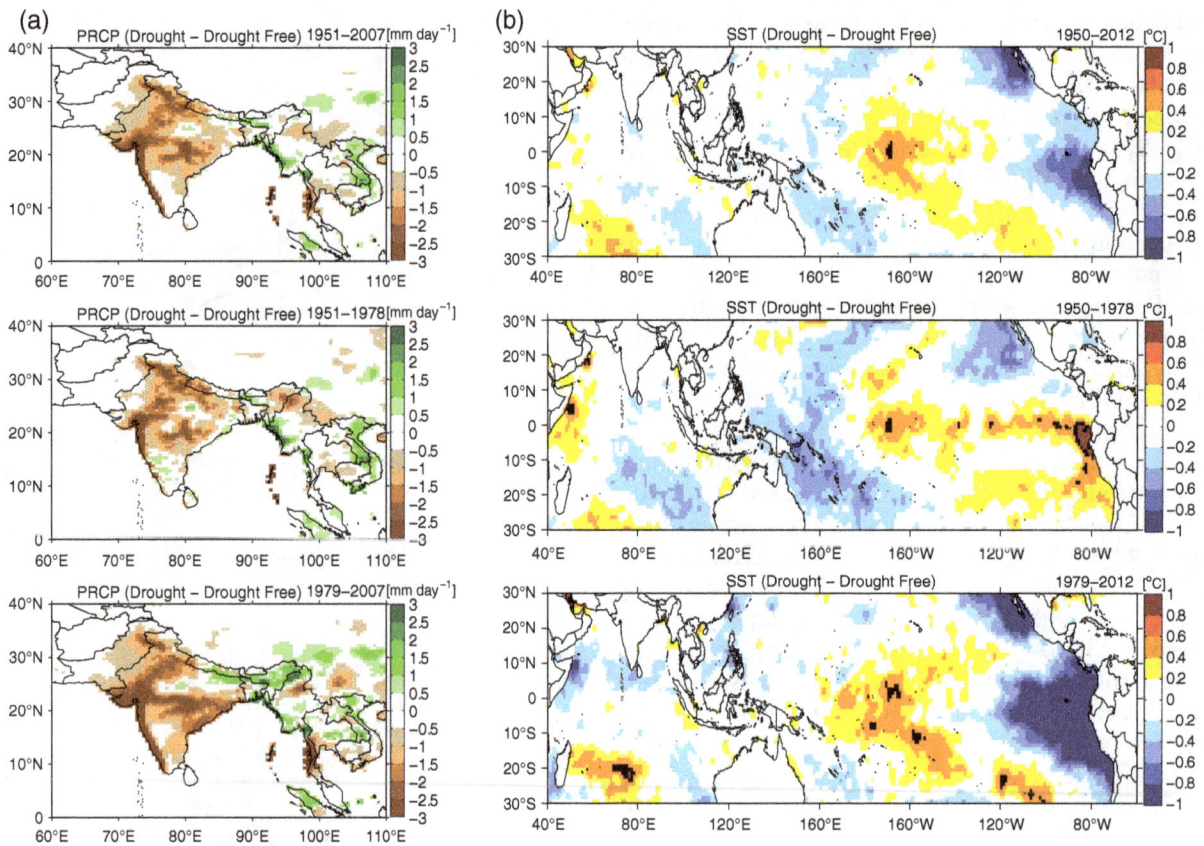

Figure 3. Composite difference patterns of (a) precipitation and (b) SST between drought (standardized AIMR index < −1.0) and drought-free (standardized AIMR index ≥ −1.0) El Niño years during the 1950−2012 period, pre-1979 subperiod and post-1979 subperiod. Precipitation patterns are based on the APHRODITE's precipitation dataset covering the period 1951−2007. SST patterns are based on the HadISST dataset over the 1950−2012 period.

(increased precipitation) in the central Pacific and intensified subsidence (decreased precipitation) over South Asia for both pre-1979 subperiod and post-1979 subperiod (Figure S3). The drought-producing El Niño events in the earlier 1950−1978 subperiod are of typical EP type with positive SST anomalies extended to the date line (e.g. 1972; Figures 4(a) and S4). Although the maximum positive SST anomalies for these typical El Niños lie in the eastern equatorial Pacific, the rising anomalies of the Walker circulation excited by sea surface warming are concentrated in the central equatorial Pacific. Correspondingly, compensatory subsidence anomalies of the Walker circulation are centered over or very close to the Indian subcontinent, leading to suppressed convection and decreased SASM precipitation (Figure 4(f)). Due to the fact that the climatological SST in the central Pacific is higher than that in the eastern Pacific, SST in the central Pacific may exceed the threshold for convection with a relatively small positive SST anomaly, resulting in greater efficiency of CP warming in driving anomalous atmospheric circulation (Graham and Barnett, 1987). It is also worth noting that positive (negative) precipitation anomalies coincide with ascending (descending) motion anomalies, indicating a consistency between convective precipitation changes and large-scale circulation variations in the tropics. Accompanied by more

frequent occurrence of the CP type of El Niño after the late 1970s, the drought-producing El Niño events in the later 1979−2012 subperiod may be of traditional EP type with warming extended westward (e.g. 1982; Figure 4(c)) or the newly emerged CP type (e.g. 1987; Figure 4(d)). Both these two types of El Niño warming patterns may lead to the Indian monsoon failure through the atmospheric teleconnection (Figures 4(h) and (i)).

In contrast, the center of anomalous rising motion in drought-free El Niño years shifts to the eastern equatorial Pacific due to the lack of CP warming (e.g. 1976; Figures 4(b) and (g)). Consequently, the anomalous sinking motion shows a prominent southeastward displacement and is centered over the Maritime Continent, leaving the South Asian region less affected by the subsidence anomalies. Despite the fact that the El Niño of 1997 was extremely strong (Niño-3 SST anomaly = 2.40 °C) with positive SST anomalies extended to the date line, the even stronger EP warming effect dominated over the CP warming effect (Figure 4(e)). Accordingly, the centers of anomalous ascending and descending motion were located over the eastern equatorial Pacific and the Maritime Continent, respectively, keeping the Indian monsoon rainfall at a normal state (Figure 4(j)).

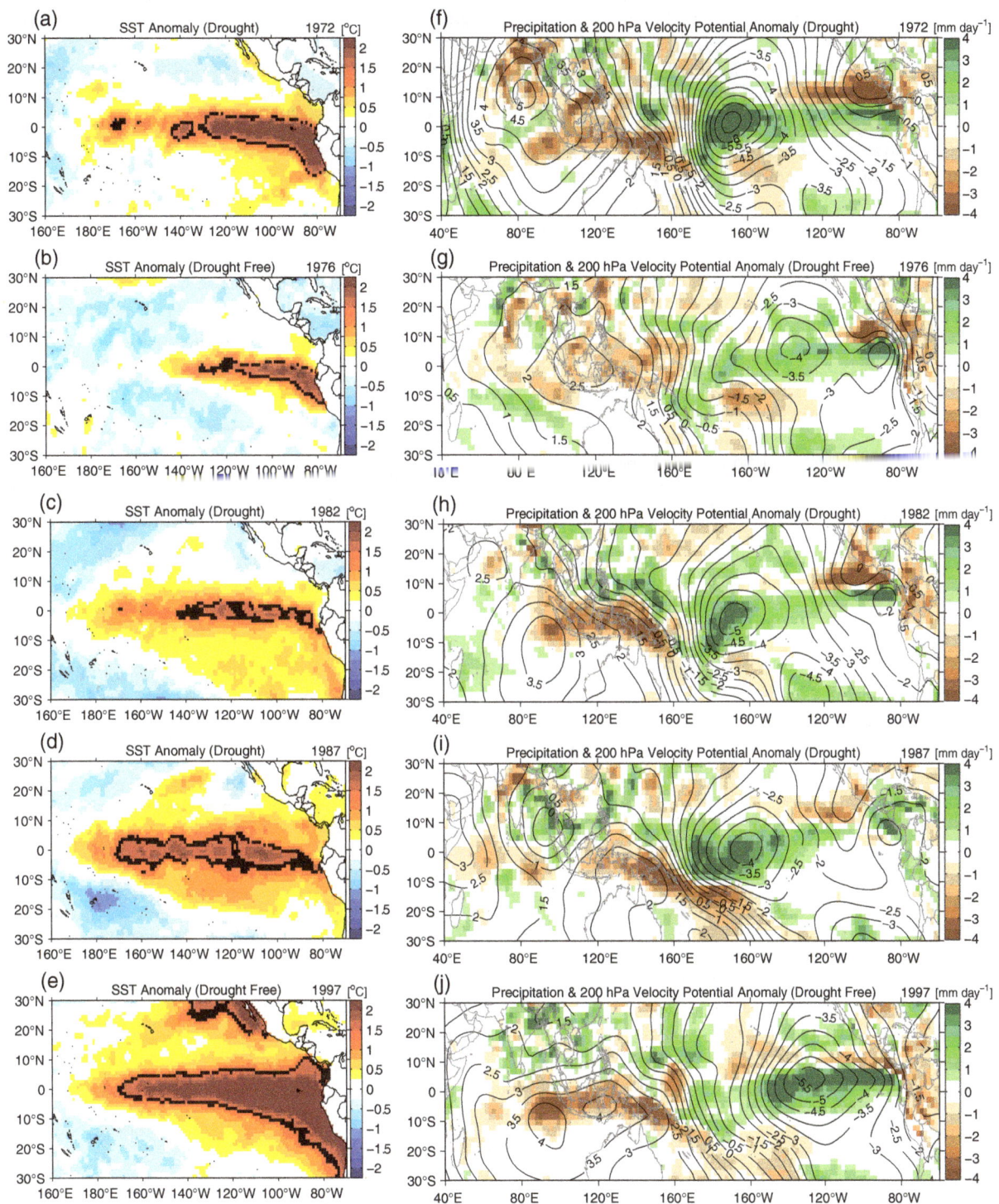

Figure 4. (a)–(e) SST anomalies and (f)–(j) 200 hPa velocity potential anomalies (contours; $\times 10^6\,m^2\,s^{-1}$) superimposed on precipitation anomalies (shaded) during summer season (June–September) in five strong El Niño years (i.e. 1972, 1976, 1982, 1987 and 1997). Precipitation and velocity potential anomalies are obtained from the Twentieth Century Reanalysis dataset during the 1950–2012 period.

4. Conclusions

We have confirmed that the CP warming plays an essential role in producing Indian monsoon droughts. However, the "flavor" of the drought-inducing El Niño events is not necessarily of the CP type. In the earlier 1950–1978 subperiod, the composite SST difference pattern between drought and drought-free El Niño years is featured by consistently enhanced warming throughout the eastern-to-central equatorial Pacific, and the Indian monsoon drought-inducing El Niño events are of canonical EP type with warming signals extended westward. By contrast, the composite SST difference pattern in the later 1979–2012 subperiod is characterized by positive SST anomalies in the central Pacific and negative SST anomalies in the eastern Pacific, and the drought-producing El Niño warming pattern may be of traditional EP type or the recently recognized CP type. These findings refine the previous view of drought-producing El Niño warming pattern and raise the possibility that the severe Indian monsoon droughts may be induced by different spatial distributions of El Niño warming associated with climate shifts.

Acknowledgements

This work was supported by the National Natural Science Foundation of China (Grant No. 41405058). Support for the Twentieth Century Reanalysis Project dataset is provided by the U.S. Department of Energy, Office of Science Innovative and Novel Computational Impact on Theory and Experiment (DOE INCITE) program, and Office of Biological and Environmental Research (BER), and by the National Oceanic and Atmospheric Administration Climate Program Office.

Supporting information

The following supporting information is available:

Figure S1. The standardized summer mean (June–September) Niño-3 SST anomaly index and the standardized All-India monsoon rainfall (AIMR) index during the 1950–2012 period. Linear regression of the AIMR index against the Niño-3 index is shown in thick black line. EP and CP El Niño events with standardized Niño-3 SST anomaly index >0.65 are denoted by red dots and green diamonds, respectively.

Figure S2. Composite difference patterns of (a) precipitation and (b) SST between drought (standardized AIMR index <−1.0) and drought-free (standardized AIMR index ≥−1.0) El Niño (standardized Niño-3 index >0.65) years during the 1950–2012 period, pre-1979 subperiod and post-1979 subperiod.

Figure S3. Composite difference patterns of precipitation (shaded) and velocity potential at 200 hPa (contours; ×10^6 m^2 s^{-1}) between drought (standardized AIMR index <−1.0) and drought-free (standardized AIMR index ≥−1.0) El Niño years during the 1950–2012 period, pre-1979 subperiod and post-1979 subperiod.

Figure S4. (a)–(c) SST anomalies and (d)–(f) 200 hPa velocity potential anomalies (contours; ×10^6 m^2 s^{-1}) superimposed on precipitation anomalies (shaded) for summer season (June–September) in drought-producing El Niño years (1951, 1965 and 1972) during the 1950–1978 subperiod.

Figure S5. (a)–(c) SST anomalies and (d)–(f) 200 hPa velocity potential anomalies (contours; ×10^6 m^2 s^{-1}) superimposed on precipitation anomalies (shaded) for summer season (June–September) in drought-free El Niño years (1957, 1963 and 1976) during the 1950–1978 subperiod.

Figure S6. (a)–(c) SST anomalies and (d)–(f) 200 hPa velocity potential anomalies (contours; ×10^6 m^2 s^{-1}) superimposed on precipitation anomalies (shaded) for summer season (June–September) in drought-producing El Niño years (1982, 1987 and 2009) during the 1979–2012 subperiod.

Figure S7. (a)–(b) SST anomalies and (c)–(d) 200 hPa velocity potential anomalies (contours; ×10^6 m^2 s^{-1}) superimposed on precipitation anomalies (shaded) for summer season (June–September) in drought-free El Niño years (1983 and 1997) during the 1979–2012 subperiod.

Table S1. Moderate to strong El Niño events (standardized Niño-3 index >0.65) and the associated All-India monsoon rainfall during the 1950–2012 period. The EP and CP types of El Niño are identified based on the EP/CP ENSO index method developed by Kao and Yu (2009).

Appendix S1. Relationship between El Niño and SASM drought.

Appendix S2. Composite difference patterns.

Appendix S3. Atmospheric responses in drought and drought-free El Niño years.

References

Ashok K, Yamagata T. 2009. The El Niño with a difference. *Nature* **461**: 481–484.

Ashok K, Behera SK, Rao SA, Weng H, Yamagata T. 2007. El Niño Modoki and its possible teleconnection. *Journal of Geophysical Research* **112**: C11007, doi: 10.1029/2006JC003798.

Compo GP, Whitaker JS, Sardeshmukh PD, Matsui N, Allan RJ, Yin X, Gleason BE Jr, Vose RS, Rutledge G, Bessemoulin P, Brönnimann S, Brunet M, Crouthamel RI, Grant AN, Groisman PY, Jones PD, Kruk MC, Kruger AC, Marshall GJ, Maugeri M, Mok HY, Nordli Ø, Ross TF, Trigo RM, Wang XL, Woodruff SD, Worley SJ. 2011. The twentieth century reanalysis project. *Quarterly Journal of the Royal Meteorological Society* **137**: 1–28.

Fan F, Mann ME, Lee S, Evans JL. 2010. Observed and modeled changes in the South Asian summer monsoon over the historical period. *Journal of Climate* **23**: 5193–5205.

Garfinkel CI, Hurwitz MM, Waugh DW, Butler AH. 2013. Are the teleconnections of Central Pacific and Eastern Pacific El Niño distinct in boreal wintertime? *Climate Dynamics* **41**: 1835–1852, doi: 10.1007/s00382-012-1570-2.

Gershunov A, Schneider N, Barnett T. 2001. Low-frequency modulation of the ENSO-Indian monsoon rainfall relationship: signal or noise? *Journal of Climate* **14**: 2486–2492.

Giese BS, Ray S. 2011. El Niño variability in simple ocean data assimilation (SODA), 1871–2008. *Journal of Geophysical Research* **116**: C02024, doi: 10.1029/2010JC006695.

Graham NE, Barnett TP. 1987. Sea surface temperature, surface wind divergence, and convection over tropical oceans. *Science* **238**: 657–659.

Grove RH. 1998. Global impact of the 1789-93 El Niño. *Nature* **393**: 318–319.

Kao H-Y, Yu J-Y. 2009. Contrasting eastern-Pacific and central-Pacific types of ENSO. *Journal of Climate* **22**: 615–632.

Krishna Kumar K, Rajagopalan B, Cane MA. 1999. On the weakening relationship between the Indian monsoon and ENSO. *Science* **284**: 2156–2159.

Krishna Kumar K, Rajagopalan B, Hoerling M, Bates G, Cane M. 2006. Unraveling the mystery of Indian monsoon failure during El Niño. *Science* **314**: 115–119.

Krishnamurthy V, Goswami BN. 2000. Indian monsoon-ENSO relationship on interdecadal timescale. *Journal of Climate* **13**: 579–595.

Kug J-S, Jin F-F, An S-I. 2009. Two types of El Niño events: cold tongue El Niño and warm pool El Niño. *Journal of Climate* **22**: 1499–1515.

Lin R-P, Zhou T-J. 2015. Reproducibility and future projections of the precipitation structure in East Asia in four Chinese GCMs that participated in the CMIP5 experiments. *Chinese Journal of Atmospheric Sciences* **39**(2): 338–356. (in Chinese).

Mantua NJ, Hare SR. 2002. The Pacific Decadal Oscillation. *Journal of Oceanography* **58**: 35–44.

Mehta VM, Lau K-M. 1997. Influence of solar irradiance on the Indian monsoon-ENSO relationship at decadal-multidecadal time scales. *Geophysical Research Letters* **24**: 159–162.

Mooley DA, Parthasarathy B. 1984. Fluctuations in All-India summer monsoon rainfall during 1871-1978. *Climatic Change* **6**: 287–301.

Palmer TN, Branković Č, Viterbo P, Miller MJ. 1992. Modeling interannual variations of summer monsoons. *Journal of Climate* **5**: 399–417.

Parthasarathy B, Munot AA, Kothawale DR. 1994. All-India monthly and seasonal rainfall series: 1871-1993. *Theoretical and Applied Climatology* **49**: 217–224.

Parthasarathy B, Munot AA, Kothawale DR. 1995. Monthly and seasonal rainfall series for all India, homogeneous regions and meteorological subdivisions: 1871-1994. Research Report No. RR-065, 113 pp. [Available from Indian Institute of Tropical Meteorology, Homi Bhabha Road, Pune 411008, India.]

Rasmusson EM, Carpenter TH. 1983. The relationship between eastern equatorial Pacific sea surface temperatures and rainfall over India and Sri Lanka. *Monthly Weather Review* **111**: 517–528.

Rayner NA, Parker DE, Horton EB, Folland CK, Alexander LV, Rowell DP, Kent EC, Kaplan A. 2003. Global analyses of sea surface temperature, sea ice, and night marine air temperature since the late nineteenth century. *Journal of Geophysical Research* **108**(D14): 4407, doi: 10.1029/2002JD002670.

Ropelewski CF, Halpert MS. 1987. Global and regional scale precipitation patterns associated with the El Niño/Southern Oscillation. *Monthly Weather Review* **115**: 1606–1626.

Trenberth KE. 1990. Recent observed interdecadal climate changes in the Northern Hemisphere. *Bulletin of the American Meteorological Society* **71**: 988–993.

Wang B, Wu R, Fu X. 2000. Pacific–East Asian teleconnection: how does ENSO affect East Asian climate? *Journal of Climate* **13**: 1517–1536.

Wang B, Wu R, Lau K-M. 2001. Interannual variability of the Asian summer monsoon: contrasts between the Indian and the western North Pacific-East Asian monsoons. *Journal of Climate* **14**: 4073–4090.

Webster PJ, Yang S. 1992. Monsoon and ENSO: selectively interactive systems. *Quarterly Journal of the Royal Meteorological Society* **118**: 877–926.

Yatagai A, Kamiguchi K, Arakawa O, Hamada A, Yasutomi N, Kitoh A. 2012. APHRODITE: constructing a long-term daily gridded precipitation dataset for Asia based on a dense network of rain gauges. *Bulletin of the American Meteorological Society* **93**: 1401–1415, doi: 10.1175/BAMS-D-11-00122.1.

Yu J-Y, Kao H-Y. 2007. Decadal changes of ENSO persistence barrier in SST and ocean heat content indices: 1958-2001. *Journal of Geophysical Research* **112**: D13106, doi: 10.1029/2006JD007654.

Yu J-Y, Zou Y, Kim ST, Lee T. 2012. The changing impact of El Niño on US winter temperatures. *Geophysical Research Letters* **39**: L15702, doi: 10.1029/2012GL052483.

Zheng F, Fang X-H, Yu J-Y, Zhu J. 2014. Asymmetry of the Bjerknes positive feedback between the two types of El Niño. *Geophysical Research Letters* **41**: 7651–7657, doi: 10.1002/2014GL062125.

28

Characteristics of wet and dry spells in the West African monsoon system

Stéphanie Froidurot and Arona Diedhiou*

Université Grenoble Alpes, IRD, CNRS, Grenoble INP, IGE, Grenoble, France

Correspondence to:
A. Diedhiou, Université Grenoble Alpes, IRD, CNRS, Grenoble INP, IGE, F-38000 Grenoble, France.
E-mail: arona.diedhiou@ird.fr

Abstract

Using 17 years (1998–2014) of daily TRMM 3B42 rainfall data, we provide a climatological characterization of wet and dry spells in West Africa, which should serve to assess the ability of climate model to simulate these high impact events. The study focuses on four subregions (Western and Central Sahel, Sudanian zone and Guinea Coast). Defining wet (dry) spells as sequences of consecutive days with precipitation higher (lower) than 1 mm, we describe the space-time variability of wet and dry spell occurrence. This climatology stresses the influence of the relief on the number and duration of these spells. The spatio-temporal variability of the wet and dry spells also appears to be closely related to the spatio-temporal variability of the West African monsoon. The number of wet spells of all durations and of 2–3 day dry spells have similar features with a maximum occurrence during the local rainy seasons and a spatial pattern similar to the mean annual rainfall with a north–south gradient. In contrast, dry spells lasting more than four days show some singularities such as a low occurrence over the Sahelian band or high occurrence along the Guinea Coast mainly from Ivory Coast to Benin. Moreover, the seasonal cycle of these longer dry spells presents higher occurrences at the beginning and the end of the rainy seasons.

Keywords: hydroclimatology; West Africa; wet spells; dry spells

1. Introduction

With a growing population, pressure on land and water for food production will increase in the coming decades. Moreover, according to the Food and Agriculture Organization of the United Nations (FAO, 2016), climate change will have a significant impact in sub-Saharan Africa where 90% of food production is from rain-fed agriculture which is highly dependent on the variability of the West African monsoon system. By 2080, 75 million hectares of land will no longer be suitable for rain-fed agriculture in sub-Saharan Africa (Turral *et al.*, 2011). Moreover, various studies suggest an evolution of rain features in West Africa over the past decades. Salack *et al.* (2016) showed higher occurrences of false start and early cessation of the rainy seasons, as well as more persistent dry spells compared to the 1950s–1960s while at the same time the probability of floods increased. In the Sahel, Sanogo *et al.* (2015) highlighted significant positive trends in the total yearly precipitation between 1980 and 2010 along with an enhancement and a prolongation of the rainy seasons. In contrast, over the Guinea Coast, no-significant trend is reported even though the second rainy season tends to be more intense suggesting a later withdrawal of rains from West Africa. In this context, it is necessary to characterize observed rainfall features of the West African monsoon as a baseline for future evolution. Crop growth are especially sensitive to the wet and dry spells occurrence and duration during the rainy season

(Gornall *et al.*, 2010). Understanding the variability of wet and dry spells is therefore important to design early warning systems and hydro-agricultural infrastructures. These spells are to be associated with water balance at the land plot level to determine the occurrence of crop water stress and its timing. However, the need for information at high space-time resolution makes this approach difficult to apply to large regions. Hence, in most regional to global studies, wet and dry spells are considered to be sequences of days with rainfall over or below a given threshold. In the literature, the choice of this threshold varies from study to study. Many authors chose a 0.85 mm or a 1 mm threshold to describe wet and dry days in various parts of the world (e.g. Barron *et al.*, 2003 and Seleshi and Camberlin, 2005 in Africa, Groisman and Knight, 2008 in the United States or Zolina *et al.*, 2013 over Europe). This amount is a good indicator of the ending of a dry spell but is sometimes considered insufficient for crop use (Sivakumar, 1992). Facing the need to take the local climatology into account, some studies are based on a variable threshold depending on the region. For example, Ratan and Venugopal (2013) proposed a climatology of wet and dry spells in several tropical regions based on a percentage (10%) of the local climatological mean rainfall to separate between wet and dry days.

Over West Africa, very few analyses of wet and dry spell features were realized, focusing essentially on dry spells and based on local in situ observations (Sivakumar, 1992; Sanogo *et al.*, 2015). In this paper,

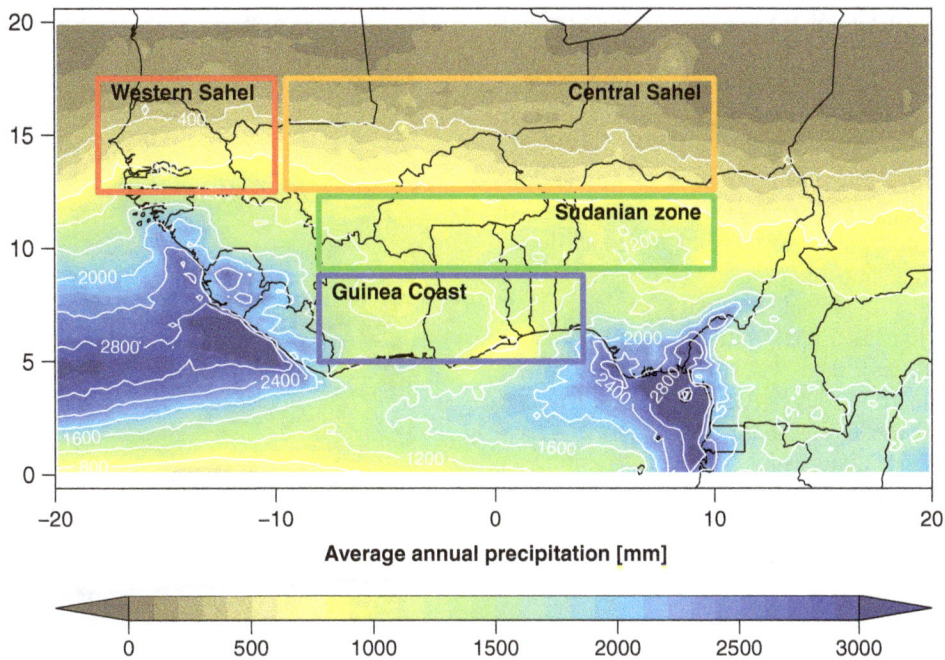

Figure 1. Studied domain: West Africa. Average annual precipitation (in mm) as observed with TRMM 3B42 data for the period 1998–2014. The four colored rectangles indicate the subregions the study focuses on: Western Sahel (red), Central Sahel (orange), Sudanian zone (green) and Guinea Coast (blue).

we propose a characterization of the main features of wet and dry spells and their space-time variability in West Africa, using TRMM 3B42 data and a 1 mm threshold to distinguish between wet and dry days. This TRMM-derived climatology of wet and dry spells will contribute to help decision making in the agricultural sector. It should also serve to assess the ability of climate models to simulate such high impact events.

2. Data and method

The study is done using TRMM 3B42 v7 rainfall product for a 17-year period from 1998 to 2014. These data are derived from a combination of calibrated microwave and infrared precipitation estimates (Kummerow et al., 1998; Huffman et al., 2007). The TRMM 3B42 rainfall product is available at temporal and spatial resolutions of 3 h and $0.25° \times 0.25°$. Several validation studies of the TRMM 3B42 rainfall data have been done over West Africa with ground-based observations and the rainfall product is generally considered reliable, especially at larger space and time scales (Nicholson et al., 2003; Pierre et al., 2011). Besides, Amekudzi et al. (2016) showed that TRMM is suitable for agricultural and other hydro-climatic impact studies over the region given its ability to capture the dry spells as well as the onset, peak and cessation of the rainy season. We consider a domain covering West Africa as presented in Figure 1 (20°W–20°E and 0–20°N). In this region, the annual precipitation follows mainly a latitudinal gradient with less than 400 mm per year in the Northern Sahel and more than 1500 mm per year along the coast of the Gulf of Guinea. The rainiest regions (more

than 2000 mm yearly) are located over the ocean at the south-west of the Gulf of Guinea and off the western Atlantic coast. Over the continent, the highest annual rain accumulation are measured over mountain ridges with elevation higher than 1000 m: the Fouta Djalon and Guinean Backbone mountains (spanning Guinea, Sierra Leone and Liberia) and the Adamawa plateau at the frontier between Nigeria and Cameroun. Other singularities compared with the latitudinal gradient can be noted. They are usually related to the topography, such as the Jos plateau (1280 m above sea level on average) in central Nigeria where annual rain accumulation reaches 1400 mm while it is below 1000 mm in this latitude band.

For the present analysis, the 3-hourly rainfall data from TRMM 3B42 have been aggregated at a daily time step. Then, we define each day as wet or dry using a threshold of 1 mm/day. Wet (respectively dry) spells are sequences of consecutive wet (dry) days, preceded and followed by dry (wet) days. Besides, a wet (dry) day that is surrounded by two dry (wet) days is defined as an isolated wet (dry) day. We analyze the spatial and temporal variability of these wet and dry events using the four subregions represented in Figure 1: Western Sahel (18°–10°W and 12.5°–17.5°N), Central Sahel (10°W–10°E and 12.5°–17.5°N), Sudanian region (8°W–10°E and 9°–12.5°N) and the Guinea Coast (8°W–4°E and 5°–9°N). The mean annual precipitation in these subregions is respectively 575, 430, 1047 and 1247 mm. Besides, these subregions differ in their seasonal cycle due to the meridional migration of the Intertropical Convergence Zone (ITCZ). Thus, the Guinea Coast displays two rainy seasons from April to July and in September–October while only one rainy season is observed from June to September to the north.

Moreover, wet and dry spells are assessed for various duration categories. Dry spells have been analyzed for durations of 2–3, 4–6, 7–15 and 16–21 days similarly to what was done by previous authors (e.g. Sivakumar, 1992). The duration categories of wet spells are chosen to correspond to the different synoptic systems causing rain in West Africa. The wet spells lasting 2–3 and 4–5 days are associated with the so-called '3–5 days' African Easterly Waves (AEWs) and the 6–9 days wet spells are chosen with reference to the '6–9 days' African Easterly Waves (Diedhiou et al. 1998, Wu et al., 2013). These AEWs are both synoptic disturbances known to influence mesoscale convective systems over West Africa. The 10–20 days is another variability mode of the African monsoon rainfall that may result from regional coupled land-atmosphere interactions (e.g. Grodsky and Carton, 2001; Mounier and Janicot, 2004).

3. Results

3.1. Wet spells

Figure 2 shows the average number of wet spells lasting 2–3, 4–5, 6–9 and 10–20 days. The shortest wet spells are the most frequent in the study region. Moreover, for all the studied durations, the spatial pattern of the number of wet spells is similar to that of the annual total precipitation (Figure 1). The number of wet spells displays a latitudinal gradient and is maximum in the same areas as those with high annual precipitation such as the Guinean backbone, the Adamawa Plateau or the Jos Plateau indicating the influence of both the relief and the latitudinal migration of the ITCZ. The yearly average of the number of the 2–3 day spells ranges from less than 10 events in the Sahel to more than 30 in the Gulf of Guinea and off the western coast of Guinea and Sierra Leone. Around 15 annual wet spells of 2–3 days are noted over the Adamawa Plateau. This small amount in comparison with the annual rainfall may be caused by the blocking role of the topography resulting in longer wet spells in this area. On average, the 4–5 day wet spells occur less than twice a year north of 12.5°N (in the Sahel) and about 7 to 10 times in the areas with high annual rain accumulation (over Guinea Coast and in mountainous areas). Concerning longer wet spells (6–9 and 10–20 days), less than 1 or 2 events occur each year over almost all the domain, except in the Fouta Djalon-Guinean backbone and the Adamawa regions where more than six (respectively three) wet spells of 6–9 days (respectively 10–20 days) occur.

For a more precise analysis of the number of wet spells and their seasonal variability, we present in Figure 3 different features of the wet spells number in the four selected subregions (see Figure 1). Figure 3(a) shows the 17-year (1998–2014) climatology of the annual number of wet spells of different durations per

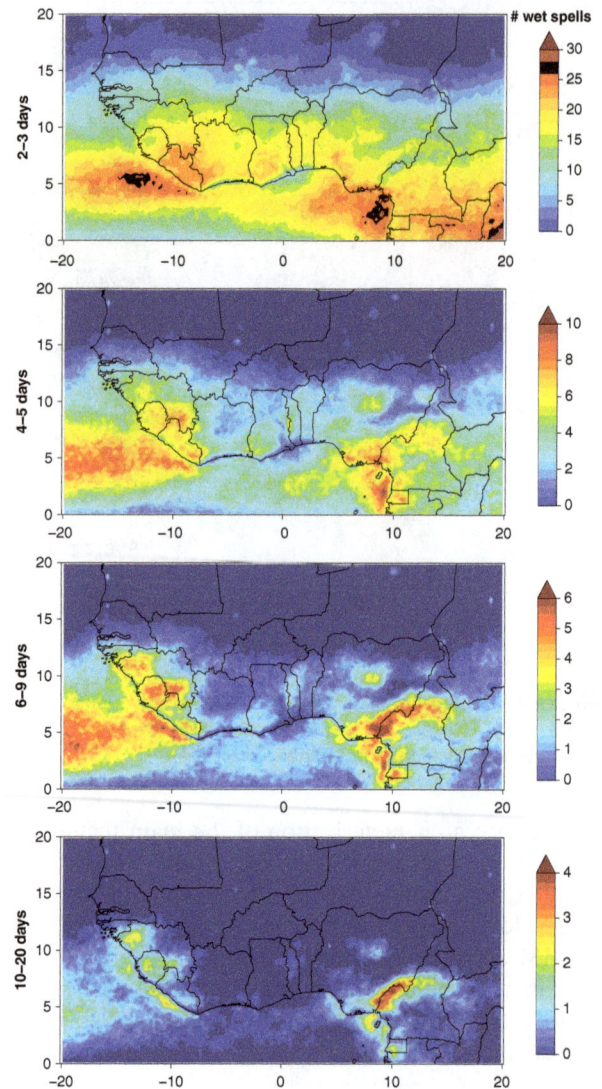

Figure 2. Mean annual number of wet spells over West Africa, for wet spells durations of 2–3, 4–5, 6–9 and 10–20 days (from top to bottom).

gridbox ($0.25° \times 0.25°$) for each of the four subregions. It is important to notice that isolated wet days are more frequent in the four subregions than wet spells of all durations and are twice as many in the Guinean or Sudanian area as in the Sahel. Moreover, the ratio of the number of isolated wet days to 2-day wet spells is higher in the driest regions. It is close to 4 in the two Sahelian regions while it is around 2.4 in the Sudanian and Guinean areas. The actual number of rainy days contributed by wet spells of different durations is another interesting quantity, which is plotted in Figure 3(b). More precisely, if there are N_d wet spells lasting d days in a season, these wet spells contribute $d \times N_d$ days to the total number of rainy days. It shows that the major contribution to all the rainy days is from the isolated wet days. Moreover, almost no wet spell longer than 10 days are observed in the selected subregions, as previously noted in Figure 2. These results are consistent with those described in Ratan and Venugopal (2013) for

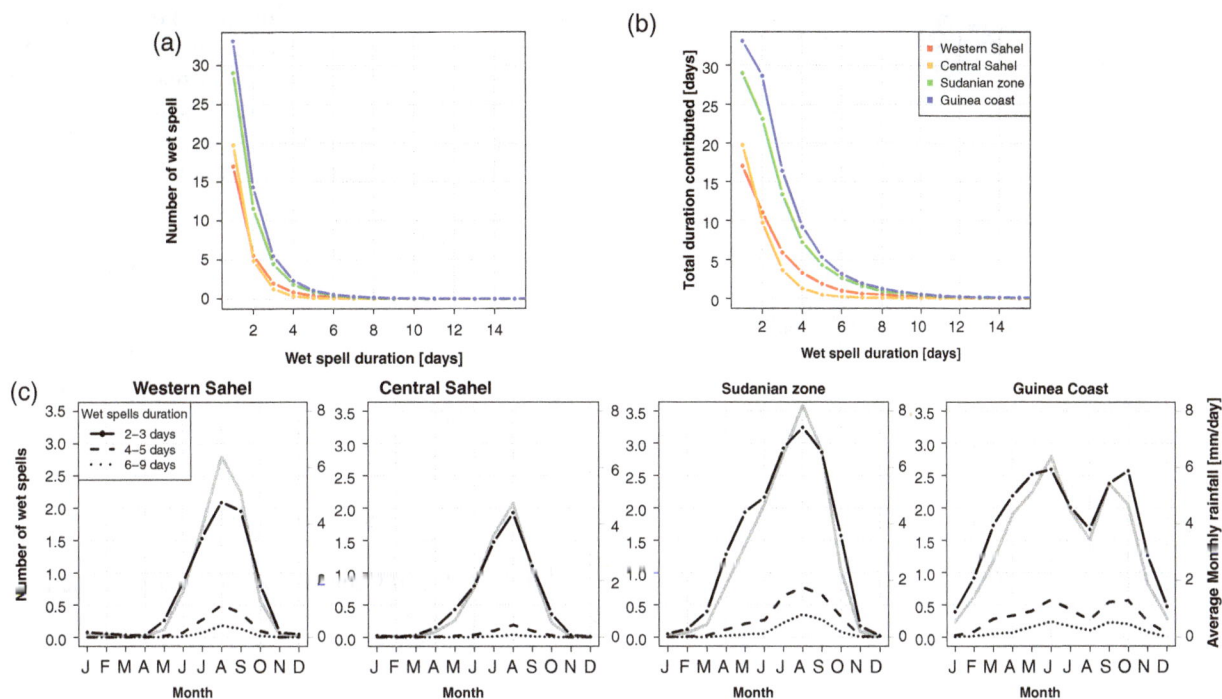

Figure 3. Wet spell characteristics for each of the four subregions. (a) Frequency histogram of wet spell duration per year and per gridpoint. (b) Average number of days each wet spell contributes to the total number of wet days. (c) Seasonal variability of the average number of wet spells of different duration (2–3, 4–5, 6–9 days, different line types). The gray line indicates the monthly mean of daily rainfall. Each of the four panels corresponds to a given subregion.

their 'arid regions' (category corresponding to the studied subregions).

The seasonal variability of the number of wet spells of different durations in each of the four subregions is presented in Figure 3(c). The 10–20 day wet spells are not shown because there are too few events. For a relevant comparison with the seasonal cycle of rainfall associated with the evolution of the West African Monsoon, the monthly mean of daily rainfall is also indicated. It appears that the number of wet spells displays the same seasonal cycle as daily rainfall, in all of the four subregions and for the different duration categories studied (2–3, 4–5 and 6–9 days). Therefore, for all the durations and in the four regions, the number of wet spells is maximum during the rainy season: between June and September in the two Sahelian regions and between April and October in the Sudanian area. Over Guinea Coast, two peaks of similar magnitude are observed corresponding to the long and short rains from April to July and from September to November, with a lower number of wet spells in August during the little dry season.

3.2. Dry spells

The characterization of the space-time variability of wet spells is repeated for dry spells. Figure 4 shows the average number of dry spells lasting 2–3, 4–6, 7–15 and 16–21 days. As expected, the longer the considered duration, the fewer the dry spells. Overall, the large-scale spatial pattern and order of magnitude of the number of the 2–3 day dry spells is similar to the number of 2–3 day wet spells, with a latitudinal gradient

and maximum occurrence off Guinea and in the Gulf of Guinea with up to 30 events each year. For longer dry spell duration, the latitudinal gradient is less pronounced and the maximum of the number of dry spells moves to the coast of Ghana and Ivory Coast with values of about 15 dry spells of 4–6 days, 10 dry spells lasting 7–15 days and more than 2 dry spells of 16–21 days on average each year. The spatial pattern of these longer dry spells differs from that of long wet spells with regions such as the Sahel or the coasts from Benin to Ivory Coast displaying large numbers of dry spells and low numbers of wet spells. Another noteworthy feature is the presence in the Sudanian band of a minimum in the number of dry spells lasting more than a week, with less than four dry spells of 7–15 days and less than one dry spell of 16–21 days on average each year. Moreover, for all durations, the mountainous areas (Fouta Djalon, Adamawa and Jos Plateaus) vary from their neighboring regions displaying fewer dry spell events. This result highlights the role of the topography in the triggering of precipitation in West Africa.

Similarly to what was shown for wet spells, Figures 5(a) and (b) present the 17-year climatology of the annual number of dry spells of different durations and the actual number of dry days contributed by these dry spells for each of the four selected subregions. In the four subregions, the isolated dry days are the more numerous than dry spells of all duration and the number of dry spells decreases with their duration. This decrease is more pronounced in the wettest regions (Sudanian zone and Guinea Coast). The

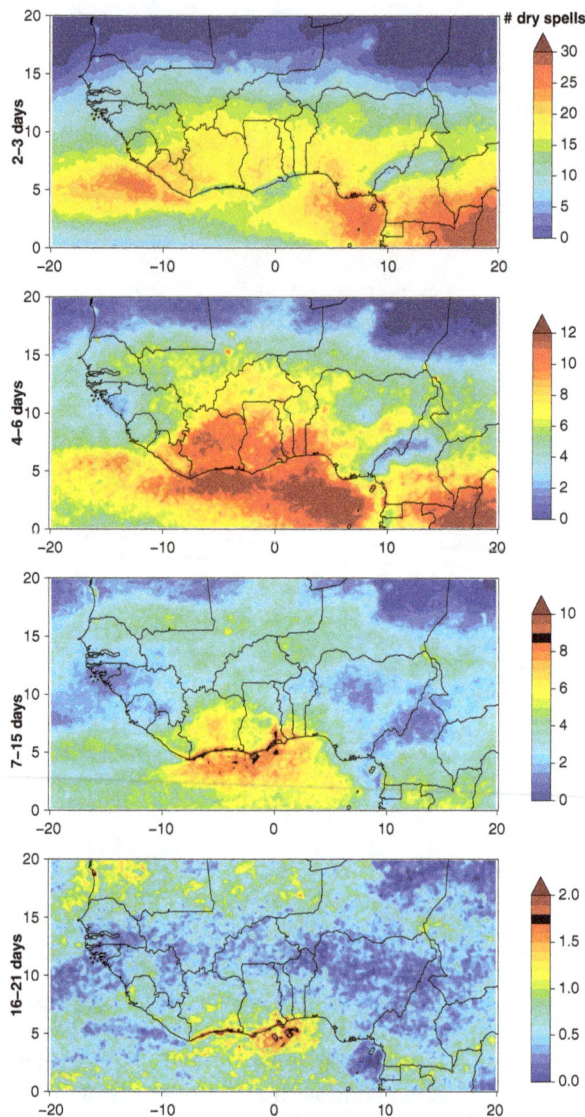

Figure 4. Mean annual number of dry spells over West Africa, for dry spells durations of 2–3, 4–6, 7–15 and 16–21 days (from top to bottom).

main contribution to the dry days comes from dry spells lasting 2 days in the four selected subregions, which is consistent with the conclusion presented, by Ratan and Venugopal (2013) in other parts of the world. The contribution of short dry spells (less than 7 days) is more important in the Sudanian zone and Guinea Coast than in the Sahel. However, the contribution of dry spells lasting more than 10 days is more important in the Western and Central Sahel than in the Sudanian band. The Guinea Coast has a little more long dry spells than the other subregions which is rather counter-intuitive and may be due to a longer rainy season leading to more possibilities to have a dry spell between two rainy days. It is noteworthy that short (1 or 2 days) wet events occur more frequently than short dry events while long (over 3 days) wet spells are fewer than long dry spells.

The seasonal variability of the number of dry spells is displayed in Figure 5(c). Unlike for wet spells, the seasonal cycle of the number of dry spells differs

depending on the considered duration of the dry spell. In the four subregions, dry spells lasting 2–3 days mostly occur during the rainy seasons with maximum occurrence in August in the two Sahelian and the Sudanian regions and in May–June and September over Guinea Coast with weaker occurrence during the little dry season in August. In the Sudanian zone and Guinea Coast, an important number of such dry spells is also noticed just before the rainy season in April and May. For longer dry spells, the seasonal cycle is less in phase with that of daily rainfall. In particular, the number of dry spells decreases in the middle of the rainy season, in August in the Central and Western Sahel and from June to September in the Sudanian region. In the Guinea Coast area, the 4–6 day dry spells are the most frequent at the beginning of the first rainy season in April and at the end of the second one in October. Regarding the 7–15 day dry spells, their occurrence is maximum at the very beginning of the first rainy season in March, during the little dry season in August and at the very end of the second rainy season in November.

4. Conclusion

This paper describes the spatial and temporal variability of wet and dry spells in West Africa based on TRMM 3B42 data, from 1998 to 2014. Thanks to its spatial resolution, this climatology of wet and dry spells should serve to assess the ability of climate models to simulate such events. It appears that the number of wet and dry spells decreases with their duration. Moreover, isolated wet days contribute the most to the total number of rainy days, while the major contribution to the dry days is from 2-day dry spells. These results confirm those of Ratan and Venugopal (2013) in other tropical regions. Despite these common specificities, the number of wet and dry spells varies spatially. The number of 2–3 day wet and dry spells show similar large-scale pattern and seasonal cycle while for events lasting more than 4 days, the main features are different whether one consider wet or dry spells. In some regions such as the south of Guinea Coast, few long wet spells are observed while long dry spells are more numerous than in the rest of the domain. Concerning the longest dry spells considered in this study (16–21 days), they occur on average once a year in the Sahel band but also over most of the Guinean region and up to twice a year along the coasts of Ivory Coast and Ghana. The relief has a particular influence leading to shorter dry spells and longer wet spells over the mountainous regions than in their neighboring area. The spatio-temporal variability of the number of wet and dry spells also appears to be linked to the spatio-temporal migration of the ITCZ. The wet spells of all duration and the short dry spells displays a large-scale spatial pattern and a diurnal cycle similar to those of the mean rainfall indicating the variability associated with the West African monsoon. These wet and dry events are more frequent in the Guinea Coast than in the Sudanian zone or Sahel and

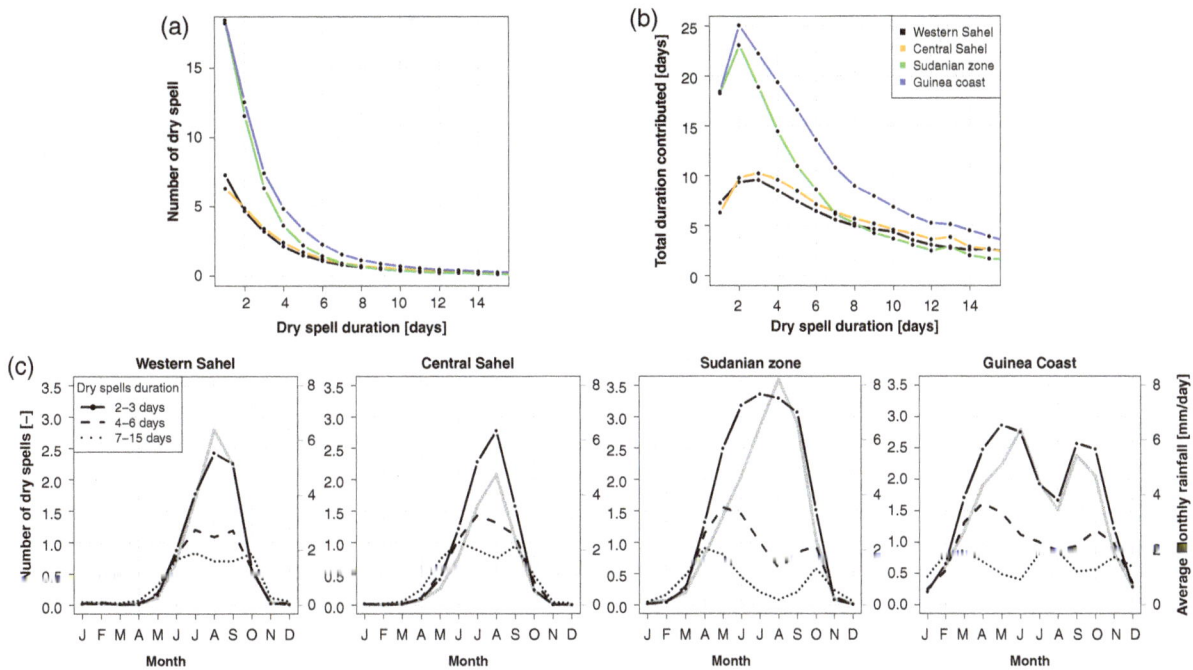

Figure 5. Dry spell characteristics for each of the four subregions. (a) Frequency histogram of dry spell duration per year and per gridpoint. (b) Average number of days each dry spell contributes to the total number of dry days. (c) Seasonal variability of the average number of dry spells of different duration (2–3, 4–6, 7–15 days, different line types). The gray line indicates the monthly mean of daily rainfall. Each of the four panels corresponds to a given subregion.

they occur mainly during the local rainy seasons. They seem to occur mainly near the core of the ITCZ, both spatially and temporally. In contrast, longer dry spells appear to be more frequent on the sides of this rainfall core: just north or south of it and at the beginning and at the end of the local rainy seasons. Therefore, wet and dry spells occurrence have a major influence on the duration of the rainy season, in particular through 'false onset' or early or late cessation of the season. Improving short-term forecasting and decadal trends of the wet and dry spells and their variability is thus a major issue for the agricultural sector in West Africa. To that purpose, further studies are needed to understand the large-scale atmospheric and surface drivers leading to these events.

Acknowledgements

The research leading to these results has received funding from the NERC/DFID Future Climate for Africa programme under the AMMA-2050 project, grant number NE/M019969/1. We also thank the TRMM science data and information system (TSDIS) and the Goddard distributed active archive center for providing us the TRMM data.

References

Amekudzi LK, Osei MA, Atiah WA, Aryee JNA, Ahiataku MA, Quansah E, Preko K, Danuor SK, Fink AH. 2016. Validation of TRMM and FEWS Satellite Rainfall Estimates with Rain Gauge Measurement over Ashanti Region, Ghana. *Atmospheric and Climate Science* **6**: 500–518.

Barron J, Rockström J, Gichuki F, Hatibu N. 2003. Dry spell analysis and maize yields for two semi-arid locations in east Africa. *Agricultural and Forest Meteorology* **117**: 23–37.

Diedhiou A, Janicot S, Viltard A, de Felice P. 1998. Evidence of two regimes of easterly waves over West Africa and the tropical Atlantic. *Geophysical Research Letters* **25**: 2805–2808.

FAO. 2016. Food and agriculture - key to achieving the 2030 agenda for sustainable development. *Food and Agriculture Organization of the United States* **I5499**: 32.

Gornall J, Betts R, Burke E, Clark R, Camp J, Willett K, Wiltshire A. 2010. Implications of climate change for agricultural productivity in the early twenty-first century. *Philosophical Transactions of the Royal Society of London. Series B: Biological Sciences* **365**: 2973–2989.

Grodsky SA, Carton JA. 2001. Coupled land/atmosphere interactions in the West African Monsoon. *Geophysical Research Letters* **28**: 1503–1506.

Groisman PY, Knight RW. 2008. Prolonged dry episodes over the conterminous United States: new tendencies emerging during the last 40 years. *Journal of Climate* **21**: 1850–1862.

Huffman GJ, Adler RF, Bolvin DT, Gu G, Nelkin EJ, Bowman KP, Hong Y, Stocker EF, Wolff DB. 2007. The TRMM multisatellite precipitation analysis (TMPA): quasi-global, multiyear, combined-sensor precipitation estimates at fine scales. *Journal of Hydrometeorology* **8**: 38–55.

Kummerow C, Barnes W, Kozu T, Shiue J, Simpson J. 1998. The tropical rainfall measuring mission (TRMM) sensor package. *Journal of Atmospheric and Ocean Technology* **15**: 809–817.

Mounier F, Janicot S. 2004. Evidence of two independent modes of convection at intraseasonal timescale in the West African summer monsoon. *Geophysical Research Letters* **31**: L16116.

Nicholson SE, Some B, McCollum J, Nelkin E, Klotter D, Berte Y, Diallo BM, Gaye I, Kpabeba G, Ndiaye O, Noukpozounkou, Tanu MM, Thiam A, Toure AA, Traore AK. 2003. Validation of TRMM and other rainfall estimates with a high-density gauge dataset for West Africa. Part II: Validation of TRMM Rainfall Products. *Journal of Applied Meteorology and Climatology* **42**: 1355–1368.

Pierre C, Bergametti G, Marticorena B, Mougin E, Lebel T, Ali A. 2011. Pluriannual comparisons of satellite-based rainfall products over the Sahelian belt for seasonal vegetation modeling. *Journal of Geophysical Research: Atmospheres* **116**: D18201.

Ratan R, Venugopal V. 2013. Wet and dry spell characteristics of global tropical rainfall. *Water Resources Research* **49**: 3830–3841.

Salack S, Klein C, Giannini A, Sarr B, Worou ON, Belko N, Bliefernicht J, Kunstman H. 2016. Global warming induced hybrid rainy seasons in the Sahel. *Environmental Research Letters: ERL* **11**: 104008.

Sanogo S, Fink AH, Omotosho JA, Ba A, Redl R, Ermert V. 2015. Spatio-temporal characteristics of the recent rainfall recovery in West Africa. *International Journal of Climatology: A Journal of the Royal Meteorological Society* **35**: 4589–4605.

Seleshi Y, Camberlin P. 2005. Recent changes in dry spell and extreme rainfall events in Ethiopia. *Theoretical and Applied Climatology* **83**: 181–191.

Sivakumar MVK. 1992. Empirical analysis of dry spells for agricultural applications in West Africa. *Journal of Climate* **5**: 532–539.

Turral H, Burke J, Faures J-M. 2011. Climate Change, water and food security. *FAO Water Reports* **36**, Food and Agriculture Organization of the United Nations, p. 15.

Wu M-LC, Reale O, Schubert SD. 2013. A characterization of African easterly waves on 2.5–6-day and 6–9-day time scales. *Journal of Climate* **26**: 6750–6774.

Zolina O, Simmer C, Belyaev K, Gulev SK, Koltermann P. 2013. Changes in the duration of European wet and dry spells during the last 60 years. *Journal of Climate* **26**: 2022–2047.

29

Interdecadal variability in the thermal difference between Western and Eastern China and its association with rainfall anomalies

Liming Wen,[1,2] Xiaoxue Yin[1,3] and Lian-Tong Zhou[1,4]*

[1]Center for Monsoon System Research, Institute of Atmospheric Physics, Chinese Academy of Sciences, Beijing, China
[2]Chengdu Meteorological Bureau, China
[3]University of Chinese Academy of Sciences, Beijing, China
[4]Key Laboratory of Global Change and Marine-Atmospheric Chemistry, Xiamen, China

*Correspondence to:
L.-T. Zhou, Center for Monsoon System Research, Institute of Atmospheric Physics, Chinese Academy of Sciences, P.O. Box 2718, Beijing 100190, China.
E-mail: zlt@mail.iap.ac.cn

Abstract

This study investigated the spring thermal difference between western and eastern China and its association with the rainfall anomalies using station and reanalysis data from 1960 to 2006. The spring thermal difference between western and eastern China underwent an obvious interdecadal shift around 1979. The thermal difference between western and eastern China was small during 1960–1978, which strengthened the southwesterly wind anomalies in line with the thermal wind. This enhanced the East Asian summer monsoon. In addition, the increase in rainfall over North China and decrease in rainfall over the Yangtze River were associated with the strong East Asian summer monsoon. However, during 1979–2006, the thermal difference between western and eastern China was large, which strengthened the northeasterly wind anomalies in line with the thermal wind. This supports the notion of weakened East Asian summer monsoon associated with the decrease in rainfall over North China and increase in rainfall over the Yangtze River.

Keywords: thermal difference; interdecadal variability; rainfall

1. Introduction

The East Asian summer monsoon is an important subsystem of the global climate system. The monsoon is related to changes in the land–sea thermal contrast. Previous studies mainly focused on the ocean thermal change and land surface processes to study the East Asian summer monsoon. In the ocean thermal change, many have studied the thermal effect of the Western Pacific warm pool and El Niño-Southern Oscillation (ENSO) cycle. East Asian summer monsoon anomalies were associated with the thermal variability of the Pacific Sea Surface Temperature (SST) anomalies (e.g. Huang *et al.*, 1999; Chang *et al.*, 2000a, 2000b; Wang, 2001; Wu and Wang, 2002; Zhou and Huang, 2003; Zhou and Chan, 2007; Ding *et al.*, 2008). In addition, the thermal state of the warm pool and its overhead convective activities play an important role in the interannual variation of the East Asian summer monsoon (Huang and Li, 1987; Nitta, 1987; Huang and Sun, 1992). In terms of land surface processes, the effect of land–air interaction on the Asian summer monsoon has been studied extensively (e.g. Webster, 1983; Yeh *et al.*, 1984; Meehl, 1994; Douville and Royer, 1996; Yang and Xu, 1994; Yang and Lau, 1998; Xu *et al.*, 2006; Wu and Kirtman, 2007; Zhou and Huang, 2010). Many studies also pointed out that the Qinghai Plateau has a huge heating effect on atmospheric and thermal anomalies, and strongly affects the

East Asian summer monsoon (e.g. Ye and Gao, 1979; Li *et al.*, 2003; Zhao and Qian, 2007). More recently, the effect of the thermal state of the arid and semi-arid Northwest China on the East Asia climate change was studied. One of the regions with the strongest surface sensible heat flux is in Northwest China, which is called a warm underlying surface, whose variability may impact local and remote climate (Zhou and Huang, 2008, 2010; Zhou *et al.*, 2009). Moreover, Gao *et al.* (2008) numerically simulated the thermal anomaly in Northwest China that affects the local and surrounding atmospheric circulation anomalies. Many ecosystems with different surface characteristics, e.g. Gobi desert and the humid monsoon regions, characterize China. Northwest China is an arid and semiarid region with strong solar radiation, sparse vegetation, and strong surface sensible heat flux. In contrast, eastern China is a monsoon region with high precipitation and weak surface sensible heat flux. Based on station data, Qu *et al.* (2011) and Wen *et al.* (2014) concluded that the surface air temperatures, precipitation, and temperature differences were not the same between eastern and western China. Zhou and Huang (2014) pointed out the significant difference in surface sensible and latent heat fluxes between western and eastern China using the ERA-40 reanalysis datasets. The surface radiation fluxes were also found to have the same characteristics based on CMIP5 data (Zhou and Du, 2016). It is well established that the monsoon is mainly caused by the

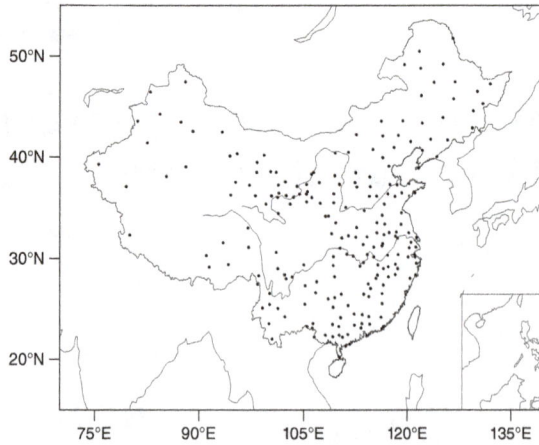

Figure 1. The locations of the 218 stations for T_s and T_a, including western China stations (red), eastern China stations (blue), and other stations (black).

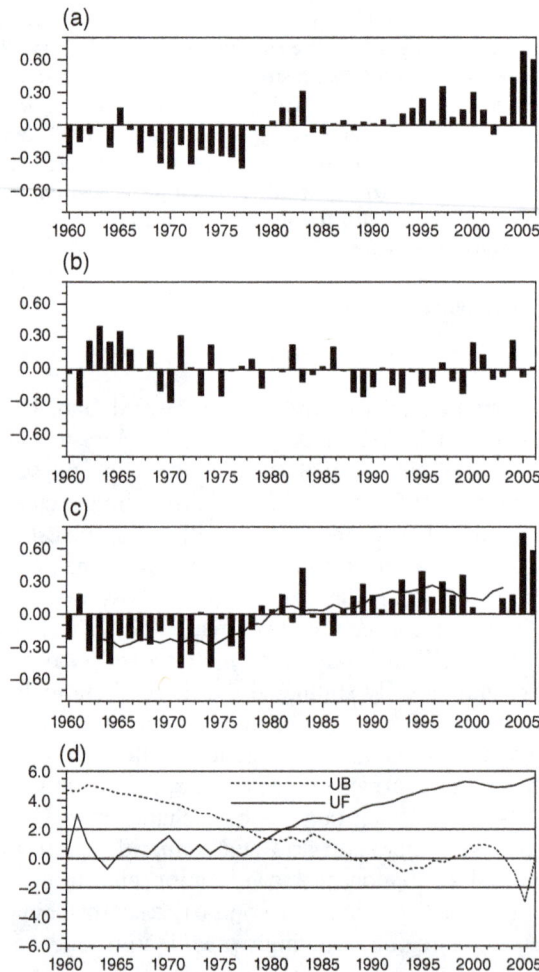

Figure 2. Time series of spring $T_s - T_a$ anomaly in western China (unit: °C) (a) and in eastern China (unit: °C) (b). (c) Time series of $\Delta(T_s - T_a)$ anomaly in spring (unit: °C). The solid curve shows the 7-year running mean. Values are the averages of the measurements from the stations in western and eastern China. The climatological mean of T_s and T_a in various months averaged for 1971–2000 is taken as their normal, respectively. (d) The forward and backward statistic rank series (solid and dashed curves) in the Mann–Kendall test of the spring $\Delta(T_s - T_a)$. The solid straight lines indicate the 0.05 confidence level of the Mann–Kendall test.

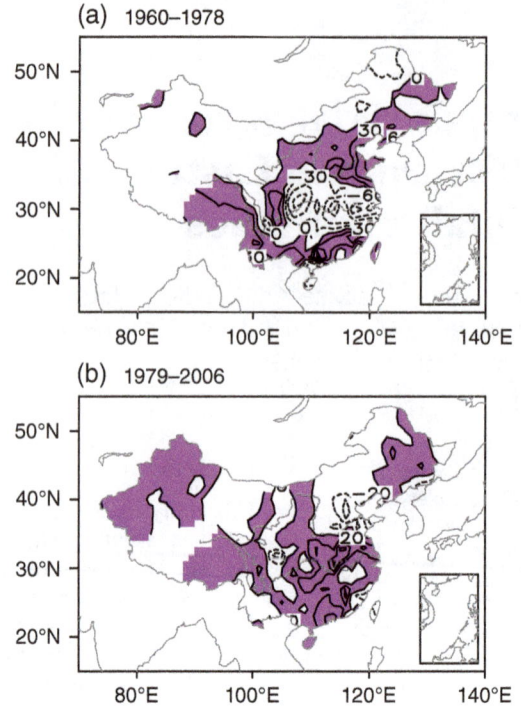

Figure 3. Distributions of the summer (JJA) rainfall (unit: mm) averaged for 1960–1978 (a), and 1979–2006 (b). The climatological monthly mean is based on the period 1971–2000. The solid and dashed lines indicate positive and negative values, respectively. Shaded regions denote positive anomalies in (a) and (b).

thermal contrast between land and sea. The thermal differences between the arid areas of Northwest China and the humid monsoon regions of eastern China inevitably contribute to the local and remote atmospheric circulation. Moreover, the temperature difference was related to the rainfall and circulation anomalies in China (Zhou and Huang, 2006; Zhou, 2015).

2. Datasets

This study used daily observations of the land surface temperature (T_s) and the surface air temperature (T_a) at 218 stations in China from 1960 to 2006. Monthly rainfall data for 1960–2006 were collected at 160 stations. The above data were provided by the Chinese Meteorological Data Center. The monthly mean wind, temperature, omega, height, sea level pressure (SLP), and special humidity were derived from the National Centers for Environmental Prediction–National Center for Atmospheric Research (NCEP–NCAR) reanalysis for 1960–2006.

3. Results

3.1. Interdecadal variability of the spring thermal difference between western and eastern China

The sensible heat flux was predominantly determined from the difference between the surface temperature

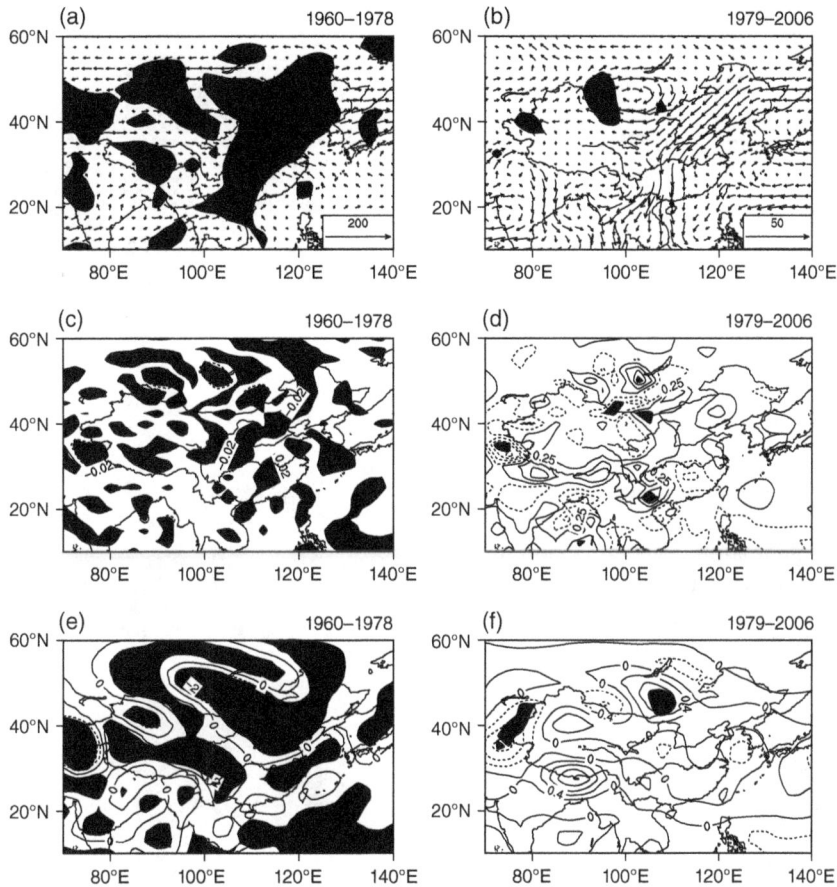

Figure 4. Distributions of the summer (JJA) water vapor flux (integration of 1000–100 hPa) anomalies (unit: kg m^{-1} s^{-1}) (a), water vapor flux (integration 1000–100 hPa) convergence anomalies (unit: g m^{-2} s^{-1}) (c), and omega anomalies at 700 hPa (unit: Pa s^{-1}) (e) averaged for 1960–1978. The (b), (d), and (f) are same as (a), (c), and (e), but for averaged for 1979–2006. The climatological monthly mean is based on the period 1971–2000. The solid and dashed lines indicate positive and negative values, respectively. The shaded regions indicate the 0.05 significance level according to the Student's t-test.

(T_s) and surface air temperature (T_a) difference, $T_s - T_a$ (Fan *et al.*, 2004; Zhou, 2009). Thus, we used this land–air temperature difference to study the thermal difference between western and eastern China and its association with the East Asian summer monsoon. Previous studies show there was an obvious difference in $T_s - T_a$ between western and eastern China (Wen *et al.*, 2014; Zhou and Wen, 2016). According to station location from Figure 1, the region average in $T_s - T_a$ of western and eastern China was calculated, respectively. Figure 2 shows the mean $T_s - T_a$ anomalies in western and eastern China in spring. The $T_s - T_a$ anomalies were mostly negative in western China before the late 1970s and largely positive after the late 1970s (Figure 2(a)). However, for eastern China (Figure 2(b)), the $T_s - T_a$ anomalies were mostly positive before the late 1970s but largely negative after the late 1970s. The data suggest that the spring thermal state in eastern and western China underwent an opposite interdecadal shift around the late 1970s. Moreover, $\Delta(T_s - T_a)$ was used to represent the thermal difference between western and eastern China, where $\Delta(T_s - T_a) = (T_s - T_a|\text{west}) - (T_s - T_a|\text{east})$.

Figure 2(c) shows that the $\Delta(T_s - T_a)$ anomalies were mainly low before the late 1970s but mostly high

thereafter. $\Delta(T_s - T_a)$ also suggests an interdecadal increase around late 1970s. The 7-year running mean (curve) in Figure 2(c) point to the interdecadal shift. The Mann–Kendall (M–K) test was used to analyze the shift in $\Delta(T_s - T_a)$. The two statistical rank series of the test are shown in Figure 2(d), which cross the year 1979 above the 95% confidence level. This suggests that the interdecadal shift of $\Delta(T_s - T_a)$ took place around 1979 and supports the appearance of the interdecadal shift around 1979. Consequently, the thermal difference between western and eastern China was stronger in 1979–2006 than 1960–1978.

3.2. Interdecadal variability of rainfall in China

The temperature difference was related to the rainfall anomalies in China (Zhou and Huang, 2006; Zhou, 2015). To investigate the association of the thermal difference anomaly with the summer rainfall variability in China, summer rainfall anomalies, water vapor flux and moisture, and vertical velocity were examined before and after 1979.

During 1960–1978, positive rainfall anomalies appeared over North China and negative anomalies appeared over the Yangtze River (Figure 3(a)). Both these anomalies were related to the East Asian summer monsoon circulation. The southwestward anomalous water vapor fluxes along the coast of eastern China substantially enhanced the moisture supply to North China from the South China Sea. This consequently increased the summer rainfall over North China (Figure 4(a)). Moreover, the increase in the rainfall was associated with water vapor flux convergence and ascending motion over North China. On the other hand, the decrease in the rainfall was associated with water vapor flux divergence and sinking motion over the Yangtze River (Figure 4(c) and (e)). However, during 1979–2006, negative rainfall anomalies also appeared over North China and positive anomalies appeared over the Yangtze River (Figure 3(b)). This was related to the weakening of the East Asian summer monsoon and the dominant northeasterly anomalies over East Asia. The northeastward anomalous water vapor fluxes were not contributed to transport into North China (Figure 4(b)), which were helpful for a decrease in rainfall over North China. Moreover, tropospheric moisture divergence and sinking motion occurred over North China (Figure 4(d) and (f)), whereas tropospheric moisture convergence and ascending motion were observed over the Yangtze River (Figure 4(d) and (f)). The latter contributed to increased rainfall in the region.

To understand the relationship between spring thermal difference between western and eastern China and the East Asian summer monsoon, the summer wind at 850 hPa, temperature at 700 hPa, and SLP were analyzed. Figure 5 shows regression of summer wind at 850 hPa, temperature at 700 hPa, and SLP with respect to the normalized spring thermal difference for the period of 1960–2006. From Figure 5(a), an anticyclonic circulation anomaly appeared over Mongolia. And the anomalous northeasterly winds occurred over eastern China. This may be related to the negative temperature anomalies appeared over Mongolia, which corresponded to positive land–air temperature difference between eastern and western China (Figure 5(b)). The negative temperature anomalies corresponded to an increased in SLP (Figure 5(c)). This enhanced in pressure difference contributed to an increase the northeasterly wind anomalies in East Asia, and a decrease in the East Asian summer monsoon.

3.3. Interdecadal variability of the atmospheric circulation

The summer rainfall anomalies in China and atmospheric circulation anomalies are closely related. In the previous section, the data suggested that the thermal difference underwent an interdecadal shift around 1979. Moreover, the thermal differences were closely related with the East Asian summer monsoon. To understand the thermal difference between western and eastern China and its association with the interdecadal variability in East Asian summer monsoon circulation anomalies, the summer 850 hPa wind, SLP, 700 hPa

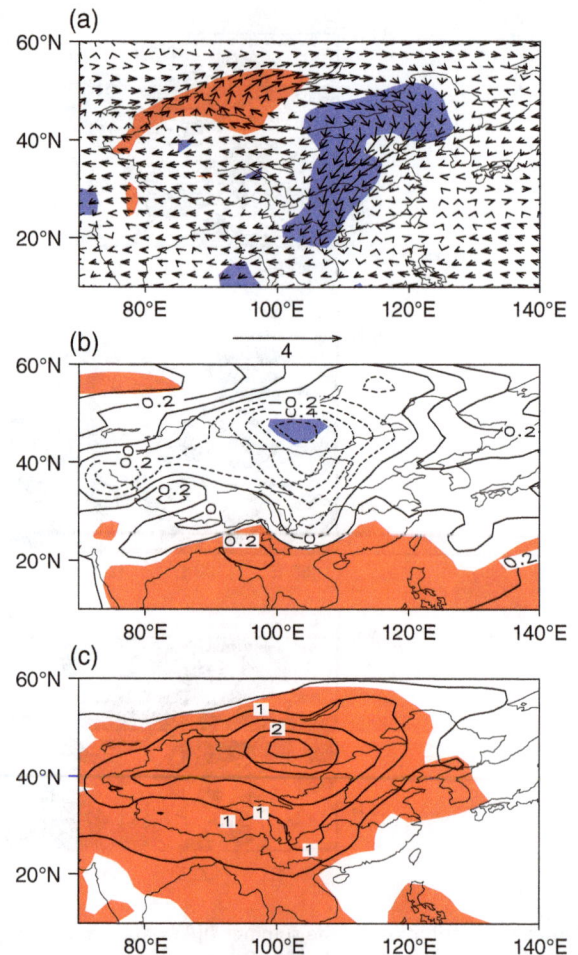

Figure 5. Regression of summer wind at 850 hPa (unit: m s⁻¹), temperature at 700 hPa (unit: °C), and sea level pressure (SLP) (unit: Pa) with respect to the normalized spring $\Delta(T_s - T_a)$ for the period of 1960–2006. The shaded regions indicate the 0.01 significance level according to the Student's t-test.

temperature, and geopotential height anomalies at 500 hPa were analyzed before and after 1979.

During 1960–1978 (Figure 6(a)), cyclonic circulation anomalies appeared over Mongolia. To the east flank of this anomalous cyclone were anomalous southwesterly winds over eastern China. These anomalous winds extended from South to North China, which helped to strengthen the East Asian summer monsoon circulation. From the above analysis, the thermal difference between eastern and western China was found to be relatively small in this period. Based on the thermal wind, the small thermal difference between western and eastern China contributed to strengthening of southwesterly wind in eastern China, which were helpful for strengthening in East Asian summer monsoon. However, during 1979–2006 (Figure 6(b)), anticyclonic circulation anomalies appeared over Mongolia. Moreover, the anomalous northeasterly winds weakened the East Asian summer monsoon circulation. The large thermal difference between western and eastern China was in 1979–2006. Therefore, according to thermal wind, it indicated that large thermal difference between western and eastern China contributed to strengthening of

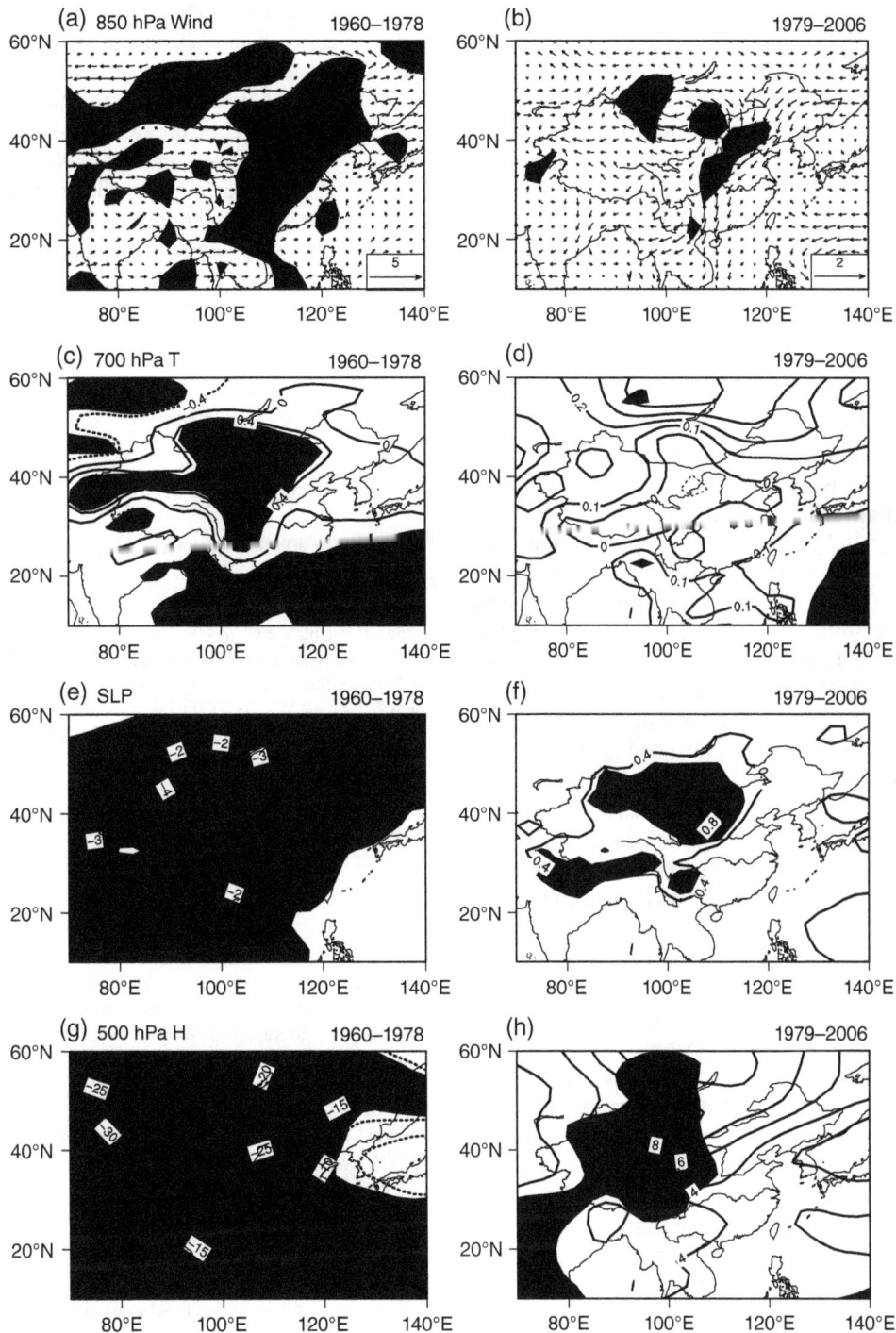

Figure 6. Distributions of the summer (JJA) wind anomalies at 850 hPa (unit: m s⁻¹) (a), temperature anomalies at 700 hPa (units: °C) (c), SLP anomalies (unit: Pa) (e), and geopotential height anomalies (unit: gpm) at 500 hPa (g) averaged for 1960–1978. The (b), (d), (f), and (h) are same as (a), (c), (e), and (g), but for averaged for 1979–2006. The climatological monthly mean is based on the period 1971–2000. The solid and dashed lines indicate positive and negative values, respectively. The shaded regions indicate the 0.05 significance level according to the Student's t-test.

northeasterly winds, which were helpful for weakening in East Asian summer monsoon. The East Asian summer monsoon anomalies are consistent with previous studies that concluded interdecadal weakening for the East Asian summer monsoon in the late 1970s (e.g. Huang *et al.*, 1999; Chang *et al.*, 2000a, 2000b; Wang, 2001; Wu and Wang, 2002; Zhou and Huang, 2003; Ding *et al.*, 2008).

To analyze in more detail, the effect of the thermal difference between western and eastern China on the East Asian summer monsoon, the temperature, SLP, and geopotential height were examined. During 1960–1978, at 700 hPa, large positive temperature anomalies appeared over Mongolia (Figure 6(c)). The positive temperature anomalies decreased the SLP (Figure 6(e)) and geopotential height anomalies

(Figure 6(g)). This enhanced the cyclonic circulation anomalies over Mongolia and increased the southwesterly wind anomalies in East Asia. However, during 1979–2006, negative temperature anomalies appeared over Mongolia (Figure 6(d)). The negative temperature anomalies increased the SLP (Figure 6(f)) and geopotential height anomalies (Figure 6(h)). This enhanced the anticyclonic circulation anomalies over Mongolia and increased the northeasterly wind anomalies in East Asia.

4. Summary and discussion

This study identified the interdecadal variability of the spring thermal difference between western and eastern China and its association with the rainfall anomalies using station and reanalysis data for 1960–2006. The spring $\Delta(T_s-T_a)$ was used to represent the thermal difference between western and eastern China. The spring $\Delta(T_s-T_a)$ data suggest an interdecadal shift around 1979 for western and eastern China. During 1960–1978, large positive temperature anomalies appeared over Mongolia. The positive temperature anomalies contributed to a decrease in pressure in this region, which contribute to an enhanced in cyclonic circulation anomalies over Mongolia. To the east flank of this anomalous cyclone were anomalous southwesterly winds over eastern China, which helped to strengthen the East Asian summer monsoon circulation. The rainfall increase over North China was associated with water vapor flux convergence and ascending motion, whereas the decrease in rainfall over the Yangtze River was associated with water vapor flux divergence and singing motion. However, during 1979–2006, negative temperature anomalies appeared over Mongolia. The negative temperature anomalies increased the pressure and the anticyclonic circulation anomalies over Mongolia. To the east flank of this anomalous anticyclone were anomalous northeasterly winds over eastern China, which were helpful for a weakening in the East Asian summer monsoon circulation. The results suggest tropospheric moisture divergence and sinking motion over North China and tropospheric moisture convergence and ascending motion over Yangtze River that contributed to increased rainfall in this region.

The interdecadal variability of the East Asian summer monsoon was affected by land surface processes and ocean thermodynamics. There are other factors, such as thermal field over the Tibetan Plateau, the snow cover, the Pacific SST, and the middle and high latitudes (e.g. Yang and Lau, 1998; Huang et al., 1999; Chang et al., 2000a, 2000b; Wang, 2001; Wu and Wang, 2002; Zhou and Huang, 2003; Ding et al., 2008), which may affect the atmospheric circulation over East Asia and the rainfall anomalies in China.

It is well known that climate shifts are characterized by changes in the background state of the Pacific Ocean and the ENSO dynamics. The interdecadal variabilities before and after the late-1970s for SST-forced

teleconnection are found in many regions (Hare and Mantua, 2000; Karspeck and Cane, 2002; Hartmann and Wendler, 2005; Rodriguez-Fonseca et al., 2011).

Because of the limited data, it cannot be determined whether the change in previous studies as well as in this study is permanent or reversible. In view of the impact of global warming on the tropical sea surface and land surface temperature, it is likely that the East Asian summer monsoon will change in the future owing to global warming.

Acknowledgements

This research was supported by the National Natural Science Foundation of China (grant nos 41175055 and 41475053), and the Fund of Key Laboratory of Global Change and Marine-Atmospheric Chemistry, SOA (GCMAC1301).

References

Chang CP, Zhang Y, Li T. 2000a. Interannual and interdecadal variations of the East Asian summer monsoon and the tropical Pacific SSTs. Part I. Roles of the subtropical ridge. *Journal of Climate* 13: 4310–4325.

Chang CP, Zhang Y, Li T. 2000b. Interannual and interdecadal variations of the East Asian summer monsoon and tropical Pacific SSTs. Part II: meridional structure of the monsoon. *Journal of Climate* 13: 4326–4340.

Ding Y, Wang Z, Sun Y. 2008. Inter-decadal variability of the summer precipitation in East China and its association with decreasing Asian summer monsoon. Part I: observed evidences. *International Journal of Climatology* 28(9): 1139–1161, doi: 10.1002/joc.1615.

Douville H, Royer JF. 1996. Sensitivity of the Asian summer monsoon to an anomalous Eurasian snow cover within theMeteo-France GCM. *Climate Dynamics* 12: 449–466.

Fan LJ, Wei ZG, Dong WJ. 2004. The characteristic of temporal and spatial distribution of the differences between ground and air temperature in the Arid Region of Northwest China. *Plateau Meteorology* 3: 360–367.

Gao R, Dong WJ, Wei ZG. 2008. Numerical simulation of the impact of abnormity of sensible heat flux in Northwest Arid Zone on precipitation in China. *Plateau Meteorology* 27(2): 320–324.

Hare SR, Mantua NJ. 2000. Empirical evidence for North Pacific regime shifts in 1977 and 1989. *Progress in Oceanography* 47: 103–145.

Hartmann B, Wendler G. 2005. The significance of the 1976 Pacific climate shift in the climatology of Alaska. *Journal of Climate* 18: 4824–4839, doi: 10.1175/JCLI3532.1.

Huang RH, Li WJ. 1987. Influence of the heat source anomaly over the tropical western Pacific on the subtropical high over East Asia. Proceedings of the International Conference on the General Circulation of East Asia, Chengdu, 40–51.

Huang RH, Sun FY. 1992. Impact of the tropical western Pacific on the East Asian summer monsoon. *Journal of the Meteorological Society of Japan* 70: 243–256.

Huang RH, Xu YH, Zhou LT. 1999. The interdecadal variation of summer precipitations in China and the drought trend in North China. *Plateau Meteorology* 18(4): 465–476.

Karspeck AR, Cane MA. 2002. Tropical Pacific 1976–77 climate shift in a linear, wind – driven model. *Journal of Physical Oceanography* 32: 2350–2360, doi: 10.1175/1520-0485.

Li DL, Wei L, Li WJ, Lv LZ, Zhong HL, Ji GL. 2003. The effect of surface sensible heat flux of the Qinghai-Xizang Plateau on general circulation over the Northern Hemisphere and climatic anomaly of China. *Climatic and Environmental Research* 8(1): 60–70.

Meehl GA. 1994. Influence of the land surface in the Asian summer monsoon: external condition versus internal feedbacks. *Journal of Climate* 8: 358–375.

Nitta TS. 1987. Convective activities in the tropical western Pacific and their impact on the Northern Hemisphere summer circulation. *Journal of the Meteorological Society of Japan* **64**: 373–390.

Qu Y, Gao X, Chen W, Hui X, Zhou C. 2011. Comparison of surface air temperatures and precipitation in Eastern and Western China during 1951–2003. *Plateau Meteorology* **27**(3): 524–529.

Rodriguez-Fonseca B, Janicot S, Mohino E, Losada T, Bader J, Caminade C, Chauvin F, Fontaine B, García-Serrano J, Gervois S, Joly M, Polo I, Ruti P, Roucou P, Voldoire A. 2011. Interannual and decadal SST – forced responses of the West African monsoon. *Atmospheric Science Letters* **12**: 67–74, doi: 10.1002/asl.308.

Wang HJ. 2001. The weakening of the Asian monsoon circulation after the end of 1970's. *Advances in Atmospheric Sciences* **18**: 378–386.

Webster PJ. 1983. Mechanisms of monsoon low-frequency variability: surface hydrological effects. *Journal of the Atmospheric Sciences* **40**: 2110–2124.

Wen LM, Zhou LT, Huang RH, Fan GZ. 2014. Characteristics of interdecadal variability in the difference between surface temperature and surface air temperature in Southeast and Northwest China (in Chinese). *Climatic and Environmental Research* **19**(5): 636–648.

Wu RG, Kirtman BP. 2007. Observed relationship of spring and summer East Asian rainfall with winter and spring Eurasian snow. *Journal of Climate* **20**: 1258–1304.

Wu R, Wang B. 2002. A contrast of the east Asian summer monsoon – ENSO relationship between 1962–77 and 1978–93. *Journal of Climate* **15**: 3266–3279, doi: 10.1175/1520-0442.

Xu M, Chang CP, Fu C, Qi Y, Robock A, Robinson D, Zhang H. 2006. Steady decline of East Asian monsoon winds, 1969-2000: evidence from direct ground measurements of wind speed. *Journal of Geophysical Research* **111**: D24111, doi: 10.1019/2006JD007337.

Yang S, Lau KM. 1998. Influences of sea surface temperature and ground wetness on Asian summer monsoon. *Journal of Climate* **11**: 3230–3246.

Yang S, Xu LZ. 1994. Linkage between Eurasian winter snow cover and regional Chinese summer rainfall. *International Journal of Climatology* **14**: 739–750, doi: 10.1002/joc.3370140704.

Ye DZ, Gao YX. 1979. *Tibet Plateau Meteorology (in Chinese)*. China Meteorological Press: Beijing; 279 pp.

Yeh TC, Wetherald RT, Manabe S. 1984. The effect of soil moisture on the short-term climate and hydrology change – a numerical experiment. *Monthly Weather Review* **112**: 474–490.

Zhao Y, Qian YF. 2007. Relationships between the surface thermal anomalies in the Tibetan Plateau and the rainfall in the Jianghuai Area in summer. *Chinese Journal of Atmospheric Sciences* **31**(1): 145–156.

Zhou LT. 2009. Difference in the interdecadal variability of spring and summer sensible heat fluxes over Northwest China. *Atmospheric and Oceanic Science Letters* **2**(2): 119–123.

Zhou LT. 2015. Influence of land–air temperature difference on spring rainfall anomalies over North China and its feedback mechanism. *International Journal of Climatology* **35**: 2676–2681, doi: 10.1002/joc.4126.

Zhou W, Chan JCL. 2007. ENSO and South China Sea summer monsoon onset. *International Journal of Climatology* **27**: 157–167, doi: 10.1002/joc.1380.

Zhou LT, Du ZC. 2016. Regional differences in the surface energy budget over China: an evaluation of a selection of CMIP5 models. *Theoretical and Applied Climatology* **25**: 241–266, doi: 10.1007/s00704-015-1407-0.

Zhou LT, Huang RH. 2003. Research on the characteristics of interdecadal variability of summer climate in China and its possible cause. *Climatic and Environmental Research* **8**: 274–290.

Zhou LT, Huang RH. 2006. Characteristics of interdecadal variability of the difference between surface temperature and surface air temperature(Tu-Ta) in spring in arid and semi-arid region of Northwest China and its impact on summer precipitation in North China. *Climatic and Environmental Research* **11**: 1–13.

Zhou LT, Huang RH. 2008. Interdecadal variability of sensible heat in arid and semi-arid regions of Northwest China and its relation to summer precipitation in China. *Chinese Journal of Atmospheric Sciences* **32**(6): 1276–1288.

Zhou LT, Huang RH. 2010. The interdecadal variability of summer rainfall in Northwest China and its possible causes. *International Journal of Climatology* **30**: 549–557.

Zhou LT, Huang RH. 2014. Regional differences in surface sensible and latent heat fluxes in China. *Theoretical and Applied Climatology* **116**: 625–637, doi: 10.1007/s00704-013-0975-0.

Zhou LT, Wen LM. 2016. Characteristics of temporal and spatial variation in land-air temperature differences in China and its association with summer rainfall. *Climatic and Environmental Research*, doi: 10.3878/j.issn.1006-9585.2016.15196 (in press).

Zhou CC, Gao XQ, Chen W, Hui XY, Li J. 2009. The impact of sensible heat flux anomaly over Central Asia on temperature and precipitation in Northwest China. *Plateau Meteorology* **28**(2): 395–401.

Permissions

All chapters in this book were first published in ASL, by John Wiley & Sons Ltd.; hereby published with permission under the Creative Commons Attribution License or equivalent. Every chapter published in this book has been scrutinized by our experts. Their significance has been extensively debated. The topics covered herein carry significant findings which will fuel the growth of the discipline. They may even be implemented as practical applications or may be referred to as a beginning point for another development.

The contributors of this book come from diverse backgrounds, making this book a truly international effort. This book will bring forth new frontiers with its revolutionizing research information and detailed analysis of the nascent developments around the world.

We would like to thank all the contributing authors for lending their expertise to make the book truly unique. They have played a crucial role in the development of this book. Without their invaluable contributions this book wouldn't have been possible. They have made vital efforts to compile up to date information on the varied aspects of this subject to make this book a valuable addition to the collection of many professionals and students.

This book was conceptualized with the vision of imparting up-to-date information and advanced data in this field. To ensure the same, a matchless editorial board was set up. Every individual on the board went through rigorous rounds of assessment to prove their worth. After which they invested a large part of their time researching and compiling the most relevant data for our readers.

The editorial board has been involved in producing this book since its inception. They have spent rigorous hours researching and exploring the diverse topics which have resulted in the successful publishing of this book. They have passed on their knowledge of decades through this book. To expedite this challenging task, the publisher supported the team at every step. A small team of assistant editors was also appointed to further simplify the editing procedure and attain best results for the readers.

Apart from the editorial board, the designing team has also invested a significant amount of their time in understanding the subject and creating the most relevant covers. They scrutinized every image to scout for the most suitable representation of the subject and create an appropriate cover for the book.

The publishing team has been an ardent support to the editorial, designing and production team. Their endless efforts to recruit the best for this project, has resulted in the accomplishment of this book. They are a veteran in the field of academics and their pool of knowledge is as vast as their experience in printing. Their expertise and guidance has proved useful at every step. Their uncompromising quality standards have made this book an exceptional effort. Their encouragement from time to time has been an inspiration for everyone.

The publisher and the editorial board hope that this book will prove to be a valuable piece of knowledge for researchers, students, practitioners and scholars across the globe.

List of Contributors

Muhammad Naveed Anjum
State Key Laboratory of Cryospheric Science, Cold and Arid Regions Environmental and Engineering Research Institute, Chinese Academy of Sciences, Lanzhou, P.R. China
Key Laboratory of Inland River Ecohydrology, Cold and Arid Regions Environmental and Engineering Research Institute, Chinese Academy of Sciences, Lanzhou, P.R. China
University of Chinese Academy of Sciences, Beijing, P.R. China

Yongjian Ding
State Key Laboratory of Cryospheric Science, Cold and Arid Regions Environmental and Engineering Research Institute, Chinese Academy of Sciences, Lanzhou, P.R. China
Key Laboratory of Inland River Ecohydrology, Cold and Arid Regions Environmental and Engineering Research Institute, Chinese Academy of Sciences, Lanzhou, P.R. China

Donghui Shangguan and Muhammad Adnan
State Key Laboratory of Cryospheric Science, Cold and Arid Regions Environmental and Engineering Research Institute, Chinese Academy of Sciences, Lanzhou, P.R. China

Adnan Ahmad Tahir
Institute of Earth Surface Dynamics, Faculty of Geosciences and Environment, University of Lausanne, Lausanne, Switzerland
Department of Environmental Sciences, COMSATS Institute of Information Technology, Abbottabad, Pakistan

Mudassar Iqbal
Key Laboratory of Land Surface Process and Climate Change in Cold and Arid Regions, Cold and Arid Regions Environmental and Engineering Research Institute, Chinese Academy of Sciences, Lanzhou, P.R. China

Hongjun Bao and Lili Wang
National Meteorological Centre, China Meteorological Administration, Beijing, China

Ke Zhang and Zhijia Li
College of Hydrology and Water Resources, Hohai University, Nanjing, China
State Key Laboratory of Hydrology-Water Resources and Hydraulic Engineering, Hohai University, Nanjing, China

Siraput Jongaramrungruang
Trinity College, University of Cambridge, UK
Physical Oceanography Department, Woods Hole Oceanographic Institution, MA, USA
Now at Division of Geological and Planetary Sciences, California Institute of Technology, CA, USA

Hyodae Seo and Caroline C. Ummenhofer
Physical Oceanography Department, Woods Hole Oceanographic Institution, MA, USA

Vinay Kumar and T. N. Krishnamurti
Department of Earth, Ocean and Atmospheric Science, Florida State University, Tallahassee, FL, USA

Tim Li
IPRC and Department of Meteorology, University of Hawaii, Honolulu, HI, USA
International Laboratory on Climate and Environment Change and Key Laboratory of Meteorological Disaster, Nanjing University of Information Science and Technology, China

Zhi Li, Weidong Yu, Kuiping Li and Yanliang Liu
Center for Ocean and Climate Research, First Institute of Oceanography, SOA, Qingdao, China

Huiqi Li
Key Laboratory of Cloud-Precipitation Physics and Severe Storms (LACS), Institute of Atmospheric Physics, Chinese Academy of Sciences, Beijing, China
College of Earth Science, University of Chinese Academy of Sciences, Beijing, China

Xiaopeng Cui
Key Laboratory of Cloud-Precipitation Physics and Severe Storms (LACS), Institute of Atmospheric Physics, Chinese Academy of Sciences, Beijing, China
Collaborative Innovation Center on Forecast and Evaluation of Meteorological Disasters, Nanjing University of Information Science and Technology, Nanjing, China

Wenlong Zhang
Institute of Urban Meteorology, China Meteorological Administration, Beijing, China

Lin Qiao
Beijing Municipal Weather Forecast Center, Beijing Meteorological Service, Beijing, China

Xiaofan Li, Peijun Zhu, Guoqing Zhai and Rui Liu
Department of Earth Sciences, Zhejiang University, Hangzhou, China

Xinyong Shen and Wei Huang
Collaborative Innovation Center on Forecast and Evaluation of Meteorological Disasters, Key Laboratory of Meteorological Disaster of Ministry of Education, Nanjing University of Information Science and Technology, China

Donghai Wang
State Key Laboratory of Severe Weather, Chinese Academy of Meteorological Sciences, Beijing, China

Gang Li and Xiaohua Jiang
Xichang Satellite Launch Center, Xichang, China

Jiepeng Chen
State Key Laboratory of Tropical Oceanography, South China Sea Institute of Oceanology, Chinese Academy of Sciences, Guangzhou, China

Xin Wang
State Key Laboratory of Tropical Oceanography, South China Sea Institute of Oceanology, Chinese Academy of Sciences, Guangzhou, China
Laboratory for Regional Oceanography and Numerical Modeling, Qingdao National Laboratory for Marine Science and Technology, Qingdao, China

Yanke Tan
College of Meteorology and Oceanography, PLA University of Science and Technology, Nanjing, China

Zhongda Lin and Riyu Lu
State Key Laboratory of Numerical Modelling for Atmospheric Sciences and Geophysical Fluid Dynamics, Institute of Atmospheric Physics, Chinese Academy of Sciences, Beijing, China

Yen-Heng Lin and Lawrence E. Hipps
Department of Plants, Soils and Climate, Utah State University, Logan, UT, USA

S.-Y. Simon Wang
Department of Plants, Soils and Climate, Utah State University, Logan, UT, USA
Utah Climate Center, Utah State University, Logan, UT, USA

Jin-Ho Yoon
School of Earth Sciences and Environmental Engineering, Gwangju Institute of Science and Technology, South Korea

Zuhan Liu
Jiangxi Province Key Laboratory for Water Information Cooperative Sensing and Intelligent Processing, Nanchang Institute of Technology, Nanchang, China
School of Information Engineering, Nanchang Institute of Technology, Nanchang, China
Key Laboratory of the Education Ministry for Poyang Lake Wetland and Watershed Research, Jiangxi Normal University, Nanchang, China

Lili Wang
School of Science, Nanchang Institute of Technology, China

Xiang Yu and Chengzhi Deng
Jiangxi Province Key Laboratory for Water Information Cooperative Sensing and Intelligent Processing, Nanchang Institute of Technology, Nanchang, China
School of Information Engineering, Nanchang Institute of Technology, Nanchang, China

Shengqian Wang
School of Information Engineering, Nanchang Institute of Technology, Nanchang, China

Jianhua Xu and Ling Bai
The Research Center for East-west Cooperation in China, East China Normal University, Shanghai, China

Zhongsheng Chen
The Research Center for East-west Cooperation in China, East China Normal University, Shanghai, China
College of Land and Resources, China West Normal University, Nanchong, China

P. V. Nagamani and M. M. Ali
National Remote Sensing Centre, ISRO, Hyderabad, India

G. J. Goni
National Oceanic and Atmospheric Administration, AOML, Washington, DC, USA

T. V. S. Udaya Bhaskar
Indian National Center for Ocean Information Services, Hyderabad, India

J. P. McCreary
University of Hawaii, Honolulu, HI, USA

R. A. Weller
Woods Hole Oceanographic Institution, Woods Hole, MA, USA

M. Rajeevan
Ministry of Earth Sciences, New Delhi, India

V. V. Gopala Krishna
National Institute of Oceanography, Goa, India

J. C. Pezzullo
Georgetown University, Washington, DC, USA

Dzung Nguyen-Le and Tomohito J. Yamada
Faculty of Engineering, Hokkaido University, Japan

Duc Tran-Anh
National Center for Hydro-Meteorological Forecasting, Hanoi, Vietnam

Edgar G. Pavia
Centro de Investigación Científica y de Educación Superior de Ensenada (CICESE), México

P. Rai and A. P. Dimri
School of Environmental Sciences, Jawaharlal Nehru University, New Delhi, India

Sajjad Saeed
Department of Earth and Environmental Sciences, KU Leuven, Belgium
Center of Excellence for Climate Change Research (CECCR), King Abdulaziz University, Jeddah, Saudi Arabia

Erwan Brisson
Goethe University, Frankfurt, Germany

Matthias Demuzere
Department of Earth and Environmental Sciences, KU Leuven, Belgium

Hossein Tabari
Hydraulics Division, Department of Civil Engineering, KU Leuven, Belgium

Patrick Willems
Hydraulics Division, Department of Civil Engineering, KU Leuven, Belgium
Department of Hydrology and Hydraulic Engineering, Vrije Universiteit Brussel, Belgium

Nicole P. M. van Lipzig
Department of Earth and Environmental Sciences, KU Leuven, Belgium

F. S. Syed
Department of Meteorology, COMSATS Institute of Information Technology, Islamabad, Pakistan

F. Kucharski
Earth System Physics Section, Abdus Salam International Centre for Theoretical Physics, Trieste, Italy
Center of Excellence for Climate Change Research/ Department of Meteorology, King Abdulaziz University, Jeddah, Saudi Arabia

Bin Wang, Xiaofan Li, Yanyan Huang and Guoqing Zhai
School of Earth Sciences, Zhejiang University, Hangzhou, China

S.-Y. Simon Wang
Utah Climate Center, Utah State University, Logan, UT, USA
Department of Plants, Soils and Climate, Utah State University, Logan, UT, USA

Yen-Heng Lin
Department of Plants, Soils and Climate, Utah State University, Logan, UT, USA

Chi-Hua Wu
Research Center for Environmental Changes, Academia Sinica, Taipei, Taiwan

Lei Wang
Center for Monsoon System Research, Institute of Atmospheric Physics, Chinese Academy of Sciences, Beijing, China
Key Laboratory of Global Change and Marine-Atmospheric Chemistry, Xiamen, China

Ronghui Huang and Renguang Wu
Center for Monsoon System Research, Institute of Atmospheric Physics, Chinese Academy of Sciences, Beijing, China

Liying Wang, Xiu-Qun Yang, Dejian Yang, Jiabei Fang and Xuguang Sun
CMA-NJU Joint Laboratory for Climate Prediction Studies and Jiangsu Collaborative Innovation Center of Climate Change, School of Atmospheric Sciences, Nanjing University, Nanjing, China

Qian Xie
College of Meteorology and Oceanography, National University of Defense Technology, Naning, China

Sudarsan Bera, G. Pandithurai and Thara V. Prabha
Indian Institute of Tropical Meteorology, Pune, India

Lijun Fan
Key Laboratory of Regional Climate-Environment Research for Temperate East Asia (RCE-TEA), Institute of Atmospheric Physics, Chinese Academy of Sciences, Beijing, China

Deliang Chen
Regional Climate Group, Department of Earth Sciences, University of Gothenburg, Sweden

Jin-Ho Yoon
Atmospheric Sciences and Global Change Division, Pacific Northwest National Laboratory, Richland, WA, USA

Wei Zhang
Key Laboratory of Meteorological Disaster of Ministry of Education, Nanjing University of Information Science and Technology, People's Republic of China
Collaborative Innovation Center on Forecast and Evaluation of Meteorological Disasters, Nanjing University of Information Science and Technology, People's Republic of China
Climate Change and Sustainability Laboratory, Shenzhen Research Institute, The Chinese University of Hong Kong, Shatin, Hong Kong, China

Ming Luo
Institute of Environment, Energy and Sustainability, The Chinese University of Hong Kong, Shatin, Hong Kong, China

Fei Xie and Cheng Sun
College of Global Change and Earth System Science, Beijing Normal University, Beijing, China

Jiankai Zhang, Wenjun Sang and Yang Li
Key Laboratory for Semi-Arid Climate Change of the Ministry of Education, College of Atmospheric Sciences, Lanzhou University, Lanzhou, China

Yulei Qi and Yang Li
School of Atmospheric Sciences, Chengdu University of Information Technology, Chengdu, China

Jianchuan Shu
Institute of Plateau Meteorology, China Meteorological Administration, Chengdu, China

Fangxing Fan, Xiao Dong, Feng Xue, Fei Zheng and Jiang Zhu
International Center for Climate and Environment Sciences, Institute of Atmospheric Physics, Chinese Academy of Sciences, Beijing, China

Xianghui Fang
Institute of Atmospheric Sciences, Fudan University, Shanghai, China

Stéphanie Froidurot and Arona Diedhiou
Université Grenoble Alpes, IRD, CNRS, Grenoble INP, IGE, Grenoble, France

Liming Wen
Center for Monsoon System Research, Institute of Atmospheric Physics, Chinese Academy of Sciences, Beijing, China
Chengdu Meteorological Bureau, China

Xiaoxue Yin
Center for Monsoon System Research, Institute of Atmospheric Physics, Chinese Academy of Sciences, Beijing, China
University of Chinese Academy of Sciences, Beijing, China

Lian-Tong Zhou
Center for Monsoon System Research, Institute of Atmospheric Physics, Chinese Academy of Sciences, Beijing, China
Key Laboratory of Global Change and Marine-Atmospheric Chemistry, Xiamen, China

Index